MW00824350

# INDUCTION
# AND ANALOGY IN
# MATHEMATICS

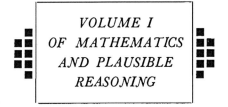

*VOLUME I
OF MATHEMATICS
AND PLAUSIBLE
REASONING*

By G. POLYA

PRINCETON UNIVERSITY PRESS
PRINCETON, NEW JERSEY

1954

*Published, 1954, by Princeton University Press*
*London: Geoffrey Cumberlege, Oxford University Press*
*L.C. Card 53–6388*

COMPOSED BY THE PITMAN PRESS, BATH, ENGLAND
PRINTED IN THE UNITED STATES OF AMERICA

# PREFACE

THIS book has various aims, closely connected with each other. In the first place, this book intends to serve students and teachers of mathematics in an important but usually neglected way. Yet in a sense the book is also a philosophical essay. It is also a continuation and requires a continuation. I shall touch upon these points, one after the other.

1. Strictly speaking, all our knowledge outside mathematics and demonstrative logic (which is, in fact, a branch of mathematics) consists of conjectures. There are, of course, conjectures and conjectures. There are highly respectable and reliable conjectures as those expressed in certain general laws of physical science. There are other conjectures, neither reliable nor respectable, some of which may make you angry when you read them in a newspaper. And in between there are all sorts of conjectures, hunches, and guesses.

We secure our mathematical knowledge by *demonstrative reasoning*, but we support our conjectures by *plausible reasoning*. A mathematical proof is demonstrative reasoning, but the inductive evidence of the physicist, the circumstantial evidence of the lawyer, the documentary evidence of the historian, and the statistical evidence of the economist belong to plausible reasoning.

The difference between the two kinds of reasoning is great and manifold. Demonstrative reasoning is safe, beyond controversy, and final. Plausible reasoning is hazardous, controversial, and provisional. Demonstrative reasoning penetrates the sciences just as far as mathematics does, but it is in itself (as mathematics is in itself) incapable of yielding essentially new knowledge about the world around us. Anything new that we learn about the world involves plausible reasoning, which is the only kind of reasoning for which we care in everyday affairs. Demonstrative reasoning has rigid standards, codified and clarified by logic (formal or demonstrative logic), which is the theory of demonstrative reasoning. The standards of plausible reasoning are fluid, and there is no theory of such reasoning that could be compared to demonstrative logic in clarity or would command comparable consensus.

2. Another point concerning the two kinds of reasoning deserves our attention. Everyone knows that mathematics offers an excellent opportunity to learn demonstrative reasoning, but I contend also that there is no subject in the usual curricula of the schools that affords a comparable opportunity to learn plausible reasoning. I address myself to all interested students of

mathematics of all grades and I say: Certainly, let us learn proving, but also *let us learn guessing.*

This sounds a little paradoxical and I must emphasize a few points to avoid possible misunderstandings.

Mathematics is regarded as a demonstrative science. Yet this is only one of its aspects. Finished mathematics presented in a finished form appears as purely demonstrative, consisting of proofs only. Yet mathematics in the making resembles any other human knowledge in the making. You have to guess a mathematical theorem before you prove it; you have to guess the idea of the proof before you carry through the details. You have to combine observations and follow analogies; you have to try and try again. The result of the mathematician's creative work is demonstrative reasoning, a proof; but the proof is discovered by plausible reasoning, by guessing. If the learning of mathematics reflects to any degree the invention of mathematics, it must have a place for guessing, for plausible inference.

There are two kinds of reasoning, as we said: demonstrative reasoning and plausible reasoning. Let me observe that they do not contradict each other; on the contrary, they complete each other. In strict reasoning the principal thing is to distinguish a proof from a guess, a valid demonstration from an invalid attempt. In plausible reasoning the principal thing is to distinguish a guess from a guess, a more reasonable guess from a less reasonable guess. If you direct your attention to both distinctions, both may become clearer.

A serious student of mathematics, intending to make it his life's work, must learn demonstrative reasoning; it is his profession and the distinctive mark of his science. Yet for real success he must also learn plausible reasoning; this is the kind of reasoning on which his creative work will depend. The general or amateur student should also get a taste of demonstrative reasoning: he may have little opportunity to use it directly, but he should acquire a standard with which he can compare alleged evidence of all sorts aimed at him in modern life. But in all his endeavors he will need plausible reasoning. At any rate, an ambitious student of mathematics, whatever his further interests may be, should try to learn both kinds of reasoning, demonstrative and plausible.

3. I do not believe that there is a foolproof method to learn guessing. At any rate, if there is such a method, I do not know it, and quite certainly I do not pretend to offer it on the following pages. The efficient use of plausible reasoning is a practical skill and it is learned, as any other practical skill, by imitation and practice. I shall try to do my best for the reader who is anxious to learn plausible reasoning, but what I can offer are only examples for imitation and opportunity for practice.

In what follows, I shall often discuss mathematical discoveries, great and small. I cannot tell the true story how the discovery did happen, because nobody really knows that. Yet I shall try to make up a likely story how the discovery could have happened. I shall try to emphasize the motives underlying the discovery, the plausible inferences that led to it, in short, everything that deserves imitation. Of course, I shall try to impress the reader; this is my duty as teacher and author. Yet I shall be perfectly honest with the reader in the point that really matters: I shall try to impress him only with things which seem genuine and helpful to me.

Each chapter will be followed by examples and comments. The comments deal with points too technical or too subtle for the text of the chapter, or with points somewhat aside of the main line of argument. Some of the exercises give an opportunity to the reader to reconsider details only sketched in the text. Yet the majority of the exercises give an opportunity to the reader to draw plausible conclusions of his own. Before attacking a more difficult problem proposed at the end of a chapter, the reader should carefully read the relevant parts of the chapter and should also glance at the neighboring problems; one or the other may contain a clue. In order to provide (or hide) such clues with the greatest benefit to the instruction of the reader, much care has been expended not only on the contents and the form of the proposed problems, but also on their *disposition*. In fact, much more time and care went into the arrangement of these problems than an outsider could imagine or would think necessary.

In order to reach a wide circle of readers I tried to illustrate each important point by an example as elementary as possible. Yet in several cases I was obliged to take a not too elementary example to support the point impressively enough. In fact, I felt that I should present also examples of historic interest, examples of real mathematical beauty, and examples illustrating the parallelism of the procedures in other sciences, or in everyday life.

I should add that for many of the stories told the final form resulted from a sort of informal psychological experiment. I discussed the subject with several different classes, interrupting my exposition frequently with such questions as: "Well, what would you do in such a situation?" Several passages incorporated in the following text have been suggested by the answers of my students, or my original version has been modified in some other manner by the reaction of my audience.

In short, I tried to use all my experience in research and teaching to give an appropriate opportunity to the reader for intelligent imitation and for doing things by himself.

4. The examples of plausible reasoning collected in this book may be put to another use: they may throw some light upon a much agitated philosophical problem: the problem of induction. The crucial question is: Are there rules for induction? Some philosophers say Yes, most scientists think No. In order to be discussed profitably, the question should be put differently. It should be treated differently, too, with less reliance on traditional verbalisms, or on new-fangled formalisms, but in closer touch with the practice of scientists. Now, observe that inductive reasoning is a

particular case of plausible reasoning.   Observe also (what modern writers almost forgot, but some older writers, such as Euler and Laplace, clearly perceived) that the role of inductive evidence in mathematical investigation is similar to its role in physical research.   Then you may notice the possibility of obtaining some information about inductive reasoning by observing and comparing examples of plausible reasoning in mathematical matters.   And so the door opens to *investigating induction inductively*.

When a biologist attempts to investigate some general problem, let us say, of genetics, it is very important that he should choose some particular species of plants or animals that lends itself well to an experimental study of his problem.   When a chemist intends to investigate some general problem about, let us say, the velocity of chemical reactions, it is very important that he should choose some particular substances on which experiments relevant to his problem can be conveniently made.   The choice of appropriate experimental material is of great importance in the inductive investigation of any problem.   It seems to me that mathematics is, in several respects, the most appropriate experimental material for the study of inductive reasoning.   This study involves psychological experiments of a sort: you have to experience how your confidence in a conjecture is swayed by various kinds of evidence.   Thanks to their inherent simplicity and clarity, mathematical subjects lend themselves to this sort of psychological experiment much better than subjects in any other field.   On the following pages the reader may find ample opportunity to convince himself of this.

It is more philosophical, I think, to consider the more general idea of plausible reasoning instead of the particular case of inductive reasoning. It seems to me that the examples collected in this book lead up to a definite and fairly satisfactory aspect of plausible reasoning.   Yet I do not wish to force my views upon the reader.   In fact, I do not even state them in Vol. I; I want the examples to speak for themselves.   The first four chapters of Vol. II, however, are devoted to a more explicit general discussion of plausible reasoning.   There I state formally the patterns of plausible inference suggested by the foregoing examples, try to systematize these patterns, and survey some of their relations to each other and to the idea of probability.

I do not know whether the contents of these four chapters deserve to be called philosophy.   If this is philosophy, it is certainly a pretty low-brow kind of philosophy, more concerned with understanding concrete examples and the concrete behavior of people than with expounding generalities.   I know still less, of course, how the final judgement on my views will turn out. Yet I feel pretty confident that my examples can be useful to any reasonably unprejudiced student of induction or of plausible reasoning, who wishes to form his views in close touch with the observable facts.

5.   This work on *Mathematics and Plausible Reasoning*, which I have always regarded as a unit, falls naturally into two parts: *Induction and Analogy in Mathematics* (Vol. I), and *Patterns of Plausible Inference* (Vol. II).   For the

convenience of the student they have been issued as separate volumes. Vol.
I is entirely independent of Vol. II, and I think many students will want to
go through it carefully before reading Vol. II. It has more of the mathematical
"meat" of the work, and it supplies "data" for the inductive investigation of
induction in Vol. II. Some readers, who should be fairly sophisticated and
experienced in mathematics, will want to go directly to Vol. II, and for
these it will be a convenience to have it separately. For ease of reference
the chapter numbering is continuous through both volumes. I have not
provided an index, since an index would tend to render the terminology
more rigid than it is desirable in this kind of work. I believe the table of
contents will provide a satisfactory guide to the book.

The present work is a continuation of my earlier book *How to Solve It*.
The reader interested in the subject should read both, but the order does not
matter much. The present text is so arranged that it can be read independently of the former work. In fact, there are only few direct references in
the present book to the former and they can be disregarded in a first reading.
Yet there are indirect references to the former book on almost every page,
and in almost every sentence on some pages. In fact, the present work
provides numerous exercises and some more advanced illustrations to the
former which, in view of its size and its elementary character, had no space
for them.

The present book is also related to a collection of problems in Analysis
by G. Szegö and the author (see Bibliography). The problems in that
collection are carefully arranged in series so that they support each other
mutually, provide cues to each other, cover a certain subject-matter jointly,
and give the reader an opportunity to practice various moves important in
problem-solving. In the treatment of problems the present book follows
the method of presentation initiated by that former work, and this link is
not unimportant.

Two chapters in Vol. II of the present book deal with the theory of
probability. The first of these chapters is somewhat related to an elementary
exposition of the calculus of probability written by the author several years
ago (see the Bibliography). The underlying views on probability and the
starting points are the same, but otherwise there is little contact.

Some of the views offered in this book have been expressed before in my
papers quoted in the Bibliography. Extensive passages of papers no. 4, 6,
8, 9, and 10 have been incorporated in the following text. Acknowledgment
and my best thanks are due to the editors of the *American Mathematical
Monthly*, *Etudes de Philosophie des Sciences en Hommage à Ferdinand Gonseth*, and
*Proceedings of the International Congress of Mathematicians 1950*, who kindly
gave permission to reprint these passages.

Most parts of this book have been presented in my lectures, some parts
several times. In some parts and in some respects, I preserved the tone of
oral presentation. I do not think that such a tone is advisable in printed

presentation of mathematics in general, but in the present case it may be appropriate, or at least excusable.

6. The last chapter of Vol. II of the present book, dealing with Invention and Teaching, links the contents more explicitly to the former work of the author and points to a possible sequel.

The efficient use of plausible reasoning plays an essential role in problem-solving. The present book tries to illustrate this role by many examples, but there remain other aspects of problem-solving that need similar illustration.

Many points touched upon here need further work. My views on plausible reasoning should be confronted with the views of other authors, the historical examples should be more thoroughly explored, the views on invention and teaching should be investigated as far as possible with the methods of experimental psychology,[1] and so on. Several such tasks remain, but some of them may be thankless.

The present book is not a textbook. Yet I hope that in time it will influence the usual presentation of the textbooks and the choice of their problems. The task of rewriting the textbooks of the more usual subjects along these lines need not be thankless.

7. I wish to express my thanks to the Princeton University Press for the careful printing, and especially to Mr. Herbert S. Bailey, Jr., Director of the Press, for understanding help in several points. I am much indebted also to Mrs. Priscilla Feigen for the preparation of the typescript, and to Dr. Julius G. Baron for his kind help in reading the proofs.

GEORGE POLYA

*Stanford University*
*May 1953*

---

[1] Exploratory work in this direction has been undertaken in the Department of Psychology of Stanford University, within the framework of a project directed by E. R. Hilgard, under O.N.R. sponsorship.

# HINTS TO THE READER

THE section 2 of chapter VII is quoted as sect. 2 in chapter VII, but as sect. 7.2 in any other chapter. The subsection (3) of section 5 of chapter XIV is quoted as sect. 5 (3) in chapter XIV, but as sect. 14.5 (3) in any other chapter. We refer to example 26 of chapter XIV as ex. 26 in the same chapter, but as ex. 14.26 in any other chapter.

Some knowledge of elementary algebra and geometry may be enough to read substantial parts of the text. Thorough knowledge of elementary algebra and geometry and some knowledge of analytic geometry and calculus, including limits and infinite series, is sufficient for almost the whole text and the majority of the examples and comments. Yet more advanced knowledge is supposed in a few incidental remarks of the text, in some proposed problems, and in several comments. Usually some warning is given when more advanced knowledge is assumed.

The advanced reader who skips parts that appear to him too elementary may miss more than the less advanced reader who skips parts that appear to him too complex.

Some details of (not very difficult) demonstrations are often omitted without warning. Duly prepared for this eventuality, a reader with good critical habits need not spoil them.

Some of the problems proposed for solution are very easy, but a few are pretty hard. Hints that may facilitate the solution are enclosed in square brackets [ ]. The surrounding problems may provide hints. Especial attention should be paid to the introductory lines prefixed to the examples in some chapters, or prefixed to the First Part, or Second Part, of such examples.

The solutions are sometimes very short: they suppose that the reader has earnestly tried to solve the problem by his own means before looking at the printed solution.

A reader who spent serious effort on a problem may profit by it even if he does not succeed in solving it. For example, he may look at the solution, try to isolate what appears to him the key idea, put the book aside, and then try to work out the solution.

At some places, this book is lavish of figures or in giving small intermediate steps of a derivation. The aim is to render visible the *evolution* of a figure or a formula; see, for instance, Fig. 16.1–16.5. Yet no book can have enough figures or formulas. A reader may want to read a passage "in first approximation" or more thoroughly. If he wants to read more thoroughly, he should have paper and pencil at hand: he should be prepared to write or draw any formula or figure given in, or only indicated by, the text. Doing so, he has a better chance to see the evolution of the figure or formula, to understand how the various details contribute to the final product, and to remember the whole thing.

# CONTENTS

.

# Volume I

# Induction and Analogy
# in Mathematics

# INDUCTION

*It will seem not a little paradoxical to ascribe a great importance to observations even in that part of the mathematical sciences which is usually called Pure Mathematics, since the current opinion is that observations are restricted to physical objects that make impression on the senses. As we must refer the numbers to the pure intellect alone, we can hardly understand how observations and quasi-experiments can be of use in investigating the nature of the numbers. Yet, in fact, as I shall show here with very good reasons, the properties of the numbers known today have been mostly discovered by observation, and discovered long before their truth has been confirmed by rigid demonstrations. There are even many properties of the numbers with which we are well acquainted, but which we are not yet able to prove; only observations have led us to their knowledge. Hence we see that in the theory of numbers, which is still very imperfect, we can place our highest hopes in observations; they will lead us continually to new properties which we shall endeavor to prove afterwards. The kind of knowledge which is supported only by observations and is not yet proved must be carefully distinguished from the truth; it is gained by induction, as we usually say. Yet we have seen cases in which mere induction led to error. Therefore, we should take great care not to accept as true such properties of the numbers which we have discovered by observation and which are supported by induction alone. Indeed, we should use such a discovery as an opportunity to investigate more exactly the properties discovered and to prove or disprove them; in both cases we may learn something useful.*—EULER[1]

**1. Experience and belief.** Experience modifies human beliefs. We learn from experience or, rather, we ought to learn from experience. To make the best possible use of experience is one of the great human tasks and to work for this task is the proper vocation of scientists.

A scientist deserving this name endeavors to extract the most correct belief from a given experience and to gather the most appropriate experience in order to establish the correct belief regarding a given question. The

---

[1] Euler, *Opera Omnia*, ser. 1, vol. 2, p. 459, Specimen de usu observationum in mathesi pura.

scientist's procedure to deal with experience is usually called *induction*. Particularly clear examples of the inductive procedure can be found in mathematical research. We start discussing a simple example in the next section.

**2. Suggestive contacts.** Induction often begins with observation. A naturalist may observe bird life, a crystallographer the shapes of crystals. A mathematician, interested in the Theory of Numbers, observes the properties of the integers 1, 2, 3, 4, 5, . . . .

If you wish to observe bird life with some chance of obtaining interesting results, you should be somewhat familiar with birds, interested in birds, perhaps you should even like birds. Similarly, if you wish to observe the numbers, you should be interested in, and somewhat familiar with, them. You should distinguish even and odd numbers, you should know the squares 1, 4, 9, 16, 25, . . . and the primes 2, 3, 5, 7, 11, 13, 17, 19, 23, 29, . . . . (It is better to keep 1 apart as "unity" and not to classify it as a prime.) Even with so modest a knowledge you may be able to observe something interesting.

By some chance, you come across the relations

$$3 + 7 = 10, \quad 3 + 17 = 20, \quad 13 + 17 = 30$$

and notice some resemblance between them. It strikes you that the numbers 3, 7, 13, and 17 are odd primes. The sum of two odd primes is necessarily an even number; in fact, 10, 20, and 30 are even. What about the *other* even numbers? Do they behave similarly? The first even number which is a sum of two odd primes is, of course,

$$6 = 3 + 3.$$

Looking beyond 6, we find that

$$8 = 3 + 5$$

$$10 = 3 + 7 = 5 + 5$$

$$12 = 5 + 7$$

$$14 = 3 + 11 = 7 + 7$$

$$16 = 3 + 13 = 5 + 11.$$

Will it go on like this forever? At any rate, the particular cases observed suggest a general statement: *Any even number greater than 4 is the sum of two odd primes.* Reflecting upon the exceptional cases, 2 and 4, which cannot be split into a sum of two odd primes, we may prefer the following more sophisticated statement: *Any even number that is neither a prime nor the square of a prime, is the sum of two odd primes.*

We arrived so at formulating a *conjecture*. We found this conjecture by *induction*. That is, it was suggested by observation, indicated by particular instances.

These indications are rather flimsy; we have only very weak grounds to believe in our conjecture. We may find, however, some consolation in the fact that the mathematician who discovered this conjecture a little more than two hundred years ago, Goldbach, did not possess much stronger grounds for it.

Is Goldbach's conjecture true? Nobody can answer this question today. In spite of the great effort spent on it by some great mathematicians, Goldbach's conjecture is today, as it was in the days of Euler, one of those "many properties of the numbers with which we are well acquainted, but which we are not yet able to prove" or disprove.

Now, let us look back and try to perceive such steps in the foregoing reasoning as might be typical of the inductive procedure.

First, we *noticed some similarity*. We recognized that 3, 7, 13, and 17 are primes, 10, 20, and 30 even numbers, and that the three equations $3 + 7 = 10$, $3 + 17 = 20$, $13 + 17 = 30$ are *analogous* to each other.

Then there was a step of *generalization*. From the examples 3, 7, 13, and 17 we passed to all odd primes, from 10, 20, and 30 to all even numbers, and then on to a possibly general relation

$$\text{even number} = \text{prime} + \text{prime}.$$

We arrived so at a clearly formulated general statement, which, however, is merely a conjecture, merely *tentative*. That is, the statement is by no means proved, it cannot have any pretension to be true, it is merely an attempt to get at the truth.

This conjecture has, however, some *suggestive points of contact* with experience, with "the facts," with "reality." It is true for the particular even numbers 10, 20, 30, also for 6, 8, 12, 14, 16.

With these remarks, we outlined roughly a first stage of the inductive process.

**3. Supporting contacts.** You should not put too much trust in any unproved conjecture, even if it has been propounded by a great authority, even if it has been propounded by yourself. You should try to prove it or to disprove it; you should *test* it.

We test Goldbach's conjecture if we examine some new even number and decide whether it is or is not a sum of two odd primes. Let us examine, for instance, the number 60. Let us perform a "quasi-experiment," as Euler expressed himself. The number 60 is even, but is it the sum of two primes? Is it true that

$$60 = 3 + \text{prime}?$$

No, 57 is not a prime. Is

$$60 = 5 + \text{prime}?$$

The answer is again "No": 55 is not a prime. If it goes on in this way, the conjecture will be exploded. Yet the next trial yields

$$60 = 7 + 53$$

and 53 is a prime. The conjecture has been verified in one more case.

The contrary outcome would have settled the fate of Goldbach's conjecture once and for all. If, trying all primes under a given even number, such as 60, you never arrive at a decomposition into a sum of two primes, you thereby explode the conjecture irrevocably. Having verified the conjecture in the case of the even number 60, you cannot reach such a definite conclusion. You certainly do not prove the theorem by a single verification. It is natural, however, to interpret such a verification as a *favorable sign*, speaking for the conjecture, rendering it *more credible*, although, of course, it is left to your personal judgement how much weight you attach to this favorable sign.

Let us return, for a moment, to the number 60. After having tried the primes 3, 5, and 7, we can try the remaining primes under 30. (Obviously, it is unnecessary to go further than 30 which equals 60/2, since one of the two primes, the sum of which should be 60, must be less than 30.) We obtain so all the decompositions of 60 into a sum of two primes:

$$60 = 7 + 53 = 13 + 47 = 17 + 43 = 19 + 41 = 23 + 37 = 29 + 31.$$

We can proceed systematically and examine the even numbers one after the other, as we have just examined the even number 60. We can *tabulate* the results as follows:

$$6 = 3 + 3$$
$$8 = 3 + 5$$
$$10 = 3 + 7 = 5 + 5$$
$$12 = 5 + 7$$
$$14 = 3 + 11 = 7 + 7$$
$$16 = 3 + 13 = 5 + 11$$
$$18 = 5 + 13 = 7 + 11$$
$$20 = 3 + 17 = 7 + 13$$
$$22 = 3 + 19 = 5 + 17 = 11 + 11$$
$$24 = 5 + 19 = 7 + 17 = 11 + 13$$
$$26 = 3 + 23 = 7 + 19 = 13 + 13$$
$$28 = 5 + 23 = 11 + 17$$
$$30 = 7 + 23 = 11 + 19 = 13 + 17.$$

The conjecture is verified in all cases that we have examined here. Each verification that lengthens the table strengthens the conjecture, renders it more credible, adds to its plausibility. Of course, no amount of such verifications could prove the conjecture.

We should examine our collected observations, we should compare and combine them, we should look for some clue that may be hidden behind them. In our case, it is very hard to discover some essential clue in the table. Still examining the table, we may realize more clearly the meaning of the conjecture. The table shows how often the even numbers listed in it can be represented as a sum of two primes (6 just once, 30 three times). The number of such representations of the even number $2n$ seems to "increase irregularly" with $n$. Goldbach's conjecture expresses the hope that the number of representations will never fall down to 0, however far we may extend the table.

Among the particular cases that we have examined we could distinguish two groups: those which preceded the formulation of the conjecture and those which came afterwards. The former suggested the conjecture, the latter supported it. Both kinds of cases provide some sort of contact between the conjecture and "the facts." The table does not distinguish between "suggestive" and "supporting" points of contact.

Now, let us look back at the foregoing reasoning and try to see in it traits typical of the inductive process.

Having conceived a conjecture, we tried to find out whether it is true or false. Our conjecture was a general statement suggested by certain particular instances in which we have found it true. We examined further particular instances. As it turned out that the conjecture is true in all instances examined, our confidence in it increased.

We did, it seems to me, only things that reasonable people usually do. In so doing, we seem to accept a principle: *A conjectural general statement becomes more credible if it is verified in a new particular case.*

Is this the principle underlying the process of induction?

**4. The inductive attitude.** In our personal life we often cling to illusions. That is, we do not dare to examine certain beliefs which could be easily contradicted by experience, because we are afraid of upsetting our emotional balance. There may be circumstances in which it is not unwise to cling to illusions, but in science we need a very different attitude, the *inductive attitude.* This attitude aims at adapting our beliefs to our experience as efficiently as possible. It requires a certain preference for what is matter of fact. It requires a ready ascent from observations to generalizations, and a ready descent from the highest generalizations to the most concrete observations. It requires saying "maybe" and "perhaps" in a thousand different shades. It requires many other things, especially the following three.

First, we should be ready to revise any one of our beliefs.

Second, we should change a belief when there is a compelling reason to change it.

Third, we should not change a belief wantonly, without some good reason.

These points sound pretty trivial. Yet one needs rather unusual qualities to live up to them.

The first point needs "intellectual courage." You need courage to revise your beliefs. Galileo, challenging the prejudice of his contemporaries and the authority of Aristotle, is a great example of intellectual courage.

The second point needs "intellectual honesty." To stick to my conjecture that has been clearly contradicted by experience just because it is *my* conjecture would be dishonest.

The third point needs "wise restraint." To change a belief without serious examination, just for the sake of fashion, for example, would be foolish. Yet we have neither the time nor the strength to examine seriously all our beliefs. Therefore it is wise to reserve the day's work, our questions, and our active doubts for such beliefs as we can reasonably expect to amend. "Do not believe anything, but question only what is worth questioning."

Intellectual courage, intellectual honesty, and wise restraint are the moral qualities of the scientist.

### EXAMPLES AND COMMENTS ON CHAPTER I.

**1.** Guess the rule according to which the successive terms of the following sequence are chosen:

$$11, 31, 41, 61, 71, 101, 131, \ldots .$$

**2.** Consider the table:

$$1 = 0 + 1$$
$$2 + 3 + 4 = 1 + 8$$
$$5 + 6 + 7 + 8 + 9 = 8 + 27$$
$$10 + 11 + 12 + 13 + 14 + 15 + 16 = 27 + 64$$

Guess the general law suggested by these examples, express it in suitable mathematical notation, and prove it.

**3.** Observe the values of the successive sums

$$1, \quad 1 + 3, \quad 1 + 3 + 5, \quad 1 + 3 + 5 + 7, \quad \ldots .$$

Is there a simple rule?

**4.** Observe the values of the consecutive sums

$$1, \quad 1 + 8, \quad 1 + 8 + 27, \quad 1 + 8 + 27 + 64, \quad \ldots \;.$$

Is there a simple rule?

**5.** The three sides of a triangle are of lengths $l$, $m$, and $n$, respectively. The numbers $l$, $m$, and $n$ are positive integers, $l \leq m \leq n$. Find the number of different triangles of the described kind for a given $n$. [Take $n = 1, 2, 3, 4, 5, \ldots$ .] Find a general law governing the dependence of the number of triangles on $n$.

**6.** The first three terms of the sequence 5, 15, 25, . . . (numbers ending in 5) are divisible by 5. Are also the following terms divisible by 5?

The first three terms of the sequence 3, 13, 23, . . . (numbers ending in 3) are prime numbers. Are also the following terms prime numbers?

**7.** By formal computation we find

$$(1 + 1!x + 2!x^2 + 3!x^3 + 4!x^4 + 5!x^5 + 6!x^6 + \ldots )^{-1}$$
$$= 1 - x - x^2 - 3x^3 - 13x^4 - 71x^5 - 461x^6 \ldots \;.$$

This suggests two conjectures about the following coefficients of the right hand power series: (1) they are all negative; (2) they are all primes. Are these two conjectures equally trustworthy?

**8.** Set

$$\left( 1 - \frac{x}{1} + \frac{x^2}{2} - \frac{x^3}{3} + \ldots \right)^{-1} = A_0 + \frac{A_1 x}{1!} + \frac{A_2 x^2}{2!} + \ldots \;.$$

We find that for

$$n = 0 \quad 1 \quad 2 \quad 3 \quad 4 \quad 5 \quad 6 \quad 7 \quad 8 \quad 9$$

$$A_n = 1 \quad 1 \quad 1 \quad 2 \quad 4 \quad 14 \quad 38 \quad 216 \quad 600 \quad 6240.$$

State a conjecture.

**9.** The great French mathematician Fermat considered the sequence

$$5, 17, 257, 65537, \ldots ,$$

the general term of which is $2^{2^n} + 1$. He observed that the first four terms (here given), corresponding to $n = 1, 2, 3$, and 4, are primes. He conjectured that the following terms are also primes. Although he did not prove it, he felt so sure of his conjecture that he challenged Wallis and other English mathematicians to prove it. Yet Euler found that the very next term, $2^{32} + 1$, corresponding to $n = 5$, is not a prime: it is divisible by 641.[2]  See the passage of Euler at the head of this chapter: "Yet we have seen cases in which mere induction led to error."

**10.** In verifying Goldbach's conjecture for $2n = 60$ we tried successively the primes $p$ under $n = 30$. We could have also tried, however, the primes $p'$ between $n = 30$ and $2n = 60$. Which procedure is likely to be more advantageous for greater $n$?

**11.** In a dictionary, you will find among the explanations for the words "induction," "experiment," and "observation" sentences like the following.

"Induction is inferring a general law from particular instances, or a production of facts to prove a general statement."

"Experiment is a procedure for testing hypotheses."

"Observation is an accurate watching and noting of phenomena as they occur in nature with regard to cause and effect or mutual relations."

Do these descriptions apply to our example discussed in sect. 2 and 3?

**12.** *Yes and No.* The mathematician as the naturalist, in testing some consequence of a conjectural general law by a new observation, addresses a question to Nature: "I suspect that this law is true. Is it true?" If the consequence is clearly refuted, the law cannot be true. If the consequence is clearly verified, there is some indication that the law may be true. Nature may answer Yes or No, but it whispers one answer and thunders the other, its Yes is provisional, its No is definitive.

**13.** *Experience and behavior.* Experience modifies human behavior. And experience modifies human beliefs. These two things are not independent of each other. Behavior often results from beliefs, beliefs are potential behavior. Yet you can see the other fellow's behavior, you cannot see his beliefs. Behavior is more easily observed than belief. Everybody knows that "a burnt child dreads the fire," which expresses just what we said: experience modifies human behavior.

Yes, and it modifies animal behavior, too.

In my neighborhood there is a mean dog that barks and jumps at people without provocation. But I have found that I can protect myself rather easily. If I stoop and pretend to pick up a stone, the dog runs away howling. All dogs do not behave so, and it is easy to guess what kind of experience gave this dog this behavior.

The bear in the zoo "begs for food." That is, when there is an onlooker around, it strikes a ridiculous posture which quite frequently prompts the onlooker to throw a lump of sugar into the cage. Bears not in captivity probably never assume such a preposterous posture and it is easy to imagine what kind of experience led to the zoo bear's begging.

A thorough investigation of induction should include, perhaps, the study of animal behavior.

---

[2] Euler, *Opera Omnia*, ser. 1, vol. 2, p. 1–5.  Hardy and Wright, *The Theory of Numbers*, p. 14–15.

**14.** *The logician, the mathematician, the physicist, and the engineer.* "Look at this mathematician," said the logician. "He observes that the first ninety-nine numbers are less than hundred and infers hence, by what he calls induction, that all numbers are less than a hundred."

"A physicist believes," said the mathematician, "that 60 is divisible by all numbers. He observes that 60 is divisible by 1, 2, 3, 4, 5, and 6. He examines a few more cases, as 10, 20, and 30, taken at random as he says. Since 60 is divisible also by these, he considers the experimental evidence sufficient."

"Yes, but look at the engineers," said the physicist. "An engineer suspected that all odd numbers are prime numbers. At any rate, 1 can be considered as a prime number, he argued. Then there come 3, 5, and 7, all indubitably primes. Then there comes 9; an awkward case, it does not seem to be a prime number. Yet 11 and 13 are certainly primes. 'Coming back to 9,' he said, 'I conclude that 9 must be an experimental error.'"

It is only too obvious that induction can lead to error. Yet it is remarkable that induction sometimes leads to truth, since the chances of error appear so overwhelming. Should we begin with the study of the obvious cases in which induction fails, or with the study of those remarkable cases in which induction succeeds? The study of precious stones is understandably more attractive than that of ordinary pebbles and, moreover, it was much more the precious stones than the pebbles that led the mineralogists to the wonderful science of crystallography.

# II

# GENERALIZATION, SPECIALIZATION, ANALOGY

---

*And I cherish more than anything else the Analogies, my most trustworthy masters. They know all the secrets of Nature, and they ought to be least neglected in Geometry.*—KEPLER

---

**1. Generalization, Specialization, Analogy, and Induction.** Let us look again at the example of inductive reasoning that we have discussed in some detail (sect. 1.2, 1.3). We started from observing the *analogy* of the three relations

$$3 + 7 = 10, \quad 3 + 17 = 20, \quad 13 + 17 = 30,$$

we *generalized* in ascending from 3, 7, 13, and 17 to all primes, from 10, 20, and 30 to all even numbers, and then we *specialized* again, came down to test particular even numbers such as 6 or 8 or 60.

This first example is extremely simple. It illustrates quite correctly the role of generalization, specialization, and analogy in inductive reasoning. Yet we should examine less meager, more colorful illustrations and, before that, we should discuss generalization, specialization, and analogy, these great sources of discovery, for their own sake.

**2. Generalization** is passing from the consideration of a given set of objects to that of a larger set, containing the given one. For example, we generalize when we pass from the consideration of triangles to that of polygons with an arbitrary number of sides. We generalize also when we pass from the study of the trigonometric functions of an acute angle to the trigonometric functions of an unrestricted angle.

It may be observed that in these two examples the generalization was effected in two characteristically different ways. In the first example, in passing from triangles to polygons with $n$ sides, we replace a constant by a variable, the fixed integer 3 by the arbitrary integer $n$ (restricted only by the inequality $n \geqq 3$). In the second example, in passing from acute angles to arbitrary angles $\alpha$, we remove a restriction, namely the restriction that $0° < \alpha < 90°$.

We often generalize in passing from just one object to a whole class containing that object.

**3. Specialization** is passing from the consideration of a given set of objects to that of a smaller set, contained in the given one. For example, we specialize when we pass from the consideration of polygons to that of regular polygons, and we specialize still further when we pass from regular polygons with $n$ sides to the regular, that is, equilateral, triangle.

These two subsequent passages were effected in two characteristically different ways. In the first passage, from polygons to regular polygons, we introduced a restriction, namely that all sides and all angles of the polygon be equal. In the second passage we substituted a special object for a variable, we put 3 for the variable integer $n$.

Very often we specialize in passing from a whole class of objects to just one object contained in the class. For example, when we wish to check some general assertion about prime numbers we pick out some prime number, say 17, and we examine whether that general assertion is true or not for just this prime 17.

**4. Analogy.** There is nothing vague or questionable in the concepts of generalization and specialization. Yet as we start discussing analogy we tread on a less solid ground.

*Analogy* is a sort of similarity. It is, we could say, similarity on a more definite and more conceptual level. Yet we can express ourselves a little more accurately. The essential difference between analogy and other kinds of similarity lies, it seems to me, in the intentions of the thinker. Similar objects agree with each other in some aspect. If you intend to reduce the aspect in which they agree to definite concepts, you regard those similar objects as *analogous*. If you succeed in getting down to clear concepts, you have *clarified* the analogy.

Comparing a young woman to a flower, poets feel some similarity, I hope, but usually they do not contemplate analogy. In fact, they scarcely intend to leave the emotional level or reduce that comparison to something measurable or conceptually definable.

Looking in a natural history museum at the skeletons of various mammals, you may find them all frightening. If this is all the similarity you can find between them, you do not see much analogy. Yet you may perceive a wonderfully suggestive analogy if you consider the hand of a man, the paw of a cat, the foreleg of a horse, the fin of a whale, and the wing of a bat, these organs so differently used, as composed of similar parts similarly related to each other.

The last example illustrates the most typical case of clarified analogy; two *systems* are analogous, if they *agree in clearly definable relations of their respective parts*.

For instance, a triangle in a plane is analogous to a tetrahedron in space. In the plane, 2 straight lines cannot include a finite figure, but 3 may include a triangle. In space, 3 planes cannot include a finite figure but 4 may include a tetrahedron. The relation of the triangle to the plane is the same as that of the tetrahedron to space in so far as both the triangle and the tetrahedron are bounded by the minimum number of simple bounding elements. Hence the analogy.

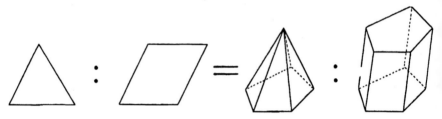

Fig. 2.1.    Analogous relations in plane and space.

One of the meanings of the Greek word "analogia," from which the word "analogy" originates, is "proportion." In fact, the system of the two numbers 6 and 9 is "analogous" to the system of the two numbers 10 and 15 in so far as the two systems agree in the ratio of their corresponding terms,

$$6 : 9 = 10 : 15.$$

Proportionality, or agreement in the ratios of corresponding parts, which we may see intuitively in geometrically similar figures, is a very suggestive case of analogy.

Here is another example. We may regard a triangle and a pyramid as analogous figures. On the one hand take a segment of a straight line, and on the other hand a polygon. Connect all points of the segment with a point outside the line of the segment, and you obtain a triangle. Connect all points of the polygon with a point outside the plane of the polygon, and you obtain a pyramid. In the same manner, we may regard a parallelogram and a prism as analogous figures. In fact, move a segment or a polygon parallel to itself, across the direction of its line or plane, and the one will describe a parallelogram, the other a prism. We may be tempted to express these corresponding relations between plane and solid figures by a sort of proportion and if, for once, we do not resist temptation, we arrive at fig. 2.1. This figure modifies the usual meaning of certain symbols (: and =) in the same way as the meaning of the word "analogia" was modified in the course of linguistic history: from "proportion" to "analogy."

The last example is instructive in still another respect. Analogy, especially incompletely clarified analogy, may be ambiguous. Thus, comparing plane and solid geometry, we found first that a triangle in a plane is analogous to a tetrahedron in space and then that a triangle is analogous to a pyramid. Now, both analogies are reasonable, each is valuable at its place. There are several analogies between plane and solid geometry and not just one privileged analogy.

Fig. 2.2 exhibits how, starting from a triangle, we may ascend to a polygon by generalization, descend to an equilateral triangle by specialization, or pass to different solid figures by analogy—there are analogies on all sides.

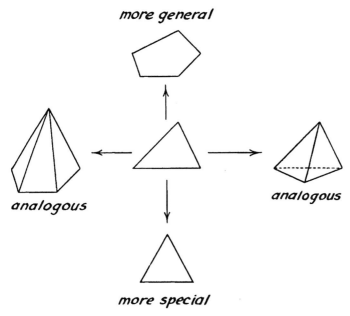

Fig. 2.2.   Generalization, specialization, analogy.

And, remember, do not neglect vague analogies. Yet, if you wish them respectable, try to clarify them.

**5. Generalization, Specialization, and Analogy** often concur in solving mathematical problems.[1] Let us take as an example the proof of the best known theorem of elementary geometry, the theorem of Pythagoras. The proof that we shall discuss is not new; it is due to Euclid himself (Euclid VI, 31).

(1) We consider a right triangle with sides $a$, $b$, and $c$, of which the first, $a$, is the hypotenuse. We wish to show that

(A) $$a^2 = b^2 + c^2.$$

---

[1] This section reproduces with slight changes a Note of the author in the *American Mathematical Monthly*, v. 55 (1948), p. 241–243.

This aim suggests that we describe squares on the three sides of our right triangle. And so we arrive at the not unfamiliar part I of our compound figure, fig. 2.3. (The reader should draw the parts of this figure as they arise, in order to see it in the making.)

(2) Discoveries, even very modest discoveries, need some remark, the recognition of some relation. We can discover the following proof by observing the *analogy* between the familiar part I of our compound figure

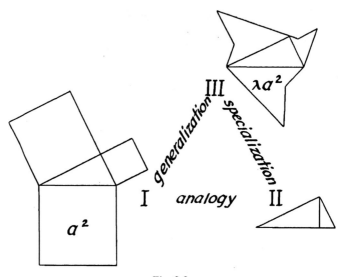

Fig. 2.3.

and the scarcely less familiar part II: the same right triangle that arises in I is divided in II into two parts by the altitude perpendicular to the hypotenuse.

(3) Perhaps, you fail to perceive the analogy between I and II. This analogy, however, can be made explicit by a common *generalization* of I and II which is expressed in III. There we find again the same right triangle, and on its three sides three polygons are described which are similar to each other but arbitrary otherwise.

(4) The area of the square described on the hypotenuse in I is $a^2$. The area of the irregular polygon described on the hypotenuse in III can be put equal to $\lambda a^2$; the factor $\lambda$ is determined as the ratio of two given areas. Yet then, it follows from the similarity of the three polygons described on the sides $a$, $b$, and $c$ of the triangle in III that their areas are equal to $\lambda a^2$, $\lambda b^2$, and $\lambda c^2$, respectively.

Now, if the equation (A) should be true (as stated by the theorem that we wish to prove), then also the following would be true:

(B)                     $$\lambda a^2 = \lambda b^2 + \lambda c^2.$$

In fact, very little algebra is needed to derive (B) from (A). Now, (B) represents a *generalization* of the original theorem of Pythagoras: *If three similar polygons are described on three sides of a right triangle, the one described on the hypotenuse is equal in area to the sum of the two others.*

It is instructive to observe that this generalization is *equivalent* to the special case from which we started. In fact, we can derive the equations (A) and (B) from each other, by multiplying or dividing by $\lambda$ (which is, as the ratio of two areas, different from 0).

(5) The general theorem expressed by (B) is equivalent not only to the special case (A), but to any other special case. Therefore, if any such special case should turn out to be obvious, the general case would be demonstrated.

Now, trying to *specialize* usefully, we look around for a suitable special case. Indeed II represents such a case. In fact, the right triangle described on its own hypotenuse is similar to the two other triangles described on the two legs, as is well known and easy to see. And, obviously, the area of the whole triangle is equal to the sum of its two parts. And so, the theorem of Pythagoras has been proved.

The foregoing reasoning is eminently instructive. A case is instructive if we can learn from it something applicable to other cases, and the more instructive the wider the range of possible applications. Now, from the foregoing example we can learn the use of such fundamental mental operations as generalization, specialization, and the perception of analogies. There is perhaps no discovery either in elementary or in advanced mathematics or, for that matter, in any other subject that could do without these operations, especially without analogy.

The foregoing example shows how we can ascend by generalization from a special case, as from the one represented by I, to a more general situation as to that of III, and redescend hence by specialization to an analogous case, as to that of II. It shows also the fact, so usual in mathematics and still so surprising to the beginner, or to the philosopher who takes himself for advanced, that the general case can be logically equivalent to a special case. Our example shows, naïvely and suggestively, how generalization, specialization, and analogy are naturally combined in the effort to attain the desired solution. Observe that only a minimum of preliminary knowledge is needed to understand fully the foregoing reasoning.

**6. Discovery by analogy.** Analogy seems to have a share in all discoveries, but in some it has the lion's share. I wish to illustrate this by an example which is not quite elementary, but is of historic interest and far more impressive than any quite elementary example of which I can think.

Jacques Bernoulli, a Swiss mathematician (1654–1705), a contemporary of Newton and Leibnitz, discovered the sum of several infinite series, but did not succeed in finding the sum of the reciprocals of the squares,

$$1 + \frac{1}{4} + \frac{1}{9} + \frac{1}{25} + \frac{1}{36} + \frac{1}{49} + \ldots \;.$$

"If somebody should succeed," wrote Bernoulli, "in finding what till now withstood our efforts and communicate it to us, we shall be much obliged to him."

The problem came to the attention of another Swiss mathematician, Leonhard Euler (1707–1783), who was born at Basle as was Jacques Bernoulli and was a pupil of Jacques' brother, Jean Bernoulli (1667–1748). He found various expressions for the desired sum (definite integrals, other series), none of which satisfied him. He used one of these expressions to compute the sum numerically to seven places (1.644934). Yet this is only an approximate value and his goal was to find the exact value. He discovered it, eventually. Analogy led him to an extremely daring conjecture.

(1) We begin by reviewing a few elementary algebraic facts essential to Euler's discovery. If the equation of degree $n$

$$a_0 + a_1 x + a_2 x^2 + \ldots + a_n x^n = 0$$

has $n$ different roots

$$\alpha_1, \; \alpha_2, \; \ldots \; \alpha_n$$

the polynomial on its left hand side can be represented as a product of $n$ linear factors,

$$a_0 + a_1 x + a_2 x^2 + \ldots + a_n x^n =$$
$$a_n(x - \alpha_1)\,(x - \alpha_2) \ldots (x - \alpha_n).$$

By comparing the terms with the same power of $x$ on both sides of this identity, we derive the well known relations between the roots and the coefficients of an equation, the simplest of which is

$$a_{n-1} = -a_n(\alpha_1 + \alpha_2 + \ldots + \alpha_n);$$

we find this by comparing the terms with $x^{n-1}$.

There is another way of presenting the decomposition in linear factors. If none of the roots $\alpha_1, \alpha_2, \ldots \alpha_n$ is equal to 0, or (which is the same) if $a_0$ is different from 0, we have also

$$a_0 + a_1 x + a_2 x^2 + \ldots + a_n x^n$$
$$= a_0 \left(1 - \frac{x}{\alpha_1}\right)\left(1 - \frac{x}{\alpha_2}\right) \ldots \left(1 - \frac{x}{\alpha_n}\right)$$

and

$$a_1 = -a_0 \left(\frac{1}{\alpha_1} + \frac{1}{\alpha_2} + \ldots + \frac{1}{\alpha_n}\right).$$

There is still another variant. Suppose that the equation is of degree $2n$, has the form

$$b_0 - b_1 x^2 + b_2 x^4 - \ldots + (-1)^n b_n x^{2n} = 0$$

and $2n$ different roots

$$\beta_1, \quad -\beta_1, \quad \beta_2, \quad -\beta_2, \quad \ldots \quad \beta_n, \quad -\beta_n.$$

Then

$$b_0 - b_1 x^2 + b_2 x^4 - \ldots + (-1)^n b_n x^{2n}$$

$$= b_0 \left(1 - \frac{x^2}{\beta_1^2}\right)\left(1 - \frac{x^2}{\beta_2^2}\right) \ldots \left(1 - \frac{x^2}{\beta_n^2}\right)$$

and

$$b_1 = b_0 \left(\frac{1}{\beta_1^2} + \frac{1}{\beta_2^2} + \ldots + \frac{1}{\beta_n^2}\right).$$

(2) Euler considers the equation

$$\sin x = 0$$

or

$$\frac{x}{1} - \frac{x^3}{1 \cdot 2 \cdot 3} + \frac{x^5}{1 \cdot 2 \cdot 3 \cdot 4 \cdot 5} - \frac{x^7}{1 \cdot 2 \cdot 3 \cdots 7} + \ldots = 0.$$

The left hand side has an infinity of terms, is of "infinite degree." Therefore, it is no wonder, says Euler, that there is an infinity of roots

$$0, \quad \pi, \quad -\pi, \quad 2\pi, \quad -2\pi, \quad 3\pi, \quad -3\pi, \quad \ldots .$$

Euler discards the root 0. He divides the left hand side of the equation by $x$, the linear factor corresponding to the root 0, and obtains so the equation

$$1 - \frac{x^2}{2 \cdot 3} + \frac{x^4}{2 \cdot 3 \cdot 4 \cdot 5} - \frac{x^6}{2 \cdot 3 \cdot 4 \cdot 5 \cdot 6 \cdot 7} + \ldots = 0$$

with the roots

$$\pi, \quad -\pi, \quad 2\pi, \quad -2\pi, \quad 3\pi, \quad -3\pi, \quad \ldots .$$

We have seen an analogous situation before, under (1), as we discussed the last variant of the decomposition in linear factors. Euler concludes, by analogy, that

$$\frac{\sin x}{x} = 1 - \frac{x^2}{2 \cdot 3} + \frac{x^4}{2 \cdot 3 \cdot 4 \cdot 5} - \frac{x^6}{2 \cdot 3 \cdots 7} + \ldots$$

$$= \left(1 - \frac{x^2}{\pi^2}\right)\left(1 - \frac{x^2}{4\pi^2}\right)\left(1 - \frac{x^2}{9\pi^2}\right) \ldots,$$

$$\frac{1}{2 \cdot 3} = \frac{1}{\pi^2} + \frac{1}{4\pi^2} + \frac{1}{9\pi^2} + \ldots,$$

$$1 + \frac{1}{4} + \frac{1}{9} + \frac{1}{16} + \ldots = \frac{\pi^2}{6}.$$

This is the series that withstood the efforts of Jacques Bernoulli—but it was a daring conclusion.

(3) Euler knew very well that his conclusion was daring. "The method was new and never used yet for such a purpose," he wrote ten years later. He saw some objections himself and many objections were raised by his mathematical friends when they recovered from their first admiring surprise.

Yet Euler had his reasons to trust his discovery. First of all, the numerical value for the sum of the series which he has computed before, agreed to the last place with $\pi^2/6$. Comparing further coefficients in his expression of sin $x$ as a product, he found the sum of other remarkable series, as that of the reciprocals of the fourth powers,

$$1 + \frac{1}{16} + \frac{1}{81} + \frac{1}{256} + \frac{1}{625} + \ldots = \frac{\pi^4}{90}.$$

Again, he examined the numerical value and again he found agreement.

(4) Euler also tested his method on other examples. Doing so he succeeded in rederiving the sum $\pi^2/6$ for Jacques Bernoulli's series by various modifications of his first approach. He succeeded also in rediscovering by his method the sum of an important series due to Leibnitz.

Let us discuss the last point. Let us consider, following Euler, the equation

$$1 - \sin x = 0.$$

It has the roots

$$\frac{\pi}{2}, \quad -\frac{3\pi}{2}, \quad \frac{5\pi}{2}, \quad -\frac{7\pi}{2}, \quad \frac{9\pi}{2}, \quad -\frac{11\pi}{2}, \quad \ldots \quad .$$

Each of these roots is, however, a double root. (The curve $y = \sin x$ does not intersect the line $y = 1$ at these abscissas, but is tangent to it. The derivative of the left hand side vanishes for the same values of $x$, but not the second derivative.) Therefore, the equation

$$1 - \frac{x}{1} + \frac{x^3}{1 \cdot 2 \cdot 3} - \frac{x^5}{1 \cdot 2 \cdot 3 \cdot 5} + \ldots = 0$$

has the roots

$$\frac{\pi}{2}, \quad \frac{\pi}{2}, \quad -\frac{3\pi}{2}, \quad -\frac{3\pi}{2}, \quad \frac{5\pi}{2}, \quad \frac{5\pi}{2}, \quad -\frac{7\pi}{2}, \quad -\frac{7\pi}{2}, \quad \ldots$$

and Euler's analogical conclusion leads to the decomposition in linear factors

$$1 - \sin x = 1 - \frac{x}{1} + \frac{x^3}{1 \cdot 2 \cdot 3} - \frac{x^5}{1 \cdot 2 \cdot 3 \cdot 4 \cdot 5} + \ldots$$

$$= \left(1 - \frac{2x}{\pi}\right)^2 \left(1 + \frac{2x}{3\pi}\right)^2 \left(1 - \frac{2x}{5\pi}\right)^2 \left(1 + \frac{2x}{7\pi}\right)^2 \ldots \quad .$$

Comparing the coefficient of $x$ on both sides, we obtain

$$-1 = -\frac{4}{\pi} + \frac{4}{3\pi} - \frac{4}{5\pi} + \frac{4}{7\pi} - \cdots ,$$

$$\frac{\pi}{4} = 1 - \frac{1}{3} + \frac{1}{5} - \frac{1}{7} + \frac{1}{9} - \frac{1}{11} + \cdots .$$

This is Leibnitz's celebrated series; Euler's daring procedure led to a known result. "For our method," says Euler, "which may appear to some as not reliable enough, a great confirmation comes here to light. Therefore, we should not doubt at all of the other things which are derived by the same method."

(5) Yet Euler kept on doubting. He continued the numerical verifications described above under (3), examined more series and more decimal places, and found agreement in all cases examined. He tried other approaches, too, and, finally, he succeeded in verifying not only numerically, but exactly, the value $\pi^2/6$ for Jacques Bernoulli's series. He found a new proof. This proof, although hidden and ingenious, was based on more usual considerations and was accepted as completely rigorous. Thus, the most conspicuous consequence of Euler's discovery was satisfactorily verified. These arguments, it seems, convinced Euler that his result was correct.[2]

**7. Analogy and induction.** We wish to learn something about the nature of inventive and inductive reasoning. What can we learn from the foregoing story?

(1) Euler's decisive step was daring. In strict logic, it was an outright fallacy: he applied a rule to a case for which the rule was not made, a rule about algebraic equations to an equation which is not algebraic. In strict logic, Euler's step was not justified. Yet it was justified by analogy, by the analogy of the most successful achievements of a rising science that he called himself a few years later the "Analysis of the Infinite." Other mathematicians, before Euler, passed from finite differences to infinitely small differences, from sums with a finite number of terms to sums with an infinity of terms, from finite products to infinite products. And so Euler passed from equations of finite degree (algebraic equations) to equations of infinite degree, applying the rules made for the finite to the infinite.

This analogy, this passage from the finite to the infinite, is beset with pitfalls. How did Euler avoid them? He was a genius, some people will answer, and of course that is no explanation at all. Euler had shrewd

---

[2] Much later, almost ten years after his first discovery, Euler returned to the subject, answered the objections, completed to some extent his original heuristic approach, and gave a new, essentially different proof. See L. Euler, *Opera Omnia*, ser. 1, vol. 14, p. 73–86, 138–155, 177–186, and also p. 156–176, containing a note by Paul Stäckel on the history of the problem.

reasons for trusting his discovery. We can understand his reasons with a little common sense, without any miraculous insight specific to genius.

(2) Euler's reasons for trusting his discovery, summarized in the fore-going,[3] are *not* demonstrative. Euler does not reexamine the grounds for his conjecture,[4] for his daring passage from the finite to the infinite; he examines only its consequences. He regards the verification of any such consequence as an argument in favor of his conjecture. He accepts both approximative and exact verifications, but seems to attach more weight to the latter. He examines also the consequences of closely related analogous conjectures[5] and he regards the verification of such a consequence as an argument for his conjecture.

Euler's reasons are, in fact, inductive. It is a typical inductive procedure to examine the consequences of a conjecture and to judge it on the basis of such an examination. In scientific research as in ordinary life, we believe, or ought to believe, a conjecture more or less according as its observable consequences agree more or less with the facts.

In short, Euler seems to think the same way as reasonable people, scientists or non-scientists, usually think. He seems to accept certain principles: *A conjecture becomes more credible by the verification of any new conse-quence.* And: *A conjecture becomes more credible if an analogous conjecture becomes more credible.*

Are the principles underlying the process of induction of this kind?

## EXAMPLES AND COMMENTS ON CHAPTER II

### First Part

**1.** *The right generalization.*

A. Find three numbers $x, y$, and $z$ satisfying the following system of equations:

$$9x - 6y - 10z = 1,$$

$$-6x + 4y + 7z = 0,$$

$$x^2 + y^2 + z^2 = 9.$$

If you have to solve A, which one of the following three generalizations does give you a more helpful suggestion, B or C or D?

B. Find three unknowns from a system of three equations.

C. Find three unknowns from a system of three equations the first two of which are linear and the third quadratic.

---

[3] Under sect. 6 (3), (4), (5). For Euler's own summary see *Opera Omnia*, ser. 1, vol. 14, p. 140.

[4] The representation of sin $x$ as an infinite product.

[5] Especially the product for $1 - \sin x$.

D. Find $n$ unknowns from a system of $n$ equations the first $n - 1$ of which are linear.

2. A point and a "regular" pyramid with hexagonal base are given in position. (A pyramid is termed "regular" if its base is a regular polygon the center of which is the foot of the altitude of the pyramid.) Find a plane that passes through the given point and bisects the volume of the given pyramid.

In order to help you, I ask you a question: What is the right generalization?

3. A. Three straight lines which are not in the same plane pass through the same point $O$. Pass a plane through $O$ that is equally inclined to the three lines.

B. Three straight lines which are not in the same plane pass through the same point. The point $P$ is on one of the lines; pass a plane through $P$ that is equally inclined to the three lines.

Compare the problems A and B. Could you use the solution of one in solving the other? What is their logical connection?

4. A.  Compute the integral

$$\int_{-\infty}^{\infty} (1 + x^2)^{-3} \, dx.$$

B.  Compute the integral

$$\int_{-\infty}^{\infty} (p + x^2)^{-3} \, dx$$

where $p$ is a given positive number.

Compare the problems A and B. Could you use the solution of one in solving the other? What is their logical connection?

5. *An extreme special case.* Two men are seated at a table of usual rectangular shape. One places a penny on the table, then the other does the same, and so on, alternately. It is understood that each penny lies flat on the table and not on any penny previously placed. The player who puts the last coin on the table takes the money. Which player should win, provided that each plays the best possible game?

This is a time-honored but excellent puzzle. I once had the opportunity to watch a really distinguished mathematician when the puzzle was proposed to him. He started by saying, "Suppose that the table is so small that it is covered by one penny. Then, obviously, the first player must win." That is, he started by picking out an *extreme special case* in which the solution is obvious.

From this special case, you can reach the full solution when you imagine the table gradually extending to leave place to more and more pennies. It may be still better to *generalize* the problem and to think of tables of various shapes and sizes. If you observe that the table has a center of symmetry and that the *right generalization* might be to consider tables with a center of symmetry, then you have got the solution, or you are at least very near to it.

**6.** Construct a common tangent to two given circles.

In order to help you, I ask you a question: Is there a more accessible extreme special case?

**7.** *A leading special case.* The area of a polygon is $A$, its plane includes with a second plane the angle $\alpha$. The polygon is projected orthogonally onto the second plane. Find the area of the projection.

Observe that the shape of the polygon is not given. Yet there is an endless variety of possible shapes. Which shape should we discuss? Which shape should we discuss first?

There is a particular shape especially easy to handle: a rectangle, the base of which is parallel to the line $l$, intersection of the plane of the projected figure with the plane of the projection. If the base of such a rectangle is $a$, its height $b$, and therefore its area is $ab$, the corresponding quantities for the projection are $a$, $b \cos \alpha$, and $ab \cos \alpha$. If the area of such a rectangle is $A$, the area of its projection is $A \cos \alpha$.

This special case of the rectangle with base parallel to $l$ is not only particularly accessible; it is a *leading special case*. The other cases follow; *the solution of the problem in the leading special case involves the solution in the general case.* In fact, starting from the rectangle with base parallel to $l$, we can extend the rule "area of the projection equals $A \cos \alpha$" successively to all other figures. First to right triangles with a leg parallel to $l$ (by bisecting the rectangle we start from); then to any triangle with a side parallel to $l$ (by combining two right triangles); finally to a general polygon (by dissecting it into triangles of the kind just mentioned). We could even pass to figures with curvilinear boundaries (by considering them as limits of polygons).

**8.** The angle at the center of a circle is double the angle at the circumference on the same base, that is, on the same arc. (Euclid III, 20.)

If the angle at the center is given, the angle at the circumference is not yet determined, but can have various positions. In the usual proof of the theorem (Euclid's proof), which is the "leading special position"?

**9.** Cauchy's theorem, fundamental in the theory of analytic functions, asserts that the integral of such a function vanishes along an arbitrary closed curve in the interior of which the function is regular. We may consider the special case of Cauchy's theorem in which the closed curve is a triangle as a leading special case: having proved the theorem for a triangle, we can easily extend it successively to polygons (by combining triangles) and to curves (by considering them as limits of polygons). Observe the analogy with ex. 7 and 8.

**10.** *A representative special case.* You have to solve some problem about polygons with $n$ sides. You draw a pentagon, solve the problem for it, study your solution, and notice that it works just as well in the general case, for any $n$, as in the special case $n = 5$. Then you may call $n = 5$ a *representative* special case: it represents to you the general case. Of course, in order to be really representative, the case $n = 5$ should have no particular simplification that could mislead you. The representative special case should *not* be simpler than the general case.

Representative special cases are often convenient in teaching. We may prove a theorem on determinants with $n$ rows in discussing carefully a determinant with just 3 rows.

**11.** *An analogous case.* The problem is to design airplanes so that the danger of skull fractures in case of accident is minimized. A medical doctor, studying this problem, experiments with eggs which he smashes under various conditions. What is he doing? He has *modified* the original problem, and is studying now an *auxiliary problem*, the smashing of eggs instead of the smashing of skulls. The link between the two problems, the original and the auxiliary, is *analogy*. From a mechanical viewpoint, a man's head and a hen's egg are roughly analogous: each consists of a rigid, fragile shell containing gelatinous material.

**12.** If two straight lines in space are cut by three parallel planes, the corresponding segments are proportional.

In order to help you to find a proof, I ask you a question: Is there a simpler analogous theorem?

**13.** The four diagonals of a parallelepiped have a common point which is the midpoint of each.

Is there a simpler analogous theorem?

**14.** The sum of any two face angles of a trihedral angle is greater than the third face angle.

Is there a simpler analogous theorem?

**15.** Consider a tetrahedron as the solid that is analogous to a triangle. List the concepts of solid geometry that are analogous to the following concepts of plane geometry: *parallelogram, rectangle, square, bisector of an angle.* State a theorem of solid geometry that is analogous to the following theorem of plane geometry: *The bisectors of the three angles of a triangle meet in one point which is the center of the circle inscribed in the triangle.*

**16.** Consider a pyramid as the solid that is analogous to a triangle. List the solids that are analogous to the following plane figures: *parallelogram, rectangle, circle.* State a theorem of solid geometry that is analogous to the following theorem of plane geometry: *The area of a circle is equal to the area of a triangle the base of which has the same length as the perimeter of the circle and the altitude of which is the radius.*

**17.** Invent a theorem of solid geometry that is analogous to the following theorem of plane geometry: *The altitude of an isosceles triangle passes through the midpoint of the base.*

What solid figure do you consider as analogous to an isosceles triangle?

**18. Great analogies.** (1) The foregoing ex. 12–17 insisted on the analogy between *plane geometry* and *solid geometry*. This analogy has many aspects and is therefore often ambiguous and not always clearcut, but it is an inexhaustible source of new suggestions and new discoveries.

(2) Numbers and figures are not the only objects of mathematics. Mathematics is basically inseparable from logic, and it deals with all objects which may be objects of an exact theory. Numbers and figures are, however, the most usual objects of mathematics, and the mathematician likes to illustrate facts about numbers by properties of figures and facts about figures by properties of numbers. Hence, there are countless aspects of the analogy between *numbers* and *figures*. Some of these aspects are very clear. Thus, in analytic geometry we study well-defined correspondences between algebraic and geometric objects and relations. Yet the variety of geometric figures is inexhaustible, and so is the variety of possible operations on numbers, and so are the possible correspondences between these varieties.

(3) The study of limits and limiting processes introduces another kind of analogy which we may call the analogy between the *infinite* and the *finite*. Thus, infinite series and integrals are in various ways analogous to the finite sums whose limits they are; the differential calculus is analogous to the calculus of finite differences; differential equations, especially linear and homogeneous differential equations, are somewhat analogous to algebraic equations, and so forth. An important, relatively recent, branch of mathematics is the theory of integral equations; it gives a surprising and beautiful answer to the question: What is the analogue, in the integral calculus, of a system of *n* linear equations with *n* unknowns? The analogy between the infinite and the finite is particularly challenging because it has characteristic difficulties and pitfalls. It may lead to discovery or error; see ex. 46.

(4) Galileo, who discovered the parabolic path of projectiles and the quantitative laws of their motion, was also a great discoverer in astronomy. With his newly invented telescope, he discovered the satellites of Jupiter. He noticed that these satellites circling the planet Jupiter are analogous to the moon circling the earth and also analogous to the planets circling the sun. He also discovered the phases of the planet Venus and noticed their similarity with the phases of the moon. These discoveries were received as a great confirmation of Copernicus's heliocentric theory, hotly debated at that time. It is strange that Galileo failed to consider the analogy between

the motion of heavenly bodies and the motion of projectiles, which can be seen quite intuitively. The path of a projectile turns its concave side towards the earth, and so does the path of the moon. Newton insisted on this analogy: " . . . a stone that is projected is by the pressure of its own weight forced out of the rectilinear path, which by the initial projection alone it should have pursued, and made to describe a curved line in the air, and . . . at last brought down to the ground; and the greater the

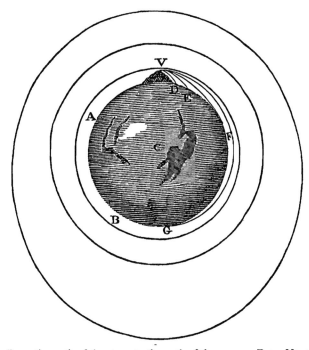

Fig. 2.4. From the path of the stone to the path of the moon. From Newton's *Principia*

velocity is with which it is projected, the farther it goes before it falls to the earth. We may therefore suppose the velocity to be so increased, that it would describe an arc of 1, 2, 5, 10, 100, 1000 miles before it arrived at the earth, till at last, exceeding the limits of the earth, it should pass into space without touching it."[6] See fig. 2.4.

Varying continuously, the path of the stone goes over into the path of the moon. And as the stone and the moon are to the earth, so are the satellites to Jupiter, or Venus and the other planets to the sun. Without visualizing this analogy, we can only very imperfectly understand Newton's discovery of universal gravitation, which we may still regard as the greatest scientific discovery ever made.

---

[6] Sir Isaac Newton's *Mathematical Principles of Natural Philosophy and his System of the World*. Translated by Motte, revised by Cajori. Berkeley, 1946; see p. 551.

**19.** *Clarified analogies.* Analogy is often vague. The answer to the question, what is analogous to what, is often ambiguous. The vagueness of analogy need not diminish its interest and usefulness; those cases, however, in which the concept of analogy attains the clarity of logical or mathematical concepts deserve special consideration.

(1) Analogy is similarity of relations. The similarity has a clear meaning if the *relations are governed by the same laws*. In this sense, the addition of numbers is analogous to the multiplication of numbers, in so far as addition and multiplication are subject to the same rules. Both addition and multiplication are commutative and associative,

$$a + b = b + a, \qquad ab = ba,$$

$$(a + b) + c = a + (b + c), \quad (ab)c = a(bc).$$

Both admit an inverse operation; the equations

$$a + x = b, \qquad ax = b$$

are similar, in so far as each admits a solution, and no more than one solution. (In order to be able to state the last rule without exceptions we must admit negative numbers when we consider addition, and we must exclude the case $a = 0$ when we consider multiplication.) In this connection subtraction is analogous to division; in fact, the solutions of the above equations are

$$x = b - a, \qquad x = \frac{b}{a},$$

respectively. Then, the number 0 is analogous to the number 1; in fact, the addition of 0 to any number, as the multiplication by 1 of any number, does not change that number,

$$a + 0 = a, \qquad a \cdot 1 = a.$$

These laws are the same for various classes of numbers; we may consider here rational numbers, or real numbers, or complex numbers. In general, *systems of objects subject to the same fundamental laws* (or axioms) may be considered as analogous to each other, and this kind of analogy has a completely clear meaning.

(2) The addition of *real* numbers is analogous to the multiplication of *positive* numbers in still another sense. Any real number $r$ is the logarithm of some positive number $p$,

$$r = \log p.$$

(If we consider ordinary logarithms, $r = -2$ if $p = 0.01$.) By virtue of this relation, to each positive number corresponds a perfectly determined real number, and to each real number a perfectly determined positive number. In this correspondence the addition of real numbers corresponds to the multiplication of positive numbers. If

$$r = \log p, \quad r' = \log p', \quad r'' = \log p'',$$

then any of the following two relations implies the other:

$$r + r' = r'', \quad pp' = p''.$$

The formula on the left and that on the right tell the same story in two different languages. Let us call one of the coordinated numbers the translation of the other; for example, let us call the real number $r$ (the logarithm of $p$) the *translation* of $p$, and $p$ the *original* of $r$. (We could have interchanged the words "translation" and "original," but we had to choose, and having chosen, we stick to our choice.) In this terminology addition appears as the translation of multiplication, subtraction as the translation of division, 0 as the translation of 1, the commutative law and associative law for the addition of real numbers are conceived as translations of these laws for the multiplication of positive numbers. The translation is, of course, different from the original, but it is a correct translation in the following sense: from any relation between the original elements, we can conclude with certainty the corresponding relation between the corresponding elements of the translation, and *vice versa*. Such a correct translation, that is a *one-to-one correspondence that preserves the laws of certain relations*, is called *isomorphism* in the technical language of the mathematician. Isomorphism is a fully clarified sort of analogy.

(3) A third sort of fully clarified analogy is what the mathematicians call in technical language *homomorphism* (or *merohedral isomorphism*). It would take too much time to discuss an example sufficiently, or to give an exact description, but we may try to understand the following approximate description. Homomorphism is a kind of *systematically abridged translation*. The original is not only translated into another language, but also abridged so that what results finally from translation and abbreviation is uniformly, systematically condensed into one-half or one-third or some other fraction of the original extension. Subtleties may be lost by such abridgement but everything that is in the original is represented by something in the translation, and, on a reduced scale, the relations are preserved.

**20.** *Quotations.*

"Let us see whether we could, by chance, conceive some other general problem that contains the original problem and is easier to solve. Thus, when we are seeking the tangent at a given point, we conceive that we are just seeking a straight line which intersects the given curve in the given point and in another point that has a given distance from the given point. After having solved this problem, which is always easy to solve by algebra, we find the case of the tangent as a special case, namely, the special case in which the given distance is minimal, reduces to a point, vanishes."
(LEIBNITZ)

"As it often happens, the general problem turns out to be easier than the special problem would be if we had attacked it directly." (P. G. Lejeune-Dirichlet, R. Dedekind)

"[It may be useful] to reduce the genus to its several species, also to a few species. Yet the most useful is to reduce the genus to just one minimal species." (Leibnitz)

"It is proper in philosophy to consider the similar, even in things far distant from each other." (Aristotle)

"Comparisons are of great value in so far as they reduce unknown relations to known relations.

"Proper understanding is, finally, a grasping of relations (un saisir de rapports). But we understand a relation more distinctly and more purely when we recognize it as the same in widely different cases and between completely heterogeneous objects." (Arthur Schopenhauer)

You should not forget, however, that there are two kinds of generalizations. One is cheap and the other is valuable. It is easy to generalize by *diluting*; it is important to generalize by *condensing*. To dilute a little wine with a lot of water is cheap and easy. To prepare a refined and condensed extract from several good ingredients is much more difficult, but valuable. Generalization by condensing compresses into one concept of wide scope several ideas which appeared widely scattered before. Thus, the Theory of Groups reduces to a common expression ideas which were dispersed before in Algebra, Theory of Numbers, Analysis, Geometry, Crystallography, and other domains. The other sort of generalization is more fashionable nowadays than it was formerly. It dilutes a little idea with a big terminology. The author usually prefers to take even that little idea from somebody else, refrains from adding any original observation, and avoids solving any problem except a few problems arising from the difficulties of his own terminology. It would be very easy to quote examples, but I don't want to antagonize people.[7]

## Second Part

The examples and comments of this second part are all connected with sect. 6 and each other. Many of them refer directly or indirectly to ex. 21, which should be read first.

**21.** *The conjecture E.* We regard the equation

$$\sin x = x \left( 1 - \frac{x^2}{\pi^2} \right) \left( 1 - \frac{x^2}{4\pi^2} \right) \left( 1 - \frac{x^2}{9\pi^2} \right) \cdots$$

as a conjecture; we call it the "conjecture *E*." Following Euler, we wish to investigate this conjecture inductively.

---

[7] Cf. G. Pólya and G. Szegő, *Aufgaben und Lehrstäze aus der Analysis*, vol. 1, p. VII.

Inductive investigation of a conjecture involves confronting its conse-quences with the facts. We shall often "predict from $E$ and verify." "Predicting from $E$" means deriving under the assumption that $E$ is true, "verifying" means deriving without this assumption. A fact "agrees with $E$" if it can be (easily) derived from the assumption that $E$ is true.

In the following we take for granted the elements of the calculus (which, from the formal side, were completely known to Euler at the time of his discovery) including the rigorous concept of limits (about which Euler never attained full clarity). We shall use only limiting processes which can be justified (most of them quite easily) but we shall not enter into detailed justifications.

**22.** We know that $\sin(-x) = -\sin x$. Does this fact agree with $E$?

**23.** Predict from $E$ and verify the value of the infinite product

$$\left(1 - \frac{1}{4}\right)\left(1 - \frac{1}{9}\right)\left(1 - \frac{1}{16}\right) \cdots \left(1 - \frac{1}{n^2}\right) \cdots .$$

**24.** Predict from $E$ and verify the value of the infinite product

$$\left(1 - \frac{4}{9}\right)\left(1 - \frac{4}{16}\right)\left(1 - \frac{4}{25}\right) \cdots \left(1 - \frac{4}{n^2}\right) \cdots .$$

**25.** Compare ex. 23 and 24, and generalize.

**26.** Predict from $E$ the value of the infinite product

$$\frac{2 \cdot 4}{3 \cdot 3} \cdot \frac{4 \cdot 6}{5 \cdot 5} \cdot \frac{6 \cdot 8}{7 \cdot 7} \cdot \frac{8 \cdot 10}{9 \cdot 9} \cdots .$$

**27.** Show that the conjecture $E$ is equivalent to the statement

$$\frac{\sin \pi z}{\pi} = \lim_{n = \infty} \frac{(z + n) \cdots (z + 1)z(z - 1) \cdots (z - n)}{(-1)^n (n!)^2} .$$

**28.** We know that $\sin(x + \pi) = -\sin x$. Does this fact agree with $E$?

**29.** The method of sect. 6 (2) leads to the conjecture

$$\cos x = \left(1 - \frac{4x^2}{\pi^2}\right)\left(1 - \frac{4x^2}{9\pi^2}\right)\left(1 - \frac{4x^2}{25\pi^2}\right) \cdots .$$

Show that this is not only analogous to, but a consequence of, the conjec-ture $E$.

**30.** We know that

$$\sin x = 2 \sin(x/2) \cos(x/2).$$

Does this fact agree with $E$?

**31.** Predict from $E$ and verify the value of the infinite product

$$\left(1 - \frac{4}{1}\right)\left(1 - \frac{4}{9}\right)\left(1 - \frac{4}{25}\right)\left(1 - \frac{4}{49}\right) \cdots .$$

**32.** Predict from $E$ and verify the value of the infinite product

$$\left(1 - \frac{16}{1}\right)\left(1 - \frac{16}{9}\right)\left(1 - \frac{16}{25}\right)\left(1 - \frac{16}{49}\right) \cdots .$$

**33.** Compare ex. 31 and 32, and generalize.

**34.** We know that $\cos(-x) = \cos x$. Does this fact agree with $E$?

**35.** We know that $\cos(x + \pi) = -\cos x$. Does this fact agree with $E$?

**36.** Derive from $E$ the product for $1 - \sin x$ conjectured in sect. 6 (4).

**37.** Derive from $E$ that

$$\cot x = \ldots + \frac{1}{x + 2\pi} + \frac{1}{x + \pi} + \frac{1}{x} + \frac{1}{x - \pi} + \frac{1}{x - 2\pi} + \cdots .$$

**38.** Derive from $E$ that

$$\cot x = \frac{1}{x} - \frac{2x}{\pi^2}\left(1 + \frac{1}{4} + \frac{1}{9} + \frac{1}{16} + \frac{1}{25} + \cdots\right)$$

$$- \frac{2x^3}{\pi^4}\left(1 + \frac{1}{16} + \frac{1}{81} + \frac{1}{256} + \frac{1}{625} + \cdots\right)$$

$$- \frac{2x^5}{\pi^6}\left(1 + \frac{1}{64} + \frac{1}{729} + \cdots\right)$$

$$- \cdots$$

and find the sum of the infinite series appearing as coefficients on the right hand side.

**39.** Derive from $E$ that

$$\frac{\cos x}{1 - \sin x} = \cot\left(\frac{\pi}{4} - \frac{x}{2}\right)$$

$$= -2\left(\frac{1}{x - \dfrac{\pi}{2}} + \frac{1}{x + \dfrac{3\pi}{2}} + \frac{1}{x - \dfrac{5\pi}{2}} + \frac{1}{x + \dfrac{7\pi}{2}} + \cdots\right)$$

$$= \frac{4}{\pi}\left(1 - \frac{1}{3} + \frac{1}{5} - \frac{1}{7} + \frac{1}{9} - \cdots\right)$$

$$+ \frac{8x}{\pi^2}\left(1 + \frac{1}{9} + \frac{1}{25} + \frac{1}{49} + \frac{1}{81} \cdots\right)$$

$$+ \frac{16x^2}{\pi^3}\left(1 - \frac{1}{27} + \frac{1}{125} - \frac{1}{343} + \cdots\right)$$

$$+ \frac{32x^3}{\pi^4}\left(1 + \frac{1}{81} + \frac{1}{625} + \cdots\right)$$

$$+ \cdots$$

and find the sum of the infinite series appearing as coefficients in the last expression.

**40.** Show that

$$1 + \frac{1}{4} + \frac{1}{9} + \frac{1}{16} + \frac{1}{25} + \cdots = \frac{4}{3}\left(1 + \frac{1}{9} + \frac{1}{25} + \frac{1}{49} + \cdots\right)$$

which yields a second derivation for the sum of the series on the left.

**41** (continued). Try to find a third derivation, knowing that

$$\arcsin x = x + \frac{1}{2}\frac{x^3}{3} + \frac{1}{2}\frac{3}{4}\frac{x^5}{5} + \frac{1}{2}\frac{3}{4}\frac{5}{6}\frac{x^7}{7} + \cdots$$

and that, for $n = 0, 1, 2, \ldots$,

$$\int_0^1 (1 - x^2)^{-1/2} x^{2n+1}\, dx = \int_0^{\pi/2} (\sin t)^{2n+1}\, dt = \frac{2 \cdot 4 \cdots 2n}{3 \cdot 5 \cdots (2n + 1)}.$$

**42** (continued). Try to find a fourth derivation, knowing that

$$(\arcsin x)^2 = x^2 + \frac{2}{3}\frac{x^4}{2} + \frac{2}{3}\frac{4}{5}\frac{x^6}{3} + \frac{2}{3}\frac{4}{5}\frac{6}{7}\frac{x^8}{4} + \cdots$$

and that, for $n = 0, 1, 2, \ldots$

$$\int_0^1 (1 - x^2)^{-1/2} x^{2n}\, dx = \int_0^{\pi/2} (\sin t)^{2n}\, dt = \frac{1}{2}\frac{3}{4} \cdots \frac{2n - 1}{2n}\frac{\pi}{2}.$$

**43.** Euler (*Opera Omnia*, ser. 1, vol. 14, p. 40–41) used the formula

$$1 + \frac{1}{4} + \frac{1}{9} + \frac{1}{16} + \cdots$$

$$= \log x \cdot \log (1 - x) + \frac{x + (1 - x)}{1} + \frac{x^2 + (1 - x)^2}{4} + \frac{x^3 + (1 - x)^3}{9}$$
$$+ \cdots,$$

valid for $0 < x < 1$, to compute numerically the sum of the series on the left hand side.

(a) Prove the formula.

(b) Which value of $x$ is the most advantageous in computing the sum on the left?

**44.** *An objection and a first approach to a proof.* There is no reason to admit *a priori* that $\sin x$ can be decomposed into linear factors corresponding to the roots of the equation

$$\sin x = 0.$$

Yet even if we should admit this, there remains an objection: Euler did *not* prove that

$$0, \quad \pi, \quad -\pi, \quad 2\pi, \quad -2\pi, \quad 3\pi, \quad -3\pi, \quad \ldots$$

are *all* the roots of this equation. We can satisfy ourselves (by discussing the curve $y = \sin x$) that there are no other real roots, yet Euler did by no means exclude the existence of complex roots.

This objection was raised by Daniel Bernoulli (a son of Jean, 1700–1788). Euler answered it by considering

$$\sin x = (e^{ix} - e^{-ix})/(2i)$$
$$= \lim_{n \to \infty} P_n(x)$$

where

$$P_n(x) = \frac{1}{2i}\left[\left(1 + \frac{ix}{n}\right)^n - \left(1 - \frac{ix}{n}\right)^n\right]$$

is a polynomial (of degree $n$ if $n$ is odd).

Show that $P_n(x)$ has no complex roots.

**45.** *A second approach to a proof.* Assuming that $n$ is odd in ex. 44, factorize $P_n(x)/x$ so that its $k$-th factor approaches

$$1 - \frac{x^2}{k^2\pi^2}$$

as $n$ tends to $\infty$, for any fixed $k$ ($k = 1, 2, 3, \ldots$).

**46.** *Dangers of analogy.* In short, the analogy between the finite and the infinite led Euler to a great discovery. Yet he skirted a fallacy. Here is an example showing the danger on a smaller scale.

The series

$$1 - \frac{1}{2} + \frac{1}{3} - \frac{1}{4} + \frac{1}{5} - \frac{1}{6} + \frac{1}{7} - \frac{1}{8} + \ldots = l$$

converges. Its sum $l$ can be roughly estimated by the first two terms:

$$1/2 < l < 1.$$

Now

$$2l = \frac{2}{1} - \frac{1}{1} + \frac{2}{3} - \frac{1}{2} + \frac{2}{5} - \frac{1}{3} + \frac{2}{7} - \frac{1}{4} + \ldots .$$

In this series, there is just one term with a given even denominator (it is negative), but two terms with a given odd denominator (one positive, and the other negative). Let us bring together the terms with the same odd denominator:

$$\frac{2}{1} - \frac{1}{2} + \frac{2}{3} - \frac{1}{4} + \frac{2}{5} - \ldots$$

$$-\frac{1}{1} \qquad -\frac{1}{3} \qquad -\frac{1}{5}$$

$$= 1 - \frac{1}{2} + \frac{1}{3} - \frac{1}{4} + \frac{1}{5} - \ldots$$

$$= l.$$

Yet $2l \neq l$, since $l \neq 0$. Where is the mistake and how can you protect yourself from repeating it?

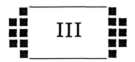

# III

## INDUCTION IN SOLID GEOMETRY

*Even in the mathematical sciences, our principal instruments to discover the truth are induction and analogy.*—LAPLACE[1]

**1. Polyhedra.** "A complicated polyhedron has many faces, corners, and edges." Some vague remark of this sort comes easily to almost anybody who has had some contact with solid geometry. Not so many people will, however, make a serious effort to deepen this remark and seek some more precise information behind it. The right thing to do is to distinguish clearly the quantities involved and to ask some definite question. Let us denote, therefore, the number of faces, the number of vertices and the number of edges of the polyhedron by $F$, $V$, and $E$, respectively (corresponding initials), and let us ask a clear question as: "Is it generally true that the number of faces increases when the number of vertices increases? Does $F$ necessarily increase with $V$?"

To begin with, we can scarcely do anything better than examine examples, particular polyhedra. Thus, for a cube (the solid I in fig. 3.1)

$$F = 6, \quad V = 8, \quad E = 12.$$

Or, for a prism with triangular base (the solid II in fig. 3.1)

$$F = 5, \quad V = 6, \quad E = 9.$$

Once launched in this direction, we naturally examine and compare various solids, for example, those exhibited in fig. 3.1 which are, besides No. I and No. II already mentioned, the following: a prism with pentagonal base (No. III), pyramids with square, triangular, or pentagonal base (Nos. IV, V, VI), an octahedron (No. VII), a "tower with roof" (No. VIII; a pyramid is placed upon the upper face of a cube as base), and a "truncated cube" (No. IX). Let us make a little effort of imagination and represent

---

[1] *Essai philosophique sur les probabilités;* see *Oeuvres complètes de Laplace,* vol. 7, p. V.

these solids, one after the other, clearly enough to count faces, vertices, and
edges.   The numbers found are listed in the following table.

| Polyhedron | F | V | E |
|---|---|---|---|
| I. cube . . . . | 6 | 8 | 12 |
| II. triangular prism . . | 5 | 6 | 9 |
| III. pentagonal prism . . | 7 | 10 | 15 |
| IV. square pyramid . . . | 5 | 5 | 8 |
| V. triangular pyramid . . | 4 | 4 | 6 |
| VI. pentagonal pyramid . . | 6 | 6 | 10 |
| VII. octahedron . . . | 8 | 6 | 12 |
| VIII. "tower" . . . . | 9 | 9 | 16 |
| IX. "truncated cube" . . | 7 | 10 | 15 |

Our fig. 3.1 has some superficial similarity with a mineralogical display,
and the above table is somewhat similar to the notebook in which the
physicist enters the results of his experiments.   We examine and compare
our figures and the numbers in our table as the mineralogist or the physicist
would examine and compare their more laboriously collected specimens
and data.   We now have something in hand that could answer our original
question: "Does $V$ increase with $F$?"   In fact, the answer is "No"; comparing the cube and the octahedron (Nos. I and VII) we see, that one has more
vertices and the other more faces.   Thus, our first attempt at establishing
a thoroughgoing regularity failed.

We can, however, try something else.   Does $E$ increase with $F$?   Or
with $V$?   To answer these questions systematically, we rearrange our table.
We dispose our polyhedra so that $E$ increases when we read down the
successive items:

| Polyhedron | F | V | E |
|---|---|---|---|
| triangular pyramid . . | 4 | 4 | 6 |
| square pyramid . . | 5 | 5 | 8 |
| triangular prism . . | 5 | 6 | 9 |
| pentagonal pyramid . | 6 | 6 | 10 |
| cube . . . . | 6 | 8 | 12 |
| octahedron . . . | 8 | 6 | 12 |
| pentagonal prism . . | 7 | 10 | 15 |
| "truncated cube" . . | 7 | 10 | 15 |
| "tower" . . . | 9 | 9 | 16 |

Looking at our more conveniently arranged data, we can easily observe
that no regularity of the suspected kind exists.   As $E$ increases from 15 to
16, $V$ drops from 10 to 9.   Again, as we pass from the octahedron to the
pentagonal prism, $E$ increases from 12 to 15 but $F$ drops from 8 to 7.   Neither
$F$ nor $V$ increases steadily with $E$.

We again failed in finding a generally valid regularity. Yet we do not like to admit that our original idea was completely wrong. Some modification of our idea may still be right. Neither $F$ nor $V$ increases with $E$, it is true, but they appear to increase "on the whole." Examining our

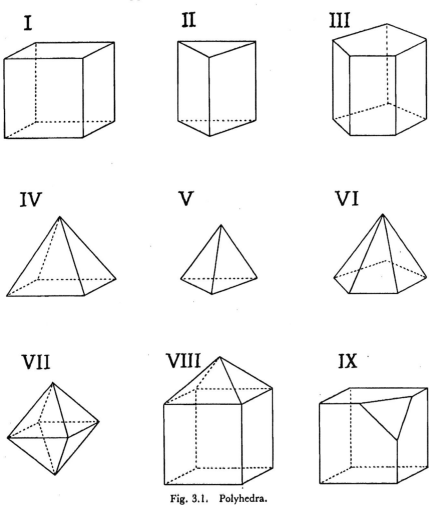

Fig. 3.1. Polyhedra.

well-arranged data, we may observe that $F$ and $V$ increase "jointly": $F + V$ *increases* as we read down the lines. And then a more precise regularity may strike us: throughout the table

$$F + V = E + 2.$$

This relation is verified in all nine cases listed in our table. It seems unlikely that such a persistent regularity should be mere coincidence.

Thus, we are led to the *conjecture* that, not only in the observed cases, but in any polyhedron the *number of faces increased by the number of vertices is equal to the number of edges increased by two.*

**2. First supporting contacts.** A well-trained naturalist does not easily admit a conjecture. Even if the conjecture appears plausible and has been verified in some cases, he will question it and collect new observations or design new experiments to test it. We may do the very same thing. We are going to examine still other polyhedra, count their faces, vertices, and edges, and compare $F + V$ to $E + 2$. These numbers may be equal or not. It will be interesting to find out which is the case.

Looking at fig. 3.1, we may observe that we have already examined three of the regular solids, the cube, the tetrahedron, and the octahedron (I, V, and VII). Let us examine the remaining two, the icosahedron and the dodecahedron.

The icosahedron has 20 faces, all triangles, and so $F = 20$. The 20 triangles have together $3 \times 20 = 60$ sides of which 2 coincide in the same edge of the icosahedron. Therefore, the number of edges is $60/2 = 30 = E$. We can find $V$ analogously. We know that 5 faces of the icosahedron are grouped around each of its vertices. The 20 triangles have together $3 \times 20 = 60$ angles, of which 5 belong to the same vertex. Therefore, the number of vertices is $60/5 = 12 = V$.

The dodecahedron has 12 faces, all pentagons, of which 3 are grouped around each vertex. We conclude hence, similarly as before, that

$$F = 12, \quad V = \frac{12 \times 5}{3} = 20, \quad E = \frac{12 \times 5}{2} = 30.$$

We can now add to our list on p. 36 two more lines:

| Polyhedron | F | V | E |
|------------|---|---|---|
| icosahedron  .  . | 20 | 12 | 30 |
| dodecahedron .  . | 12 | 20 | 30 |

Our conjecture, that $F + V = E + 2$, is verified in both cases.

**3. More supporting contacts.** Thanks to the foregoing verifications, our conjecture became perceptibly more plausible; but is it proved now? By no means. In a similar situation, a conscientious naturalist would feel satisfaction over the success of his experiments, but would go on devising further experiments. Which polyhedron should we test now?

The point is that our conjecture is so well verified by now that verification in just one more instance would add only little to our confidence, so little perhaps that it would be scarcely worth the trouble of choosing a polyhedron and counting its parts. Could we find some more worthwhile way of testing our conjecture?

Looking at fig. 3.1, we may observe that all solids in the first line are of the same nature; they are prisms. Again, all solids in the second line are pyramids. Our conjecture is certainly true of the three prisms and the three pyramids shown in fig. 3.1; but is it true of *all* prisms and pyramids? If a prism has $n$ lateral faces, it has $n + 2$ faces in all, $2n$ vertices and $3n$ edges. A pyramid with $n$ lateral faces has $n + 1$ faces in all, $n + 1$ vertices and $2n$ edges. Thus, we may add two more lines to our list on p. 36:

| Polyhedron | F | V | E |
|---|---|---|---|
| Prism with $n$ lateral faces . . | $n + 2$ | $2n$ | $3n$ |
| Pyramid with $n$ lateral faces . | $n + 1$ | $n + 1$ | $2n$ |

Our conjecture asserting that $F + V = E + 2$ turned out to be true not only for one or two more polyhedra but for two unlimited series of polyhedra.

**4. A severe test.** The last remark adds considerably to our confidence in our conjecture, but does not prove it, of course. What should we do? Should we go on testing further particular cases? Our conjecture seems to withstand simple tests fairly well. Therefore we should submit it to some severe, searching test that stands a good chance to refute it.

Let us look again at our collection of polyhedra (fig. 3.1). There are prisms (I, II, III), pyramids (IV, V, VI), regular solids (I, V, VII); yet we have already considered all these kinds of solids exhaustively. What else is there? Fig. 3.1 contains also the "tower" (No. VIII) which is obtained by placing a "roof" on the top of a cube. Here we may perceive the possibility of a generalization. We take any polyhedron instead of the cube, choose any face of the polyhedron, and place a "roof" on it. Let the original polyhedron have $F$ faces, $V$ vertices, and $E$ edges, and let its face chosen have $n$ sides. We place on this face a pyramid with $n$ lateral faces and so obtain a new polyhedron. How many faces, vertices, and edges has the new "roofed" polyhedron? One face (the chosen one) is lost in the process, and $n$ new ones are won (the $n$ lateral faces of the pyramid) so that the new polyhedron has $F - 1 + n$ faces. All vertices of the polyhedron belong also to the new one, but one vertex is added (the summit of the pyramid) and so the new polyhedron has $V + 1$ vertices. Again, all edges of the old polyhedron belong also to the new one, but $n$ edges are added (the lateral edges of the pyramid) and so the new polyhedron has $E + n$ edges.

Let us summarize. The original polyhedron had $F$, $V$, and $E$ faces, vertices, and edges, respectively, whereas the new "roofed" polyhedron has

$$F + n - 1, \quad V + 1, \quad \text{and} \quad E + n$$

parts of the corresponding kind. Are these facts consistent with our conjecture?

*If* the relation $F + V = E + 2$ holds, then, obviously,

$$(F + n - 1) + (V + 1) = (E + n) + 2$$

holds *also*. That is, if our conjecture happens to be verified in the case of the original polyhedron, it must be verified also in the case of the new "roofed" polyhedron. Our conjecture survives the "roofing," and so it has passed a very severe test, indeed. There is such an inexhaustible variety of polyhedra which we can derive from those already examined by repeated "roofing," and we have proved that our conjecture is true for all of them.

By the way, the last solid of our fig. 3.1, the "truncated cube" (IX), opens the way to a similar consideration. Instead of the cube, let us "truncate" any polyhedron, cutting off an arbitrarily chosen vertex. Let the original polyhedron have

$$F, V, \text{ and } E$$

faces, vertices, and edges, respectively, and let $n$ be the number of the edges radiating from the vertex we have chosen. Cutting off this vertex we introduce 1 new face (which has $n$ sides), $n$ new edges, and also $n$ new vertices, but we lose 1 vertex. To sum up, the new "truncated" polyhedron has

$$F + 1, \quad V + n - 1, \quad \text{and} \quad E + n$$

faces, vertices, and edges, respectively. Now, from

$$F + V = E + 2$$

follows

$$(F + 1) + (V + n - 1) = (E + n) + 2.$$

That is, our conjecture is tenacious enough to survive the "truncating." It has passed another severe test.

It is natural to regard the foregoing remarks as a very strong argument for our conjecture. We can perceive in them even something else: the first hint of a proof. Starting from some simple polyhedron, as the tetrahedron or the cube, for which the conjecture holds, we can derive by roofing and truncating a vast variety of other polyhedra for which the conjecture also holds. Could we derive *all* polyhedra? Then we would have a proof! Besides, there may be other operations which, like truncating and roofing, preserve the conjectural relation.

**5. Verifications and verifications.** The mental procedures of the trained naturalist are not essentially different from those of the common man, but they are more thorough. Both the common man and the scientist are led to conjectures by a few observations and they are both paying attention to later cases which could be in agreement or not with the conjecture. A case in agreement makes the conjecture more likely, a conflicting case disproves it, and here the difference begins: Ordinary people are usually more apt to look for the first kind of cases, but the scientist looks for the second kind. The reason is that everybody has a little vanity, the common man as the scientist, but different people take pride in different things. Mr. Anybody does not like to confess, even to himself, that he was mistaken and so he does not like conflicting cases, he avoids them, he is even inclined to explain them away when they present themselves. The scientist, on the contrary, is ready enough to recognize a mistaken conjecture, but he does not like to leave questions undecided. Now, a case in agreement does not settle the question definitively, but a conflicting case does. The scientist, seeking a definitive decision, looks for cases which have a chance to upset the conjecture, and the more chance they have, the more they are welcome. There is an important point to observe. If a case which threatens to upset the conjecture turns out, after all, to be in agreement with it, the conjecture emerges greatly strengthened from the test. The more danger, the more honor; passing the most threatening examination grants the highest recognition, the strongest experimental evidence to the conjecture. There are instances and instances, verifications and verifications. An instance which is *more likely to be conflicting* brings the conjecture in any case nearer to decision than an instance which is less so, and this explains the preference of the scientist.

Now, we may get down to our own particular problem and see how the foregoing remarks apply to the "experimental research on polyhedra" that we have undertaken. Each new case in which the relation $F + V = E + 2$ is verified adds to the confidence that this relation is generally true. Yet we soon get tired of a monotonous sequence of verifications. A case little different from the previously examined cases, if it agrees with the conjecture, adds to our confidence, of course, but it adds little. In fact we easily believe, before the test, that the case at hand will behave as the previous cases from which it differs but little. We desire not only another verification, but a *verification of another kind*. In fact, looking back at the various phases of our research (sect. 2, 3, and 4), we may observe that each one yielded a kind of verification that surpassed essentially those obtained in the foregoing. Each phase verified the conjecture for a *more extensive variety of cases than the foregoing*.

**6. A very different case.** Variety being important, let us look for some polyhedron very different from those heretofore examined. Thus, we may hit upon the idea of regarding a picture frame as a polyhedron. We take a very long triangular rod, we cut four pieces of it, we adjust these pieces at the ends, and fit them together to a framelike polyhedron. Fig. 3.2 suggests that the frame is placed on a table so that the edges which have already been on the uncut rod all lie horizontally. There are 4 times 3, that is, 12, horizontal edges, and also 4 times 3 non-horizontal edges, so that the total number of edges is $E = 12 + 12 = 24$. Counting the faces and vertices, we find that $F = 4 \times 3 = 12$, and $V = 4 \times 3 = 12$. Now, $F + V = 24$ is different from $E + 2 = 26$. Our conjecture, taken in full generality, turned out to be false!

We can say, of course, that we have never intended to state the proposition in such generality, that we meant all the time polyhedra which are convex or, so to say, "sphere-shaped" and not polyhedra which are "doughnut-shaped" as the picture frame. But these are excuses. In fact, we have to shift our position and modify our original statement. It is quite possible that the blow that we received will be beneficial in the end and lead us eventually to an amended and more precise statement of our conjecture. Yet it was a blow to our confidence, anyway.

Fig. 3.2. A doughnut-shaped polyhedron.

**7. Analogy.** The example of the "picture frame" killed our conjecture in its original form but it can be promptly revived in a revised (and, let us hope, improved) form, with an important restriction.

The tetrahedron is convex, and so is the cube, and so are the other polyhedra in our collection (fig. 3.1), and so are all the polyhedra that we can derive from them by truncating and by "moderate" roofing (by placing *sufficiently flat* roofs on their various faces). At any rate, there is no danger that these operations could lead from a convex or "sphere-shaped" polyhedron to a "doughnut-shaped" solid.

Observing this, we introduce some much-needed precision. We conjecture that in any *convex* polyhedron the relation $F + V = E + 2$ holds between the numbers of faces, vertices, and edges. (The restriction to "sphere-shaped" polyhedra may be even preferable, but we do not wish to stop to define here the meaning of the term.)

This conjecture has some chance to be true. Nevertheless, our confidence was shaken and we look around for some new support for our conjecture. We cannot hope for much help from further verifications. It seems that we have exhausted the most obvious sources. Yet we may still hope for some help from analogy. Is there any simpler analogous case that could be instructive?

Polygons are analogous to polyhedra. A polygon is a part of a plane as a polyhedron is a part of space. A polygon has a certain number, $V$, of vertices (the vertices of its angles) and a certain number, $E$, of edges (or sides). Obviously

$$V = E.$$

Yet this relation, valid for convex polygons, appears too simple and throws little light on the more intricate relation

$$F + V = E + 2$$

which we suspect to be valid for all convex polyhedra.

If we are genuinely concerned in the question, we naturally try to bring these two relations nearer to each other. There is an ingenious way of doing so. We have to bring first the various numbers into a natural order. The polyhedron is 3-dimensional; its faces (polygons) are 2-dimensional, its edges 1-dimensional, and its vertices (points) 0-dimensional, of course. We may now rewrite our equations, arranging the quantities in the order of increasing dimensions. The relation for polygons, written in the form

$$V - E + 1 = 1,$$

becomes comparable to the relation for polyhedra, written in the form

$$V - E + F - 1 = 1.$$

The 1, on the left hand side of the equation for polygons, stands for the only two-dimensional element concerned, the interior of the polygon. The 1, on the left hand side of the equation for polyhedra, stands for the only three-dimensional element concerned, the interior of the polyhedron. The numbers, on the left hand side, counting elements of 0, 1, 2, and 3 dimensions, respectively, are disposed in this natural order, and have alternating signs. The right hand side is the same in both cases; the analogy seems complete. As the first equation, for polygons, is obviously true, the analogy adds to our confidence in the second equation, for polyhedra, which we have conjectured.

**8. The partition of space.** We pass now to another example of inductive research in solid geometry. In our foregoing example, we started from a general, somewhat vague remark. Our point of departure now is a particular clear-cut problem. We consider a simple but not too familiar problem of solid geometry: *Into how many parts is space divided by 5 planes?*

This question is easily answered if the five given planes are all parallel to each other, in which case space is visibly divided into 6 parts. This case, however, is too particular. If our planes are in a "general position," no two among them will be parallel and there will be considerably more parts than 6. We have to restate our problem more precisely, adding an essential clause: *Into how many parts is space divided by 5 planes, provided that these planes are in a general position?*

The idea of a "general position" is quite intuitive; planes are in such a position when they are not linked by any particular relation, when they are given independently, chosen at random. It would not be difficult to clear up the term completely by a technical definition but we shall not do so, for two reasons. First, this presentation should not be too technical. Second, leaving the notion somewhat hazy, we come nearer to the mental attitude of the naturalist who is often obliged to start with somewhat hazy notions, but clears them up as he goes ahead.

**9. Modifying the problem.** Let us concentrate upon our problem. We are given 5 planes in general position. They cut space into a certain number of partitions. (We may think of a cheese sliced into pieces by 5 straight cuts with a sharp knife.) We have to find the number of these partitions. (Into how many pieces is the cheese cut?)

It seems difficult to see at once all the partitions effected by the 5 planes. (It may be impossible to "see" them. At any rate, do not overstrain your geometric imagination; rather, try to think. Your reason may carry you farther than your imagination.) But why just 5 planes? Why not any number of planes? In how many parts is space divided by 4 planes? By 3 planes? Or by 2 planes? Or by just 1 plane?

We reach here cases which are accessible to our geometric intuition. One plane divides space obviously into 2 parts. Two planes divide space into 3 parts if they are parallel. We have to discard, however, this particular position; 2 planes in a general position intersect, and divide space into 4 parts. Three planes in a general position divide space into 8 parts. In order to realize this last, more difficult, case, we may think of 2 vertical walls inside a building, crossing each other, and of a horizontal layer, supported by beams, crossing both walls and forming around the point where it crosses both the ceiling of 4 rooms and the floor of 4 other rooms.

**10. Generalization, specialization, analogy.** Our problem is concerned with 5 planes but, instead of considering 5 planes, we first played with 1, 2, and 3 planes. Have we squandered our time? Not at all. We have prepared ourselves for our problem by examining *simpler analogous cases*. We have tried our hand at these simpler cases; we have clarified the intervening concepts and familiarized ourselves with the kind of problem we have to face.

Even the way that led us to those simpler analogous problems is typical and deserves our attention. First, we passed from the case of 5 planes to the case of any number of planes, let us say, to $n$ planes: we *generalized*. Then, from $n$ planes, we passed back to 4 planes, to 3 planes, to 2 planes, to just 1 plane, that is, we put $n = 4, 3, 2, 1$ in the general problem: we *specialized*. But the problem about dividing space by, let us say, 3 planes is *analogous* to our original question involving 5 planes. Thus, we have reached analogy in a typical manner, by *introductory generalization and subsequent specialization*.

**11. An analogous problem.** What about the next case of 4 planes? Four planes, in general position, determine various portions of space, one of which is limited, contained by four triangular faces, and is called a tetrahedron (see fig. 3.3). This configuration reminds us of three straight lines in a plane, in general position, which determine various portions of the plane, one of which is limited, contained by three line-segments, and is a triangle (see fig. 3.4). We have to ascertain the number of portions of space determined by the four planes. Let us try our hand at the simpler analogous problem: Into how many portions is the plane divided by three lines?

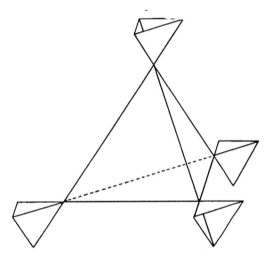

Fig. 3.3. Space divided by four planes.

Many of us will see the answer immediately, even without drawing a figure, and anybody may see it, by using a rough sketch (see fig. 3.4). The required number of parts is 7.

We have found the solution of the simpler analogous problem; but can we use this solution for our original problem? Yes, we can, if we handle the analogy of the two configurations intelligently. We ought to consider the dissection of the plane by 3 straight lines so that we may apply afterwards the same consideration to the dissection of space by 4 planes.

Thus, let us look again at the dissection of the plane by 3 lines, bounding a triangle. One division is finite, it is the interior of the triangle. And the infinite divisions have either a common side with the triangle (there are 3 such divisions), or a common vertex (there are also 3 of this kind). Thus the number of all divisions is $1 + 3 + 3 = 7$.

Now, we consider the dissection of space by 4 planes bounding a tetrahedron. One division is finite, it is the interior of the tetrahedron. An infinite division may have a common face (a 2-dimensional part of the boundary) with the tetrahedron (there are 4 such divisions), or a common edge (1-dimensional part of the boundary; there are 6 divisions of this kind) or a common vertex (0-dimensional part of the boundary; there are 4 divisions of this kind, emphasized in fig. 3.3). Thus the number of all divisions is $1 + 4 + 6 + 4 = 15$.

We have reached this result by analogy, and we have used analogy in a typical, important way. First, we devised an easier analogous problem and we solved it. Then, in order to solve the original, more difficult problem (about the tetrahedron), we used the new easier analogous problem

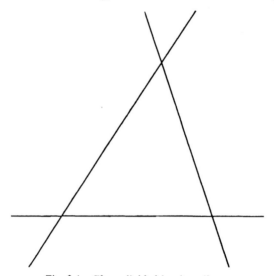

Fig. 3.4.    Plane divided by three lines.

(about the triangle) as a *model*; in solving the more difficult problem we followed the pattern of the solution of the easier problem. But before doing this, we had to reconsider the solution of the easier problem. We rearranged it, remade it into a new pattern fit for imitation.

To single out an analogous easier problem, to solve it, to remake its solution so that it may serve as a model and, at last, to reach the solution of the original problem by following the model just created—this method may seem roundabout to the uninitiated, but is frequently used in mathematical and non-mathematical scientific research.

**12. An array of analogous problems.** Yet our original problem is still unsolved. It is concerned with the dissection of space by 5 planes. What is the analogous problem in two dimensions? Dissection by 5 straight lines? Or by 4 straight lines? It may be better for us to consider these problems in full generality, the dissection of space by $n$ planes, and the dissection of a plane by $n$ straight lines. These dissecting straight lines must be, of course, in general position (no 2 are parallel and no 3 meet in the same point).

If we are accustomed to use geometrical analogy, we may go one step further and consider also the division of the straight line by $n$ different points. Although this problem seems to be rather trivial, it may be instructive. We easily see that a straight line is divided by 1 point into 2 parts, by 2 points into 3, by 3 points into 4, and, generally, by $n$ points into $n + 1$ different parts.

Again, if we are accustomed to pay attention to extreme cases, we may consider the undivided space, plane or line, and regard it as a "division effected by 0 dividing elements."

Let us set up the following table that exhibits all our results obtained hitherto.

| Number of dividing elements | Number of divisions of | | |
|:---:|:---:|:---:|:---:|
| | space by planes | plane by straight lines | line by points |
| 0 | 1 | 1 | 1 |
| 1 | 2 | 2 | 2 |
| 2 | 4 | 4 | 3 |
| 3 | 8 | 7 | 4 |
| 4 | 15 | | 5 |
| ... | ... | ... | ... |
| $n$ | | | $n + 1$ |

**13. Many problems may be easier than just one.** We started out to solve a problem, that about the dissection of space by 5 planes. We have not yet solved it, but we set up many new problems. Each unfilled case of our table corresponds to an open question.

This procedure of heaping up new problems may seem foolish to the uninitiated. But some experience in solving problems may teach us that many problems together may be easier to solve than just one of them—if the many problems are well coordinated, and the one problem by itself is isolated. Our original problem appears now as one in an array of unsolved problems. But the point is that all these unsolved problems form an array: they are well disposed, grouped together, in close analogy with each other and with a few problems solved already. If we compare the present position of our question, well inserted into an array of analogous questions, with its original position, as it was still completely isolated, we are naturally inclined to believe that some progress has been made.

**14. A conjecture.** We look at the results displayed in our table as a naturalist looks at the collection of his specimens. This table is a challenge to our inventive ability, to our faculties of observation. Can we discover any connection, any regularity?

Looking at the second column (division of space by planes) we may notice the sequence 1, 2, 4, 8—there is a clear regularity; we see here the successive powers of 2. Yet, what a disappointment! The next term in the column is 15, and not 16 as we have expected. Our first guess was not so good; we must look for something else.

Eventually, we may chance upon adding two juxtaposed numbers and observe that their sum is in the table. We recognize a peculiar connection; we obtain a number of the table by adding two others, the number above it and the number to the right of the latter. For example,

$$8 \qquad 7$$

$$15$$

are linked by the relation

$$8 + 7 = 15.$$

This is a remarkable connection, a striking clue. It seems unlikely that this connection which we can observe throughout the whole table so far calculated should result from mere chance.

Thus, the situation suggests that the regularity observed extends beyond the limits of our observation, that the numbers of the table not yet calculated are connected in the same way as those already calculated, and so we are led to conjecture that the law we chanced upon is generally valid.

*If* this is so, however, we can solve our original problem. By adding juxtaposed numbers we can extend our table till we reach the number we wished to obtain:

| | | | |
|---|---|---|---|
| 0 | 1 | 1 | 1 |
| 1 | 2 | 2 | 2 |
| 2 | 4 | 4 | 3 |
| 3 | 8 | 7 | 4 |
| 4 | 15 | **11** | 5 |
| 5 | **26** | | |

In the table as it is reprinted here two new numbers appear in heavy type, computed by addition, $11 = 7 + 4$, $26 = 15 + 11$. If our guess is correct, 26 should be the number of portions into which space is dissected by 5 planes in general position. We have solved the proposed problem, it seems. Or, at least, we have succeeded in hitting on a plausible conjecture supported by all the evidence heretofore collected.

**15. Prediction and verification.** In the foregoing we have followed exactly the typical procedure of the naturalist. If a naturalist observes a striking regularity, which cannot be reasonably attributed to mere chance, he conjectures that the regularity extends beyond the limits of his actual observations. Making such a conjecture is often the decisive step in inductive research.

The following step may be prediction. On the basis of his former observations and their concordance with conjectural law, the naturalist predicts the result of his next observation. Much depends on the outcome of that next observation. Will the prediction turn out to be true or not? We are very much in the same position. We have found, or, rather, guessed or

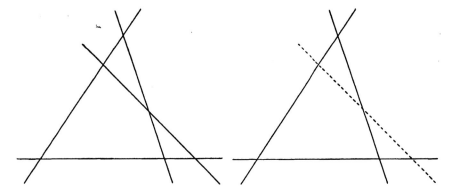

Fig. 3.5. Plane divided by four lines.   Fig. 3.6. Transition from three lines to four.

predicted that 11 should be the number of regions into which a plane is dissected by 4 straight lines in general position. Is that so? Is our prediction correct?

Examining a rough sketch (see fig. 3.5) we can convince ourselves that our guess was good, that 11 *is* actually the correct number. This confirmation of our prediction yields inductive evidence in favor of the rule on the basis of which we made our prediction. Having passed the test successfully, our conjecture comes out strengthened.

**16. Again and better.** We have verified that number 11 by looking at the figure and counting. Yes, 4 lines in general position seem to divide the plane into 11 portions. But let us do it again and do it better. We have counted those portions in some way. Let us count them again and count them so that we should be certain of avoiding confusion and miscounts and traps set by special positions.

Let us start from the fact that 3 lines determine exactly 7 portions of the plane. We have some reasons to believe that 4 lines determine 11 portions. Why just 4 portions more? Why does the number 4 intervene in this connection? Why does the introduction of a new line increase the number of portions just by 4?

We emphasize one line in fig. 3.5, we redraw it in short strokes (see fig. 3.6). The new figure does not look very different but it expresses a very different conception. We regard the emphasized line as *new* and the three other lines as *old*. The old lines cut the plane into 7 portions. What happens when a new line is added?

The new line, drawn at random, must intersect each old line and each one in a different point. That makes 3 points. These 3 points divide the new line into 4 segments. Now each segment bisects an old division of the plane, makes two new divisions out of an old one. Taken together, the 4 segments of the new line create 8 new divisions and abolish 4 old divisions —the *number* of divisions increases by just 4. *This* is the reason that the number of divisions now is just 4 more than it was before: $7 + 4 = 11$.

This way of arriving at the number 11 is convincing and illuminating. We may begin now to see a reason for the regularity which we have observed and on which we have based our prediction of that number 11. We begin to suspect an explanation behind the facts and our confidence in the general validity of the observed regularity is greatly strengthened.

**17. Induction suggests deduction; the particular case suggests the general proof.** We have been careful all along to point out the parallelism between our reasoning and the procedures of the naturalist. We started from a special problem as the naturalist may start from a puzzling observation. We advanced by tentative generalizations, by noticing accessible special cases, by observing instructive analogies. We tried to guess some regularity and failed, we tried again and did better. We succeeded in conjecturing a general law which is supported by all experimental evidence at our disposal. We put one more special case to the test and found concordance with the conjectured law, the authority of which gained by such verification. At last, we began to see a reason for that general law, a sort of explanation, and our confidence was greatly strengthened. A naturalist's research may pass through exactly the same phases.

There is, however, a parting of the ways at which the mathematician turns sharply away from the naturalist. Observation is the highest authority for the naturalist, but not for the mathematician. Verification in many well-chosen instances is the only way of confirming a conjectural law in the natural sciences. Verification in many well-chosen instances may be very helpful as an encouragement, but can never prove a conjectural law in the mathematical sciences. Let us consider our own concrete case. By examining various special cases and comparing them, we have been led to conjecturing a general rule from which it would follow that the solution of our originally proposed problem is 26. Are all our observations and verifications sufficient to prove the general rule? Or can they prove the special result that the solution of our problem is actually 26? Not in the least. For a mathematician with rigid standards, the number 26 is just a clever guess and no amount of experimental verification could demonstrate the suspected general rule. Induction renders its results probable, it never proves them.

It may be observed, however, that inductive research may be useful in mathematics in another respect that we have not yet mentioned. The careful observation of special cases that leads us to a general mathematical result may also *suggest its proof.* From the intent examination of a particular case a general insight may emerge.

In fact, this actually happened to us already in the foregoing section. The general rule that we have discovered by induction is concerned with two juxtaposed numbers in our table, such as 7 and 4, and with their sum, which is 11 in the case at hand. Now, in the foregoing section we have visualized the geometrical significance of 7, 4, and 11 in our problem and, in doing so, we have understood *why* the relation $7 + 4 = 11$ arises there. We dealt, in fact, with the passage from 3 lines dividing the plane to 4 such lines. Yet there is no particular virtue in the numbers 3 and 4; we could pass just as well from any number to the following, from $n$ to $n + 1$. The special case discussed may represent to us the general situation (ex. 2.10). I leave to the reader the pleasure of fully extracting the general idea from the particular observation of the foregoing section. In doing so, he may give a formal proof for the rule discovered inductively, at least as far as the last two columns are concerned.

Yet, in order to complete the proof, we have to consider not only the dissection of a plane by straight lines, but also the dissection of space by planes. We may hope, however, that if we are able to clear up the dissection of a plane, analogy will help us to clear up the dissection of space. Again, I leave to the reader the pleasure of profiting from the advice of analogy.

**18. More conjectures.** We have not yet exhausted the subject of plane and space partitions. There are a few more little discoveries to make and they are well accessible to inductive reasoning. We may be easily led to them by careful observation and understanding combination of particular instances.

We may wish to find a *formula* for the number of divisions of a plane by $n$ lines in general position. In fact, we have already a formula in a simpler analogous case: $n$ different points divide a straight line into $n + 1$ segments. This analogous formula, the particular cases listed in our table, our inductively discovered general rule (which we have almost proved), all our results hitherto obtained may help us to solve this new problem. I do not enter into details. I just note the solution which we may find, following the foregoing hints, in various manners.

$1 + n$ is the number of portions into which a straight line is divided by $n$ different points.

$1 + n + \dfrac{n(n-1)}{2}$ is the number of portions into which a plane is divided by $n$ straight lines in general position.

The reader may derive the latter formula or at least he can check it in the simplest cases, for $n = 0, 1, 2, 3, 4$. I leave also to the reader the pleasure of discovering a third formula of the same kind, for the number of space partitions. In making this little discovery, the reader may broaden his experience of inductive reasoning in mathematical matters and enjoy the help that analogy lends us in the solution of problems little or great.

### EXAMPLES AND COMMENTS ON CHAPTER III

The formula $F + V = E + 2$, conjectured in sect. 1, is due to Leonhard Euler. We call it "Euler's formula," regard it as a conjecture, and examine it in various ways, sometimes inductively and sometimes with a view to finding a proof, in ex. 1–10. We return to it in ex. 21–30 and ex. 31–41. Before attempting any example in these two divisions, read ex. 21 and ex. 31, respectively.

**1.** Two pyramids, standing on opposite sides of their common base, form jointly a "double pyramid." An octahedron is a particular double pyramid; the common base is a square. Does Euler's formula hold for the general double pyramid?

**2.** Take a convex polyhedron with $F$ faces, $V$ vertices, and $E$ edges, choose a point $P$ in its interior (its centroid, for example), describe a sphere with center $P$ and project the polyhedron from the center $P$ onto the surface of the sphere. This projection transforms the $F$ faces into $F$ regions or "countries" on the surface of the sphere, it transforms any of the $E$ edges into a boundary line separating two neighboring countries and any of the $V$ vertices into a "corner" or a common boundary point of three or more countries (a "three-country corner" or a "four-country corner," etc.). This projection yields boundary lines of particularly simple nature (arcs of great circles) but, obviously, the validity of Euler's formula for such a *subdivision of the surface of the sphere into countries* is independent of the precise form of the boundary lines; the numbers $F$, $V$, and $E$ are not influenced by continuous deformation of these lines.

(1) A *meridian* is one half of a great circle connecting the two poles, South and North. A *parallel circle* is the intersection of the globe's surface with a plane parallel to the equator. The earth's surface is divided by $m$ meridians and $p$ parallel circles into $F$ countries. Compute $F$, $V$, and $E$. Does Euler's formula hold?

(2) The projection of the octahedron from its center $P$ onto the surface of the sphere is a special case of the situation described in (1). For which values of $m$ and $p$?

3. Chance plays a rôle in discovery. Inductive discovery obviously depends on the observational material. In sect. 1 we came across certain polyhedra, but we could have chanced upon others. Probably we would not have missed the regular solids, but our list could have come out thus:

| Polyhedron | F | V | E |
|---|---|---|---|
| tetrahedron | 4 | 4 | 6 |
| cube | 6 | 8 | 12 |
| octahedron | 8 | 6 | 12 |
| pentagonal prism | 7 | 10 | 15 |
| pentagonal double pyramid | 10 | 7 | 15 |
| dodecahedron | 12 | 20 | 30 |
| icosahedron | 20 | 12 | 30 |

Do you observe some regularity? Can you explain it? What is the connection with Euler's formula?

4. Try to generalize the relation between two polyhedra observed in the table of ex. 3. [The relation described in the solution of ex. 3 under (2) is too "narrow," too "detailed." Take, however, the cube and the octahedron in the situation there described, color the edges of one in red, those of the other in blue, and project them from their common center $P$ onto a sphere as described in ex. 2. Then generalize.]

5. It would be sufficient to prove Euler's formula in a particular case: for convex polyhedra that have only triangular faces. Why? [Sect. 4.]

6. It would be sufficient to prove Euler's formula in a particular case: for convex polyhedra that have only three-edged vertices. Why? [Sect. 4.]

7. In proving Euler's formula we can restrict ourselves to figures in a plane. In fact, imagine that $F - 1$ faces of the polyhedron are made of cardboard, but one face is made of glass; we call this face the "window." You look through the window into the interior of the polyhedron, holding your eyes so close to the window that you see the whole interior. (This may be impossible if the polyhedron is not convex.) You can interpret what you see as a plane figure drawn on the window pane: you see a subdivision of the window into smaller polygons.

In this subdivision there are $N_2$ polygons, $N_1$ straight boundary lines (some outer, some inner) and $N_0$ vertices.

(1) Express $N_0$, $N_1$, $N_2$ in terms of $F$, $V$, $E$.

(2) If Euler's formula holds for $F$, $V$, and $E$, which formula holds for $N_0$, $N_1$, and $N_2$?

8. A rectangle is $l$ inches long and $m$ inches wide; $l$ and $m$ are integers. The rectangle is subdivided into $lm$ equal squares by straight lines parallel to its sides.

(1) Express $N_0$, $N_1$, and $N_2$ (defined in ex. 7) in terms of $l$ and $m$.

(2) Is the relation ex. 7 (2) valid in the present case?

**9.** Ex. 5 and 7 suggest that we should examine the subdivision of a triangle into $N_2$ triangles with $N_0 - 3$ vertices in the interior of the subdivided triangle. In computing the sum of all the angles in those $N_2$ triangles in two different ways, you may prove Euler's formula.

**10.** Sect. 7 suggests the extension of Euler's formula to four and more dimensions. How can we make such an extension tangible? How can we visualize it?

Ex. 7 shows that the case of polyhedra can be reduced to the subdivision of a plane polygon. Analogy suggests that the case of four dimensions may be reduced to the subdivision of a polyhedron in our visible three-dimensional space. If we wish to proceed inductively, we have to examine some example of such a subdivision. By analogy, ex. 8 suggests the following.

A box (that is, a rectangular parallelepiped) has the dimensions $l$, $m$, and $n$; these three numbers are integers. The box is subdivided into $lmn$ equal cubes by planes parallel to its faces. Let $N_0$, $N_1$, $N_2$, and $N_3$ denote the number of vertices, edges, faces, and polyhedra (cubes) forming the subdivision, respectively.

(1) Express $N_0$, $N_1$, $N_2$, and $N_3$ in terms of $l$, $m$, and $n$.

(2) Is there a relation analogous to equation (2) in the solution of ex. 7?

**11.** Let $P_n$ denote the number of parts into which the plane is divided by $n$ straight lines in general position. Prove that $P_{n+1} = P_n + (n + 1)$.

**12.** Let $S_n$ denote the number of parts into which space is divided by $n$ planes in general position. Prove that $S_{n+1} = S_n + P_n$.

**13.** Verify the conjectural formula

$$P_n = 1 + n + \frac{n(n - 1)}{2}$$

for $n = 0, 1, 2, 3, 4$.

**14.** Guess a formula for $S_n$ and verify it for $n = 0, 1, 2, 3, 4, 5$.

**15.** How many parts out of the 11 into which the plane is divided by 4 straight lines in general position are finite? [How many are infinite?]

**16.** Generalize the foregoing problem.

**17.** How many parts out of the 26 into which space is divided by 5 planes in general position are infinite?

**18.** Five planes pass through the center of a sphere, but in other respects their position is general. Find the number of the parts into which the surface of the sphere is divided by the five planes.

**19.** Into how many parts is the plane divided by 5 mutually intersecting circles in general position?

**20.** Generalize the foregoing problems.

**21.** *Induction: adaptation of the mind, adaptation of the language.* Induction results in adapting our mind to the facts. When we compare our ideas to the observations, there may be agreement or disagreement. If there is agreement, we feel more confident of our ideas; if there is disagreement, we modify our ideas. After repeated modification our ideas may fit the facts somewhat better. Our first ideas about any new subject are almost bound to be wrong, at least in part; the inductive process gives us a chance to correct them, to adapt them to reality. Our examples show this process on a small scale, but pretty clearly. In sect. 1, after two or three wrong conjectures, we arrived eventually at the right conjecture. We arrived at it by accident, you may say. "Yet such accidents happen only to people who deserve them," as Lagrange said once when an incomparably greater discovery, by Newton, was discussed.

Adaptation of the mind may be more or less the same thing as adaptation of the language; at any rate, one goes hand in hand with the other. The progress of science is marked by the progress of terminology. When the physicists started to talk about "electricity," or the physicians about "contagion," these terms were vague, obscure, muddled. The terms that the scientists use today, such as "electric charge," "electric current," "fungus infection," "virus infection," are incomparably clearer and more definite. Yet what a tremendous amount of observation, how many ingenious experiments lie between the two terminologies, and some great discoveries too. Induction changed the terminology, clarified the concepts. We can illustrate also this aspect of the process, the inductive clarification of concepts, by a suitable small-scale mathematical example. The situation, not infrequent in mathematical research, is this: A theorem has been already formulated, but we have to give a more precise meaning to the terms in which it is formulated in order to render it strictly correct. This can be done conveniently by an inductive process, as we shall see.

Let us look back at ex. 2 and its solution. We talked about the "subdivision of the sphere into countries" without proposing a formal definition of this term. We hoped that Euler's formula remains valid if $F$, $V$, and $E$ denote the number of countries, boundary lines, and corners in such a subdivision. Yet again, we relied on examples and a rough description and did not give formal definitions for $F$, $V$, and $E$. In what exact sense should we take these terms to render Euler's formula strictly correct? This is our question.

Let us say that a subdivision of the sphere (that is, of the spherical *surface*) with a corresponding interpretation of $F$, $V$, and $E$ is "right" if Euler's formula holds, and is "wrong" if this formula does not hold. Propose examples of subdivisions which could help us to discover some clear and simple distinction between "right" and "wrong" cases.

**22.** The whole surface of the globe consists of just one country.   Is that right?   (We mean "right" from the viewpoint of Euler's formula.)

**23.** The globe's surface is divided into just two countries, the western hemisphere and the eastern hemisphere, separated by a great circle.   Is that wrong?

**24.** Two parallel circles divide the sphere into three countries.   Is it right or wrong?

**25.** Three meridians divide the sphere into three countries.   Is it right or wrong?

**26.** Call the division of the sphere by $m$ meridians and $p$ parallel circles the "division $(m,p)$"; cf. ex. 2 (1).   Is the extreme case $(0,p)$ right or wrong?

**27.** Is the extreme case $(m,0)$ right or wrong?   (Cf. ex. 26.)

**28.** Which subdivisions $(m,p)$ (cf. ex. 26) can be generated by the process described in ex. 2?   (Projection of a convex polyhedron onto the sphere, followed by continuous shifting of the boundaries which leaves the number of countries and the number of the boundary lines around each country unaltered.)   Which conditions concerning $m$ and $p$ characterize such subdivisions?

**29.** What is wrong with the examples in which Euler's formula fails? Which geometrical conditions, rendering more precise the meaning of $F$, $V$, and $E$, would ensure the validity of Euler's formula?

**30.** Propose more examples to illustrate the answer to ex. 29.

**31.** *Descartes' work on polyhedra.*   Among the manuscripts left by Descartes there were brief notes on the general theory of polyhedra.   A copy of these notes (by the hand of Leibnitz) was discovered and published in 1860, more than two hundred years after Descartes' death; cf. Descartes' *Oeuvres*, vol. 10, pp. 257–276.   These notes treat of subjects closely related to Euler's theorem: although the notes do not state the theorem explicitly, they contain results from which it immediately follows.

We consider, with Descartes, a convex polyhedron.   Let us call any angle of any face of the polyhedron a *surface angle*, and let $\Sigma\alpha$ stand for the sum of all surface angles.   Descartes computes $\Sigma\alpha$ in two different manners, and Euler's theorem results immediately from the comparison of the two expressions.

The following examples give the reader an opportunity to reconstruct some of Descartes' conclusions.   We shall use the following notation:

$F_n$ denotes the number of faces with $n$ edges,

$V_n$ the number of vertices in which $n$ edges end, so that

$$F_3 + F_4 + F_5 + \ldots = F,$$
$$V_3 + V_4 + V_5 + \ldots = V.$$

We continue to call $E$ the number of all edges of the polyhedron.

**32.** Express the number of all surface angles in three different ways: in terms of $F_3$, $F_4$, $F_5$, . . . , of $V_3$, $V_4$, $V_5$, . . . , and of $E$, respectively.

**33.** Compute $\Sigma\alpha$ for the five regular solids: the tetrahedron, the cube, the octahedron, the dodecahedron, and the icosahedron.

**34.** Express $\Sigma\alpha$ in terms of $F_3$, $F_4$, $F_5$, . . . .

**35.** Express $\Sigma\alpha$ in terms of $E$ and $F$.

**36.** *Supplementary solid angles, supplementary spherical polygons.* We call *solid angle* what is more usually called a polyhedral angle.

Two convex solid angles have the same number of faces and the same vertex, but no other point in common. To each face of one solid angle corresponds an edge of the other, and the face is perpendicular to the corresponding edge. (This relation between the two solid angles is reciprocal: the edge $e$, intersection of two contiguous faces of the first solid angle, corresponds to the face $f'$, of the second solid angle, if $f'$ is bounded by the two edges corresponding to the two above mentioned faces.) Two solid angles in this mutual relation are called *supplementary* solid angles. (This name is not usual, but two ordinary supplementary angles can be brought into an analogous mutual position.) Each of two supplementary solid angles is called the supplement of the other.

The sphere with radius 1, described about the common vertex of two supplementary solid angles as center, is intersected by these in two spherical polygons: also these polygons are called supplementary.

We consider two supplementary spherical polygons. Let $a_1$, $a_2$, . . . $a_n$ denote the sides of the first polygon, $\alpha_1$, $\alpha_2$, . . . $\alpha_n$ its angles, $A$ its area, $P$ its perimeter, and let $a'_1$, $a'_2$, . . . $a'_n$, $\alpha'_1$, $\alpha'_2$, . . . $\alpha'_n$, $A'$, $P'$ stand for the analogous parts of the other polygon. Then, if the notation is appropriately chosen,

$$a_1 + \alpha'_1 = a_2 + \alpha'_2 = \ldots = a_n + \alpha'_n = \pi,$$

$$a'_1 + \alpha_1 = a'_2 + \alpha_2 = \ldots = a'_n + \alpha_n = \pi;$$

this is well known and easily seen.
    Prove that

$$P + A' = P' + A = 2\pi.$$

[Assume as known that the area of a spherical triangle with angles $\alpha$, $\beta$, and $\gamma$ is the "spherical excess" $\alpha + \beta + \gamma - \pi$.]

**37.** "As in a plane figure all exterior angles jointly equal 4 right angles, so in a solid figure all exterior solid angles jointly equal 8 right angles." Try to interpret this sentence found in Descartes' notes as a theorem which you can prove. [See fig. 3.7.]

**38.** Express $\Sigma\alpha$ in terms of $V$.

**39.** Prove Euler's theorem.

**40.** The initial remark of sect. 1 is vague, but can suggest several precise statements. Here is one that we have not considered in sect. 1: "If any one of the three quantities $F$, $V$, and $E$ tends to $\infty$, also the other two must tend to $\infty$." Prove the following inequalities which hold generally for convex polyhedra and give still more precise information:

$$2E \geq 3F, \quad 2V \geq F + 4, \quad 3V \geq E + 6,$$
$$2E \geq 3V, \quad 2F \geq V + 4, \quad 3F \geq E + 6,$$

Fig. 3.7. Exterior angles of a polygon.

Can the case of equality be attained in these inequalities? For which kind of polyhedra can it be attained?

**41.** There are convex polyhedra all faces of which are polygons of the same kind, that is, polygons with the same number of sides. For example, all faces of a tetrahedron are triangles, all faces of a parallelepiped quadrilaterals, all faces of a regular dodecahedron pentagons. "And so on," you may be tempted to say. Yet such simple induction may be misguiding: there exists *no* convex polyhedron with faces which are all hexagons. Try to prove this. [Ex. 31.]

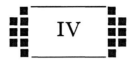

# IV

# INDUCTION IN THE THEORY
# OF NUMBERS

---

*In the Theory of Numbers it happens rather frequently that, by some unexpected luck, the most elegant new truths spring up by induction.*
—GAUSS[1]

---

**1. Right triangles in integers.**[2] The triangle with sides 3, 4, and 5 is a right triangle since

$$3^2 + 4^2 = 5^2.$$

This is the simplest example of a right triangle of which the sides are measured by integers. Such "right triangles in integers" have played a rôle in the history of the Theory of Numbers; even the ancient Babylonians discovered some of their properties.

One of the more obvious problems about such triangles is the following: *Is there a right triangle in integers, the hypotenuse of which is a given integer n?*

We concentrate upon this question. We seek a triangle the hypotenuse of which is measured by the given integer $n$ and the legs by some integers $x$ and $y$. We may assume that $x$ denotes the longer of the two legs. Therefore, being given $n$, we seek two integers $x$ and $y$ such that

$$n^2 = x^2 + y^2, \quad 0 < y \leqq x < n.$$

We may attack the problem inductively and, unless we have some quite specific knowledge, we cannot attack it any other way. Let us take an example. We choose $n = 12$. Therefore, we seek two positive integers $x$ and $y$, such that $x \geqq y$ and

$$144 = x^2 + y^2.$$

---

[1] *Werke*, vol. 2, p. 3.

[2] Parts of this chapter appeared already under the title "Let us teach guessing" in the volume *Études de philosophie des sciences en hommage à Ferdinand Gonseth*. Éditions du Griffon, 1950; see pp. 147–154.

Which values are available for $x^2$? The following:

$$1, 4, 9, 16, 25, 36, 49, 64, 81, 100, 121.$$

Is $x^2 = 121$? That is, is

$$144 - x^2 = 144 - 121 = y^2$$

a square? No, 23 is not a square. We should now try other squares but, in fact, we need not try too many of them. Since $y \leq x$,

$$144 = x^2 + y^2 \leq 2x^2$$

$$x^2 \geq 72.$$

Therefore, $x^2 = 100$ and $x^2 = 81$ are the only remaining possibilities. Now, neither of the numbers

$$144 - 100 = 44, \quad 144 - 81 = 63$$

is a square and hence the answer: there is no right triangle in integers with hypotenuse 12.

We treat similarly the hypotenuse 13. Of the three numbers

$$169 - 144 = 25, \quad 169 - 121 = 48, \quad 169 - 100 = 69$$

only one is a square and so there is just one right triangle in integers with hypotenuse 13:

$$169 = 144 + 25.$$

Proceeding similarly, we can examine with a little patience all the numbers under a given not too high limit, such as 20. We find only five "hypotenuses" less than 20, the numbers 5, 10, 13, 15, and 17:

$$25 = 16 + 9$$
$$100 = 64 + 36$$
$$169 = 144 + 25$$
$$225 = 144 + 81$$
$$289 = 225 + 64.$$

By the way, the cases 10 and 15 are not very interesting. The triangle with sides 10, 8, and 6 is similar to the simpler triangle with sides 5, 4, and 3, and the same is true of the triangle with sides 15, 12, and 9. The remaining three right triangles, with hypotenuse 5, 13, and 17, respectively, are essentially different, none is similar to another among them.

We may notice that all three numbers 5, 13, and 17 are *odd primes*. They are, however, not all the odd primes under 20; none of the other odd primes, 3, 7, 11, and 19 is a hypotenuse. Why that? What is the difference between the two sets? *When, under which circumstances, is an odd prime the hypotenuse of some right triangle in integers, and when is it not?* This is a modification of our original question. It may appear more hopeful; at any rate, it is new. Let us investigate it—again, inductively. With a little patience, we construct the following table (the dash indicates that there is no right triangle with hypotenuse $p$).

| Odd prime $p$ | Right triangles with hypotenuse $p$ |
|:---:|:---:|
| 3 | — |
| 5 | $25 = 16 + 9$ |
| 7 | — |
| 11 | — |
| 13 | $169 = 144 + 25$ |
| 17 | $289 = 225 + 64$ |
| 19 | — |
| 23 | — |
| 29 | $841 = 441 + 400$ |
| 31 | — |

When is a prime a hypotenuse; when is it not? What is the difference between the two cases? A physicist could easily ask himself some very similar questions. For instance, he investigates the double refraction of crystals. Some crystals do show double refraction; others do not. Which crystals are doubly refracting, which are not? What is the difference between the two cases?

The physicist looks at his crystals and we look at our two sets of primes

$$5, 13, 17, 29, \ldots \quad \text{and} \quad 3, 7, 11, 19, 23, 31, \ldots .$$

We are looking for some characteristic difference between the two sets. The primes in both sets increase by irregular jumps. Let us look at the lengths of these jumps, at the successive differences:

| 5 | 13 | 17 | 29 | | 3 | 7 | 11 | 19 | 23 | 31 |
|:---:|:---:|:---:|:---:|:---:|:---:|:---:|:---:|:---:|:---:|:---:|
| 8 | | 4 | 12 | | | 4 | 4 | 8 | 4 | 8 |

Many of these differences are equal to 4, and, as it is easy to notice, *all are divisible by 4*. The primes in the first set, led by 5, leave the remainder 1 when divided by 4, are of the form $4n + 1$ with integral $n$. The primes in the second set, led by 3, are of the form $4n + 3$. Could this be the characteristic difference we are looking for? If we do not discard this possibility from the outset, we are led to the following conjecture: *A prime of the form $4n + 1$ is the hypotenuse of just one right triangle in integers; a prime of the form $4n + 3$ is the hypotenuse of no such triangle.*

**2. Sums of squares.** The problem of the right triangles in integers, one aspect of which we have just discussed (in sect. 1), played, as we have said, an important rôle in the history of the Theory of Numbers. It leads on, in fact, to many further questions. Which numbers, squares or not, can be decomposed into two squares? What about the numbers which cannot be decomposed into two squares? Perhaps, they are decomposable into three squares; but what about the numbers which are not decomposable into three squares?

We could go on indefinitely, but, and this is highly remarkable, we need not. Bachet de Méziriac (author of the first printed book on mathematical recreations) remarked that *any number* (that is, positive integer) *is either a square, or the sum of two, three, or four squares.* He did not pretend to possess a proof. He found indications pointing to his statement in certain problems of Diophantus and verified it up to 325.

In short, Bachet's statement was just a conjecture, found inductively. It seems to me that his main achievement was to put the question: *HOW MANY squares are needed to represent all integers?* Once this question is clearly put, there is not much difficulty in discovering the answer inductively. We construct a table beginning with

$$1 = 1$$
$$2 = 1 + 1$$
$$3 = 1 + 1 + 1$$
$$4 = 4$$
$$5 = 4 + 1$$
$$6 = 4 + 1 + 1$$
$$7 = 4 + 1 + 1 + 1$$
$$8 = 4 + 4$$
$$9 = 9$$
$$10 = 9 + 1.$$

This verifies the statement up to 10. Only the number 7 requires four squares; the others are representable by one or two or three. Bachet went on tabulating up to 325 and found many numbers requiring four squares and none requiring more. Such inductive evidence satisfied him, it seems, at least to a certain degree, and he published his statement. He was lucky. His conjecture turned out to be true and so he became the discoverer of the "four-square theorem" which we can state also in the form: The equation

$$n = x^2 + y^2 + z^2 + w^2$$

where $n$ is any given positive integer has a solution in which $x$, $y$, $z$, and $w$ are non-negative integers.

The decomposition of a number into a sum of squares has still other aspects. Thus, we may investigate the number of solutions of the equation

$$n = x^2 + y^2$$

in integers $x$ and $y$. We may admit only positive integers, or all integers, positive, negative, and 0. If we choose the latter conception of the problem and take as example $n = 25$, we find 12 solutions of the equation

$$25 = x^2 + y^2,$$

namely the following

$$25 = 5^2 + 0^2 = (-5)^2 + 0^2 = 0^2 + 5^2 = 0^2 + (-5)^2$$
$$= 4^2 + 3^2 = (-4)^2 + 3^2 = 4^2 + (-3)^2 = (-4)^2 + (-3)^2$$
$$= 3^2 + 4^2 = (-3)^2 + 4^2 = 3^2 + (-4)^2 = (-3)^2 + (-4)^2.$$

By the way, these solutions have an interesting geometric interpretation, but we need not discuss it now. See ex. 2.

**3. On the sum of four odd squares.** Of the many problems concerned with sums of squares I choose one that looks somewhat far-fetched, but will turn out to be exceptionally instructive.

*Let $u$ denote a positive odd integer. Investigate inductively the number of the solutions of the equation*

$$4u = x^2 + y^2 + z^2 + w^2$$

*in positive odd integers $x$, $y$, $z$, and $w$.*

For example, if $u = 1$ we have the equation

$$4 = x^2 + y^2 + z^2 + w^2$$

and there is obviously just one solution, $x = y = z = w = 1$. In fact, we do not regard

$$x = -1, \quad y = 1, \quad z = 1, \quad w = 1$$

or

$$x = 2, \quad y = 0, \quad z = 0, \quad w = 0$$

as a solution, since we admit only positive odd numbers for $x, y, z$, and $w$. If $u = 3$, the equation is

$$12 = x^2 + y^2 + z^2 + w^2$$

and the following two solutions:

$$x = 3, \quad y = 1, \quad z = 1, \quad w = 1$$
$$x = 1, \quad y = 3, \quad z = 1, \quad w = 1$$

are different.

In order to emphasize the restriction laid upon the values of $x, y, z$, and $w$, we shall avoid the term "solution" and use instead the more specific description: "representation of $4u$ as a sum of four odd squares." As this description is long, we shall abbreviate it in various ways, sometimes even to the one word "representation."

**4. Examining an example.** In order to familiarize ourselves with the meaning of our problem, let us consider an example. We choose $u = 25$. Then $4u = 100$, and we have to find all representations of 100 as a sum of four odd squares. Which odd squares are available for this purpose? The following:

$$1, \quad 9, \quad 25, \quad 49, \quad 81.$$

If 81 is one of the four squares the sum of which is 100, then the sum of the three others must be

$$100 - 81 = 19.$$

The only odd squares less than 19 are 1 and 9, and $19 = 9 + 9 + 1$ is evidently the only possibility to represent 19 as a sum of 3 odd squares if the terms are arranged in order of magnitude. We obtain

$$100 = 81 + 9 + 9 + 1.$$

We find similarly

$$100 = 49 + 49 + 1 + 1,$$
$$100 = 49 + 25 + 25 + 1,$$
$$100 = 25 + 25 + 25 + 25.$$

Proceeding systematically, by splitting off the largest square first, we may convince ourselves that we have exhausted all possibilities, provided that the 4 squares are arranged in descending order (or rather in non-ascending order). But there are more possibilities if we take into account, as we should, all arrangements of the terms. For example,

$$
\begin{aligned}
100 &= 49 + 49 + 1 + 1 \\
&= 49 + 1 + 49 + 1 \\
&= 49 + 1 + 1 + 49 \\
&= 1 + 49 + 49 + 1 \\
&= 1 + 49 + 1 + 49 \\
&= 1 + 1 + 49 + 49.
\end{aligned}
$$

These 6 sums have the same terms, but the order of the terms is different; they are to be considered, according to the statement of our problem, as 6 different representations; the one representation

$$100 = 49 + 49 + 1 + 1$$

with non-increasing terms is a source of 5 other representations, of 6 representations in all. We have similarly

| *Non-increasing terms* | *Number of arrangements* |
|---|---|
| $81 + 9 + 9 + 1$ | 12 |
| $49 + 49 + 1 + 1$ | 6 |
| $49 + 25 + 25 + 1$ | 12 |
| $25 + 25 + 25 + 25$ | 1 |

To sum up, we found in our case where $u = 25$ and $4u = 100$

$$12 + 6 + 12 + 1 = 31$$

representations of $4u = 100$ as a sum of 4 odd squares.

**5. Tabulating the observations.** The special case $u = 25$ where $4u = 100$ and the number of representations is 31 has shown us clearly the meaning of the problem.    We may now explore systematically the simplest cases, $u = 1, 3, 5, \ldots$ up to $u = 25$.    We construct Table I.    (See below; the reader should construct the table by himself, or at least check a few items.)

Table I

| $u$ | $4u$ | Non-increasing | Arrangements | Representations |
|---|---|---|---|---|
| 1  | 4   | $1 + \ 1 + \ 1 + \ 1$ | 1  | 1  |
| 3  | 12  | $9 + \ 1 + \ 1 + \ 1$ | 4  | 4  |
| 5  | 20  | $9 + \ 9 + \ 1 + \ 1$ | 6  | 6  |
| 7  | 28  | $25 + \ 1 + \ 1 + \ 1$ | 4  | 8  |
|    |     | $9 + \ 9 + \ 9 + \ 1$ | 4  |    |
| 9  | 36  | $25 + \ 9 + \ 1 + \ 1$ | 12 | 13 |
|    |     | $9 + \ 9 + \ 9 + \ 9$ | 1  |    |
| 11 | 44  | $25 + \ 9 + \ 9 + \ 1$ | 12 | 12 |
| 13 | 52  | $49 + \ 1 + \ 1 + \ 1$ | 4  | 14 |
|    |     | $25 + 25 + \ 1 + \ 1$ | 6  |    |
|    |     | $25 + \ 9 + \ 9 + \ 9$ | 4  |    |
| 15 | 60  | $49 + \ 9 + \ 1 + \ 1$ | 12 | 24 |
|    |     | $25 + 25 + \ 9 + \ 1$ | 12 |    |
| 17 | 68  | $49 + \ 9 + \ 9 + \ 1$ | 12 | 18 |
|    |     | $25 + 25 + \ 9 + \ 9$ | 6  |    |
| 19 | 76  | $49 + 25 + \ 1 + \ 1$ | 12 | 20 |
|    |     | $49 + \ 9 + \ 9 + \ 9$ | 4  |    |
|    |     | $25 + 25 + 25 + \ 1$ | 4  |    |
| 21 | 84  | $81 + \ 1 + \ 1 + \ 1$ | 4  | 32 |
|    |     | $49 + 25 + \ 9 + \ 1$ | 24 |    |
|    |     | $25 + 25 + 25 + \ 9$ | 4  |    |
| 23 | 92  | $81 + \ 9 + \ 1 + \ 1$ | 12 | 24 |
|    |     | $49 + 25 + \ 9 + \ 9$ | 12 |    |
| 25 | 100 | $81 + \ 9 + \ 9 + \ 1$ | 12 | 31 |
|    |     | $49 + 49 + \ 1 + \ 1$ | 6  |    |
|    |     | $49 + 25 + 25 + \ 1$ | 12 |    |
|    |     | $25 + 25 + 25 + 25$ | 1  |    |

**6. What is the rule?**    Is there any recognizable law, any simple connection between the odd number $u$ and the number of different representations of $4u$ as a sum of four odd squares?

This question is the kernel of our problem.  We have to answer it on the basis of the observations collected and tabulated in the foregoing section. We are in the position of the naturalist trying to extract some rule, some general formula from his experimental data.  Our experimental material available at this moment consists of two parallel series of numbers

| 1 | 3 | 5 | 7 | 9 | 11 | 13 | 15 | 17 | 19 | 21 | 23 | 25 |
|---|---|---|---|---|----|----|----|----|----|----|----|----|
| 1 | 4 | 6 | 8 | 13 | 12 | 14 | 24 | 18 | 20 | 32 | 24 | 31. |

The first series consists of the successive odd numbers, but what is the rule governing the second series?

As we try to answer this question, our first feeling may be close to despair. That second series looks quite irregular, we are puzzled by its complex origin, we can scarcely hope to find any rule.  Yet, if we forget about the complex origin and concentrate upon what is before us, there is a point easy enough to notice.  It happens rather often that a term of the second series exceeds the corresponding term of the first series by just one unit. Emphasizing these cases by heavy print in the first series, we may present our experimental material as follows:

| 1 | **3** | **5** | 7 | 9 | **11** | **13** | 15 | **17** | **19** | 21 | **23** | 25 |
|---|---|---|---|---|----|----|----|----|----|----|----|----|
| 1 | 4 | 6 | 8 | 13 | 12 | 14 | 24 | 18 | 20 | 32 | 24 | 31. |

The numbers in heavy print attract our attention.  It is not difficult to recognize them: they are *primes*.  In fact, they are *all* the primes in the first row as far as our table goes.  This remark may appear very surprising if we remember the origin of our series.  We considered squares, we made no reference whatever to primes.  Is it not strange that the prime numbers play a rôle in our problem?  It is difficult to avoid the impression that our observation is significant, that there is something remarkable behind it.

What about those numbers of the first series which are not in heavy print? They are odd numbers, but not primes.  The first, 1, is unity, the others are composite

$$9 = 3 \times 3, \quad 15 = 3 \times 5, \quad 21 = 3 \times 7, \quad 25 = 5 \times 5.$$

What is the nature of the corresponding numbers in the second series?

If the odd number $u$ is a prime, the corresponding number is $u + 1$; if $u$ is not a prime, the corresponding number is not $u + 1$.  This we have observed already.  We may add one little remark.  If $u = 1$, the corresponding number is also 1, and so *less* than $u + 1$, but in all other cases in which $u$ is not a prime the corresponding number is *greater* than $u + 1$.  That is, the number corresponding to $u$ is less than, equal to, or greater than $u + 1$ accordingly as $u$ is unity, a prime, or a composite number.  There is some regularity.

Let us concentrate upon the composite numbers in the upper line and the corresponding numbers in the lower line:

$$3 \times 3 \quad 3 \times 5 \quad 3 \times 7 \quad 5 \times 5$$
$$13 \qquad 24 \qquad 32 \qquad 31 \quad .$$

There is something strange.    Squares in the first line correspond to primes in the second line.    Yet we have too few observations; probably we should not attach too much weight to this remark.    Still, it is true that, conversely, under the composite numbers in the first line which are not squares, we find numbers in the second line which are not primes:

$$3 \times 5 \quad 3 \times 7$$
$$4 \times 6 \quad 4 \times 8.$$

Again, there is something strange.    Each factor in the second line exceeds the corresponding factor in the first line by just one unit.    Yet we have too few observations; we had better not attach too much weight to this remark. Still, our remark shows some parallelism with a former remark.    We noticed before ·

$$p$$
$$p + 1$$

and we notice now

$$pq$$
$$(p + 1)(q + 1)$$

where $p$ and $q$ are primes.    There is some regularity.

Perhaps we shall see more clearly if we write the entry corresponding to $pq$ differently:

$$(p + 1)(q + 1) = pq + p + q + 1.$$

What can we see there?    What are these numbers $pq$, $p$, $q$, $1$?    At any rate, the cases

$$9 \qquad 25$$
$$13 \qquad 31$$

remain unexplained.    In fact, the entries corresponding to 9 and 25 are greater than $9 + 1$ and $25 + 1$, respectively, as we have already observed:

$$13 = 9 + 1 + 3 \qquad 31 = 25 + 1 + 5 \ .$$

What are these numbers?

If one more little spark comes from somewhere, we may succeed in combining our fragmentary remarks into a coherent whole, our scattered indications into an illuminating view of the full correspondence:

$$p \qquad\qquad pq \qquad\qquad 9 \qquad\qquad 25 \qquad\qquad 1$$
$$p + 1 \quad pq + p + q + 1 \quad 9 + 3 + 1 \quad 25 + 5 + 1 \quad 1 \ .$$

DIVISORS! The second line shows the divisors of the numbers in the first line. This may be the desired rule, and a discovery, a real discovery: *To each number in the first line corresponds the sum of its divisors.*

And so we have been led to a conjecture, perhaps to one of those "most elegant new truths" of Gauss: *If u is an odd number, the number of representations of 4u as a sum of four odd squares is equal to the sum of the divisors of u.*

**7. On the nature of inductive discovery.** Looking back at the foregoing sections (3 to 6) we may find many questions to ask.

*What have we obtained?* Not a proof, not even the shadow of a proof, just a conjecture: a simple description of the facts within the limits of our experimental material, and a certain hope that this description may apply beyond the limits of our experimental material.

*How have we obtained our conjecture?* In very much the same manner that ordinary people, or scientists working in some non-mathematical field, obtain theirs. We collected relevant observations, examined and compared them, noticed fragmentary regularities, hesitated, blundered, and eventually succeeded in *combining the scattered details into an apparently meaningful whole.* Quite similarly, an archaeologist may reconstitute a whole inscription from a few scattered letters on a worn-out stone, or a palaeontologist may reconstruct the essential features of an extinct animal from a few of its petrified bones. In our case the meaningful whole appeared at the same moment when we recognized the appropriate unifying concept (the divisors).

**8. On the nature of inductive evidence.** There remain a few more questions.

*How strong is the evidence?* Your question is incomplete. You mean, of course, the inductive evidence for our conjecture stated in sect. 6 that we can derive from Table I of sect. 5; this is understood. Yet what do you mean by "strong"? The evidence is strong if it is convincing; it is convincing if it convinces somebody. Yet you did not say whom it should convince—me, or you, or Euler, or a beginner, or whom?

Personally, I find the evidence pretty convincing. I feel sure that Euler would have thought very highly of it. (I mention Euler because he came very near to discovering our conjecture; see ex. 6.24.) I think that a beginner who knows a little about the divisibility of numbers ought to find the evidence pretty convincing, too. A colleague of mine, an excellent mathematician who however was not familiar with this corner of the Theory of Numbers, found the evidence "hundred per cent convincing."

*I am not concerned with subjective impressions. What is the precise, objectively evaluated degree of rational belief, justified by the inductive evidence?* You give me one thing (A), you fail to give me another thing (B), and you ask me a third thing (C).

(A) You give me exactly the inductive evidence: the conjecture has been verified in the first thirteen cases, for the numbers 4, 12, 20, . . . , 100. This is perfectly clear.

(B) You wish me to evaluate the degree of rational belief justified by this evidence. Yet such belief must depend, if not on the whims and the temperament, certainly on the *knowledge* of the person receiving the evidence. He may know a proof of the conjectural theorem or a counter-example exploding it. In these two cases the degree of his belief, already firmly established, will remain unchanged by the inductive evidence. Yet if he knows something that comes very close to a complete proof, or to a complete refutation, of the theorem, his belief is still capable of modification and will be affected by the inductive evidence here produced, although different degrees of belief will result from it according to the kind of knowledge he has. Therefore, if you wish a definite answer, you should specify a definite level of knowledge on which the proposed inductive evidence (A) should be judged. You should give me a definite set of relevant known facts (an explicit list of known elementary propositions in the Theory of Numbers, perhaps).

(C) You wish me to evaluate the degree of rational belief justified by the inductive evidence exactly. Should I give it to you perhaps expressed in percentages of "full credence"? (We may agree to call "full credence" the degree of belief justified by a complete mathematical proof of the theorem in question.) Do you expect me to say that the given evidence justifies a belief amounting to 99% or to 2.875% or to .000001% of "full credence"?

In short, you wish me to solve a problem: Given (A) the inductive evidence and (B) a definite set of known facts or propositions, compute the percentage of full credence rationally resulting from both (C).

To solve this problem is much more than I can do. I do not know anybody who could do it, or anybody who would dare to do it. I know of some philosophers who promise to do something of this sort in great generality. Yet, faced with the concrete problem, they shrink and hedge and find a thousand excuses why not to do just this problem.

Perhaps the problem is one of those typical philosophical problems about which you can talk a lot in general, and even worry genuinely, but which fade into nothingness when you bring them down to concrete terms.

*Could you compare the present case of inductive inference with some standard case and so arrive at a reasonable estimate of the strength of the evidence?* Let us compare the inductive evidence for our conjecture with Bachet's evidence for his conjecture.

Bachet's conjecture was: For $n = 1, 2, 3, \ldots$ the equation

$$n = x^2 + y^2 + z^2 + w^2$$

has at least one solution in non-negative integers $x, y, z$, and $w$. He verified this conjecture for $n = 1, 2, 3, \ldots, 325$. (See sect. 2, especially the short table.)

Our conjecture is: For a given odd $u$, the number of solutions of the equation

$$4u = x^2 + y^2 + z^2 + w^2$$

in positive odd integers $x, y, z$, and $w$ is equal to the sum of the divisors of $u$. We verified this conjecture for $u = 1, 3, 5, 7, \ldots, 25$ (13 cases). (See sect. 3 to 6.)

I shall compare these two conjectures and the inductive evidence yielded by their respective verifications in three respects.

*Number of verifications.* Bachet's conjecture was verified in 325 cases, ours in 13 cases only. The advantage in this respect is clearly on Bachet's side.

*Precision of prediction.* Bachet's conjecture predicts that the number of solutions is $\geq 1$; ours predicts that the number of solutions is exactly equal to such and such a quantity. It is obviously reasonable to assume, I think, that *the verification of a more precise prediction carries more weight* than that of a less precise prediction. The advantage in this respect is clearly on our side.

*Rival conjectures.* Bachet's conjecture is concerned with the maximum number of squares, say $M$, needed in representing an arbitrary positive integer as sum of squares. In fact, Bachet's conjecture asserts that $M = 4$. I do not think that Bachet had any *a priori* reason to prefer $M = 4$ to, say, $M = 5$, or to any other value, as $M = 6$ or $M = 7$; even $M = \infty$ is not excluded *a priori*. (Naturally, $M = \infty$ would mean that there are larger and larger integers demanding more and more squares. On the face, $M = \infty$ could appear as the most likely conjecture.) In short, Bachet's conjecture has many obvious rivals. Yet ours has none. Looking at the irregular sequence of the numbers of representations (sect. 6) we had the impression that we might not be able to find any rule. Now we did find an admirably clear rule. We hardly expect to find any other rule.

It may be difficult to choose a bride if there are many desirable young ladies to choose from; if there is just one eligible girl around, the decision may come much quicker. It seems to me that our attitude toward conjectures is somewhat similar. Other things being equal, a conjecture that has many obvious rivals is more difficult to accept than one that is unrivalled. If you think as I do, you should find that in this respect the advantage is on the side of our conjecture, not on Bachet's side.

Please observe that the evidence for Bachet's conjecture is stronger in one respect and the evidence for our conjecture is stronger in other respects, and do not ask unanswerable questions.

### EXAMPLES AND COMMENTS ON CHAPTER IV

**1.** *Notation.* We assume that $n$ and $k$ are positive integers and consider the Diophantine equation

$$n = x_1^2 + x_2^2 + \ldots + x_k^2.$$

We say that two solutions $x_1, x_2, \ldots x_k$ and $x_1', x_2', \ldots x_k'$ are equal if, and only if, $x_1 = x_1', x_2 = x_2', \ldots x_k = x_k'$. If we admit for $x_1, x_2, \ldots x_k$ *all* integers, positive, negative, or null, we call the number of solutions $R_k(n)$. If we admit only *positive odd* integers, we call the number of solutions $S_k(n)$. This notation is important in the majority of the following problems.

Bachet's conjecture (sect. 2) is expressed in this notation by the inequality

$$R_4(n) > 0 \text{ for } n = 1, 2, 3, \ldots .$$

The conjecture that we discovered in sect. 6 affirms that $S_4(4(2n - 1))$ equals the sum of the divisors of $2n - 1$, for $n = 1, 2, 3, \ldots .$

Find $R_2(25)$ and $S_3(11)$.

**2.** Let $x$ and $y$ be rectangular coordinates in a plane. The points for which both $x$ and $y$ are integers are called the "lattice points" of the plane. Lattice points in space are similarly defined.

Interpret $R_2(n)$ and $R_3(n)$ geometrically, in terms of lattice points.

**3.** Express the conjecture encountered in sect. 1 in using the symbol $R_2(n)$.

**4.** When is an odd prime the sum of two squares? Try to answer this question inductively, by examining the table

$$
\begin{array}{rl}
3 & - \\
5 = & 4 + 1 \\
7 & - \\
11 & - \\
13 = & 9 + 4 \\
17 = & 16 + 1 \\
19 & - \\
23 & - \\
29 = & 25 + 4 \\
31 & -
\end{array}
$$

Extend this table if necessary and compare it with the table in sect. 1.

**5.** Could you verify by mathematical deduction some part of your answer to ex. 4 obtained by induction? After such a verification, would it be reasonable to change your confidence in the conjecture?

**6.** Verify Bachet's conjecture (sect. 2) up to 30 inclusively. Which numbers require actually four squares?

**7.** In order to understand better Table I in sect. 5, let $a^2$, $b^2$, $c^2$, and $d^2$ denote four different odd squares and consider the sums

$$(1) \quad a^2 + b^2 + c^2 + d^2$$
$$(2) \quad a^2 + a^2 + b^2 + c^2$$
$$(3) \quad a^2 + a^2 + b^2 + b^2$$
$$(4) \quad a^2 + a^2 + a^2 + b^2$$
$$(5) \quad a^2 + a^2 + a^2 + a^2 .$$

How many different representations (in the sense of sect. 3) can you derive from each by permuting the terms?

**8.** The number of representations of $4u$, as a sum of four odd squares is odd if, and only if, $u$ is a square. (Following the notation of sect. 3, we assume that $u$ is odd.)   Prove this statement and show that it agrees with the conjecture of sect. 6.   How does this remark influence your confidence in the conjecture?

**9.** Now, let $a$, $b$, $c$, and $d$ denote different positive integers (odd or even). Consider the five sums mentioned in ex. 7 and also the following:

$$(6) \quad a^2 + b^2 + c^2 \qquad (9) \quad a^2 + b^2$$
$$(7) \quad a^2 + a^2 + b^2 \qquad (10) \quad a^2 + a^2$$
$$(8) \quad a^2 + a^2 + a^2 \qquad (11) \quad a^2 .$$

Find in each of these eleven cases the contribution to $R_4(n)$.   You derive from each sum all possible representations by the following obvious operations:  you add $0^2$ as many times as necessary to bring the number of terms to 4, you change the arrangement, and you replace some (or none, or all) of the numbers $a$, $b$, $c$, $d$ by $-a$, $-b$, $-c$, $-d$, respectively.   (Check examples in Table II.)

**10.** Investigate inductively the number of solutions of the equation $n = x^2 + y^2 + z^2 + w^2$ in integers $x$, $y$, $z$, and $w$, positive, negative, or 0. Start by constructing a table analogous to Table I.

**11** (continued).   Try to use the method, or the result, of sect. 6.

**12** (continued).   Led by the analogy of sect. 6 or by your observation of Table II, distinguish appropriate classes of integers and investigate each class by itself.

**13** (continued).   Concentrate upon the most stubborn class.

**14** (continued).   Try to summarize all fragmentary regularities and express the law in one sentence.

**15** (continued).    Check the rule found in the first three cases not contained in Table II.

**16.** Find $R_8(5)$ and $S_8(40)$.

**17.** Check at least two entries of Table III, p. 75, not yet given in Tables I and II.

**18.** Using Table III, investigate inductively $R_8(n)$ and $S_8(8n)$.

**19** (continued).    Try to use the method, or the result, of sect. 6 and ex. 10–15.

**20** (continued).    Led by analogy or observation, distinguish appropriate classes of integers and investigate each class by itself.

**21** (continued).    Try to discover a cue in the most accessible case.

**22** (continued).    Try to find some unifying concept that could summarize the fragmentary regularities.

**23** (continued).    Try to express the law in one sentence.

**24.** Which numbers can and which numbers cannot be expressed in the form $3x + 5y$, where $x$ and $y$ are non-negative integers?

**25.** Try to guess the law of the following table:

| $a$ | $b$ | Last integer not expressible in form $ax + by$ |
|-----|-----|-----|
| 2 | 3 | 1 |
| 2 | 5 | 3 |
| 2 | 7 | 5 |
| 2 | 9 | 7 |
| 3 | 4 | 5 |
| 3 | 5 | 7 |
| 3 | 7 | 11 |
| 3 | 8 | 13 |
| 4 | 5 | 11 |
| 5 | 6 | 19 |

It is understood that $x$ and $y$ are non-negative integers.    Check a few items and extend the table, if necessary.    [Observe the change in the last column when just one of the two numbers $a$ and $b$ changes.]

**26.** *Dangers of induction.*    Examine inductively the following assertions:

(1) $(n - 1)! + 1$ is divisible by $n$ when $n$ is a prime, but not divisible by $n$ when $n$ is composite.

(2) $2^{n-1} - 1$ is divisible by $n$ when $n$ is an odd prime, but not divisible by $n$ when $n$ is composite.

Table II

| $n$ | Non-increasing | Repres. | $R_4(n)/8$ |
|---|---|---|---|
| 1 | 1 | $4 \times 2$ | 1 |
| 2 | $1 + 1$ | $6 \times 4$ | 3 |
| 3 | $1 + 1 + 1$ | $4 \times 8$ | 4 |
| 4 | 4<br>$1 + 1 + 1 + 1$ | $4 \times 2$<br>$1 \times 16$ | 3 |
| 5 | $4 + 1$ | $12 \times 4$ | 6 |
| 6 | $4 + 1 + 1$ | $12 \times 8$ | 12 |
| 7 | $4 + 1 + 1 + 1$ | $4 \times 16$ | 8 |
| 8 | $4 + 4$ | $. \ 6 \times 4$ | 3 |
| 9 | 9<br>$4 + 4 + 1$ | $4 \times 2$<br>$12 \times 8$ | 13 |
| 10 | $9 + 1$<br>$4 + 4 + 1 + 1$ | $12 \times 4$<br>$6 \times 16$ | 18 |
| 11 | $9 + 1 + 1$ | $12 \times 8$ | 12 |
| 12 | $9 + 1 + 1 + 1$<br>$4 + 4 + 4$ | $4 \times 16$<br>$4 \times 8$ | 12 |
| 13 | $9 + 4$<br>$4 + 4 + 4 + 1$ | $12 \times 4$<br>$4 \times 16$ | 14 |
| 14 | $9 + 4 + 1$ | $24 \times 8$ | 24 |
| 15 | $9 + 4 + 1 + 1$ | $12 \times 16$ | 24 |
| 16 | 16<br>$4 + 4 + 4 + 4$ | $4 \times 2$<br>$1 \times 16$ | 3 |
| 17 | $16 + 1$<br>$9 + 4 + 4$ | $12 \times 4$<br>$12 \times 8$ | 18 |
| 18 | $16 + 1 + 1$<br>$9 + 9$<br>$9 + 4 + 4 + 1$ | $12 \times 8$<br>$6 \times 4$<br>$12 \times 16$ | 39 |
| 19 | $16 + 1 + 1 + 1$<br>$9 + 9 + 1$ | $4 \times 16$<br>$12 \times 8$ | 20 |
| 20 | $16 + 4$<br>$9 + 9 + 1 + 1$ | $12 \times 4$<br>$6 \times 16$ | 18 |
| 21 | $16 + 4 + 1$<br>$9 + 4 + 4 + 4$ | $24 \times 8$<br>$4 \times 16$ | 32 |
| 22 | $16 + 4 + 1 + 1$<br>$9 + 9 + 4$ | $12 \times 16$<br>$12 \times 8$ | 36 |
| 23 | $9 + 9 + 4 + 1$ | $12 \times 16$ | 24 |
| 24 | $16 + 4 + 4$ | $12 \times 8$ | 12 |
| 25 | 25<br>$16 + 9$<br>$16 + 4 + 4 + 1$ | $4 \times 2$<br>$12 \times 4$<br>$12 \times 16$ | 31 |

Table II (continued)

| n | Non-increasing | Repres. | $R_4(n)/8$ |
|---|---|---|---|
| 26 | 25 + 1<br>16 + 9 + 1<br>9 + 9 + 4 + 4 | 12 × 4<br>24 × 8<br>6 × 16 | 42 |
| 27 | 25 + 1 + 1<br>16 + 9 + 1 + 1<br>9 + 9 + 9 | 12 × 8<br>12 × 16<br>4 × 8 | 40 |
| 28 | 25 + 1 + 1 + 1<br>16 + 4 + 4 + 4<br>9 + 9 + 9 + 1 | 4 × 16<br>4 × 16<br>4 × 16 | 24 |
| 29 | 25 + 4<br>16 + 9 + 4 | 12 × 4<br>24 × 8 | 30 |
| 30 | 25 + 4 + 1<br>16 + 9 + 4 + 1 | 24 × 8<br>24 × 16 | 72 |

Table III

| n | $R_4(n)/8$ | $R_8(n)/16$ | $S_8(8n)$ | $S_4(4(2n-1))$ | $2n-1$ |
|---|---|---|---|---|---|
| 1 | 1 | 1 | 1 | 1 | 1 |
| 2 | 3 | 7 | 8 | 4 | 3 |
| 3 | 4 | 28 | 28 | 6 | 5 |
| 4 | 3 | 71 | 64 | 8 | 7 |
| 5 | 6 | 126 | 126 | 13 | 9 |
| 6 | 12 | 196 | 224 | 12 | 11 |
| 7 | 8 | 344 | 344 | 14 | 13 |
| 8 | 3 | 583 | 512 | 24 | 15 |
| 9 | 13 | 757 | 757 | 18 | 17 |
| 10 | 18 | 882 | 1008 | 20 | 19 |
| 11 | 12 | 1332 | 1332 | 32 | 21 |
| 12 | 12 | 1988 | 1792 | 24 | 23 |
| 13 | 14 | 2198 | 2198 | 31 | 25 |
| 14 | 24 | 2408 | 2752 | 40 | 27 |
| 15 | 24 | 3528 | 3528 | 30 | 29 |
| 16 | 3 | 4679 | 4096 | 32 | 31 |
| 17 | 18 | 4914 | 4914 | 48 | 33 |
| 18 | 39 | 5299 | 6056 | 48 | 35 |
| 19 | 20 | 6860 | 6860 | 38 | 37 |
| 20 | 18 | 8946 | 8064 | 56 | 39 |

# MISCELLANEOUS EXAMPLES
# OF INDUCTION

---

*When you have satisfied yourself that the theorem is true, you start proving it.*—THE TRADITIONAL MATHEMATICS PROFESSOR[1]

---

**1. Expansions.** In dealing with problems of any kind, we need inductive reasoning of some kind. In various branches of mathematics there are certain problems which call for inductive reasoning in a typical manner. The present chapter illustrates this point by a few examples. We begin with a relatively simple example.

*Expand into powers of $x$ the function $1/(1 - x + x^2)$.*

This problem can be solved in many ways. The following solution is somewhat clumsy, but is based on a sound principle and may occur naturally to an intelligent beginner who knows little, yet knows at least the sum of the geometric series:

$$1 + r + r^2 + r^3 + \ldots = \frac{1}{1 - r}.$$

There is an opportunity to use this formula in our problem:

$$\frac{1}{1 - x + x^2} = \frac{1}{1 - x(1 - x)}$$
$$= 1 + x(1 - x) + x^2(1 - x)^2 + x^3(1 - x)^3 + \ldots$$
$$= 1 + x - x^2$$
$$\qquad + x^2 - 2x^3 + x^4$$
$$\qquad\qquad + x^3 - 3x^4 + 3x^5 - x^6$$
$$\qquad\qquad\qquad + x^4 - 4x^5 + 6x^6 - 4x^7 + x^8$$
$$\qquad\qquad\qquad\qquad + x^5 - 5x^6 + 10x^7 - 10x^8 + \ldots$$
$$\qquad\qquad\qquad\qquad\qquad + x^6 - 6x^7 + 15x^8 - \ldots$$
$$\qquad\qquad\qquad\qquad\qquad\qquad + x^7 - 7x^8 + \ldots$$
$$\qquad\qquad\qquad\qquad\qquad\qquad\qquad + x^8 - \ldots$$
$$\qquad\qquad\qquad\qquad\qquad\qquad\qquad\qquad \ldots$$
$$= 1 + x \qquad - x^3 - x^4 \qquad + x^6 + x^7 \qquad \ldots\ .$$

---

[1] This dictum of the well-known pedagogue (*How to Solve It*, p. 181) is sometimes preceded by the following exhortation: "If you have to prove a theorem, do not rush. First of all, understand fully what the theorem says, try to see clearly what it means. Then check the theorem; it could be false. Examine its consequences, verify as many particular instances as are needed to convince yourself of its truth. When . . ."

The result is striking. Any non-vanishing coefficient has either the value 1 or the value $-1$. The succession of the coefficients seems to show some regularity which becomes more apparent if we compute more terms:

$$\frac{1}{1-x+x^2} = 1 + x - x^3 - x^4 + x^6 + x^7 - x^9 - x^{10} + x^{12} + x^{13} - \dots .$$

Periodic! The sequence of coefficients appears to be periodical with the period 6:

$$1, 1, 0, -1, -1, 0 \mid 1, 1, 0, -1, -1, 0 \mid 1, 1, \dots .$$

We naturally expect that the periodicity observed extends beyond the limit of our observations. Yet this is an inductive conclusion, or a mere guess, which we should regard with due skepticism. The guess, however, is based on facts, and so it deserves serious examination. Examining it means, among other things, restating it. There is an interesting way of restating our conjecture:

$$\frac{1}{1-x+x^2} = 1 - x^3 + x^6 - x^9 + x^{12} - \dots$$
$$+ x - x^4 + x^7 - x^{10} + x^{13} - \dots .$$

Now, we may easily notice two geometric series on the right hand side, both with the same ratio $-x^3$, which we can sum. And so our conjecture boils down to

$$\frac{1}{1-x+x^2} = \frac{1}{1+x^3} + \frac{x}{1+x^3} = \frac{1+x}{1+x^3},$$

which is, of course, true. We proved our conjecture.

Our example, simple as it is, is typical in many respects. If we have to expand a given function, we can often obtain the first few coefficients without much trouble. Looking at these coefficients, we should try, as we did here, to guess the law governing the expansion. Having guessed the law, we should try, as we did here, to prove it. It may be a great advantage, as it was here, to work out the proof backwards, starting from an appropriate, clear statement of the conjecture.

By the way, our example is quite rewarding (which is also typical). It leads to a curious relation between binomial coefficients.

It is not superfluous to add that the problem of expanding a given function in a series frequently arises in various branches of mathematics. See the next section, and the Examples and Comments on Chapter VI.

**2. Approximations.**[2] Let $E$ denote the length of the perimeter of an ellipse with semiaxes $a$ and $b$. There is no simple expression for $E$ in terms of $a$ and $b$, but several approximate expressions have been proposed among which the following two are perhaps the most obvious:

$$P = \pi(a + b), \quad P' = 2\pi(ab)^{1/2};$$

---

[2] Cf. Putnam, 1949.

$P$ is a proximate, $P'$ another proximate, $E$ an exact, expression for the same quantity, the length of the perimeter of the ellipse. When $a$ coincides with $b$, the ellipse becomes a circle, and both $P$ and $P'$ coincide with $E$.

How closely do $P$ and $P'$ approximate $E$ when $a$ is different from $b$? Which one comes nearer to the truth, $P$ or $P'$? Questions of this kind frequently arise in all branches of applied mathematics and there is a widely accepted procedure to deal with them which we may describe roughly as follows. *Expand $(P - E)/E$, the relative error of the approximation, in powers of a suitable small quantity and base your judgement upon the initial term* (the first non-vanishing term) *of the expansion.*

Let us see what this means and how the procedure works when applied to our case. First, we should choose a "suitable small quantity." We try $\varepsilon$, the numerical eccentricity of the ellipse, defined by the formula

$$\varepsilon = \frac{(a^2 - b^2)^{1/2}}{a};$$

we take $a$ as the major, and $b$ as the minor, semiaxis. When $a$ becomes $b$ and the ellipse a circle, $\varepsilon$ vanishes. When the ellipse is not very different from a circle, $\varepsilon$ is small. Therefore, let us expand the relative error into powers of $\varepsilon$. We obtain (let us skip the details here)

$$\frac{P - E}{E} = -\frac{1}{64}\varepsilon^4 + \cdots, \qquad \frac{P' - E}{E} = -\frac{3}{64}\varepsilon^4 + \cdots.$$

We computed only the initial term which, in both cases, is of order 4, contains $\varepsilon^4$. We omitted in both expansions the terms of higher order, containing $\varepsilon^5$, $\varepsilon^6$, .... The terms omitted are negligible in comparison with the initial terms when $\varepsilon$ is very small (infinitely small), that is, when the ellipse is almost circular. Therefore, for almost circular ellipses, $P$ comes nearer to the true value $E$ than $P'$. (In fact, the ratio of the errors becomes 1 : 3 as $\varepsilon$ tends to 0.) Both $P$ and $P'$ approximate $E$ from below:

$$E > P > P'.$$

All this holds for very small $\varepsilon$, for almost circular ellipses. We do not know yet how much of these results remains valid when $\varepsilon$ is not so small. In fact, at this moment we know only limit relations, valid for $\varepsilon \to 0$. We do not yet know anything definitive about the error of our approximations when $\varepsilon = 0.5$ or $\varepsilon = 0.1$. Of course, what we need in practice is information about such concrete cases.

In such circumstances, practical people test their formulas numerically. We may follow them, but which case should we test first? It is advisable not to forget the extreme cases. The numerical eccentricity $\varepsilon$ varies between the extreme values 0 and 1. When $\varepsilon = 0$, $b = a$ and the ellipse becomes a circle. Yet we know this case fairly well by now and we turn rather to the other extreme case. When $\varepsilon = 1$, $b = 0$, the ellipse becomes a line segment of length $2a$, and the length of the perimeter is $4a$. We have

$$E = 4a, \quad P = \pi a, \quad P' = 0 \quad \text{when } \varepsilon = 1.$$

It may be worth noticing that in both extreme cases, for $\varepsilon = 1$ just as for very small $\varepsilon$, $E > P > P'$. Are these inequalities generally valid? For the second inequality, the answer is easy. In fact, we have, for $a > b$,

$$P = \pi(a + b) > 2\pi(ab)^{1/2} = P'$$

since this is equivalent to

$$(a + b)^2 > 4ab$$

or to

$$(a - b)^2 > 0.$$

We focus our attention upon the remaining question. Is the inequality $E > P$ generally valid? It is natural to conjecture that what we found true in the extreme cases ($\varepsilon$ small, and $\varepsilon = 1$) remains true in the intermediate cases (for all values of $\varepsilon$ between 0 and 1). Our conjecture is not supported by many observations, that is true, but it is supported by analogy. A similar question (concerning $P > P'$) which we asked in the same breath and based on similar grounds has been answered in the affirmative.

Let us test a case numerically. We know a little more about the case where $\varepsilon$ is nearly 0 than about the case where it is nearly 1. We choose a simple value for $\varepsilon$, nearer to 1 than to 0: $a = 5$, $b = 3$, $\varepsilon = 4/5$. We find for this $\varepsilon$ (using appropriate tables)

$$E = 2\pi \times 4.06275, \quad P = 2\pi \times 4.00000.$$

The inequality $E > P$ is verified. This numerical verification of our conjecture comes from a new side, from a different source, and therefore carries some weight. Let us note also that

$$(P - E)/E = -0.0155, \quad -\varepsilon^4/64 = -0.0064.$$

The relative error is about 1.5%. It is considerably larger than the initial term of its expansion, but has the same sign. As $\varepsilon = 4/5 = 0.8$ is not too small, our remark fits into the whole picture and tends to increase our confidence in the conjecture.

Approximate formulas play an important role in applied mathematics. Trying to judge such a formula, we often adopt in practice the procedure followed in this section. We compute the initial term in the expansion of the relative error and supplement the information so gained by numerical tests, considerations of analogy, etc., in short, by inductive, non-demonstrative reasoning.

**3. Limits.** In order to see inductive reasoning at work in still another domain, we consider the following problem.[3]

---

[3] See Putnam, 1948.

*Let $a_1, a_2, \ldots, a_n, \ldots$ be an arbitrary sequence of positive numbers.  Show that*

$$\lim_{n \to \infty} \sup \left( \frac{a_1 + a_{n+1}}{a_n} \right)^n \geq e.$$

This problem requires some preliminary knowledge, especially familiarity with the concept of "lim sup" or "upper limit of indetermination."[4]  Yet even if you are thoroughly familiar with this concept, you may experience some difficulty in finding a proof.  My congratulations to any undergraduate who can do the problem by his own means in a few hours.

If you have struggled with the problem yourself a little while, you may follow with more sympathy the struggle described in the following sections.

**4. Trying to disprove it.**  We begin with the usual questions.

*What is the hypothesis?*  Just $a_n > 0$, nothing else.

*What is the conclusion?*  That inequality with $e$ on the right and that complicated limit on the left.

*Do you know a related theorem?*  No, indeed.  It is very different from anything I know.

*Is it likely that the theorem is true?  Or is it more likely that it is false?*  False, of course.  In fact I cannot believe that such a precise consequence can be derived from such a broad hypothesis, just $a_n > 0$.

*What are you required to do?*  To prove the theorem.  Or to disprove it.  I am very much for disproving it.

*Can you test any particular case of the theorem?*  Yes, that is what I am about to do.

[In order to simplify the formulas, we set

$$\left( \frac{a_1 + a_{n+1}}{a_n} \right)^n = b_n$$

and write $b_n \to b$ for $\lim_{n \to \infty} b_n = b$.]

I try $a_n = 1$, for $n = 1, 2, 3, \ldots$ .  Then

$$b_n = \left( \frac{1 + 1}{1} \right)^n = 2^n \to \infty.$$

In this case, the assertion of the theorem is verified.

Yet I could set $a_1 = 0$, $a_n = 1$ for $n = 2, 3, 4, \ldots$ .  Then

$$b_n = \left( \frac{0 + 1}{1} \right)^n = 1^n \to 1 < e.$$

The theorem is exploded!  No, it is not.  The hypothesis allows $a_1 = 0.00001$, but it prohibits $a_1 = 0$.  What a pity!

---

[4] See e.g., G. H. Hardy, *Pure Mathematics*, sect. 82.

Let me try something else. Let $a_n = n$. Then

$$b_n = \left(\frac{1 + (n + 1)}{n}\right)^n = \left(1 + \frac{2}{n}\right)^n \to e^2.$$

Again verified.

Now, let $a_n = n^2$. Then

$$b_n = \left(\frac{1 + (n + 1)^2}{n^2}\right)^n = \left(1 + \frac{2}{n}\frac{n + 1}{n}\right)^n \to e^2.$$

Again verified. And again $e^2$. Should $e^2$ stand on the right hand side in the conclusion instead of $e$? That would improve the theorem.

Let me *introduce a parameter.* Let me take . . . . Yes, let me take $a_1 = c$, where I can dispose of $c$, but $a_n = n$ for $n = 2, 3, 4, \ldots$ . Then

$$b_n = \left(\frac{c + (n + 1)}{n}\right)^n = \left(1 + \frac{1 + c}{n}\right)^n \to e^{1+c}.$$

This is always $> e$, since $c = a_1 > 0$. Yet it can come as close to $e$ as we please, since $c$ can be arbitrarily small. I cannot disprove it, I cannot prove it.

Just one more trial. Let me take $a_n = n^c$. Then [we skip some computations]

$$b_n = \left[\frac{1 + (n + 1)^c}{n^c}\right]^n \to \begin{cases} \infty & \text{if } 0 < c < 1, \\ e^2 & \text{if } c = 1, \\ e^c & \text{if } c > 1. \end{cases}$$

Again, the limit can come as close to $e$ as we please, but remains always superior to $e$. I shall never succeed in bringing down this . . . limit below $e$. It is time to turn round.

**5. Trying to prove it.** In fact, the indications for a *volte-face* are quite strong. In the light of the accumulated inductive evidence the prospects of disproving the theorem appear so dim that the prospects of proving it look relatively bright.

Therefore, nothing remains but to start reexamining the theorem, its statement, its hypothesis, its conclusion, the concepts involved, etc.

*Can you relax the hypothesis?* No, I cannot. If I admit $a_n = 0$, the conclusion is no more valid, the theorem becomes false ($a_1 = 0, a_2 = a_3 = a_4 = \ldots = 1$).

*Can you improve the conclusion?* I certainly cannot improve it by substituting some greater number for $e$, since then the conclusion is no more valid, the theorem becomes false (examples in the foregoing sect. 4).

*Have you taken into account all essential notions involved in the problem?* No, I have not. That may be the trouble.

*What have you failed to take into account?* The definition of lim sup. The definition of the number $e$.

What is lim sup $b_n$? It is the upper limit of indetermination of $b_n$ as $n$ tends to infinity.

What is $e$? I could define $e$ in various ways. The above examples suggest that the most familiar definition of $e$ may be the best:

$$e = \lim_{n \to \infty} \left(1 + \frac{1}{n}\right)^n.$$

*Could you restate the theorem?* . . . . . . . . . . . . . . . . .

*Could you restate the theorem in some more accessible form?* . . . . . . . . . .

*Could you restate the conclusion? What is the conclusion?* The conclusion contains $e$. What is $e$? (I failed to ask this before.) Oh, yes—the conclusion is

$$\limsup_{n \to \infty} \left(\frac{a_1 + a_{n+1}}{a_n}\right)^n \geq \lim_{n \to \infty} \left(1 + \frac{1}{n}\right)^n$$

or, which is the same,

$$\limsup_{n \to \infty} \left[\frac{n(a_1 + a_{n+1})}{(n+1)a_n}\right]^n \geq 1.$$

This looks much better!

*Can the conclusion be false, when the hypothesis is fulfilled?* Yes, that is the question. Let me see it. Let me look squarely at the negation of the assertion, at the exactly opposite assertion. Let me write it down:

$$(?) \qquad \limsup_{n \to \infty} \left[\frac{n(a_1 + a_{n+1})}{(n+1)a_n}\right]^n < 1.$$

I put a query in front of it, because just this point is in doubt. Let me call it the "formula (?)." What does (?) mean? It certainly implies that there is an $N$ such that

$$\left[\frac{n(a_1 + a_{n+1})}{(n+1)a_n}\right]^n < 1 \text{ for } n \geq N.$$

It follows hence that

$$\frac{n(a_1 + a_{n+1})}{(n+1)a_n} < 1 \text{ for } n \geq N.$$

It follows further . . . . Let me try something. Yes, I can write it neatly! It follows further from (?) that

$$\frac{a_1}{n+1} + \frac{a_{n+1}}{n+1} < \frac{a_n}{n}$$

or

$$\frac{a_{n+1}}{n+1} - \frac{a_n}{n} < -\frac{a_1}{n+1} \text{ for } n \geq N.$$

Let me write this out broadly. It follows that

$$\frac{a_n}{n} - \frac{a_{n-1}}{n-1} < -\frac{a_1}{n}$$

$$\frac{a_{n-1}}{n-1} - \frac{a_{n-2}}{n-2} < -\frac{a_1}{n-1}$$

$$\cdot \quad \cdot \quad \cdot \quad \cdot \quad \cdot \quad \cdot \quad \cdot \quad \cdot \quad \cdot \quad \cdot \quad \cdot$$

$$\frac{a_{N+1}}{a_{N+1}} - \frac{a_N}{N} < -\frac{a_1}{N+1}$$

and so

$$\frac{a_n}{n} < \frac{a_N}{N} - a_1 \left( \frac{1}{N+1} + \frac{1}{N+2} + \cdots + \frac{1}{n-1} + \frac{1}{n} \right)$$

$$= C - a_1 \left( 1 + \frac{1}{2} + \frac{1}{3} + \cdots + \frac{1}{n} \right)$$

where $C$ is a constant, independent of $n$ provided that $n \geq N$. It does not really matter but, in fact,

$$C = \frac{a_N}{N} + a_1 \left( 1 + \frac{1}{2} + \cdots + \frac{1}{N} \right).$$

It matters, however, that $n$ can be arbitrarily large, and that the harmonic series diverges. It follows, therefore, that

$$\lim_{n \to \infty} \frac{a_n}{n} = -\infty.$$

Now, this contradicts flatly the hypothesis that $a_n > 0$ for $n = 1, 2, 3, \ldots$ . Yet this contradiction follows faultlessly from the formula (?). Therefore, in fact, (?) must be responsible for the contradiction; (?) is incompatible with the hypothesis $a_n > 0$; the opposite to (?) must be true—the theorem is proved!

**6. The role of the inductive phase.** Looking back at the foregoing solution superficially, we could think that the first, inductive, phase of the solution (sect. 4) is not used at all in the second, demonstrative, phase (sect. 5). Yet this is not so. The inductive phase was useful in several respects.

First, examining concrete particular cases of the theorem, we understood it thoroughly, and realized its full meaning. We satisfied ourselves that its hypothesis is essential, its conclusion sharp. This information was helpful in the second phase: we knew that we must use the whole hypothesis and that we must take into account the precise value of the constant $e$.

Second, having verified the theorem in several particular cases, we gathered strong inductive evidence for it. The inductive phase overcame our initial suspicion and gave us a strong confidence in the theorem. Without such confidence we would have scarcely found the courage to undertake the proof which did not look at all a routine job. "When you have satisfied yourself that the theorem is true, you start proving it"—the traditional mathematics professor is quite right.

Third, the examples in which the familiar limit formula for $e$ popped up again and again, gave us reasonable ground for introducing that limit formula into the statement of the theorem. And introducing it turned out the crucial step toward the solution.

On the whole, it seems natural and reasonable that the inductive phase precedes the demonstrative phase. First guess, then prove.

### EXAMPLES AND COMMENTS ON CHAPTER V.

**1.** By multiplying the series

$$(1 - x^2)^{-1/2} = 1 + \frac{1}{2} x^2 + \frac{1}{2}\frac{3}{4} x^4 + \ldots$$

$$\arcsin x = \frac{x}{1} + \frac{1}{2}\frac{x^3}{3} + \frac{1}{2}\frac{3}{4}\frac{x^5}{5} + \ldots$$

you find the first terms of the expansion

$$y = (1 - x^2)^{-1/2} \arcsin x = x + \frac{2}{3} x^3 + \ldots .$$

(a) Compute a few more terms and try to guess the general term.

(b) Show that $y$ satisfies the differential equation

$$(1 - x^2)y' - xy = 1$$

and use this equation to prove your guess.

**2.** By multiplying the series

$$e^{x^2/2} = 1 + \frac{x^2}{2} + \frac{x^4}{2 \cdot 4} + \ldots$$

$$\int_0^x e^{-t^2/2} \, dt = \frac{x}{1} - \frac{1}{2}\frac{x^3}{3} + \frac{1}{2 \cdot 4}\frac{x^5}{5} - \ldots$$

you find the first terms of the expansion

$$y = e^{x^2/2} \int_0^x e^{-t^2/2} \, dt = x + \frac{1}{3} x^3 + \ldots .$$

(a) Compute a few more terms and try to guess the general term.

(b) Your guess, if correct, suggests that $y$ satisfies a simple differential equation. By establishing this equation, prove your guess.

**3.** The functional equation

$$f(x) = \frac{1}{1+x} f\left(\frac{2\sqrt{x}}{1+x}\right)$$

is satisfied by the power series

$$f(x) = 1 + \frac{1}{4} x^2 + \frac{9}{64} x^4 + \frac{25}{256} x^6 + \frac{1225}{16384} x^8 + \cdots.$$

Verify these coefficients, derive a few more, if necessary, and try to guess the general term.

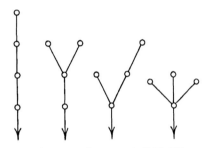

Fig. 5.1.   Compounds $C_4H_9OH$.

**4.** The functional equation

$$f(x) = 1 + \frac{x}{6} [f(x)^3 + 3f(x)f(x^2) + 2f(x^3)]$$

is satisfied by the power series

$$f(x) = 1 + x + x^2 + 2x^3 + 4x^4 + \ldots + a_n x^n + \cdots.$$

It is asserted that $a_n$ is the number of the structurally different chemical compounds (aliphatic alcohols) having the same chemical formula $C_nH_{2n+1}OH$. In the case $n = 4$, the answer is true. There are $a_4 = 4$ alcohols $C_4H_9OH$; they are represented in fig. 5.1, each compound as a "tree," each C as a little circle or "knot," and the radical —OH as an arrow; the $H$'s are dropped. Test other values of $n$.

**5.**        $$\sum_{k=0}^{n/2} (-1)^k \binom{n-k}{k} = 1, 1, 0, -1, -1, 0$$

according as $n \equiv 0, 1, 2, 3, 4, 5$ (mod. 6).

**6.** An ellipse describes a prolate, or an oblate, spheroid according as it rotates about its major, or minor, axis.

For the area of the surface of the prolate spheriod

$$E = 2\pi ab[(1 - \varepsilon^2)^{1/2} + (\arcsin \varepsilon)/\varepsilon], \quad P = 4\pi(a^2 + 2b^2)/3$$

are the exact, and a proximate expression, respectively ($a$, $b$, and $\varepsilon$ as in sect. 2). Find

(a) the initial term of the relative error

(b) the relative error when $b = 0$.

What about the sign of the relative error?

**7.** For the area of the surface of the oblate ellipsoid

$$E = 2\pi a^2 \left[ 1 + \frac{1 - \varepsilon^2}{2\varepsilon} \log \frac{1 + \varepsilon}{1 - \varepsilon} \right], \qquad P = \frac{4\pi(2a^2 + b^2)}{3}$$

are the exact, and a proximate, expression, respectively. Find

(a) the initial term of the relative error

(b) the relative error when $b = 0$.

What about the sign of the relative error?

**8.** Comparing ex. 6 and ex. 7, which approximate formula would you propose for the area of the surface of the general ellipsoid with semiaxes $a$, $b$, and $c$?

What about the sign of the error?

**9.** [Sect. 2.] Starting from the parametric representation of the ellipse, $x = a \sin t$, $y = b \cos t$, show that

$$E = 4a \int_0^{\pi/2} (1 - \varepsilon^2 \sin^2 t)^{1/2}\, dt$$

$$= 2\pi a \left[ 1 - \sum_1^\infty \left( \frac{1}{2} \frac{3}{4} \cdots \frac{2n - 1}{2n} \right)^2 \frac{\varepsilon^{2n}}{2n - 1} \right]$$

and derive hence the initial terms given without proof in sect. 2.

**10** (continued). Using the expansions in powers of $\varepsilon$, prove that $E > P$ for $0 < \varepsilon \leq 1$.

**11.** [Sect. 2.] Determine the number $\alpha$ so that the expression

$$P'' = \alpha P + (1 - \alpha)P'$$

should yield the best possible approximation to $E$ for small $\varepsilon$. (That is, the order of the initial term of $(P'' - E)/E$ should be as high as possible.)

**12** (continued). Investigate the approximation by $P''$ following the method of sect. 2. (Inductively!)

**13.** Given a positive integer $p$ and a sequence of positive numbers $a_1, a_2, a_3, \ldots, a_n, \ldots$ . Show that

$$\limsup_{n \to \infty} \left( \frac{a_1 + a_{n+p}}{a_n} \right)^n \geq e^p.$$

**14** (continued). Point out a sequence $a_1, a_2, a_3, \ldots$ for which equality is attained.

**15.** *Explain the observed regularities.* A discovery in physics is often attained in two steps. First a certain regularity is noticed in the data of observation. Then this regularity is explained as a consequence of some general law. Different persons may take the two steps which may be separated by a long interval of time. A great example is that of Kepler and Newton: the regularities in the motion of the planets observed by Kepler have been explained by the law of gravitation discovered by Newton. Something similar may happen in mathematical research, and here is a neat example which requires little preliminary knowledge.

The usual table of four-place common logarithms lists 900 mantissas, those of the logarithms of the integers from 100 to 999. We may be inclined to think, before observation, that the ten figures, 0, 1, . . . , 9 are about equally frequent in these tables, but this is not so: they certainly do not turn up equally often as the *first* figure of the mantissa. By counting the mantissas that have the same first figure, we obtain Table I. (Check it!)

Table I.   Mantissas with the same first figure in four-place logarithms

| First figure | Nr. of mantissas | Ratios |
|:---:|:---:|:---:|
| 0 | 26 | |
| | | 1.269 |
| 1 | 33 | |
| | | 1.242 |
| 2 | 41 | |
| | | 1.268 |
| 3 | 52 | |
| | | 1.250 |
| 4 | 65 | |
| | | 1.262 |
| 5 | 82 | |
| | | 1.256 |
| 6 | 103 | |
| | | 1.252 |
| 7 | 129 | |
| | | 1.271 |
| 8 | 164 | |
| | | 1.250 |
| 9 | 205 | |
| | Total 900 | |

Inspecting the second column of Table I we may notice that any two consecutive numbers in it have approximately the same ratio. This induces us to compute these ratios to a few decimals: they are listed in the last column of Table I.

Why are these ratios approximately equal? Try to perceive some precise regularity behind the observed approximate regularity. The numbers in the second column of Table I are approximately the terms of a geometric progression. Could you discover an exact geometric progression to which the terms of the approximate progression are simply related? [The ratio of the exact progression should be, perhaps, some kind of average of the ratios listed in the last column of Table I.]

**16.** *Classify the observed facts.* A great part of the naturalist's work is aimed at describing and classifying the objects that he observes. Such work was

Fig. 5.2. Symmetries of friezes.

predominant for a long time after Linnaeus when the main activity of the naturalists consisted in describing new species and genera of plants and animals, and in reclassifying the known species and genera. Not only plants and animals are described and classified by the naturalists, but also other objects, especially minerals; the classification of crystals is based on their symmetry. A good classification is important; it reduces the observable variety to relatively few clearly characterized and well ordered types. The mathematician has not often opportunity to indulge in description and classification, but it may happen.

If you are acquainted with a few simple notions of plane geometry (line of symmetry, center of symmetry) you may have fun with ornaments. Fig. 5.2

exhibits fourteen ornamental bands each of which is generated by a simple figure, repeated periodically along a (horizontal) straight line.    Let us call such a band a "frieze."    Match each frieze on the left-hand side of fig. 5.2 (marked with a numeral) with a frieze on the right-hand side (marked with a letter) so that the two friezes matched have the *same type of symmetry*.    Moreover, examine ornamental bands which you can find on all sorts of objects, or in older architectural works, and try to match each with a frieze in fig. 5.2.    Finally, give a complete list of the various types of symmetry that a frieze may have and an exhaustive description of each type of symmetry.    [Consider a frieze as infinitely long, in both directions, and the generating figure as periodically repeated an infinity of times.    Observe that the term "type of symmetry" was not formally defined: to arrive at an appropriate interpretation of this term is an important part of your task.]

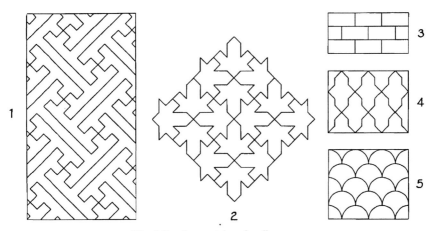

Fig. 5.3.    Symmetries of wallpapers.

**17.**    Find two ornaments in fig. 5.3 that have the same type of symmetry. Each ornament must be conceived as covering the whole plane with its repeated patterns.

**18.**    *What is the difference?*    The twenty-six capital letters are divided into five batches as follows:

A M T U V W Y
B C D E K
N S Z
H I O X
F G J L P Q R.

What is the difference?    What could be a simple basis for the exhibited classification?    [Look at the five equations:

$$y = x^2, \quad y^2 = x, \quad y = x^3, \quad x^2 + 2y^2 = 1, \quad y = x + x^4.$$

What is the difference?]

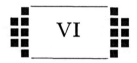

# VI

# A MORE GENERAL STATEMENT

*He [Euler] preferred instructing his pupils to the little satisfaction of amazing them. He would have thought not to have done enough for science if he should have failed to add to the discoveries, with which he enriched science, the candid exposition of the ideas that led him to those discoveries.*
—CONDORCET

**1. Euler.** Of all mathematicians with whose work I am somewhat acquainted, Euler seems to be by far the most important for our inquiry. A master of inductive research in mathematics, he made important discoveries (on infinite series, in the Theory of Numbers, and in other branches of mathematics) by induction, that is, by observation, daring guess, and shrewd verification. In this respect, however, Euler is not unique; other mathematicians, great and small, used induction extensively in their work.

Yet Euler seems to me almost unique in one respect: he takes pains to present the relevant inductive evidence carefully, in detail, in good order. He presents it convincingly but honestly, as a genuine scientist should do. His presentation is "the candid exposition of the ideas that led him to those discoveries" and has a distinctive charm. Naturally enough, as any other author, he tries to impress his readers, but, as a really good author, he tries to impress his readers only by such things as have genuinely impressed himself.

The next section brings a sample of Euler's writing. The memoir chosen can be read with very little previous knowledge and is entirely devoted to the exposition of an inductive argument.

**2. Euler's memoir** is given here, in English translation, *in extenso*, except for a few unessential alterations which should make it more accessible to a modern reader.[1]

---

[1] The original is in French; see Euler's *Opera Omnia*, ser. 1, vol. 2, p. 241–253. The alterations consist in a different notation (footnote 2), in the arrangement of a table (explained in footnote 3), in slight changes affecting a few formulas, and in dropping a repetition of former arguments in the last No. 13 of the memoir. The reader may consult the easily available original.

DISCOVERY OF A MOST EXTRAORDINARY LAW OF THE NUMBERS
CONCERNING THE SUM OF THEIR DIVISORS

1. Till now the mathematicians tried in vain to discover some order in the sequence of the prime numbers and we have every reason to believe that there is some mystery which the human mind shall never penetrate. To convince oneself, one has only to glance at the tables of the primes, which some people took the trouble to compute beyond a hundred thousand, and one perceives that there is no order and no rule. This is so much more surprising as the arithmetic gives us definite rules with the help of which we can continue the sequence of the primes as far as we please, without noticing, however, the least trace of order. I am myself certainly far from this goal, but I just happened to discover an extremely strange law governing the sums of the divisors of the integers which, at the first glance, appear just as irregular as the sequence of the primes, and which, in a certain sense, comprise even the latter. This law, which I shall explain in a moment, is, in my opinion, so much more remarkable as it is of such a nature that we can be assured of its truth without giving it a perfect demonstration. Nevertheless, I shall present such evidence for it as might be regarded as almost equivalent to a rigorous demonstration.

2. A prime number has no divisors except unity and itself, and this distinguishes the primes from the other numbers. Thus 7 is a prime, for it is divisible only by 1 and itself. Any other number which has, besides unity and itself, further divisors, is called composite, as for instance, the number 15, which has, besides 1 and 15, the divisors 3 and 5. Therefore, generally, if the number $p$ is prime, it will be divisible only by 1 and $p$; but if $p$ was composite, it would have, besides 1 and $p$, further divisors. Therefore, in the first case, the sum of its divisors will be $1 + p$, but in the latter it would exceed $1 + p$. As I shall have to consider the sum of divisors of various numbers, I shall use[2] the sign $\sigma(n)$ to denote the sum of the divisors of the number $n$. Thus, $\sigma(12)$ means the sum of all the divisors of 12, which are 1, 2, 3, 4, 6, and 12; therefore, $\sigma(12) = 28$. In the same way, one can see that $\sigma(60) = 168$ and $\sigma(100) = 217$. Yet, since unity is only divisible by itself, $\sigma(1) = 1$. Now, 0 (zero) is divisible by all numbers. Therefore, $\sigma(0)$ should be properly infinite. (However, I shall assign to it later a finite value, different in different cases, and this will turn out serviceable.)

3. Having defined the meaning of the symbol $\sigma(n)$, as above, we see clearly that if $p$ is a prime $\sigma(p) = 1 + p$. Yet $\sigma(1) = 1$ (and not $1 + 1$); hence we see that 1 should be excluded from the sequence of the primes; 1 is the beginning of the integers, neither prime nor composite. If, however, $n$ is composite, $\sigma(n)$ is greater than $1 + n$.

---

[2] Euler was the first to introduce a symbol for the sum of the divisors; he used $\int n$, not the modern $\sigma(n)$ of the text.

In this case we can easily find $\sigma(n)$ from the factors of $n$. If $a, b, c, d, \ldots$ are different primes, we see easily that

$$\sigma(ab) = 1 + a + b + ab = (1 + a)(1 + b) = \sigma(a)\sigma(b),$$
$$\sigma(abc) = (1 + a)(1 + b)(1 + c) = \sigma(a)\sigma(b)\sigma(c),$$
$$\sigma(abcd) = \sigma(a)\sigma(b)\sigma(c)\sigma(d)$$

and so on. We need particular rules for the powers of primes, as

$$\sigma(a^2) = 1 + a + a^2 = \frac{a^3 - 1}{a - 1}$$

$$\sigma(a^3) = 1 + a + a^2 + a^3 = \frac{a^4 - 1}{a - 1}$$

and, generally,

$$\sigma(a^n) = \frac{a^{n+1} - 1}{a - 1}.$$

Using this, we can find the sum of the divisors of any number, composite in any way whatever. This we see from the formulas

$$\sigma(a^2 b) = \sigma(a^2)\sigma(b)$$
$$\sigma(a^3 b^2) = \sigma(a^3)\sigma(b^2)$$
$$\sigma(a^3 b^4 c) = \sigma(a^3)\sigma(b^4)\sigma(c)$$

and, generally,

$$\sigma(a^\alpha b^\beta c^\gamma d^\delta e^\varepsilon) = \sigma(a^\alpha)\sigma(b^\beta)\sigma(c^\gamma)\sigma(d^\delta)\sigma(e^\varepsilon).$$

For instance, to find $\sigma(360)$ we set, since 360 factorized is $2^3 \cdot 3^2 \cdot 5$,

$$\sigma(360) = \sigma(2^3)\sigma(3^2)\sigma(5) = 15 \cdot 13 \cdot 6 = 1170.$$

4. In order to show the sequence of the sums of the divisors, I add the following table[3] containing the sums of the divisors of all integers from 1 up to 99.

| $n$ | 0 | 1 | 2 | 3 | 4 | 5 | 6 | 7 | 8 | 9 |
|---|---|---|---|---|---|---|---|---|---|---|
| 0 | — | 1 | 3 | 4 | 7 | 6 | 12 | 8 | 15 | 13 |
| 10 | 18 | 12 | 28 | 14 | 24 | 24 | 31 | 18 | 39 | 20 |
| 20 | 42 | 32 | 36 | 24 | 60 | 31 | 42 | 40 | 56 | 30 |
| 30 | 72 | 32 | 63 | 48 | 54 | 48 | 91 | 38 | 60 | 56 |
| 40 | 90 | 42 | 96 | 44 | 84 | 78 | 72 | 48 | 124 | 57 |
| 50 | 93 | 72 | 98 | 54 | 120 | 72 | 120 | 80 | 90 | 60 |
| 60 | 168 | 62 | 96 | 104 | 127 | 84 | 144 | 68 | 126 | 96 |
| 70 | 144 | 72 | 195 | 74 | 114 | 124 | 140 | 96 | 168 | 80 |
| 80 | 186 | 121 | 126 | 84 | 224 | 108 | 132 | 120 | 180 | 90 |
| 90 | 234 | 112 | 168 | 128 | 144 | 120 | 252 | 98 | 171 | 156 |

[3] The number in the intersection of the row marked 60 and the column marked 7, that is, 68, is $\sigma(67)$. If $p$ is prime, $\sigma(p)$ is in heavy print. This arrangement of the table is a little more concise than the arrangement in the original.

If we examine a little the sequence of these numbers, we are almost driven to despair. We cannot hope to discover the least order. The irregularity of the primes is so deeply involved in it that we must think it impossible to disentangle any law governing this sequence, unless we know the law governing the sequence of the primes itself. It could appear even that the sequence before us is still more mysterious than the sequence of the primes.

5. Nevertheless, I observed that this sequence is subject to a completely definite law and could even be regarded as a *recurring* sequence. This mathematical expression means that each term can be computed from the foregoing terms, according to an invariable rule. In fact, if we let $\sigma(n)$ denote any term of this sequence, and $\sigma(n-1)$, $\sigma(n-2)$, $\sigma(n-3)$, $\sigma(n-4)$, $\sigma(n-5)$, ... the preceding terms, I say that the value of $\sigma(n)$ can always be combined from some of the preceding as prescribed by the following formula:

$$\sigma(n) = \sigma(n-1) + \sigma(n-2) - \sigma(n-5) - \sigma(n-7)$$

$$+ \sigma(n-12) + \sigma(n-15) - \sigma(n-22) - \sigma(n-26)$$

$$+ \sigma(n-35) + \sigma(n-40) - \sigma(n-51) - \sigma(n-57)$$

$$+ \sigma(n-70) + \sigma(n-77) - \sigma(n-92) - \sigma(n-100)$$

$$+ \dots .$$

On this formula we must make the following remarks.

I. In the sequence of the signs $+$ and $-$, each arises twice in succession.

II. The law of the numbers 1, 2, 5, 7, 12, 15, ... which we have to subtract from the proposed number $n$, will become clear if we take their differences:

| Nrs. | 1, | 2, | 5, | 7, | 12, | 15, | 22, | 26, | 35, | 40, | 51, | 57, | 70, | 77, | 92, | 100, ... |
|------|----|----|----|----|-----|-----|-----|-----|-----|-----|-----|-----|-----|-----|-----|----------|
| Diff. | 1, | 3, | 2, | 5, | 3, | 7, | 4, | 9, | 5, | 11, | 6, | 13, | 7, | 15, | 8, ... | |

In fact, we have here, alternately, all the integers 1, 2, 3, 4, 5, 6, ... and the odd numbers 3, 5, 7, 9, 11, ... , and hence we can continue the sequence of these numbers as far as we please.

III. Although this sequence goes to infinity, we must take, in each case, only those terms for which the numbers under the sign $\sigma$ are still positive and omit the $\sigma$ for negative values.

IV. If the sign $\sigma(0)$ turns up in the formula, we must, as its value in itself is indeterminate, substitute for $\sigma(0)$ the number $n$ proposed.

6. After these remarks it is not difficult to apply the formula to any given particular case, and so anybody can satisfy himself of its truth by as many examples as he may wish to develop. And since I must admit that I am not in a position to give it a rigorous demonstration, I will justify it by a sufficiently large number of examples.

$\sigma(1) = \sigma(0)$ $= 1$ $= 1$

$\sigma(2) = \sigma(1) + \sigma(0)$ $= 1 + 2$ $= 3$

$\sigma(3) = \sigma(2) + \sigma(1)$ $= 3 + 1$ $= 4$

$\sigma(4) = \sigma(3) + \sigma(2)$ $= 4 + 3$ $= 7$

$\sigma(5) = \sigma(4) + \sigma(3) - \sigma(0)$ $= 7 + 4 - 5$ $= 6$

$\sigma(6) = \sigma(5) + \sigma(4) - \sigma(1)$ $= 6 + 7 - 1$ $= 12$

$\sigma(7) = \sigma(6) + \sigma(5) - \sigma(2) - \sigma(0)$ $= 12 + 6 - 3 - 7$ $= 8$

$\sigma(8) = \sigma(7) + \sigma(6) - \sigma(3) - \sigma(1)$ $= 8 + 12 - 4 - 1$ $= 15$

$\sigma(9) = \sigma(8) + \sigma(7) - \sigma(4) - \sigma(2)$ $= 15 + 8 - 7 - 3$ $= 13$

$\sigma(10) = \sigma(9) + \sigma(8) - \sigma(5) - \sigma(3)$ $= 13 + 15 - 6 - 4$ $= 18$

$\sigma(11) = \sigma(10) + \sigma(9) - \sigma(6) - \sigma(4)$ $= 18 + 13 - 12 - 7$ $= 12$

$\sigma(12) = \sigma(11) + \sigma(10) - \sigma(7) - \sigma(5) + \sigma(0)$ $= 12 + 18 - 8 - 6 + 12$ $= 28$

$\sigma(13) = \sigma(12) + \sigma(11) - \sigma(8) - \sigma(6) + \sigma(1)$ $= 28 + 12 - 15 - 12 + 1$ $= 14$

$\sigma(14) = \sigma(13) + \sigma(12) - \sigma(9) - \sigma(7) + \sigma(2)$ $= 14 + 28 - 13 - 8 + 3$ $= 24$

$\sigma(15) = \sigma(14) + \sigma(13) - \sigma(10) - \sigma(8) + \sigma(3) + \sigma(0)$ $= 24 + 14 - 18 - 15 + 4 + 15 = 24$

$\sigma(16) = \sigma(15) + \sigma(14) - \sigma(11) - \sigma(9) + \sigma(4) + \sigma(1)$ $= 24 + 24 - 12 - 13 + 7 + 1 = 31$

$\sigma(17) = \sigma(16) + \sigma(15) - \sigma(12) - \sigma(10) + \sigma(5) + \sigma(2)$ $= 31 + 24 - 28 - 18 + 6 + 3 = 18$

$\sigma(18) = \sigma(17) + \sigma(16) - \sigma(13) - \sigma(11) + \sigma(6) + \sigma(3)$ $= 18 + 31 - 14 - 12 + 12 + 4 = 39$

$\sigma(19) = \sigma(18) + \sigma(17) - \sigma(14) - \sigma(12) + \sigma(7) + \sigma(4)$ $= 39 + 18 - 24 - 28 + 8 + 7 = 20$

$\sigma(20) = \sigma(19) + \sigma(18) - \sigma(15) - \sigma(13) + \sigma(8) + \sigma(5)$ $= 20 + 39 - 24 - 14 + 15 + 6 = 42$

I think these examples are sufficient to discourage anyone from imagining that it is by mere chance that my rule is in agreement with the truth.

7. Yet somebody could still doubt whether the law of the numbers 1, 2, 5, 7, 12, 15, . . . which we have to subtract is precisely that one which I have indicated, since the examples given imply only the first six of these numbers. Thus, the law could still appear as insufficiently established and, therefore, I will give some examples with larger numbers.

I. Given the number 101, find the sum of its divisors. We have

$$\sigma(101) = \sigma(100) + \sigma(99) - \sigma(96) - \sigma(94)$$
$$+ \sigma(89) + \sigma(86) - \sigma(79) - \sigma(75)$$
$$+ \sigma(66) + \sigma(61) - \sigma(50) - \sigma(44)$$
$$+ \sigma(31) + \sigma(24) - \sigma(9) - \sigma(1)$$
$$= 217 + 156 - 252 - 144$$
$$+ 90 + 132 - 80 - 124$$
$$+ 144 + 62 - 93 - 84$$
$$+ 32 + 60 - 13 - 1$$
$$= 893 - 791$$
$$= 102$$

and hence we could conclude, if we would not have known it before, that 101 is a prime number.

II. Given the number 301, find the sum of its divisors. We have

$$
\begin{aligned}
\text{diff.} \quad & \qquad 1 \qquad\quad 3 \qquad\quad 2 \qquad\quad 5 \\
\sigma(301) = {}& \sigma(300) + \sigma(299) - \sigma(296) - \sigma(294) + \\
& \qquad 3 \qquad\quad 7 \qquad\quad 4 \qquad\quad 9 \\
+{}& \sigma(289) + \sigma(286) - \jmath(279) - \sigma(275) + \\
& \qquad 5 \qquad\quad 11 \qquad\quad 6 \qquad\quad 13 \\
+{}& \sigma(266) + \sigma(261) - \sigma(250) - \sigma(244) + \\
& \qquad 7 \qquad\quad 15 \qquad\quad 8 \qquad\quad 17 \\
+{}& \sigma(231) + \sigma(224) - \sigma(209) - \sigma(201) + \\
& \qquad 9 \qquad\quad 19 \qquad\quad 10 \qquad\quad 21 \\
+{}& \sigma(184) + \sigma(175) - \sigma(156) - \sigma(146) + \\
& \qquad 11 \qquad\quad 23 \qquad\quad 12 \qquad\quad 25 \\
+{}& \sigma(125) + \sigma(114) - \ \sigma(91) - \ \ \sigma(79) + \\
& \qquad 13 \qquad\quad 27 \qquad\quad 14 \\
+{}& \ \ \sigma(54) + \ \ \sigma(41) - \ \ \sigma(14) - \ \ \sigma(0).
\end{aligned}
$$

We see by this example how we can, using the differences, continue the formula as far as is necessary in each case. Performing the computations, we find

$$\sigma(301) = 4939 - 4587 = 352.$$

We see hence that 301 is not a prime. In fact, $301 = 7 \cdot 43$ and we obtain

$$\sigma(301) = \sigma(7)\sigma(43) = 8 \cdot 44 = 352$$

as the rule has shown.

8. The examples that I have just developed will undoubtedly dispel any qualms which we might have had about the truth of my formula. Now, this beautiful property of the numbers is so much more surprising as we do not perceive any intelligible connection between the structure of my formula and the nature of the divisors with the sum of which we are here concerned. The sequence of the numbers 1, 2, 5, 7, 12, 15, . . . does not seem to have any relation to the matter in hand. Moreover, as the law of these numbers is "interrupted" and they are in fact a mixture of two sequences with a regular law, of 1, 5, 12, 22, 35, 51, . . . and 2, 7, 15, 26, 40, 57, . . . , we would not expect that such an irregularity can turn up in Analysis. The lack of demonstration must increase the surprise still more, since it seems wholly impossible to succeed in discovering such a property without being guided by some reliable method which could take the place of a perfect proof. I confess that I did not hit on this discovery by mere chance, but another proposition opened the path to this beautiful property—another proposition of the same nature which must be accepted as true although I am unable to

prove it. And although we consider here the nature of integers to which the Infinitesimal Calculus does not seem to apply, nevertheless I reached my conclusion by differentiations and other devices. I wish that somebody would find a shorter and more natural way, in which the consideration of the path that I followed might be of some help, perhaps.

9. In considering the partitions of numbers, I examined, a long time ago, the expression

$$(1 - x) (1 - x^2) (1 - x^3) (1 - x^4) (1 - x^5) (1 - x^6) (1 - x^7) (1 - x^8) \ldots ,$$

in which the product is assumed to be infinite. In order to see what kind of series will result, I multiplied actually a great number of factors and found

$$1 - x - x^2 + x^5 + x^7 - x^{12} - x^{15} + x^{22} + x^{26} - x^{35} - x^{40} + \ldots .$$

The exponents of $x$ are the same which enter into the above formula; also the signs $+$ and $-$ arise twice in succession. It suffices to undertake this multiplication and to continue it as far as it is deemed proper to become convinced of the truth of this series. Yet I have no other evidence for this, except a long induction which I have carried out so far that I cannot in any way doubt the law governing the formation of these terms and their exponents. I have long searched in vain for a rigorous demonstration of the equation between the series and the above infinite product $(1 - x) (1 - x^2) (1 - x^3) \ldots ,$ and I have proposed the same question to some of my friends with whose ability in these matters I am familiar, but all have agreed with me on the truth of this transformation of the product into a series, without being able to unearth any clue of a demonstration. Thus, it will be a known truth, but not yet demonstrated, that if we put

$$s = (1 - x) (1 - x^2) (1 - x^3) (1 - x^4) (1 - x^5) (1 - x^6) \ldots$$

the same quantity $s$ can also be expressed as follows:

$$s = 1 - x - x^2 + x^5 + x^7 - x^{12} - x^{15} + x^{22} + x^{26} - x^{35} - x^{40} + \ldots .$$

For each of us can convince himself of this truth by performing the multiplication as far as he may wish; and it seems impossible that the law which has been discovered to hold for 20 terms, for example, would not be observed in the terms that follow.

10. As we have thus discovered that those two infinite expressions are equal even though it has not been possible to demonstrate their equality, all the conclusions which may be deduced from it will be of the same nature, that is, true but not demonstrated. Or, if one of these conclusions could be demonstrated, one could reciprocally obtain a clue to the demonstration of that equation; and it was with this purpose in mind that I maneuvered those two expressions in many ways, and so I was led among other discoveries to that which I explained above; its truth, therefore, must be as certain as

that of the equation between the two infinite expressions. I proceeded as follows. Being given that the two expressions

I. $s = (1 - x)(1 - x^2)(1 - x^3)(1 - x^4)(1 - x^5)(1 - x^6)(1 - x^7)\ldots$

II. $s = 1 - x - x^2 + x^5 + x^7 - x^{12} - x^{15} + x^{22} + x^{26} - x^{35} - x^{40} + \ldots$

are equal, I got rid of the factors in the first by taking logarithms

$$\log s = \log(1 - x) + \log(1 - x^2) + \log(1 - x^3) + \log(1 - x^4) + \ldots$$

In order to get rid of the logarithms, I differentiate and obtain the equation

$$\frac{1}{s}\frac{ds}{dx} = -\frac{1}{1 - x} - \frac{2x}{1 - x^2} - \frac{3x^2}{1 - x^3} - \frac{4x^3}{1 - x^4} - \frac{5x^4}{1 - x^5} - \ldots$$

or

$$-\frac{x}{s}\frac{ds}{dx} = \frac{x}{1 - x} + \frac{2x^2}{1 - x^2} + \frac{3x^3}{1 - x^3} + \frac{4x^4}{1 - x^4} + \frac{5x^5}{1 - x^5} + \ldots$$

From the second expression for $s$, as infinite series, we obtain another value for the same quantity

$$-\frac{x}{s}\frac{ds}{dx} = \frac{x + 2x^2 - 5x^5 - 7x^7 + 12x^{12} + 15x^{15} - 22x^{22} - 26x^{26} + \ldots}{1 - x - x^2 + x^5 + x^7 - x^{12} - x^{15} + x^{22} + x^{26} - \ldots}.$$

11. Let us put

$$-\frac{x}{s}\frac{ds}{dx} = t.$$

We have above two expressions for the quantity $t$. In the first expression, I expand each term into a geometric series and obtain

$$
\begin{aligned}
t = x + \;\; &x^2 + \;\; x^3 + \;\; x^4 + \;\; x^5 + \;\; x^6 + \;\; x^7 + \;\; x^8 + \ldots \\
+ \;\; &2x^2 \qquad\quad + 2x^4 \qquad\quad + 2x^6 \qquad\quad + 2x^8 + \ldots \\
&\qquad\quad + 3x^3 \qquad\qquad\qquad + 3x^6 \qquad\qquad\qquad + \ldots \\
&\qquad\qquad\qquad + 4x^4 \qquad\qquad\qquad\qquad\quad + 4x^8 + \ldots \\
&\qquad\qquad\qquad\qquad\quad + 5x^5 \qquad\qquad\qquad\qquad\qquad + \ldots \\
&\qquad\qquad\qquad\qquad\qquad\qquad + 6x^6 \qquad\qquad\qquad\quad + \ldots \\
&\qquad\qquad\qquad\qquad\qquad\qquad\qquad\qquad + 7x^7 \qquad\qquad + \ldots \\
&\qquad\qquad\qquad\qquad\qquad\qquad\qquad\qquad\qquad\qquad + 8x^8 + \ldots.
\end{aligned}
$$

Here we see easily that each power of $x$ arises as many times as its exponent has divisors, and that each divisor arises as a coefficient of the same power of $x$. Therefore, if we collect the terms with like powers, the coefficient of each power of $x$ will be the sum of the divisors of its exponent. And, therefore, using the above notation $\sigma(n)$ for the sum of the divisors of $n$, I obtain

$$t = \sigma(1)x + \sigma(2)x^2 + \sigma(3)x^3 + \sigma(4)x^4 + \sigma(5)x^5 + \ldots.$$

The law of the series is manifest. And, although it might appear that some induction was involved in the determination of the coefficients, we can easily satisfy ourselves that this law is a necessary consequence.

12. By virtue of the definition of $t$, the last formula of No. 10 can be written as follows:

$$t(1 - x - x^2 + x^5 + x^7 - x^{12} - x^{15} + x^{22} + x^{26} - \ldots)$$
$$- x - 2x^2 + 5x^5 + 7x^7 - 12x^{12} - 15x^{15} + 22x^{22} + 26x^{26} - \ldots = 0.$$

Substituting for $t$ the value obtained at the end of No. 11, we find

$$0 = \sigma(1)x + \sigma(2)x^2 + \sigma(3)x^3 + \sigma(4)x^4 + \sigma(5)x^5 + \sigma(6)x^6 + \ldots$$
$$- x - \sigma(1)x^2 - \sigma(2)x^3 - \sigma(3)x^4 - \sigma(4)x^5 - \sigma(5)x^6 - \ldots$$
$$- 2x^2 - \sigma(1)x^3 - \sigma(2)x^4 - \sigma(3)x^5 - \sigma(4)x^6 - \ldots$$
$$+ 5x^5 + \sigma(1)x^6 + \ldots .$$

Collecting the terms, we find the coefficient for any given power of $x$. This coefficient consists of several terms. First comes the sum of the divisors of the exponent of $x$, and then sums of divisors of some preceding numbers, obtained from that exponent by subtracting successively 1, 2, 5, 7, 12, 15, 22, 26, .... Finally, if it belongs to this sequence, the exponent itself arises. We need not explain again the signs assigned to the terms just listed. Therefore, generally, the coefficient of $x^n$ is

$$\sigma(n) - \sigma(n - 1) - \sigma(n - 2) + \sigma(n - 5) + \sigma(n - 7) - \sigma(n - 12)$$
$$- \sigma(n - 15) + \ldots .$$

This is continued as long as the numbers under the sign $\sigma$ are not negative. Yet, if the term $\sigma(0)$ arises, we must substitute $n$ for it.

13. Since the sum of the infinite series considered in the foregoing No. 12 is 0, whatever the value of $x$ may be, the coefficient of each single power of $x$ must necessarily be 0. Hence we obtain the law that I explained above in No. 5; I mean the law that governs the sum of the divisors and enables us to compute it recursively for all numbers. In the foregoing development, we may perceive some reason for the signs, some reason for the sequence of the numbers

$$1, 2, 5, 7, 12, 15, 22, 26, 35, 40, 51, 57, 70, 77, \ldots$$

and, especially, a reason why we should substitute for $\sigma(0)$ the number $n$ itself, which could have appeared the strangest feature of my rule. This reasoning, although still very far from a perfect demonstration, will certainly lift some doubts about the most extraordinary law that I explained here.

**3. Transition to a more general viewpoint.** Euler's foregoing text is extraordinarily instructive. We can learn from it a great deal about mathematics, or the psychology of invention, or inductive reasoning. The examples and comments at the end of this chapter provide for opportunity to examine some of Euler's mathematical ideas, but now we wish to concentrate on his inductive argument.

The theorem investigated by Euler is remarkable in several respects and is of great mathematical interest even today. However, we are concerned here not so much with the mathematical content of this theorem, but rather with the reasons which induced Euler to believe in the theorem when it was still unproved. In order to understand better the nature of these reasons, I shall ignore the mathematical content of Euler's memoir and give a schematic outline of it, emphasizing a certain general aspect of his inductive argument.

As we shall disregard the mathematical content of the various theorems that we must discuss, we shall find it advantageous to designate them by letters, as $T$, $T^*$, $C_1$, $C_2$, $\ldots$ , $C_1^*$, $C_2^*$, $\ldots$ . The reader may ignore the meaning of these letters completely. Yet, in case he wishes to recognize them in Euler's text, here is the key.

$T$ is the theorem

$$(1 - x)(1 - x^2)(1 - x^3) \ldots = 1 - x - x^2 + x^5 + x^7 - x^{12} - x^{15} + \ldots .$$

The law of the numbers $1, 2, 5, 7, 12, 15, \ldots$ is explained in sect. 2, No. 5, II.

$C_n$ is the assertion that the coefficient of $x^n$ is the same on both sides of the foregoing equation. For example, $C_6$ asserts that expanding the product on the left hand side, we shall find that the coefficient of $x^6$ is 0. Observe that $C_n$ is a consequence of the theorem $T$.

$C_n^*$ is the equation

$$\sigma(n) = \sigma(n - 1) + \sigma(n - 2) - \sigma(n - 5) - \sigma(n - 7) + \ldots$$

explained at length in sect. 2, No. 5. For example, $C_6^*$ asserts that

$$\sigma(6) = \sigma(5) + \sigma(4) - \sigma(1).$$

$T^*$ is the "most extraordinary law," asserting that $C_1^*, C_2^*, C_3^*, \ldots$ are all true. Observe that $C_n^*$ is a consequence (a particular case) of the theorem $T^*$.

**4. Schematic outline of Euler's memoir.**[4] *Theorem $T$ is of such a nature that we can be assured of its truth without giving it a perfect demonstration. Nevertheless, I shall present such evidence for it as might be regarded as almost equivalent to a rigorous demonstration.*

*Theorem $T$ includes an infinite number of particular cases: $C_1$, $C_2$, $C_3$, $\ldots$ . Conversely, the infinite set of these particular cases $C_1$, $C_2$, $C_3$, $\ldots$ is equivalent to theorem $T$. We can find out by a simple calculation whether $C_1$ is true or not.*

---

[4] This outline was first published in my paper, "Heuristic Reasoning and the Theory of Probability," *Amer. Math. Monthly*, vol. 48, 1941, p. 450–465. The italics indicate phrases which are not due to Euler.

*Another simple calculation determines whether* $C_2$ *is true or not, and similarly for* $C_3$, *and so on. I have made these calculations and I find that* $C_1$, $C_2$, $C_3$, . . . , $C_{40}$ *are all true.* It suffices to undertake *these calculations* and to continue *them* as far as is deemed proper to become convinced of the truth of this sequence *continued indefinitely.* Yet I have no other evidence for this, except a long induction which I have carried out so far that I cannot in any way doubt the law of which $C_1$, $C_2$, . . . *are the particular cases.* I have long searched in vain for a rigorous demonstration of *theorem* $T$, and I have proposed the same question to some of my friends with whose ability in these matters I am familiar, but all have agreed with me on the truth of *theorem* $T$ without being able to unearth any clue of a demonstration. Thus it will be a known truth, but not yet demonstrated; for each of us can convince himself of this truth by the *actual calculation of the cases* $C_1$, $C_2$, $C_3$, . . . as far as he may wish; and it seems impossible that the law which has been discovered to hold for 20 terms, for example, would not be observed in the terms that follow.

As we have thus discovered the truth of *theorem* $T$ even though it has not been possible to demonstrate it, all the conclusions which may be deduced from it will be of the same nature, that is, true but not demonstrated. Or, if one of these conclusions could be demonstrated, one could reciprocally obtain a clue to the demonstration of *theorem* $T$; and it was with this purpose in mind that I maneuvered *theorem* $T$ in many ways and so discovered among others *theorem* $T^*$ whose truth must be as certain as that of *theorem* $T$.

*Theorems* $T$ *and* $T^*$ *are equivalent; they are both true or false; they stand or fall together. Like* $T$, *theorem* $T^*$ *includes an infinity of particular cases* $C_1^*, C_2^*, C_3^*, \ldots,$ *and this sequence of particular cases is equivalent to theorem* $T^*$. *Here again, a simple calculation shows whether* $C_1^*$ *is true or not. Similarly, it is possible to determine whether* $C_2^*$ *is true or not, and so on.* It is not difficult to apply theorem $T^*$ to any given particular case, and so anybody can satisfy himself of its truth by as many examples as he may wish to develop. And since I must admit that I am not in a position to give it a rigorous demonstration, I will justify it by a sufficiently large number of examples, by $C_1^*$, $C_2^*$, . . . , $C_{20}^*$. I think these examples are sufficient to discourage anyone from imagining that it is by mere chance that my rule is in agreement with the truth.

If one still doubts that the law is precisely that one which I have indicated, I will give some examples with larger numbers. *By examination, I find that* $C_{101}^*$ *and* $C_{301}^*$ *are true, and so I find that theorem* $T^*$ *is valid even for these cases which are far removed from those which I examined earlier.* These examples which I have just developed undoubtedly will dispel any qualms which we might have had about the truth of *theorems* $T$ *and* $T^*$.

### EXAMPLES AND COMMENTS ON CHAPTER VI

In discovering his "Most Extraordinary Law of the Numbers" Euler "reached his conclusion by differentiations and other devices" although "the

Infinitesimal Calculus does not seem to apply to the nature of integers." In order to understand Euler's method, we apply it to similar examples. We begin by giving a name to his principal "device" or mathematical tool.

**1. Generating functions.** We restate the result of No. 11 of Euler's memoir in modern notation:

$$\sum_{n=1}^{\infty} \frac{nx^n}{1-x^n} = \sigma(1)x + \sigma(2)x^2 + \ldots + \sigma(n)x^n + \ldots .$$

The right hand side is a power series in $x$. The coefficient of $x^n$ in this power series is $\sigma(n)$, the sum of the divisors of $n$. Both sides of the equation represent the same function of $x$. The expansion of this function in powers of $x$ "generates" the sequence $\sigma(1)$, $\sigma(2)$, $\ldots \sigma(n)$, $\ldots$ and so we call this function the *generating function* of $\sigma(n)$. Generally, if

$$f(x) = a_0 + a_1x + a_2x^2 + \ldots + a_nx^n + \ldots$$

we say that $f(x)$ is the generating function of $a_n$, or the function generating the sequence $a_0$, $a_1$, $a_2$, $\ldots a_n$, $\ldots$ .

The name "generating function" is due to Laplace. Yet, without giving it a name, Euler used the device of generating functions long before Laplace, in several memoirs of which we have seen one in sect. 2. He applied this mathematical tool to several problems in Combinatory Analysis and the Theory of Numbers.

A generating function is a device somewhat similar to a bag. Instead of carrying many little objects detachedly, which could be embarrassing, we put them all in a bag, and then we have only one object to carry, the bag. Quite similarly, instead of handling each term of the sequence $a_0$, $a_1$, $a_2$, $\ldots$ $a_n$, $\ldots$ individually, we put them all in a power series $\Sigma a_n x^n$, and then we have only one mathematical object to handle, the power series.

**2.** Find the generating function of $n$. Or, what is the same, find the sum of the series $\Sigma nx^n$.

**3.** Being given that $f(x)$ generates the sequence $a_0$, $a_1$, $a_2$, $\ldots a_n$, $\ldots$ find the function generating the sequence

$$0a_0, 1a_1, 2a_2, \ldots na_n, \ldots .$$

**4.** Being given that $f(x)$ generates the sequence $a_0$, $a_1$, $a_2$, $\ldots a_n$, $\ldots$ , find the function generating the sequence

$$0, a_0, a_1, \ldots a_{n-1}, \ldots .$$

**5.** Being given that $f(x)$ is the generating function of $a_n$, find the generating function of

$$s_n = a_0 + a_1 + a_2 + \ldots + a_n.$$

**6.** Being given that $f(x)$ and $g(x)$ are the generating functions of $a_n$ and $b_n$, respectively, find the generating function of

$$c_n = a_0 b_n + a_1 b_{n-1} + a_2 b_{n-2} + \ldots + a_n b_0.$$

**7.** *A combinatorial problem in plane geometry.* A convex polygon with $n$ sides is dissected into $n - 2$ triangles by $n - 3$ diagonals; see fig. 6.1. Call $D_n$ the number of different dissections.

Find $D_n$ for $n = 3, 4, 5, 6$.

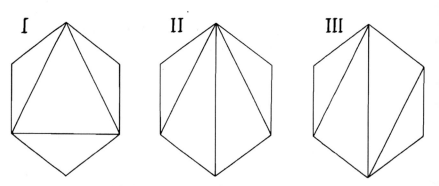

Fig. 6.1.   Three types of dissection for a hexagon.

**8** (continued). It is not easy to guess a general, explicit expression for $D_n$ on the basis of the numerical values considered in ex. 7. Yet the sequence $D_3, D_4, D_5, \ldots$ is a "recurring" sequence in the following, very general sense: each term can be computed from the foregoing terms according to an invariable rule, a "recursion formula." (See Euler's memoir, No. 5.)

Define $D_2 = 1$, and show that for $n \geqq 3$

$$D_n = D_2 D_{n-1} + D_3 D_{n-2} + D_4 D_{n-3} + \ldots + D_{n-1} D_2.$$

[Check the first cases.   Refer to fig. 6.2.]

**9** (continued). The derivation of an explicit expression for $D_n$ from the recursion formula of ex. 8 is not obvious. Yet consider the generating function

$$g(x) = D_2 x^2 + D_3 x^3 + D_4 x^4 + \ldots + D_n x^n + \ldots .$$

Show that $g(x)$ satisfies a quadratic equation and derive hence that for $n = 3, 4, 5, 6, \ldots$

$$D_n = \frac{2}{2} \frac{6}{3} \frac{10}{4} \frac{14}{5} \ldots \frac{4n - 10}{n - 1}.$$

**10.** *Sums of squares.* Recall the definition of $R_k(n)$ (ex. 4.1), extend it to $n = 0$ in setting $R_k(0) = 1$ (a reasonable extension), introduce the generating function

$$\sum_{n=0}^{\infty} R_k(n)x^n = R_k(0) + R_k(1)x + R_k(2)x^2 + \dots ,$$

and show that

$$\sum_{n=0}^{\infty} R_3(n)x^n = (1 + 2x + 2x^4 + 2x^9 + \dots)^3.$$

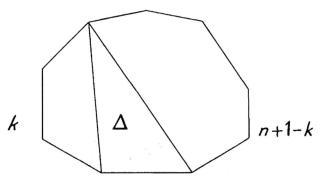

Fig. 6.2.   Starting the dissection of a polygon with $n$ sides.

[What is $R_3(n)$?   The number of solutions of the equation

$$u^2 + v^2 + w^2 = n$$

in integers $u$, $v$, and $w$, positive, negative, or 0.

What may be the rôle of the series on the right-hand side of the equation that you are required to prove?

$$1 + 2x + 2x^4 + 2x^9 + \dots = \sum_{u=-\infty}^{\infty} x^{u^2} = \sum_{n=0}^{\infty} R_1(n)x^n.$$

How should you conceive the right-hand side of the equation you are aiming at?   Perhaps so:

$$\Sigma x^{u^2} \cdot \Sigma x^{u^2} \cdot \Sigma x^{u^2}.]$$

**11.** Generalize the result of ex. 10.

**12.** Recall the definition of $S_k(n)$ (ex. 4.1) and express the generating function

$$\sum_{n=1}^{\infty} S_k(n)x^n.$$

**13.** Use ex. 11 to prove that, for $n \geq 1$, $R_2(n)$ is divisible by 4, $R_4(n)$ by 8, and $R_8(n)$ by 16.   (The result was already used in ch. IV, Tables II and III.)

**14.** Use ex. 12 to prove that

$$S_2(n) = 0 \text{ if } n \text{ is not of the form } 8m + 2,$$
$$S_4(n) = 0 \text{ if } n \text{ is not of the form } 8m + 4,$$
$$S_8(n) = 0 \text{ if } n \text{ is not of the form } 8m.$$

**15.** Use ex. 11 to prove that

$$R_{k+l}(n) = R_k(0)R_l(n) + R_k(1)R_l(n-1) + \ldots + R_k(n)R_l(0).$$

**16.** Prove that

$$S_{k+l}(n) = S_k(1)S_l(n-1) + S_k(2)S_l(n-2) + \ldots + S_k(n-1)S_l(1).$$

**17.** Propose a simple method for computing Table III of ch. IV from the Tables I and II of the same chapter.

**18.** Let $\sigma_k(n)$ stand for the sum of the $k$th powers of the divisors of $n$. For example,

$$\sigma_3(15) = 1^3 + 3^3 + 5^3 + 15^3 = 3528;$$

$\sigma_1(n) = \sigma(n)$.

(1)  Show that the conjectures found in sect. 4.6 and ex. 4.23 imply

$$\sigma(1)\sigma(2u-1) + \sigma(3)\sigma(2u-3) + \ldots + \sigma(2u-1)\sigma(1) = \sigma_3(u)$$

where $u$ denotes an odd integer.

(2)  Test particular cases of the relation found in (1) numerically.

(3)  How does such a verification influence your confidence in the conjectures from which the relation verified has been derived?

**19.** *Another recursion formula.*  We consider the generating functions

$$G = \sum_{m=1}^{\infty} S_1(m)x^m, \qquad H = \sum_{m=1}^{\infty} S_4(m)x^m.$$

We set

$$S_4(4u) = s_u$$

where $u$ is an odd integer.  Then

$$G = x + x^9 + x^{25} + \ldots + x^{(2n-1)^4} + \ldots ,$$
$$H = s_1 x^4 + s_3 x^{12} + s_5 x^{20} + \ldots + s_{2n-1} x^{8n-4} + \ldots ,$$
$$G^4 = H$$

by ex. 14 and 12.  We derive from the last equation, by taking the logarithms and differentiating, that

$$4 \log G = \log H,$$
$$\frac{4G'}{G} = \frac{H'}{H},$$
$$G \cdot xH' = 4 \cdot xG' \cdot H,$$

$$(x + x^9 + x^{25} + \ldots)(4s_1 x^4 + 12s_3 x^{12} + 20s_5 x^{20} + \ldots)$$
$$= 4(x + 9x^9 + 25x^{25} + \ldots)(s_1 x^4 + s_3 x^{12} + s_5 x^{20} + \ldots).$$

Comparing the coefficients of $x^5$, $x^{13}$, $x^{21}$, . . . on both sides of the foregoing equation, we find, after some elementary work, the following relations:

$$0s_1 = 0$$
$$1s_3 - 4s_1 = 0$$
$$2s_5 - 3s_3 = 0$$
$$3s_7 - 2s_5 - 12s_1 = 0$$
$$4s_9 - 1s_7 - 11s_3 = 0$$
$$5s_{11} \qquad - 10s_5 = 0$$
$$6s_{13} + 1s_{11} - 9s_7 - 24s_1 = 0$$
$$7s_{15} + 2s_{13} - 8s_9 - 23s_3 = 0$$
$$8s_{17} + 3s_{15} - 7s_{11} - 22s_5 = 0$$
$$9s_{19} + 4s_{17} - 6s_{13} - 21s_7 = 0$$
$$10s_{21} + 5s_{19} - 5s_{15} - 20s_9 - 40s_1 = 0$$
$$11s_{23} + 6s_{21} - 4s_{17} - 19s_{11} - 39s_3 = 0$$
$$\cdot \quad \cdot \quad \cdot \quad \cdot \quad \cdot \quad \cdot \quad \cdot \quad \cdot \quad \cdot \quad \cdot \quad \cdot \quad \cdot \quad \cdot \quad \cdot$$

The very first equation of this system is vacuous and is displayed here only to emphasize the general law. Yet we know that $s_1 = 1$. Knowing this, we obtain from the next equation $s_3$. Knowing $s_3$, we obtain from the following equation $s_5$. And so on, we can compute from the system the terms of the sequence $s_1$, $s_3$, $s_5$, . . . as far as we wish, one after the other, *recurrently*.

The system has a remarkable structure. There is 1 equation containing 1 of the quantities $s_1$, $s_3$, $s_5$, . . . , 2 equations containing 2 of them, 3 equations containing 3 of them, and so on. The coefficients in each column are increased by 1 and the subscripts by 2 as we pass from one row to the next. The subscript at the head of each column is 1 and the coefficient is $-4$ multiplied by the first coefficient in the same row.

We can concentrate the whole system in one equation (recursion formula); write it down.

**20.** *Another Most Extraordinary Law of the Numbers Concerning the Sum of their Divisors.* If the conjecture of sect. 4.6 stands

$$s_{2n-1} = S_4(4(2n-1)) = \sigma(2n-1)$$

and so ex. 19 yields a recursion formula connecting the terms of the sequence $\sigma(1)$, $\sigma(3)$, $\sigma(5)$, $\sigma(7)$, . . . which is in many ways strikingly similar to Euler's formula.

Write out in detail and verify numerically the first cases of the indicated recursion formula.

**21.** For us there is also a heuristic similarity between Euler's recursion formula for $\sigma(n)$ (sect. 2) and the foregoing recursion formula for $\sigma(2n-1)$ (ex. 20). For us this latter is a conjecture. We derived this conjecture, as Euler has derived his, "by differentiation and other devices" from another conjecture.

Show that the recursion formula for $\sigma(2n-1)$ indicated by ex. 20 is equivalent to the equation

$$S_4(4(2n-1)) = \sigma(2n-1)$$

to which we arrived in sect. 4.6. That is, if one of the two assertions is true, the other is necessarily also true.

**22.** Generalize ex. 19.

**23.** Devise a method for computing $R_8(n)$ independently of $R_4(n)$.

**24.** *How Euler missed a discovery.* The method illustrated by ex. 19 and ex. 23, and generally stated in ex. 22, is due to Euler.[5] In inventing his method, Euler aimed at the problem of four squares and some related problems. In fact, he applied his method to the problem of four squares and investigated inductively the number of representations, but failed to discover the remarkable law governing $R_4(n)$, which is after all not so difficult to discover inductively (ex. 4.10–4.15). How did it happen?

In examining the equation

$$n = x^2 + y^2 + z^2 + w^2$$

we may choose various standpoints, especially the following:

(1) We admit for $x, y, z$, and $w$ only non-negative integers.

(2) We admit for $x, y, z$, and $w$ all integers, positive, negative, and null.

The second standpoint may be less obvious, but leads to $R_4(n)$ and to the remarkable connection between $R_4(n)$ and the divisors of $n$. The first standpoint is more obvious, but the number of solutions does not seem to have any simple remarkable property. Euler chose the standpoint (1), not the standpoint (2), he applied his method explained in ex. 22 to

$$(1 + x + x^4 + x^9 + \ldots)^4,$$

not to

$$(1 + 2x + 2x^4 + 2x^9 + \ldots)^4,$$

and so he bypassed a great discovery. It is instructive to compare two lines of inquiry which look so much alike at the outset, but one of which is wonderfully fruitful and the other almost completely barren.

---

[5] *Opera Omnia,* ser. 1, vol. 4, p. 125–135.

The properties of $R_4(n)$, $S_4(n)$, $R_8(n)$, and $S_8(n)$ investigated in ch. IV (ex. 4.10–4.15, sect. 4.3–4.6, ex. 4.18–4.23) have been discovered by Jacobi, not inductively, but as incidental corollaries of his researches on elliptic functions. Several proofs of these theorems have been found since, but no known proof is quite elementary and straightforward.[6]

**25.** *A generalization of Euler's theorem on $\sigma(n)$.* Given $k$, set

$$\prod_{n=1}^{\infty} (1 - x^n)^k = 1 - \sum_{n=1}^{\infty} a_n x^n$$

and show that, for $n = 1, 2, 3, \ldots$

$$\sigma(n) = \sum_{m=1}^{n-1} a_m \sigma(n - m) + na_n/k.$$

Which particular case yields Euler's theorem of sect. 2?

---

[6] See also for further references G. H. Hardy and E. M. Wright, *An introduction to the theory of numbers*, Oxford, 1938, chapter XX.

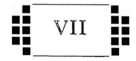

# VII

## MATHEMATICAL INDUCTION

---

*Jacques Bernoulli's method is important also to the naturalist. We find what seems to be a property A of the concept B by observing the cases $C_1$, $C_2$, $C_3$, . . . . We learn from Bernoulli's method that we should not attribute such a property A, found by incomplete, non-mathematical induction, to the concept B, unless we perceive that A is linked to the characteristics of B and is independent of the variation of the cases. As in many other points, mathematics offers here a model to natural science.*—ERNST MACH[1]

---

**1. The inductive phase.** Again, we begin with an example. There is little difficulty in finding the sum of the first $n$ integers. We take here for granted the formula

$$1 + 2 + 3 + \ldots + n = \frac{n(n + 1)}{2}$$

which can be discovered and proved in many ways.[2] It is harder to find a formula for the sum of the first $n$ squares

$$1 + 4 + 9 + 16 + \ldots + n^2.$$

There is no difficulty in computing this sum for small values of $n$, but it is not so easy to disentangle a rule. It is quite natural, however, to seek some sort of parallelism between the two sums and to observe them together:

| $n$ | 1 | 2 | 3 | 4 | 5 | 6 | ... |
|---|---|---|---|---|---|---|---|
| $1 + 2 + \ldots + n$ | 1 | 3 | 6 | 10 | 15 | 21 | ... |
| $1^2 + 2^2 + \ldots + n^2$ | 1 | 5 | 14 | 30 | 55 | 91 | ... . |

How are the last two rows related? We may hit upon the idea of examining their ratio:

| $n$ | 1 | 2 | 3 | 4 | 5 | 6 | ... |
|---|---|---|---|---|---|---|---|
| $\dfrac{1^2 + 2^2 + \ldots + n^2}{1 + 2 + \ldots + n}$ | 1 | $\dfrac{5}{3}$ | $\dfrac{7}{3}$ | 3 | $\dfrac{11}{3}$ | $\dfrac{13}{3}$ | ... |

---

[1] *Erkenntnis und Irrtum*, 4th ed., 1920, p. 312.
[2] See *How to Solve It*, p. 107.

Here the rule is obvious and it is almost impossible to miss it if the foregoing ratios are written as follows:

$$\frac{3}{3} \quad \frac{5}{3} \quad \frac{7}{3} \quad \frac{9}{3} \quad \frac{11}{3} \quad \frac{13}{3}.$$

We can hardly refrain from formulating the conjecture that

$$\frac{1^2 + 2^2 + \ldots + n^2}{1 + 2 + \ldots + n} = \frac{2n + 1}{3}.$$

Using the value of the denominator on the left-hand side, which we took for granted, we are led to restating our conjecture in the form

$$1^2 + 2^2 + 3^2 + \ldots + n^2 = \frac{n(n + 1)(2n + 1)}{6}.$$

Is this true? That is, is it generally true? The formula is certainly true in the particular cases $n = 1, 2, 3, 4, 5, 6$ which suggested it. Is it true also in the next case $n = 7$? The conjecture leads us to predicting that

$$1 + 4 + 9 + 16 + 25 + 36 + 49 = \frac{7 \cdot 8 \cdot 15}{6}$$

and, in fact, both sides turn out to be equal to 140.

We could, of course, go on to the next case $n = 8$ and test it, but the temptation is not too strong. We are inclined to believe anyhow that the formula will be verified in the next case too, and so this verification would add but little to our confidence—so little that going through the computation is hardly worth while. How could we test the conjecture more efficiently?

If the conjecture is true at all, it should be independent of the variation of the cases, it should hold good in the transition from one case to another. Supposedly,

$$1 + 4 + \ldots + n^2 = \frac{n(n + 1)(2n + 1)}{6}.$$

Yet, if this formula is generally true, it *should hold also in the next case*: we should have

$$1 + 4 + \ldots + n^2 + (n + 1)^2 = \frac{(n + 1)(n + 2)(2n + 3)}{6}.$$

Here is an opportunity to check efficiently the conjecture: by subtracting the upper line from the lower we obtain

$$(n + 1)^2 = \frac{(n + 1)(n + 2)(2n + 3)}{6} - \frac{n(n + 1)(2n + 1)}{6}.$$

Is this consequence of the conjecture true?

An easy rearrangement of the right hand side yields

$$\frac{n+1}{6}\left[(n+2)(2n+3) - n(2n+1)\right]$$

$$= \frac{n+1}{6}\left[2n^2 + 3n + 4n + 6 - 2n^2 - n\right]$$

$$= \frac{n+1}{6}\left[6n + 6\right]$$

$$= (n+1)^2.$$

The consequence examined is incontestably true, the conjecture passed a severe test.

**2. The demonstrative phase.** The verification of any consequence increases our confidence in the conjecture, but the verification of the consequence just examined can do more: it can *prove* the conjecture. We need only a little change of our viewpoint and a little reshuffling of our remarks.

It is *supposedly* true that

$$1^2 + 2^2 + \ldots + n^2 = \frac{n(n+1)(2n+1)}{6}.$$

It is *incontestably* true that

$$(n+1)^2 = \frac{(n+1)(n+2)(2n+3)}{6} - \frac{n(n+1)(2n+1)}{6}.$$

It is *consequently* true that

$$1^2 + 2^2 + \ldots + n^2 + (n+1)^2 = \frac{(n+1)(n+2)(2n+3)}{6} - -$$

(we added the two foregoing equations). This means: If our conjecture is true for a certain integer $n$, it remains necessarily true for the next integer $n + 1$.

Yet we know that the conjecture is true for $n = 1, 2, 3, 4, 5, 6, 7$. Being true for 7, it must be true also for the next integer 8; being true for 8, it must be true for 9; since true for 9, also true for 10, and so also for 11, and so on. The conjecture is true for all integers; we succeeded in proving it in full generality.

**3. Examining transitions.** The last reasoning of the foregoing section can be simplified a little. It is enough to know two things about the conjecture:

It is true for $n = 1$.

Being true for $n$, it is also true for $n + 1$.

Then the conjecture is true for all integers: true for 1, therefore, also for 2; true for 2, therefore, also for 3; and so on.

We have here a fundamentally important procedure of demonstration. We could call it "passing from $n$ to $n + 1$," but it is usually called" mathematical induction." This usual designation is a very inappropriate name for a procedure of demonstration, since induction (in the meaning in which the term is most frequently used) yields only a plausible, and not a demonstrative, inference.

Has mathematical induction anything to do with induction? Yes, it has, and we consider it here for this reason and not only for its name.

In our foregoing example, the demonstrative reasoning of sect. 2 naturally completes the inductive reasoning of sect. 1, and this is typical. The demonstration of sect. 2 appears as a "mathematical complement to induction," and if we take "mathematical induction" as an abbreviation in this sense, the term may appear quite appropriate, after all. (Therefore, let us take it in this sense—no use quarreling with established technical terms.) Mathematical induction often arises as the finishing step, or last phase, of an inductive research, and this last phase often uses suggestions which turned up in the foregoing phases.

Another and still better reason to consider mathematical induction in the present context is hinted by the passage quoted from Ernst Mach at the beginning of this chapter.[3] Examining a conjecture, we investigate the various cases to which the conjecture is supposed to apply. We wish to see whether the relation asserted by the conjecture is *stable*, that is, independent of, and undisturbed by, the variation of the cases. Our attention turns so naturally to the *transition* from one such case to another. "That by means of centripetal forces the planets may be retained in certain orbits, we may easily understand, if we consider the motions of projectiles" says Newton, and then he imagines a stone that is projected with greater and greater initial velocity till its path goes round the earth as the path of the moon; see ex. 2.18 (4). Thus Newton visualizes a continuous transition from the motion of a projectile to the motion of a planet. He considers the transition between two cases to which the law of universal gravitation, that he undertook to prove, should equally apply. Any beginner, who uses mathematical induction in proving some elementary theorem, acts like Newton in this respect: he considers the transition from $n$ to $n + 1$, the transition between two cases to which the theorem that he undertook to prove should equally apply.

**4. The technique of mathematical induction.** To be a good mathematician, or a good gambler, or good at anything, you must be a good guesser. In order to be a good guesser, you should be, I would think, naturally clever to begin with. Yet to be naturally clever is certainly not

---

[3] Mach believed that Jacques Bernoulli invented the method of mathematical induction, but most of the credit for its invention seems to be due to Pascal. Cf. H. Freudenthal, *Archives internationales d'histoire des sciences*, no. 22, 1953, p. 17–37. Cf. also *Jacobi Bernoulli Basileensis Opera*, Geneva 1744, vol. I, p. 282–283.

enough.   You should examine your guesses, compare them with the facts, modify them if need be, and so acquire an extensive (and intensive) experience with guesses that failed and guesses that came true.   With such an experience in your background, you may be able to judge more competently which guesses have a chance to turn out correct and which have not.

Mathematical induction is a demonstrative procedure often useful in verifying mathematical conjectures at which we arrived by some inductive procedure.   Therefore, if we wish to acquire some experience in inductive mathematical research, some acquaintance with the technique of mathematical induction is desirable.

The present section and the following examples and comments may give a little help in acquiring this technique.

(1) *The inductive phase.*   We begin with an example very similar to that discussed in sect. 1 and 2.   We wish to express in some shorter form another sum connected with the first $n$ squares,

$$\frac{1}{3} + \frac{1}{15} + \frac{1}{35} + \ldots + \frac{1}{4n^2 - 1} =$$

$$\frac{1}{4 \cdot 1^2 - 1} + \frac{1}{4 \cdot 2^2 - 1} + \frac{1}{4 \cdot 3^2 - 1} + \ldots + \frac{1}{4n^2 - 1}.$$

We compute this sum in the first few cases and tabulate the results:

$$n \qquad\qquad = 1, \; 2, \; 3, \; 4, \; \ldots$$

$$\frac{1}{3} + \frac{1}{15} + \ldots + \frac{1}{4n^2 - 1} = \frac{1}{3}, \frac{2}{5}, \frac{3}{7}, \frac{4}{9}, \ldots \; .$$

There is an obvious guess:

$$\frac{1}{3} + \frac{1}{15} + \frac{1}{35} + \ldots + \frac{1}{4n^2 - 1} = \frac{n}{2n + 1}.$$

Profiting from our experience with our former similar problem, we test our conjecture right away as efficiently as we can: we test the transition from $n$ to $n + 1$.   If our conjecture is generally true, it must be true both for $n$ and for $n + 1$:

$$\frac{1}{3} + \frac{1}{15} + \ldots + \frac{1}{4n^2 - 1} \qquad\qquad = \frac{n}{2n + 1},$$

$$\frac{1}{3} + \frac{1}{15} + \ldots + \frac{1}{4n^2 - 1} + \frac{1}{4(n + 1)^2 - 1} = \frac{n + 1}{2n + 3}.$$

By subtracting we obtain

$$\frac{1}{4(n + 1)^2 - 1} = \frac{n + 1}{2n + 3} - \frac{n}{2n + 1}.$$

Is this consequence of our conjecture true? We transform both sides, trying to bring them nearer to each other:

$$\frac{1}{(2n+2)^2 - 1} = \frac{2n^2 + 3n + 1 - 2n^2 - 3n}{(2n+3)(2n+1)}.$$

Very little algebra is enough to see that the two sides of the last equation are actually identical. The consequence examined is incontestably true.

(2) *The demonstrative phase.* Now we reshuffle our remarks, as in our foregoing example, sect. 2.

Supposedly

$$\frac{1}{3} + \frac{1}{15} + \ldots + \frac{1}{4n^2 - 1} = \frac{n}{2n+1}.$$

Incontestably

$$\frac{1}{4(n+1)^2 - 1} = \frac{n+1}{2n+3} - \frac{n}{2n+1}.$$

Consequently

$$\frac{1}{3} + \frac{1}{15} + \ldots + \frac{1}{4n^2 - 1} + \frac{1}{4(n+1)^2 - 1} = \frac{n+1}{2n+3}.$$

The conjecture, supposed to be true for $n$, turns out to be true, in consequence of this supposition, also for $n+1$. As it is true for $n = 1$, it is generally true.

(3) *Shorter.* We could have spent a little less time on the inductive phase of our solution. Having conceived the conjecture, we could have suspected that mathematical induction may be appropriate to prove it. Then, without any testing, we could have tried to apply mathematical induction directly, as follows.

Supposedly

$$\frac{1}{3} + \frac{1}{15} + \ldots + \frac{1}{4n^2 - 1} = \frac{n}{2n+1}.$$

Consequently

$$\frac{1}{3} + \frac{1}{15} + \ldots + \frac{1}{4n^2 - 1} + \frac{1}{4(n+1)^2 - 1} = \frac{n}{2n+1} + \frac{1}{4(n+1)^2 - 1}$$

$$= \frac{n}{2n+1} + \frac{1}{(2n+2)^2 - 1}$$

$$= \frac{n(2n+3) + 1}{(2n+1)(2n+3)}$$

$$= \frac{2n^2 + 3n + 1}{(2n+1)(2n+3)}$$

$$= \frac{(2n+1)(n+1)}{(2n+1)(2n+3)}.$$

$$= \frac{n+1}{2n+3}$$

and so we succeeded in deriving for $n + 1$ the relation that we have supposed for $n$. This is exactly what we were required to do, and so we proved the conjecture.

This variant of the solution is less repetitious, but perhaps also a trifle less natural, than the first, presented under (1) and (2).

(4) *Still shorter.* We can see the solution almost at a glance if we notice that

$$\frac{1}{4n^2 - 1} = \frac{1}{(2n - 1)(2n + 1)} = \frac{1}{2}\left(\frac{1}{2n - 1} - \frac{1}{2n + 1}\right).$$

(We are led to this formula quite naturally, if we are familiar with the decomposition of rational functions in partial fractions.)  Putting $n = 1$, 2, 3, . . . , $n$ and adding, we obtain

$$\frac{1}{4 - 1} + \frac{1}{16 - 1} + \frac{1}{36 - 1} + \cdots + \frac{1}{4n^2 - 1}$$

$$= \frac{1}{1 \cdot 3} + \frac{1}{3 \cdot 5} + \frac{1}{5 \cdot 7} + \cdots + \frac{1}{(2n - 1)(2n + 1)}$$

$$= \frac{1}{2}\left[\left(\frac{1}{1} - \frac{1}{3}\right) + \left(\frac{1}{3} - \frac{1}{5}\right) + \left(\frac{1}{5} - \frac{1}{7}\right) + \cdots + \left(\frac{1}{2n - 1} - \frac{1}{2n + 1}\right)\right]$$

$$= \frac{1}{2}\left[1 - \frac{1}{2n + 1}\right]$$

$$= \frac{n}{2n + 1}.$$

What has just happened happens not infrequently.  A theorem proved by mathematical induction can often be proved more shortly by some other method.  Even the careful examination of the proof by mathematical induction may lead to such a shortcut.

(5) *Another example.* We examine two numbers, $a$ and $b$, subject to the inequalities

$$0 < a < 1. \quad 0 < b < 1.$$

Then, obviously,

$$(1 - a)(1 - b) = 1 - a - b + ab > 1 - a - b.$$

A natural generalization leads us to suspect the following statement: *If $n \geq 2$ and $0 < a_1 < 1, 0 < a_2 < 1, \ldots 0 < a_n < 1$, then*

$$(1 - a_1)(1 - a_2) \ldots (1 - a_n) > 1 - a_1 - a_2 - \ldots - a_n.$$

We use mathematical induction to prove this.  We have seen that the inequality is true in the first case to which it is asserted to apply, for $n = 2$. Therefore, supposing it to be true for $n$, where $n \geq 2$, we have to derive it for $n + 1$.

Supposedly

$$(1 - a_1) \ldots (1 - a_n) > 1 - a_1 - \ldots - a_n$$

and we know that

$$0 < a_{n+1} < 1.$$

Consequently

$$(1 - a_1) \ldots (1 - a_n) (1 - a_{n+1}) > (1 - a_1 - \ldots - a_n) (1 - a_{n+1})$$

$$= 1 - a_1 - \ldots - a_n - a_{n+1} + (a_1 + \ldots + a_n)a_{n+1}$$

$$> 1 - a_1 - \ldots - a_n - a_{n+1}.$$

We derived for $n + 1$ what we have supposed for $n$: the proof is complete.

Fig. 7.1. From $n$ to $n + 1$.      Fig. 7.2. The case $n = 1$.

Let us note that mathematical induction may be used to prove propositions which apply, not to all positive integers absolutely, but to all positive integers from a certain integer onwards. For example, the theorem just proved is concerned only with values $n \geq 2$.

(6) *What is n?* We discuss now a theorem of plane geometry.

*If the polygon P is convex and contained in the polygon Q, the perimeter of P is shorter than the perimeter of Q.*

That the area of the inner polygon $P$ is less than the area of the outer polygon $Q$ is obvious. Yet the theorem stated is not quite so obvious; without the restriction that $P$ is convex, it would be false.

Fig. 7.1 shows the essential idea of the proof. We cut off the shaded piece from the outer polygon $Q$; there remains a new polygon $Q'$, a part of $Q$, which has two properties:

First, $Q'$ still contains the convex polygon $P$ which, being convex, lies fully on one side of the straight line $A'B'$ into which the side $AB$ of $P$ has been produced.

Second, the perimeter of $Q'$ is shorter than that of $Q$. In fact, the perimeter of $Q'$ differs from that of $Q$ in so far as the former contains the straight line-segment joining the points $A'$ and $B'$, and the latter contains a broken line instead, joining the same points (on the far side of the shaded piece). Yet the straight line is the shortest distance between the points $A'$ and $B'$.

As we passed from $Q$ to $Q'$, so we can pass from $Q'$ to another polygon $Q''$. We thus obtain a sequence of polygons $Q$, $Q'$, $Q''$, .... Each polygon is included in, and has a shorter perimeter than, the foregoing, and the last polygon in this sequence is $P$. Therefore, the perimeter of $P$ is shorter than that of $Q$.

We should recognize the nature of the foregoing proof: it is, in fact, a proof by mathematical induction. But what is $n$? With respect to which quantity is the induction performed?

This question is serious. Mathematical induction is used in various domains and sometimes in very difficult and intricate questions. Trying to find a hidden proof, we may face a crucial decision: What should be $n$? With respect to what quantity should we try mathematical induction?

In the foregoing proof it is advisable to choose as $n$ the *number of those sides of the inner convex polygon which do not belong entirely to the perimeter of the outer polygon.* Fig. 7.2 illustrates the case $n = 1$. I leave to the reader to find out what is advisable to call $n$ in fig. 7.1.

### EXAMPLES AND COMMENTS ON CHAPTER VII

**1.** Observe that

$$
\begin{aligned}
1 &= 1 \\
1 - 4 &= -(1 + 2) \\
1 - 4 + 9 &= 1 + 2 + 3 \\
1 - 4 + 9 - 16 &= -(1 + 2 + 3 + 4).
\end{aligned}
$$

Guess the general law suggested by these examples, express it in suitable mathematical notation, and prove it.

**2.** Prove the explicit formulas for $P_n$, $S_n$ and $S_n^{\infty}$ guessed in ex. 3.13, 3.14 and 3.20, respectively. [Ex. 3.11, 3.12.]

**3.** Guess an expression for

$$
1^3 + 2^3 + 3^3 + \ldots + n^3
$$

and prove it by mathematical induction. [Ex. 1.4.]

**4.** Guess an expression for

$$
\left(1 - \frac{1}{4}\right)\left(1 - \frac{1}{9}\right)\left(1 - \frac{1}{16}\right) \cdots \left(1 - \frac{1}{n^2}\right)
$$

valid for $n \geq 2$ and prove it by mathematical induction.

**5.** Guess an expression for

$$\left(1 - \frac{4}{1}\right)\left(1 - \frac{4}{9}\right)\left(1 - \frac{4}{25}\right) \cdots \left(1 - \frac{4}{(2n-1)^2}\right)$$

valid for $n \geq 1$ and prove it by mathematical induction.

**6.** Generalize the relation

$$\frac{x}{1+x} + \frac{2x^2}{1+x^2} + \frac{4x^4}{1+x^4} + \frac{8x^8}{1+x^8} = \frac{x}{1-x} - \frac{16x^{16}}{1-x^{16}}$$

and prove your generalization by mathematical induction.

**7.** We consider the operation that consists in passing from the sequence

$$a_1, a_2, a_3, \ldots, a_n, \ldots$$

to the sequence

$$s_1, s_2, s_3, \ldots, s_n, \ldots$$

with the general term

$$s_n = a_1 + a_2 + a_3 + \ldots + a_n.$$

We shall call this operation (the formation of the sequence $s_1, s_2, s_3, \ldots$ ) "summing the sequence $a_1, a_2, a_3, \ldots$ ". With this terminology, we can express a fact already observed (in ex. 1.3) as follows.

You can pass from the sequence of all positive integers 1, 2, 3, 4, . . . to the sequence of the squares 1, 4, 9, 16 . . . in two steps: (1) leave out each second term (2) sum the remaining sequence. In fact, see the table:

| 1 | 2 | 3 | 4 | 5 | 6 | 7 | 8 | 9 | 10 | 11 | 12 | 13 | . . . |
|---|---|---|---|---|---|---|---|---|----|----|----|----|-------|
| 1 |   | 3 |   | 5 |   | 7 |   | 9 |    | 11 |    | 13 | . . . |
| 1 |   | 4 |   | 9 |   | 16 |  | 25 |   | 36 |    | 49 | . . . |

Prove this assertion by mathematical induction.

**8** (continued). You can pass from the sequence of all positive integers 1, 2, 3, 4, . . . to the sequence of the cubes 1, 8, 27, 64, . . . in four steps: (1) leave out each third term (2) sum the remaining sequence (3) leave out each second term (4) sum the remaining sequence. Prove this by mathematical induction, after having examined the table:

| 1 | 2 | 3 | 4 | 5 | 6 | 7 | 8 | 9 | 10 | 11 | 12 | 13 | . . . |
|---|---|---|---|---|---|---|---|---|----|----|----|----|-------|
| 1 | 2 |   | 4 | 5 |   | 7 | 8 |   | 10 | 11 |    | 13 | . . . |
| 1 | 3 |   | 7 | 12 |  | 19 | 27 | | 37 | 48 |    | 61 | . . . |
| 1 |   |   | 7 |   |   | 19 |   |  | 37 |    |    | 61 | . . . |
| 1 |   |   | 8 |   |   | 27 |   |  | 64 |    |    | 125 | . . . |

**9** (continued).　You can pass from the sequence of all positive integers 1, 2, 3, 4, . . . to the sequence of the fourth powers 1, 16, 81, 256, . . . in six steps, visible from the table:

| 1 | 2 | 3 | 4 | 5 | 6 | 7 | 8 | 9 | 10 | 11 | 12 | 13 | . . . |
|---|---|---|---|---|---|---|---|---|----|----|----|----|-------|
| 1 | 2 | 3 |   | 5 | 6 | 7 |   | 9 | 10 | 11 |    | 13 | . . . |
| 1 | 3 | 6 |   | 11 | 17 | 24 |  | 33 | 43 | 54 |   | 67 | . . . |
| 1 | 3 |   |   | 11 | 17 |    |   | 33 | 43 |    |    | 67 | . . . |
| 1 | 4 |   |   | 15 | 32 |    |   | 65 | 108 |   |    | 175 | . . . |
| 1 |   |   |   | 15 |    |    |   | 65 |    |    |    | 175 | . . . |
| 1 |   |   |   | 16 |    |    |   | 81 |    |    |    | 256 | . . . |

What do these facts suggest?

**10.** Observing that

$$
\begin{aligned}
1 &= 1 \\
1 - 5 &= -4 \\
1 - 5 + 10 &= 6 \\
1 - 5 + 10 - 10 &= -4 \\
1 - 5 + 10 - 10 + 5 &= 1 \\
1 - 5 + 10 - 10 + 5 - 1 &= 0
\end{aligned}
$$

we are led to the general statement

$$
\binom{n}{0} - \binom{n}{1} + \binom{n}{2} - \binom{n}{3} + \cdots + (-1)^k \binom{n}{k} = (-1)^k \binom{n-1}{k}
$$

for $0 < k < n$, $n = 1, 2, 3, \ldots$ .

In proving this by mathematical induction, would you prefer to proceed from $n$ to $n + 1$ or rather from $k$ to $k + 1$?

**11.** In a tennis tournament there are $2n$ participants.　In the first round of the tournament each participant plays just once, so there are $n$ games, each occupying a pair of players.　Show that the pairing for the first round can be arranged in exactly

$$1 \cdot 3 \cdot 5 \cdot 7 \cdots (2n - 1)$$

different ways.

**12.** *To prove more may be less trouble.*　Let

$$\frac{1}{1 - x} = f_0(x)$$

and define the sequence $f_0(x), f_1(x), f_2(x), \ldots$ by the condition that

$$f_{n+1}(x) = x \frac{df_n(x)}{dx}$$

for $n = 0, 1, 2, 3, \ldots$ . (Such a definition is called *recursive*: to find $f_{n+1}$ we have to go back to $f_n$.) Observing that

$$f_1(x) = \frac{x}{(1-x)^2}, \quad f_2(x) = \frac{x+x^2}{(1-x)^3}, \quad f_3(x) = \frac{x+4x^2+x^3}{(1-x)^4},$$

prove by mathematical induction that, for $n \geq 1$, the numerator of $f_n(x)$ is a polynomial, the constant term of which is 0 and the other coefficients positive integers.

**13** (continued). Find by induction and prove by mathematical induction further properties of $f_n(x)$.

**14.** *Balance your theorem.* The typical proposition $A$ accessible to proof by mathematical induction has an infinity of cases $A_1, A_2, A_3, \ldots A_n, \ldots$ . The case $A_1$ is often easy; at any rate, $A_1$ has to be handled by specific means. Once $A_1$ is established, we have to prove $A_{n+1}$ assuming $A_n$. A proposition $A'$ stronger than $A$ may be easier to prove than $A$.[4] In fact, let $A'$ consist of the cases $A'_1 \ A'_2, \ldots, A'_n, \ldots$ . In passing from $A$ to $A'$ we make the burden of the proof heavier: we have to prove the stronger $A'_{n+1}$ instead of $A_{n+1}$. Yet we make also the support of the proof stronger: we may use the more informative $A'_n$ instead of $A_n$.

The solution of ex. 12 provides an illustration. Yet we would have made this solution uselessly cumbersome by including the materials treated in ex. 13 which are more conveniently handled by additional remarks, as a corollary.

In general, in trying to devise a proof by mathematical induction, you may fail for two opposite reasons. You may fail because you try to prove too much: your $A_{n+1}$ is too heavy a burden. Yet you may also fail because you try to prove too little: your $A_n$ is too weak a support. You have to balance the statement of your theorem so that the support is just enough for the burden. And so the machinery of the proof edges you towards a more balanced, better adapted view of the facts. This may be typical of the rôle of proofs in building up science.

**15.** *Outlook.* More intricate problems in more difficult domains demand a more sophisticated technique of mathematical induction and lead to various modifications of this important method of proof. The theory of groups provides some of the most remarkable examples. An interesting variant is the "backward mathematical induction" or "inference from $n$ to

---

[4] This is the "inventor's paradox"; see *How to Solve It*, p. 110.

$n-1$"; for an interesting elementary example cf. G. H. Hardy, J. E. Littlewood, and G. Pólya, *Inequalities*, p. 17 and p. 20.

**16.** Being given that $Q_1 = 1$ and

$$Q_{n-1}Q_n = \frac{1}{2^{n(n-1)}} \frac{n!(n+1)!(n+2)! \ldots (2n-1)!}{0! \quad 1! \quad 2! \quad \ldots \quad (n-1)!}$$

for $n = 2, 3, \ldots$, find, and prove, a general expression for $Q_n$.

**17.** *Are any n numbers equal?* You would say, No. Yet we can try to prove the contrary by mathematical induction. It may be more attractive however, to prove the assertion: "Any $n$ girls have eyes of the same color."

For $n = 1$ the statement is obviously (or "vacuously") true. It remains to pass from $n$ to $n + 1$. For the sake of concreteness, I shall pass from 3 to 4 and leave the general case to you.

Let me introduce you to any four girls, Ann, Berthe, Carol, and Dorothy, or $A$, $B$, $C$, and $D$, for short. Allegedly ($n = 3$) the eyes of $A$, $B$, and $C$ are of the same color. Also the eyes of $B$, $C$, and $D$ are of the same color, allegedly ($n = 3$). Consequently, the eyes of all four girls, $A$, $B$, $C$, and $D$, must be of the same color; for the sake of full clarity, you may look at the diagram:

$$A, \quad B, \quad C, \quad D.$$

This proves the point for $n + 1 = 4$, and the passage from 4 to 5, for example, is, obviously, not more difficult.

Explain the paradox. You may try the experimental approach by looking into the eyes of several girls.

**18.** If parallel lines are regarded as intersecting (at infinity) the statement "Any $n$ lines in a plane have a common point" is true for $n = 1$ (vacuously) and for $n = 2$ (thanks to our interpretation). Construct a (paradoxical) proof by mathematical induction.

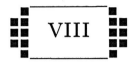

# VIII

## MAXIMA AND MINIMA

---

*Since the fabric of the world is the most perfect and was established by the wisest Creator, nothing happens in this world in which some reason of maximum or minimum would not come to light.*—EULER

---

**1. Patterns.** Problems concerned with greatest and least values, or maximum and minimum problems, are more attractive, perhaps, than other mathematical problems of comparable difficulty, and this may be due to a quite primitive reason. Everyone of us has his personal problems. We may observe that these problems are very often maximum or minimum problems of a sort. We wish to obtain a certain object at the lowest possible price, or the greatest possible effect with a certain effort, or the maximum work done within a given time and, of course, we wish to run the minimum risk. Mathematical problems on maxima and minima appeal to us, I think, because they idealize our everyday problems.

We are even inclined to imagine that Nature acts as we would like to act, obtaining the greatest effect with the least effort. The physicists succeeded in giving clear and useful forms to ideas of this sort; they describe certain physical phenomena in terms of "minimum principles." The first dynamical principle of this kind (the "Principle of Least Action" which usually goes under the name of Maupertuis) was essentially developed by Euler; his words, quoted at the beginning of this chapter, describe vividly a certain aspect of the problems on minima and maxima which may have appealed to many scientists in his century.

In the next chapter we shall discuss a few problems on minima and maxima arising in elementary physics. The present chapter prepares us for the next. The Differential Calculus provides a general method for solving problems on minima and maxima. We shall not use this method here. It will be more instructive to develop a few "patterns" of our own instead.

Having solved a problem with real insight and interest, you acquire a precious possession: a pattern, a model, that you can imitate in solving similar problems. You develop this pattern if you try to follow it, if you score a success in following it, if you reflect upon the reasons of your success, upon the analogy of the problems solved, upon the relevant circumstances that make a problem accessible to this kind of solution, etc. Developing such a pattern, you may finally attain a real discovery. At any rate, you have a chance to acquire some well ordered and readily available knowledge.

**2. Example.** *Given two points and a straight line, all in the same plane, both points on the same side of the line. On the given straight line, find a point from which the segment joining the two given points is seen under the greatest possible angle.*

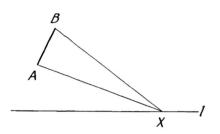

Fig. 8.1.  Looking for the best view.     Fig. 8.2.  The variation of the angle may look like this.

This is the problem that we wish to solve. We draw a figure (fig. 8.1) and introduce suitable notation. Let

*A* and *B* denote the two given points,

*l* the given straight line,

*X* a variable point of the line *l*.

We consider $\angle AXB$, the angle subtended by the given segment *AB* at the variable point *X*. We are required to find that position of *X* on the given line *l* for which this angle attains its maximum.

Imagine that *l* is a straight road. If from some point of the road *l* you wish to fire a shot on a target stretching from *A* to *B*, you should choose the point that we are seeking; it gives you the best chance to hit. If you have the more peaceful intention to take, from the road *l*, a snapshot of a façade the corners of which are at *A* and at *B*, you should again choose the point that we are seeking; it gives you the most extensive view.

The solution of our problem is not quite immediate. But, even if we do not know yet *where* the maximum is attained, we do not doubt that it *is* attained somewhere. Why is this so plausible?

We can account for the plausibility if we visualize the variation of the angle the maximum of which we are trying to find. Let us imagine that we walk along the line *l* and look at the segment *AB*. · Let us start from the point at which the line *l* and the line through *A* and *B* intersect and proceed to the right. At the start, the angle under which *AB* appears vanishes; then the angle increases; yet finally, when we are very far from *AB*, it must decrease again since it vanishes at infinite distance.[1] Between the two extreme cases in which the angle vanishes it must become a maximum somewhere. But where?

This question is not easy to answer, although we could point out long stretches of the line $l$ where the maximum is probably not attained. Let us choose any point on the line and call it $X$. This point, chosen at random, is very likely not in the maximum position that we are trying to find. How could we decide quite clearly whether it is in the maximum position or not?

There is a fairly easy remark.[2] If a point is *not* in the maximum position, there must be another point, on the other side of the maximum position, at which the angle in question has the same value. Is there any other point $X'$ on the line $l$ seen from which the segment $AB$ appears under the same angle as it does from $X$? Here is, at last, a question that we can readily answer: both $X$ and $X'$ (if there is an $X'$) must be on the same circle passing through the points $AB$ by virtue of a familiar property of the angles inscribed in a circle (Euclid III, 21).

And now, the idea may appear. Let us draw several circles passing through the given points $A$ and $B$. If such a circle intersects the line in two points, as $X$ and $X'$ in fig. 8.3, the segment $AB$ is seen from both points $X$ and $X'$ under the same angle, but this angle is not the greatest possible: a circle that intersects $l$ between $X$ and $X'$ yields a greater angle. Intersecting circles cannot do the trick: the vertex of the maximum angle is the point at which a circle through $A$ and $B$ *touches* the line $l$ (the point $M$ in fig. 8.3).

**3. The pattern of the tangent level line.** Let us look back at the solution that we have just found. What can we learn from it? What is essential in it? Which features are capable of an appropriate generalization?

The step which appears the most essential after some reflection is not too conspicuous. I think that the decisive step was to broaden our viewpoint. to step out of the line $l$, to consider the values of the quantity to be maximized (the angle subtended by $AB$) at points of the plane outside $l$. We considered the variation of this angle when its vertex moved in the plane, we considered the dependence of this angle on the position of its vertex. In short, we conceived this angle as the *function of a variable point* (its vertex), and regarded this point (the vertex) as varying in the *plane*.

The angle remains unchanged when its vertex moves along an arc of circle joining the points $A$ and $B$. Let us call such an arc of circle a *level line*. This expression underscores the general viewpoint that we are about to attain. The lines along which a function of a variable point remains constant are usually called the level lines of that function.

---

[1] If we consider $\angle AXB$ as function of the distance measured along the line $l$, we can graph it (represent it in rectangular coordinates) in the usual manner. Fig. 8.2 gives a qualitative sketch of the graph; $\angle AXB$ is represented by the ordinate $XY$.

[2] Very easy if we look at fig. 8.2.

Yet, let us not forget the unknown of our problem. We were required to find the maximum of the angle (of this function of a variable point) when its vertex (the variable point) cannot move freely in the plane, but is restricted to a *prescribed path*, the line *l*. At which point of the prescribed path is the maximum attained?

We know the answer already, but let us understand it better, let us look at it from a more general viewpoint. Let us consider an analogous, fairly general and very intuitive, example.

 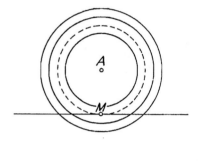

Fig. 8.3.  A tangent level line.                    Fig. 8.4.  Another tangent level line.

You know what "level lines" or "contour lines" are on a map or in the terrain (let us think of a hilly country) that the map represents. They are the lines of constant elevation; a level line connects those points on the map which represent points upon the earth's surface of equal height above sea level. If you imagine the sea rising 100 feet, a new coastline with bays intruding into the valleys appears at the new sea level. This new coastline is the level line of elevation 100. The map-maker plots only a few level lines, at constant intervals, for example, at elevations 100, 200, 300, . . . ; yet we can think a level line, there *is* a level line, at each elevation, through each point of the terrain. The function of the variable point that is important for the map-maker, or for you when you move about in the terrain, is the elevation above sea level; this function remains constant along each level line.

Now, here is a problem analogous to our problem just discussed (in sect. 2). You walk along a road, a prescribed path. At which point of the road do you attain the maximum elevation?

It is very easy to say where you do not attain it. A point which you pass going up or down is certainly not the point of maximum elevation, nor the point of minimum elevation. At such a point, your road crosses a level line: *the maximum* (or the minimum) *can* NOT *be attained at a point where the prescribed path crosses a level line.*

With this essential remark, let us return to our example (sect. 2, fig. 8.1, 8.2, 8.3). Let us consider the whole prescribed path: the line $l$ from its intersection with the line through $A$ and $B$ to infinite distance (to the right). At each of its points, this prescribed path intersects a level line (an arc of circle with endpoints $A$ and $B$) except in just one point, where it is tangent to a level line (to such a circle). If the maximum is anywhere, it must be at that point: *at the point of maximum the prescribed path is tangent to a level line.*

This hints very strongly the general idea behind our example. Yet let us examine this hint. Let us apply it to a simple analogous case and see how it works. Here is an easy example.

*On a given straight line find the point which is at minimum distance from a given point.*

Let us introduce suitable notation:

$A$ is the given point,

$a$ is the given straight line.

It is understood that the given point $A$ is not on the given line $a$. We have to find the shortest distance from $A$ to $a$.

Everybody knows the solution. Imagine that you are swimming in a calm sea; in this moment you are at the point $A$; the line $a$ marks a straight uniform beach. Suddenly you are scared, you wish to reach firm ground as quickly as possible. Where is the nearest point of the beach? You know it without reflection. A dog would know it. A dog or a cow, thrown into the water, would start swimming without delay along the perpendicular from $A$ to $a$.

Yet our purpose is not just to find the solution, but to examine a general idea in finding it. The quantity that we wish to minimize is the distance of a variable point from the given point $A$. This distance depends on the position of the variable point. The level lines of the distance are obviously concentric circles with $A$ as their common center. The "prescribed path" is the given straight line $a$. The minimum is not attained at a point where the prescribed path crosses a level line. In fact, it is attained at the (only) point at which the prescribed path is tangent to a level line (at the point $M$ of fig. 8.4). The shortest distance from the point $A$ to the line $a$ is the radius of the circle with center $A$ that is tangent to $a$—as we knew from the start. Still, we learned something. The general idea appears now more clearly and it may be left to the reader to clarify it completely.

With the essential common features of the foregoing problems clearly in mind, we are naturally looking out for analogous problems to which we could apply the same pattern of solution. In the foregoing, we considered a point variable in a plane and sought the minimum or maximum of a function of such a point along a prescribed path. Yet we could consider a point variable in space and seek the minimum or maximum of a function of such a point along a prescribed path, or on a prescribed surface. In the plane, the tangent level lines played a special rôle. Analogy prompts us to expect that the tangent level surfaces will play a similar rôle in space.

**4. Examples.** We discuss two examples which can be treated by the same method, but have very little in common otherwise.

(1) *Find the minimum distance between two given skew straight lines.*

Let us call

*a* and *b* the two given skew lines,

*X* a variable point on *a*,

*Y* a variable point on *b*;

see fig. 8.5. We are required to determine that position of the line-segment *XY* in which it is the shortest.

The distance *XY* depends on the position of its two end-points, *X* and *Y*, which are both variable. There are two variable points, not just one, and this may be the characteristic difficulty of the problem. If one of the two points were given, fixed, invariable, and only the other variable, the problem would be easy. In fact, it would not even be new; it would be identical with a problem just solved (sect. 3).

Let us *fix for a moment* one of the originally variable points, say, *Y*. Then the segment *XY* is in the plane through the fixed point *Y* and the given line *a*, and only one of its end points, *X*, is variable, running along the line *a*. Obviously, *XY* becomes a minimum when it becomes perpendicular to *a* (by sect. 3, fig. 8.4).

Yet we could interchange the rôles of the two points, *X* and *Y*. Let us now, for a change, fix *X* and make *Y* alone variable. Obviously, the segment *XY* becomes shortest when perpendicular to *b*.

The minimum position of *XY*, however, is independent of our whims and of our choice of rôles for *X* and *Y*, and so we are led to suspect that it is perpendicular both to *a* and to *b*. Yet let us look more closely at the situation.

In fact, the foregoing argument shows directly where the minimum position can *not* be (and only indirectly where it should be). I say that a position in which the segment *XY* is not perpendicular to the line *a* at the point *X*, is *not* the minimum position. In fact, I fix the point *Y* and move *X* to another position where *XY* becomes perpendicular to *a* and so doing I make *XY* shorter (by sect. 3). This reasoning obviously applies just as well to *Y* as to *X*, and so we see: *the length of the segment XY cannot be a minimum unless this segment is perpendicular both to a and to b.* If there is a shortest distance, it must be along the common perpendicular to the two given lines.

We need not take anything for granted. In fact, we can see at a glance that the common perpendicular is actually the shortest distance. Let us assume that, in fig. 8.5, the plane of the drawing is parallel to both given lines *a* and *b* (*a* above, *b* below). We may consider any point or line in space as represented in fig. 8.5 by its orthogonal projection. The true length of the segment *XY* is the hypotenuse of a right triangle of which another side is the orthogonal projection of *XY* seen in fig. 8.5; the third side is the shortest distance of two parallel planes, one through $a$, the other through $b$, both parallel to the plane of the drawing to which that third side is perpendicular. Therefore, the shorter the projection of $XY$ shown in fig. 8.5, the shorter is $XY$ itself. The projection of $XY$ reduces to a point, its length to nil, and so the length of $XY$ is a minimum if, and only if, $XY$ is perpendicular to the plane of the drawing and so to both lines $a$ and $b$.

And so we have verified directly what we discovered before by another method.

(2) *Find the maximum of the area of a polygon inscribed in a given circle and having a given number of sides.*

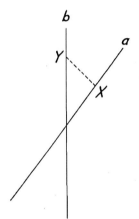

Fig. 8.5. Two skew lines.

The circle is given. On this circle, we have to choose the $n$ vertices $U, \ldots, W, X, Y$, and $Z$ of a polygon so that the area becomes a maximum. Just as in the foregoing problem, under (1), the main difficulty seems to be that there are many variables (the vertices $U, \ldots, W, X, Y$, and $Z$). We should, perhaps, try the method that worked in the foregoing problem. What is the essential point of this method?

Let us take the problem as *almost solved*. Let·us imagine that we have obtained already the desired position of all vertices except one, say, $X$. The $n-1$ other points, $U, \ldots, W, Y$, and $Z$, are already fixed, each in the position where it should be, but we have yet to choose $X$ so that the area becomes a maximum. The whole area consists, however, of two parts: the polygon $U \ldots WYZ$ with $n-1$ fixed vertices which is independent of $X$ and the triangle $WXY$ which depends on $X$. We focus our attention upon this triangle which must become a maximum when the whole area becomes a maximum; see fig. 8.6. The base $WY$ of $\triangle WXY$ is fixed. If the vertex $X$ moves along a parallel to the base $WY$, the area remains constant: such parallels to $WY$ are the level lines. We pick out the tangent level line: the tangent to the circle parallel to $WY$. Its point of contact is obviously the position of $X$ that renders the area of $\triangle WXY$ a maximum. With $X$ in this position the triangle is isosceles, $WX = XY$. These two adjacent sides must be equal, if the area of the polygon is a maximum. Yet the same reasoning applies to any pair of adjacent sides: all sides must be equal when the maximum of the area is attained, and so the inscribed polygon with maximum area must be *regular*.

**5. The pattern of partial variation.** Comparing the two examples discussed in the foregoing section (sect. 4), we easily recognize some common features and a common pattern of solution. In both problems we seek the extremum (minimum or maximum) of a quantity depending on *several* variable elements. In both solutions we fix for a moment all originally

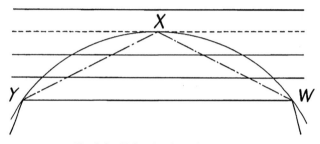

Fig. 8.6.   Triangle of maximum area.

variable elements except one and study the effect of the variation of this single element. The simultaneous variation of all variable elements, or *total* variation, is not so easy to survey. We treated our example with good results in studying the *partial* variation when only a single element varies and the others are fixed. The principle underlying our procedure seems to be: *a function f of several variables cannot attain a maximum with respect to all its variables jointly, unless it attains a maximum with respect to each single variable.*

This statement is fairly general, although unnecessarily restricted in one respect: it clings too closely to the foregoing examples in which we varied just one element at a time. Yet we can imagine that in other examples it may be advantageous to vary just two elements at a time and fix the others, or perhaps just three, and so on. In such cases we could still speak appropriately of "partial variation." The general idea appears now fairly clearly and after one more example the reader may undertake by himself to clarify it completely.

*A line of length l is divided into n parts. Find the maximum of the product of these n parts.*

Let $x_1, x_2, \ldots x_n$ denote the lengths of the $n$ parts; $x_1, x_2, \ldots x_n$ are positive numbers with a given sum

$$x_1 + x_2 + \ldots + x_n = l.$$

We are required to make $x_1 x_2 \ldots x_n$ a maximum.

We examine first the simplest special case: Being given the sum $x_1 + x_2$ of two positive quantities, find the maximum of their product $x_1 x_2$. We can interpret $x_1$ and $x_2$ as adjacent sides of a rectangle and restate the problem in the following more appealing form: Being given $L$, the length of the perimeter of a rectangle, find the maximum of its area. In fact, the sum of the two sides just mentioned is given:

$$x_1 + x_2 = \frac{L}{2}.$$

There is an obvious guess: the area becomes a maximum when the rectangle becomes a square. This guess cannot be hard to verify. Each side of the square with perimeter $L$ is equal to

$$\frac{L}{4} = \frac{x_1 + x_2}{2}.$$

We have to verify that the area of the square is greater than the area of the rectangle, or, which is the same, that their difference

$$\left(\frac{x_1 + x_2}{2}\right)^2 - x_1 x_2$$

is positive. Is that so? Very little algebra is enough to see that

$$\left(\frac{x_1 + x_2}{2}\right)^2 - x_1 x_2 = \left(\frac{x_1 - x_2}{2}\right)^2.$$

This formula shows the whole situation at a glance. The right hand side is positive, unless $x_1 = x_2$ and the rectangle is a square.

In short, the area of a rectangle with given perimeter becomes a maximum when the rectangle is a square; the product of two positive quantities with given sum becomes a maximum when these two quantities are equal.

Let us try to use the special case just solved as a stepping stone to the solution of the general problem. Let us take the problem as almost solved. Let us imagine that we have obtained already the desired values of all parts, except those of the first two, $x_1$ and $x_2$. Thus, we regard $x_1$ and $x_2$ as variables, but $x_3$, $x_4$, ... $x_n$ as constants. The sum of the two variable parts is constant,

$$x_1 + x_2 = l - x_3 - x_4 - \ldots - x_n.$$

Now, the product of all parts

$$x_1 x_2 (x_3 x_4 \ldots x_n)$$

cannot become a maximum unless the product $x_1 x_2$ of the first two parts becomes a maximum, too. This requires, however, that $x_1 = x_2$. Yet there is no reason why any other pair of parts should behave differently.

The desired maximum of the product cannot be attained unless all the quantities with given sum are equal. We quote Colin Maclaurin (1698–1746) to whom the foregoing reasoning is due: "If the Line *AB* is divided into any Number of Parts *AC*, *CD*, *DE*, *EB*, the Product of all those Parts multiplied into one another will be a *Maximum* when the Parts are equal amongst themselves."

The reader can learn a great deal in clarifying the foregoing proof. Is it quite satisfactory?

**6. The theorem of the arithmetic and geometric means and its first consequences.** Let us reconsider the result of the foregoing section: If

$$x_1 + x_2 + x_3 + \ldots + x_n = l$$

then

$$x_1 x_2 x_3 \ldots x_n < \left(\frac{l}{n}\right)^n$$

unless $x_1 = x_2 = x_3 \ldots = x_n = l/n$. Eliminating $l$, we can restate this result in the form:

$$x_1 x_2 \ldots x_n < \left(\frac{x_1 + x_2 + \ldots + x_n}{n}\right)^n$$

or

$$\sqrt[n]{x_1 x_2 \ldots x_n} < \frac{x_1 + x_2 + \ldots + x_n}{n}$$

*unless all the positive quantities $x_1$, $x_2$, ... $x_n$ are equal*; if these quantities are equal, the inequality becomes an equation. The left-hand side of the above inequality is called the geometric, the right-hand side the arithmetic, mean of $x_1$, $x_2$, ... $x_n$. The theorem just stated is sometimes called the "theorem of the arithmetic and geometric means" or, shortly, the "theorem of the means."

The theorem of the means is interesting and important in many respects. It is worth mentioning that it can be stated in two different ways:

*The product of n positive quantities with a given sum becomes a maximum when these quantities are all equal.*

*The sum of n positive quantities with a given product becomes a minimum when these quantities are all equal.*

The first statement is concerned with a maximum, the second with the corresponding minimum. The derivation of the foregoing section is aimed at the first statement. Changing this derivation systematically we could arrive at the second statement. It is simpler, however, to observe that the inequality between the means yields impartially both statements: to obtain one or the other, we have to regard one or the other side of the inequality as given. We may call these two (essentially equivalent) statements *conjugate statements*.

The theorem of the means yields the solution of many geometric problems on minima and maxima. We discuss here just one example (several others can be found at the end of this chapter).

*Being given the area of the surface of a box, find the maximum of its volume.*

I use the word "box" instead of "rectangular parallelepiped" because "box" is expressive enough and so much shorter than the official term.

The solution is easily foreseen and, once foreseen, it is easily reduced to the theorem of the means as follows. Let

$a$, $b$, $c$ denote the lengths of the three edges of the box drawn from the same vertex,

$S$ the area of the surface,

$V$ the volume.

Obviously

$$S = 2(ab + ac + bc), \qquad V = abc.$$

Observing that the sum of the three quantities $ab$, $ac$, and $bc$ is $S/2$ and their product $V^2$, we naturally think of the theorem of the means which yields

$$V^2 = (abc)^2 < \left(\frac{ab + ac + bc}{3}\right)^3 = \left(\frac{S}{6}\right)^3$$

unless

$$ab = ac = bc$$

or, which is the same,

$$a = b = c.$$

That is

$$V < (S/6)^{3/2}$$

unless the box is a cube, when equality occurs. We can express the result in two different (although essentially equivalent) forms.

*Of all boxes with a given surface area the cube has the maximum volume.*

*Of all boxes with a given volume the cube has the minimum surface area.*

As above, we may call these two statements conjugate statements. As above, one of the two conjugate statements is concerned with a maximum, the other with a minimum.

The preceding application of the theorem of the means has its merits. We may regard it as a pattern and collect cases to which the theorem of the means can be similarly applied.

### EXAMPLES AND COMMENTS ON CHAPTER VIII

### First Part

**1.** *Minimum and maximum distances in plane geometry.* Find the minimum distance between (1) two points, (2) a point and a straight line, (3) two parallel straight lines.

Find the minimum, and the maximum, distance between (4) a point and a circle, (5) a straight line and a circle, (6) two circles.

The solution is obvious in all cases. Recall the elementary proof at least in some cases.

**2.** *Minimum and maximum distances in solid geometry.* Find the minimum distance between (1) two points, (2) a point and a plane, (3) two parallel planes, (4) a point and a straight line, (5) a plane and a parallel straight line, (6) two skew straight lines.

Find the minimum, and the maximum, distance between (7) a point and a sphere, (8) a plane and a sphere, (9) a straight line and a sphere, (10) two spheres.

**3.** *Level lines in a plane.* Consider the distance of a variable point from a given (1) point, (2) straight line, (3) circle. What are the level lines?

**4.** *Level surfaces in space.* Consider the distance of a variable point from a given (1) point, (2) plane, (3) straight line, (4) sphere. What are the level surfaces?

**5.** Answer the questions of ex. 1 using level lines.

**6.** Answer the questions of ex. 2 using level surfaces.

**7.** Given two sides of a triangle, find the maximum of the area using level lines.

**8.** Given one side and the length of the perimeter of a triangle, find the maximum of the area using level lines.

**9.** Given the area of a rectangle, find the minimum of the perimeter using level lines. (In a rectangular coordinate system, let $(0,0)$, $(x,0)$, $(0,y)$, $(x,y)$ be the four vertices of the rectangle and use analytic geometry.)

**10.** Examine the following statement: "The shortest distance from a given point to a given curve is perpendicular to the given curve."

**11.** *The principle of the crossing level line.* We consider a function $f$ of a point which varies in a plane, the maximum and the minimum of $f$ along a prescribed path, and a level line of $f$ which separates two regions of the plane; in one of these regions $f$ takes higher, in the other lower, values than on the level line itself.

If the prescribed path crosses the level line, neither the maximum nor the minimum of $f$ is attained at the point of crossing.

**12.** The contour-map of fig. 8.7 shows a peak $P$ and a pass (or saddle-point with horizontal tangent plane) $S$. Hiking through such a country, do you necessarily attain the highest point of your path at a point where the path is tangent to a level line?.

**13.** Let $A$ and $B$ denote two given points and $X$ a variable point in a plane. The angle at $X$ subtended by the segment $AB$ ($\angle AXB$) which can take any value between 0° and 180° (limits inclusively) is a function of the variable point $X$.

(1) Give a full description of the level lines.

(2) Of two different level lines, which one corresponds to a higher value of the angle?

You may use figs. 8.1 and 8.3, but you should realize that now you may look at the segment $AB$ from *both* sides.

**14.** Consider figs. 8.1, 8.2, 8.3, take $\angle AXB$ as in ex. 13, and find its minimum along $l$. Does the result conform to the principle of ex. 11?

Fig. 8.7. Level lines on a contour map.

**15.** Given the volume of a box (rectangular parallelepiped), find the minimum of the area of its surface, using partial variation.

**16.** Of all triangles with a given perimeter, which one has the maximum area? [Ex. 8.]

**17.** Of all tetrahedra inscribed in a given sphere, which one has the maximum volume? [Do you know a related problem?]

**18.** Given $a$, $b$, and $c$, the lengths of three edges of a tetrahedron drawn from the same vertex, find the maximum of the volume of the tetrahedron. [Do you know an analogous problem?]

**19.** Find the shortest distance between a sphere and a cylinder. (Cylinder means: infinite cylinder of revolution.)

**20.** Find the shortest distance between two cylinders with skew axes.

**21.** Examine the following statement: "The shortest distance between two given surfaces is perpendicular to both."

**22.** *The principle of partial variation.* The function $f(X,Y,Z,\ldots)$ of several variables $X,Y,Z,\ldots$ attains its maximum for $X = A$, $Y = B$, $Z = C$, $\ldots$. Then the function $f(X,B,C,\ldots)$ of the single variable $X$ attains its maximum for $X = A$; and the function $f(X,Y,C,\ldots)$ of the two variables $X$ and $Y$ attains its maximum for $X = A$, $Y = B$; and so on.

*A function of several variables cannot attain a maximum with respect to all its variables jointly, unless it attains a maximum with respect to any subset of variables.*

**23.** *Existence of the extremum.* Both the principle of the level line and that of partial variation give usually only "negative information." They show directly in which points a proposed function $f$ can *not* attain a maximum, and we have to infer hence where $f$ *may* attain one. That $f$ *must* attain a maximum somewhere, cannot be derived from these principles alone. Yet the existence of the maximum can sometimes be derived by some modification of the reasoning. Moreover, the existence of the maximum can often be derived from general theorems on continuous functions of several variables.[3] At any rate, whenever the existence of the maximum appears obvious from the intuitive standpoint, we have a good reason to hope that some special device or some general theorem will apply and prove the existence.

**24.** *A modification of the pattern of partial variation: An infinite process.* Find the maximum of $xyz$, being given that $x + y + z = l$.

It is understood that $x$, $y$, and $z$ are positive, and that $l$ is given. The present problem is a particular case of the problem of sect. 5. Following the method used there, we keep one of the three numbers $x$, $y$, and $z$ fixed and change the other two so that they become equal, which increases their product. Let us start from any given system $(x, y, z)$; performing the change indicated we pass to another system $(x_1, y_1, z_1)$; then we pass to still another $(x_2, y_2, z_2)$ and hence to $(x_3, y_3, z_3)$, and so on. Let us leave the three terms unchanged in turn: first the $x$-term, then $y$, then $z$, then again $x$, then $y$, then $z$, then again $x$, and so on. Thus, we set

$$x_1 = x, \quad y_1 = z_1 = \frac{y + z}{2},$$

$$y_2 = y_1, \quad z_2 = x_2 = \frac{z_1 + x_1}{2},$$

$$z_3 = z_2, \quad x_3 = y_3 = \frac{x_2 + y_2}{2},$$

$$x_4 = x_3, \quad y_4 = z_4 = \frac{y_3 + z_3}{2},$$

. . . . . . . . . . . . . .

---

[3] A function of several variables, continuous in a closed set, attains there its lower and upper bounds. This generalizes G. H. Hardy, *Pure Mathematics*, p. 194, theorem 2.

Each step leaves the sum unchanged, but increases the product:

$$x + y + z = x_1 + y_1 + z_1 = x_2 + y_2 + z_2 = \cdots$$

$$xyz < x_1 y_1 z_1 < x_2 y_2 z_2 < \cdots .$$

We assumed that $y \neq z$ and that $x_1 \neq z_1$. (This is the non-exceptional case; in the exceptional case we attain our end more easily.) We naturally expect that the three numbers $x_n, y_n$, and $z_n$ become less and less different from each other as $n$ increases. *If* we can prove that finally

$$\lim_{n \to \infty} x_n = \lim_{n \to \infty} y_n = \lim_{n \to \infty} z_n$$

we can immediately infer that

$$xyz < \lim_{n \to \infty} x_n y_n z_n = (l/3)^3.$$

We obtain so this result at considerable expense, yet *without assuming the existence of the maximum* from the outset.

Prove that $\lim\limits_{n \to \infty} x_n = \lim\limits_{n \to \infty} y_n = \lim\limits_{n \to \infty} z_n$.

**25.** *Another modification of the pattern of partial variation: A finite process.* We are still concerned with the problem of ex. 24, but we use now a more sophisticated modification of the method of sect. 5.

Let $l = 3A$; thus, $A$ is the arithmetic mean of $x, y$, and $z$, and we have

$$(x - A) + (y - A) + (z - A) = 0.$$

It may happen that $x = y = z$. If this is not the case, one of the differences on the left hand side of our equation must be negative and another positive. Let us choose the notation so that

$$y < A < z.$$

We pass now from the system $(x, y, z)$ to $(x', y', z')$, putting

$$x' = x, \quad y' = A, \quad z' = y + (z - A);$$

we left the first quantity unchanged. Then

$$x + y + z = x' + y' + z'$$

and

$$y'z' - yz = A(y + z - A) - yz$$

$$= (A - y)(z - A) > 0$$

so that

$$xyz < x'y'z'.$$

It may happen that $x' = y' = z'$. If this is not the case, we pass from $(x', y', z',)$ to $(x'', y'', z'')$, putting

$$y'' = y', \qquad z'' = x'' = \frac{z' + x'}{2}$$

which gives

$$x'' = y'' = z'' = A$$

and again increases the product (as we know from sect. 5) so that

$$xyz < x'y'z' < x''y''z'' = A^3$$

$$= \left( \frac{x + y + z}{3} \right)^3.$$

Fig. 8.8. Triangular coordinates.

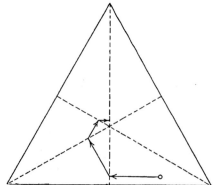

Fig. 8.9. Successive steps approaching the center.

We proved the desired result, *without assuming the existence of the maximum and without considering limits.*

By a suitable extension of this procedure, prove the theorem of the means (sect. 6) generally for $n$ quantities.

**26. Graphic comparison.** Let $P$ be a point in the interior of an equilateral triangle with altitude $l$, and $x$, $y$, and $z$ the distances of $P$ from the three sides of the triangle; see fig. 8.8. Then

$$x + y + z = l.$$

(Why?) The numbers $x$, $y$, and $z$ are the *triangular coordinates* of the point $P$. Any system of three positive numbers $x$, $y$, and $z$ with the sum $l$ can be interpreted as the triangular coordinates of a uniquely determined point inside the triangle.

The sequence

$$(x, y, z), \quad (x_1, y_1, z_1), \quad (x_2, y_2, z_2), \quad \ldots$$

considered in ex. 24 is represented by a sequence of points in fig. 8.9. The segments joining the consecutive points are parallel to the various sides of the triangle in succession, to the first, the second, and the third side, respectively, then again to the first side, and so on; each segment ends on an altitude of the triangle. (Why?) The procedure of ex. 25 is represented by three points and two segments. (How?)

**27.** Reconsider the argument of sect. 4(2) and modify it, taking first ex. 24, then ex. 25 as a model.

**28.** A necessary condition for a maximum or a minimum value of a function $f(x,y,z)$ at the point $(a,b,c)$ is that the partial derivatives

$$\frac{\partial f}{\partial x}, \quad \frac{\partial f}{\partial y}, \quad \frac{\partial f}{\partial z}$$

vanish for $x = a$, $y = b$, $z = c$.

The usual proof of this theorem exemplifies one of our patterns. Which one?

**29.** State the well-known necessary condition (in terms of partial derivatives) for a maximum or minimum value of the function $f(x,y)$ under the side-condition (or subsidiary condition) that $x$ and $y$ are linked by the equation $g(x,y) = 0$. Explain the connection with the pattern of the tangent level line.

**30.** Reexamine the cases mentioned in the solution of ex. 12 in the light of the condition mentioned in ex. 29. Is there any contradiction?

**31.** State the well-known necessary condition for a maximum or minimum value of a function $f(x,y,z)$ under the side-condition that $g(x,y,z) = 0$. Explain the connection with the pattern of the tangent level surface.

**32.** State the well-known necessary condition for a maximum or minimum value of a function $f(x,y,z)$ under the two simultaneous side-conditions that $g(x,y,z) = 0$ and $h(x,y,z) = 0$. Explain the connection with the pattern of the tangent level surface.

### Second Part

The terminology and notation used in the following are explained in ex. 33, which should be read first.

**33.** *Polygons and polyhedra.* *Area and perimeter.* *Volume and surface.* Dealing with polygons, we shall use the following notation most of the time:
$A$ for the area, and
$L$ for the length of the perimeter.
Dealing with polyhedra, we let
$V$ denote the volume, and
$S$ the area of the surface.

We shall discuss problems of maxima and minima concerned with $A$ and $L$, or $V$ and $S$.   Such problems were known to the ancient Greeks.[4]   We shall discuss mainly problems treated by Simon Lhuilier and Jacob Steiner.[5] Elementary algebraic inequalities, especially the theorem of the means (sect. 6), will turn out useful in solving the majority of the following problems.

Most of the time these problems deal only with the simplest polygons (triangles and quadrilaterals) and the simplest polyhedra (prisms and pyramids).   We have to learn a few less usual terms.

Two pyramids, standing on opposite sides of their common base, form jointly a *double pyramid*.   If the base has $n$ sides, the double pyramid has $2n$ faces, $n + 2$ vertices, and $3n$ edges.   The base is *not* a face of the double pyramid.

If all lateral faces are perpendicular to the base, we call the prism a *right prism*.

If the base of a pyramid is circumscribed about a circle and the altitude meets the base at the center of this circle, we call the pyramid a *right pyramid*.

If the two pyramids forming a double pyramid are right pyramids and symmetrical to each other with respect to their common base, we call the double pyramid a *right double pyramid*.

If a prism, pyramid, or double pyramid is not "right," we call it "oblique." Among the five regular solids, there is just one prism, just one pyramid, and just one double pyramid: the cube, the tetrahedron, and the octahedron, respectively.   Each of these three is a "right" solid of its kind.

We shall consider also cylinders, cones, and double cones; if there is no remark to the contrary, their bases are supposed to be circles.

**34.** *Right prism with square base.*   Of all right prisms with square base having a given volume, the cube has the minimum surface.

Prove this special case of a theorem already proved (sect. 6, ex. 15) directly, using the theorem of the means.

You may be tempted to proceed as follows.   Let $V$, $S$, $x$, and $y$ denote the volume, the surface area, the side of the base, and the altitude of the prism, respectively.   Then

$$V = x^2 y, \qquad S = 2x^2 + 4xy.$$

Applying the theorem of the means, we obtain

$$(S/2)^2 = [(2x^2 + 4xy)/2]^2 \geq 2x^2 \cdot 4xy = 8x^3 y.$$

Yet this has no useful relation to $V = x^2 y$—the theorem of the means does not seem to be applicable.

This was, however, a rash, thoughtless, unprofessional application of the theorem.   Try again.   [What is the desired conclusion?]

---

[4] Pappus, *Collectiones*, Book V.

[5] Simon Lhuilier, *Polygonométrie et Abrégé d'Isopérimétrie élémentaire*, Genève, 1789.   J. Steiner, *Gesammelte Werke*, vol. 2, p. 177–308.

**35.** *Right cylinder.* Observe that, of all the prisms considered in ex. 34, only the cube is circumscribed about a sphere and prove: Of all right cylinders having a given volume, the cylinder circumscribed about a sphere has the minimum surface. [What is the desired conclusion?]

**36.** *General right prism.* Of a right prism, given the volume and the shape (but not the size) of the base. When the area of the surface is a minimum, which fraction of it is the area of the base? [Do you know a related problem?]

**37.** *Right double pyramid with square base.* Prove: Of all right double pyramids with square base having a given volume, the regular octahedron has the minimum surface.

**38.** *Right double cone.* Observe that the inscribed sphere touches each face of the regular octahedron at its center, which divides the altitude of the face in the ratio 1 : 2 and prove: Of all right double cones having a given volume, the minimum of the surface is attained by the double cone the generatrices of which are divided in the ratio 1 : 2 by the points of contact with the inscribed sphere.

**39.** *General right double pyramid.* Of a right double pyramid, given the volume and the shape (but not the size) of the base. When the area of the surface is a minimum, which fraction of it is the area of the base?

**40.** Given the area of a triangle, find the minimum of its perimeter. [Could you predict the result? If you wish to try the theorem of the means, you may need the expression of the area in terms of the sides.]

**41.** Given the area of a quadrilateral, find the minimum of its perimeter. [Could you predict the result? Call $a$, $b$, $c$, and $d$ the sides of the quadrilateral, $\varepsilon$ the sum of two opposite angles, and express the area $A$ in terms of $a$, $b$, $c$, $d$, and $\varepsilon$. This is a generalization of the problem solved by Heron's formula.]

**42.** A right prism and an oblique prism have the same volume and the same base. Then the right prism has the smaller surface.

A right pyramid and an oblique pyramid have the same volume and the same base. Then the right pyramid has the smaller surface.

A right double pyramid and an oblique double pyramid have the same volume and the same base. Then the right double pyramid has the smaller surface.

In all three statements, the bases of the two solids compared agree both in shape and in size. (The volumes, of course, agree only in size.) Choose the statement that seems to you the most accessible of the three and prove it.

**43.** *Applying geometry to algebra.* Prove: If $u_1, u_2, \ldots u_n, v_1, v_2, \ldots v_n$ are real numbers,

$$\sqrt{u_1^2 + v_1^2} + \sqrt{u_2^2 + v_2^2} + \ldots + \sqrt{u_n^2 + v_n^2}$$
$$\geq \sqrt{(u_1 + u_2 + \ldots + u_n)^2 + (v_1 + v_2 + \ldots + v_n)^2}$$

and equality is attained if, and only if,

$$u_1 : v_1 = u_2 : v_2 = \ldots = u_n : v_n.$$

[Consider $n + 1$ points $P_0$, $P_1$, $P_2$, $\ldots P_n$ in a rectangular coordinate system and the length of the broken line $P_0 P_1 P_2 \ldots P_n$.]

**44.** Prove the inequality of ex. 43 independently of geometric considerations. [In the geometric proof of the inequality, the leading special case is $n = 2$.]

**45.** *Applying algebra to geometry.* Prove: Of all triangles with given base and area, the isosceles triangle has the minimum perimeter. [Ex. 43.]

**46.** Let $V$, $S$, $A$, and $L$ denote the volume, the area of the whole surface, the area of the base, and the length of the perimeter of the base of a pyramid $P$, respectively. Let $V_0$, $S_0$, $A_0$, and $L_0$ stand for the corresponding quantities connected with another pyramid $P_0$. Supposing that

$$V = V_0, \quad A = A_0, \quad L \geqq L_0$$

and that $P_0$ is a right pyramid, prove that

$$S \geqq S_0.$$

Equality is attained if, and only if, $L = L_0$ and $P$ is also a right pyramid. [Ex. 43.]

**47.** Let $V$, $S$, $A$, and $L$ denote the volume, the area of the surface, the area of the base, and the perimeter of the base of a double pyramid $D$, respectively. Let $V_0$, $S_0$, $A_0$, and $L_0$ stand for the corresponding quantities connected with another double pyramid $D_0$. Supposing that

$$V = V_0, \quad A = A_0, \quad L \geqq L_0 .$$

and that $D_0$ is a right double pyramid, prove that

$$S \geqq S_0.$$

Equality is attained if, and only if, $L = L_0$ and $D$ is also a right double pyramid. [Ex. 45, 46.]

**48.** Prove: Of all quadrilateral prisms with a given volume, the cube has the minimum surface. [Compare with ex. 34; which statement is stronger?]

**49.** Prove: Of all quadrilateral double pyramids with a given volume, the regular octahedron has the minimum surface. [Compare with ex. 37; which statement is stronger?]

**50.** Prove: Of all triangular pyramids with a given volume, the regular tetrahedron has the minimum surface.

**51.** *Right pyramid with square base.* Prove: Of all right pyramids with square base having a given volume, the pyramid in which the base is $\frac{1}{4}$ of the total surface has the minimum surface.

**52.** *Right cone.* Of all right cones having a given volume, the cone in which the base is $\frac{1}{4}$ of the total surface has the minimum surface.

**53.** *General right pyramid.* Of a right pyramid, given the volume and the shape (but not the size) of the base. When the area of the surface is a minimum, which fraction of it is the area of the base? [Do you know a special case?]

**54.** Looking back at our various examples dealing with prisms, pyramids, and double pyramids, observe their mutual relations and arrange them in a table so that the analogy of the *results* becomes conspicuous. Point out the gaps which you expect to fill out with further results.

**55.** *The box with the lid off.* Given $S_5$, the sum of the areas of five faces of a box. Find the maximum of the volume $V$. [Do you know a related problem? Could you use the result, or the method?]

**56.** *The trough.* Given $S_4$, the sum of the areas of four faces of a right triangular prism; the missing face is a *lateral* face. Find the maximum of the volume $V$.

**57.** *A fragment.* In a right prism with triangular base, given $S_3$, the sum of the areas of three mutually adjacent faces (that is, of two lateral faces and one base). Show that these three faces are of equal area and perpendicular to each other when the volume $V$ attains its maximum. [A fragment—of what?]

**58.** Given the area of a sector of a circle. Find the value of the angle at the center when the perimeter is a minimum.

**59.** In a triangle, given the area and an angle. Find the minimum (1) of the sum of the two sides including the given angle, (2) of the side opposite the given angle, (3) of the whole perimeter.

**60.** Given in position an angle and a point in the plane of the angle, inside the angle. A variable straight line, passing through the given point, cuts off a triangle from the angle. Find the minimum of the area of this triangle.

**61.** Given $E$, the sum of the lengths of the 12 edges of a box, find the maximum (1) of its volume $V$, (2) of its surface $S$.

**62.** *A post office problem.* Find the maximum of the volume of a box, being given that the length and girth combined do not exceed $l$ inches.

**63.** *A problem of Kepler.* Given $d$, the distance from the center of a generator of a right cylinder to the farthermost point of the cylinder. Find the maximum of the volume of the cylinder.

# IX

## PHYSICAL MATHEMATICS

*The science of physics does not only give us [mathematicians] an opportunity to solve problems, but helps us also to discover the means of solving them, and it does this in two ways: it leads us to anticipate the solution and suggests suitable lines of argument.*—HENRI POINCARÉ[1]

**1. Optical interpretation.** Mathematical problems are often inspired by nature, or rather by our interpretation of nature, the physical sciences. Also, the solution of a mathematical problem may be inspired by nature; physics provides us with clues with which, left alone, we had very little chance to provide ourselves. Our outlook would be too narrow without discussing mathematical problems suggested by physical investigation and solved with the help of physical interpretation. Here follows a first, very simple problem of this kind.

(1) *Nature suggests a problem.* The straight line is the shortest path between two given points. Light, travelling through the air from one point to another, chooses this shortest path, so at least our everyday experience seems to show. But what happens when light travels from one point to another not directly, but undergoing a reflection on an interposed mirror? Will light again choose the shortest path? What is the shortest path in these circumstances? By considerations on the propagation of light we are led to the following purely geometrical problem:

*Given two points and a straight line, all in the same plane, both points on the same side of the line. On a given straight line, find a point such that the sum of its distances from the two given points be a minimum.*

Let (see fig. 9.1)

*A* and *B* denote the two given points,

*l* the given straight line,

*X* a variable point of the line *l*.

We consider $AX + XB$, the sum of two distances or, which is the same, the

[1] *La valeur de la science*, p. 152.

length of the path leading from $A$ to $X$ and hence to $B$. We are required to find that position of $X$ on the given line $l$ for which the length of this path attains its minimum.

We have seen a very similar problem before (sect. 8.2, figs. 8.1, 8.2, 8.3). In fact, both problems have exactly the same data, and even the unknown is of the same nature: Here, as there, we seek the position of a point on a given line for which a certain extremum is attained. The two problems differ only in the nature of this extremum: Here we seek to minimize the sum of two lines; there we sought to maximize the angle included by those two lines.

Still, the two problems are so closely related that it is natural to try the same method. In solving the problem of sect. 8.2, we used level lines; let us use them again.

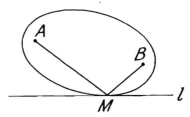

Fig. 9.1. Which path is the shortest?      Fig. 9.2. A tangent level line.

Consider a point $X$ that is not bound to the prescribed path but is free to move in the whole plane. If the quantity $AX + XB$ (which we wish to minimize) has a constant value, how can $X$ move? Along an ellipse with foci $A$ and $B$. Therefore, the level lines are "confocal" ellipses, that is, ellipses with the same foci (the given points $A$ and $B$). *The desired minimum is attained at the point of contact of the prescribed path $l$ with an ellipse the foci of which are the given points $A$ and $B$* (see fig. 9.2).

(2) *Nature suggests a solution.* We have found the solution, indeed. Yet, unless we know certain geometric properties of the ellipse, our solution is not of much use. Let us make a fresh start and seek a more informative solution.

Let us visualize the physical situation that suggested our problem. The point $A$ is a source of light, the point $B$ the eye of an observer, and $l$ marks the position of a reflecting plane surface; we may think of the horizontal surface of a quiet pool (which is perpendicular to the plane of fig. 9.1, and inter-sects it in the line $l$). The broken line $AXB$ represents, if the point $X$ is correctly chosen, the path of light. We know this path fairly well, by experience. We suspect that the length of the broken line $AXB$ is a minimum when it represents the actual path of the reflected light.

Your eye is in the position $B$ and you look down at the reflecting pool observing in it the image of $A$. The ray of light that you perceive does not come directly from the object $A$, but appears to come from a point under the surface of the pool. From which point? From the point $A^*$, the mirror image of the object $A$, symmetrical to $A$ with respect to the line $l$.

Introduce the point $A^*$, suggested by your physical experience, into the figure! This point $A^*$ changes the face of the problem. We see a host of new relations (fig. 9.3) which we proceed to order and to exploit rapidly. Obviously

$$AX = A^*X.$$

($A^*X$ is the mirror image of $AX$. You can also argue from the congruence of the triangles $\triangle ACX$, $\triangle A^*CX$; the line $l$ is the perpendicular bisector of the segment $AA^*$.) Therefore,

$$AX + XB = A^*X + XB.$$

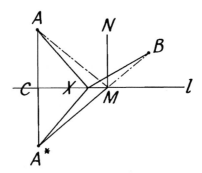

Fig. 9.3. A more informative solution.

Both sides of this equation are minimized by the same position of $X$. Yet the right-hand side is obviously a *minimum when $A^*$, $X$, and $B$ are on the same straight line.* The straight line is the shortest.

This is the solution (see fig. 9.3). The point $M$, the minimum position of $X$, is obtained as intersection of the line $l$ and of the line joining $A^*$ and $B$. Obviously, $AM$ and $MB$ include the same angle with $l$. Introducing the line $MN$, normal to $l$ (parallel to $A^*A$), we see that

$$\angle AMN = \angle BMN.$$

The equality of these two angles characterizes the shortest path. Yet the very same equality

angle of incidence = angle of reflection

characterizes the actual path of light, as we know by experience. Therefore, in fact, the reflected ray of light takes the shortest possible course between the object and the eye. This discovery is due to Heron of Alexandria.

(3) *Comparing two solutions.* It is often useful to look back at the completed solution. In the present case it should be doubly useful, since we have two solutions which we can compare with each other (under (1) and (2)). Both

methods of solving the problem (figs. 9.2 and 9.3) must yield the same result (imagine the two figures superposed). We can obtain the point $M$, the solution of our minimum problem, by means of an ellipse tangent to $l$, or by means of two rays equally inclined to $l$. Yet these two constructions must agree, whatever the relative position of the data (the points $A$ and $B$ and the line $l$) may be. The agreement of the two constructions involves a geometric property of the ellipse: *The two straight lines, joining the two foci of an ellipse to any point on the periphery of the ellipse, are equally inclined to the tangent of the ellipse at the point where they meet.*

If we conceive the ellipse as a mirror and take into account the law of reflection (which we have just discussed), we can restate the geometric property in intuitive optical interpretation: *Any ray of light coming from a focus of an elliptic mirror is reflected into the other focus.*

(4) *An application.* Simple as it is, Heron's discovery deserves a place in the history of science. It is the first example of the use of a minimum principle in describing physical phenomena. It is a suggestive example of interrelations between mathematical and physical theories. Much more general minimum principles have been discovered after Heron and mathematical and physical theories have been interrelated on a much grander scale, but the first and simplest examples are in some respects the most impressive.

Looking back at the impressively successful solution under (2), we should ask: Can you use it? Can you use the result? Can you use the method? In fact, there are several openings. We could examine the reflection of light in a curved mirror, or successive reflections in a series of plane mirrors, or combine the result with methods that we learned before, and so on.

Let us discuss here just one example, the problem of the "traffic center." Three towns intend to construct three roads to a common traffic center which should be chosen so that the total cost of road construction is a minimum. If we take all this in utmost simplification, we have the following purely geometric problem: *Given three points, find a fourth point so that the sum of its distances from the three given points is a minimum.*

Let $A$, $B$, and $C$ denote the three given points (towns) and $X$ a variable point in the plane determined by $A$, $B$, and $C$. We seek the minimum of $AX + BX + CX$.

This problem seems to be related to Heron's problem. We should bring the two problems together, work out the closest possible relation between them. If, for a moment, we take the distance $CX$ as fixed ($= r$, say), the relation appears very close indeed: Here, as there, we have to find the minimum of $AX + BX$, the sum of the distances of one variable point from two fixed points. The difference is that $X$ is obliged to move along a circle here (with radius $r$ and center $C$), and along a straight line there. The former problem was about reflection in a plane mirror, the present problem is about reflection in a circular mirror.

Let us rely on the light: it is clever enough to find by itself the shortest path from $A$ to the circular mirror and hence to $B$. Yet the light moves so that the angle of incidence is equal to the angle of reflection. Therefore, in the desired minimum position, $\angle AXB$ must be *bisected* by the straight line passing through $C$ and $X$ (see fig. 9.4). By the principle of partial variation and the symmetry of the situation, $\angle AXC$ and $\angle BXC$ must be similarly bisected. The three straight lines joining $X$ to $A$, $B$, and $C$ dissect the plane into six angles, the common vertex of which is $X$. Focussing our attention upon the pairs of vertical angles in fig. 9.5, we easily see that all six angles are equal and, therefore, each of them is equal to 60°. *The three roads diverging from the traffic center are equally inclined to each other;* the angle between any two is 120°.

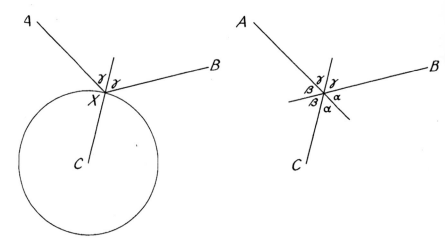

Fig. 9.4. Traffic center and circular mirror.    Fig. 9.5. The traffic center.

(If we remember that the method of partial variation which we have used is subject to certain limitations, we may find a critical reexamination of our solution advisable.)

**2. Mechanical interpretation.** Mathematical problems and their solutions can be suggested by any sector of our experience, by optical, mechanical, or other phenomena. We shall discuss next how simple mechanical principles can help us to discover the solution.

(1) *A string, of which both extremities are fixed, passes through a heavy ring. Find the position of equilibrium.*

It is understood that the string is perfectly flexible and inextensible, its weight is negligible, the ring slides along the string without friction, and the dimensions of the ring are so small that it can be regarded as a mathematical point.

Let $A$ and $B$ denote the fixed endpoints of the string and $X$ any position of the ring. The string forms the broken line $AXB$ on fig. 9.6.

The proposed problem can be solved by two different methods. First, the ring must hang as low as possible. (In fact, the ring is heavy; it "wants" to come as close to the ground, or to the center of the earth, as possible.) Both parts of the inextensible string, $AX$ and $BX$, are stretched, and so the ring, sliding along the string, describes an ellipse with foci $A$ and $B$. Obviously, the position of equilibrium is at the lowest point $M$ of the ellipse where the tangent is *horizontal*.

Second, the forces acting on the point $M$ of the string must be in equilibrium. The weight of the ring and the tensions in the string act on the point $M$. The tensions in both parts of the string, $MA$ and $MB$, are equal and directed along the string to $A$ and $B$, respectively.

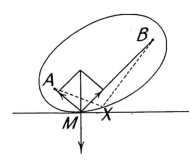

Fig. 9.6. Two conditions of equilibrium.

Their resultant bisects the $\angle AMB$ and, being opposite to the weight of the ring, is *vertical*.

The two solutions, however, must agree. Therefore, the lines $MA$ and $MB$, equally inclined to the vertical normal of the ellipse, are also equally inclined to its horizontal tangent: *The two straight lines, joining the two foci of an ellipse to any point $M$ on the periphery, are equally inclined to the tangent at the point $M$.* (By keeping the length $AB$ but changing its angle of inclination to the horizontal, we can bring $M$ in any desired position on one half of the ellipse.)

We derived a former result (sect. 1(3)) by a new method which may be capable of further applications.

(2) We seem to have a surplus of knowledge. Without having learned too much mechanics, we know enough of it, it seems, not only to find a solution of a proposed mechanical problem, but to find two solutions, based on two different principles. These two solutions, compared, led us to an interesting geometrical fact. Could we divert some more of this overflow of mechanical knowledge into other channels?

With a little luck, we can imagine a mechanism to solve the problem of the traffic center considered above (sect. 1 (4)): Three pulleys turn around axles (nails) fixed in a vertical wall at the points $A$, $B$, and $C$; see fig. 9.7.

Three strings, $XAP$, $XBQ$, and $XCR$ in fig. 9.7, pass over the pulleys at $A$, $B$, and $C$, respectively. At their common endpoint $X$ the three strings are attached together and each carries a weight, $P$, $Q$, and $R$, respectively, at its other end. These weights $P$, $Q$, and $R$ are equal. Our problem is to find the position of equilibrium.

Of course, this problem must be understood with the usual simplifications: The strings are perfectly flexible and inextensible, the friction, the weight of the strings, and the dimensions of the pulleys are negligible (the pulleys are treated as points). As under (1) we can solve the problem by two different methods.

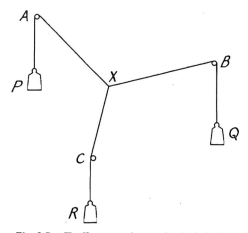

Fig. 9.7.  Traffic center by mechanical device.

First, the three weights must hang jointly as low as possible. That is, the sum of their distances from a given horizontal level (the ground) must be a minimum. (That is, the potential energy of the system must be a minimum; remember that the three weights are equal.) Therefore $AP + BQ + CR$ must be a maximum. Therefore, since the length of each string is invariable, $AX + BX + CX$ must be a minimum and so our problem turns out to be identical with the problem of the traffic center of sect. 1 (4), fig. 9.4, 9.5.

Second, the forces acting at the point $X$ must be in equilibrium. The three equal weights pull, each at its own string, with equal force and these forces are transmitted undiminished by the frictionless pulleys. Three equal forces acting at $X$ along the lines $XA$, $XB$, and $XC$, respectively, must be in equilibrium. Obviously, by symmetry, they must be equally inclined to each other; the angle between any two of the three strings meeting at $X$ is 120°. (The triangle formed by the three forces is equilateral, its exterior angles are 120°.)

This confirms the solution of sect. 1 (4). (On the other hand, the mechanical interpretation may emphasize the necessity of some restriction concerning the configuration of the three points $A$, $B$, and $C$.)

**3. Reinterpretation.** A stick, half immersed in water, appears sharply bent. We conclude hence that the light that follows a straight course in the water as in the air undergoes an abrupt change of direction in emerging from the water into the air. This is the phenomenon of refraction—a phenomenon apparently more complicated and more difficult to understand than reflection. The law of refraction, after unsuccessful efforts by Kepler and others, was finally discovered by Snellius (about 1621) and published by Descartes. Still later came Fermat (1601–1665) who took up the thread of ideas started by Heron.

The light, proceeding from an object $A$ under water to an eye $B$ above the water describes a broken line with an angular point on the surface that separates the air from the water; see fig 9.8. The straight line is, however, the shortest path between $A$ and $B$, and so the light, in its transition from one medium into another, fails to obey Heron's principle. This is disappointing; we do not like to admit that a simple rule that holds good in two cases (direct propagation and reflection) fails in a third case (refraction). Fermat hit upon an expedient. He was familiar with the idea that the light takes time to travel from one point to another, that it travels with a certain (finite) speed; in fact, Galileo proposed a method for measuring the velocity of light. Perhaps the light that travels with a certain velocity through the air travels with another velocity through the water; such a difference in velocity could explain, perhaps, the phenomenon of refraction. As long as it travels at constant speed, the light, in choosing the shortest course, chooses also the *fastest* course. If the velocity depends on the medium traversed, the shortest course is no more necessarily the fastest. Perhaps the light chooses *always* the fastest course, also in proceeding from the water into the air.

This train of ideas leads to a clear problem of minimum (see fig. 9.8): *Given two points $A$ and $B$, a straight line $l$ separating $A$ from $B$, and two velocities $u$ and $v$, find the minimum time needed in travelling from $A$ to $B$; you are supposed to travel from $A$ to $l$ with the velocity $u$, and from $l$ to $B$ with the velocity $v$.*

Obviously, it is the quickest to follow a straight line from $A$ to a certain point $X$ on $l$, and some other straight line from $X$ to $B$. The problem consists essentially in finding the point $X$. Now, in uniform motion time equals distance divided by speed. Therefore, the time spent in travelling from $A$ to $X$ and hence to $B$ is

$$\frac{AX}{u} + \frac{XB}{v}.$$

This quantity should be made a minimum by the suitable choice of the point $X$ on $l$. We have to find $X$, being given $A$, $B$, $u$, $v$, and $l$.

To solve this problem without differential calculus is not too easy. Fermat solved it by inventing a method that eventually led to the differential calculus. We rather follow the lead given by the examples of the foregoing section. With a little luck, we can succeed in imagining a mechanism that solves for us the proposed problem of minimum; see fig. 9.9.

A ring $X$ can slide along a fixed horizontal rod $l$ that passes through it. Two strings $XAP$ and $XBQ$ are attached to the ring $X$. Each of these strings passes over a pulley (at $A$ and at $B$, respectively) and carries a weight at its other end (at $P$ and at $Q$, respectively). The main point is to choose the weights. These weights cannot be equal; if they were, the line $AXB$ would be straight in the equilibrium position (this seems plausible at least) and so $AXB$ would not be fit to represent the path of refracted light. Let us postpone the choice of the weights, but let us introduce a suitable notation. We

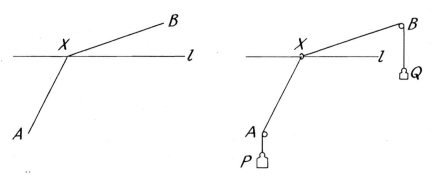

Fig. 9.8.  Refraction.          Fig. 9.9.  Refraction by mechanical device.

call $p$ the weight at the endpoint $P$ of the first string and $q$ the weight at the endpoint $Q$ of the second string. And now we have to find the position of equilibrium. (We assume the usual simplifications: the rod is perfectly inflexible, the strings perfectly flexible, but also inextensible; we disregard the friction, the weight and stiffness of the strings, the dimensions of the pulleys, and those of the ring.) As in sect. 2, we solve our problem by two different methods.

First, the two weights must hang jointly as low as possible. (That is, the potential energy of the system must be a minimum.) This implies that

$$AP \cdot p + BQ \cdot q$$

must be a maximum. Therefore, since the length of each string is invariable,

$$AX \cdot p + XB \cdot q$$

must be a minimum.

This is very close to Fermat's problem, but not exactly the same. Yet the two problems, the optical and the mechanical, coincide mathematically if we choose

$$p = 1/u, \qquad q = 1/v.$$

Then the problem of equilibrium (fig. 9.9) requires, just as Fermat's problem of the fastest travel, that

$$\frac{AX}{u} + \frac{XB}{v}$$

be a minimum. This we found in looking at the equilibrium of the mechanical system in fig. 9.9 from a first viewpoint.

Second, the forces acting at the point $X$ must be in equilibrium. The pull of the weights is transmitted undiminished by the frictionless pulleys. Two forces, of magnitude $1/u$ and $1/v$, respectively, act on the ring, each pulling it in the direction of the respective string. They cannot move it in the

Fig. 9.10. The law of refraction.

vertical direction, because the rod $l$ passing through the ring is perfectly rigid (there is a vertical reaction of unlimited amount due to the rod). Yet the *horizontal components* of the two pulls, which are opposite in direction, must cancel, must be *equal in magnitude*. In order to express this relation, we introduce the angles $\alpha$ and $\beta$ between the vertical through the point $X$ and the two strings; see fig. 9.10. The equality of the horizontal components is expressed by

$$\frac{1}{u} \sin \alpha = \frac{1}{v} \sin \beta$$

or

$$\frac{\sin \alpha}{\sin \beta} = \frac{u}{v}.$$

This is the *condition of minimum*.

Now, let us return to the optical interpretation. The angle $\alpha$ between the incoming ray and the normal to the refracting surface is called the angle of incidence, and $\beta$ between the outgoing ray and the normal the angle of refraction. The ratio $u/v$ of the velocities depends on the two media, water and air, but not on geometric circumstances, as the situation of the points $A$ and $B$. Therefore, the condition of minimum requires that *the sines of the angles of incidence and refraction are in a constant ratio depending on the two media alone* (called nowadays the refractive index). Fermat's "principle of least time" leads to Snellius' law of refraction, confirmed by countless observations.

We reconstructed as well as we could the birth of an important discovery. The procedure of solution (that we used instead of Fermat's) is also worth noticing. Our problem had from the start a physical (optical) interpretation. Yet, in order to solve it, we invented *another* physical (mechanical) interpretation. Our solution was a *solution by reinterpretation.* Such solutions may reveal new analogies between different physical phenomena and have a peculiar artistic quality.

**4. Jean Bernoulli's discovery of the Brachistochrone.** A heavy material point starts from rest at the point A and glides without friction along an inclined plane to a lower point B. The material point starting

Fig. 9.11.  Path of a material point.

from rest could also swing from A to B along a circular arc, as the bob of a pendulum. Which motion takes less time, that along the straight line or that along the circular arc? Galileo thought that the descent along the circular arc is faster. Jean Bernoulli imagined an arbitrary curve in the vertical plane through A and B connecting these two points. There are infinitely many such curves, and he undertook to find the curve that makes the time of descent a minimum; this curve is called the "curve of fastest descent" or the "brachistochrone." We wish to understand Jean Bernoulli's wonderfully imaginative solution of his problem.

We place an arbitrary curve descending from A to B in a coordinate system; see fig. 9.11. We choose A as the origin, the x-axis horizontal, and the y-axis vertically downward. We focus the moment when the material point sliding down the curve passes a certain point $(x, y)$ with a certain velocity $v$. We have the relation

$$v^2/2 = gy$$

which was perfectly familiar to Bernoulli; we derive it today from the conservation of energy. That is, whatever the path of the descent may be, $v$, the velocity attained depends only on $y$, the depth of the descent:

(1) $$v = (2gy)^{1/2}.$$

What does this mean? Let us try to see intuitively the significance of this basic fact.

We draw horizontal lines (see fig. 9.11) dividing the plane in which the material point descends into thin horizontal layers. The descending material point crosses these layers one after the other. Its velocity does not depend on the path that it took, but depends only on the layer that it is just crossing; its velocity varies from layer to layer. Where have we seen such a situation? When the light of the sun comes down to us it crosses several layers of air each of which is of a different density; therefore, the velocity of light varies from layer to layer. The proposed mechanical problem admits an *optical reinterpretation*.

We see now fig. 9.11 in a new context. We regard this figure as representing an optically inhomogeneous medium. This medium is stratified, has strata of different quality; the velocity of light in the horizontal layer at the depth $y$ is $(2gy)^{1/2}$. The light crossing this medium from $A$ to $B$ (from one of the given points to the other) *could* travel along various curves. Yet the light chooses the fastest course; it travels actually along the curve that renders the time of travel a minimum. Therefore, *the actual path of light, traversing the described inhomogeneous, stratified medium from $A$ to $B$,* **is** *the brachistochrone!* Yet the actual path of light is governed by Snellius' law of refraction: the solution suddenly appears within reach. Jean Bernoulli's imaginative reinterpretation renders accessible a problem that seemed entirely novel and inaccessible.

There still remains some work to do, but it demands incomparably less originality. In order to make Snellius' law applicable in its familiar form (which we have discussed in the preceding sect. 3) we change again our interpretation of fig. 9.11, slightly: the velocity $v$ should not vary continuously with $y$ in infinitesimal steps, but discontinuously, in small steps. We imagine several horizontal layers of transparent matter (several plates of glass) each somewhat different optically from its neighbors. Let $v$, $v'$, $v''$, $v'''$, . . . be the velocity of light in the successive layers, and let the light crossing them successively include the angle $\alpha$, $\alpha'$, $\alpha''$, $\alpha'''$, . . . with the vertical, respectively; see fig. 9.12. By the law of Snellius (see sect. 3)

$$\frac{\sin \alpha}{v} = \frac{\sin \alpha'}{v'} = \frac{\sin \alpha''}{v''} = \frac{\sin \alpha'''}{v'''} = \cdots .$$

Now we may return from the medium consisting of thin plates to the stratified medium in which $v$ varies continuously with the depth. (Let the plates become infinitely thin.)   We see that

(2)                  $$\frac{\sin \alpha}{v} = \text{const.}$$

along the path of light.

Let $\beta$ be the angle included by the tangent to the curve with the horizontal. Then

$$\alpha + \beta = 90°, \qquad \tan \beta = dy/dx = y'$$

and so

(3)                  $$\sin \alpha = \cos \beta = (1 + y'^2)^{-1/2}.$$

Fig. 9.12.   Path of light.

We combine the equations (1), (2), and (3) (derived from mechanics, optics, and the calculus, respectively) introduce a suitable notation for the constant arising in (2), and obtain so

$$y(1 + y'^2) = c;$$

$c$ is a positive constant.   We obtained a differential equation of the first order for the brachistochrone.   Finding the curves satisfying such an equation was a problem familiar to Bernoulli.   We need not go into detail here (see, however, ex. 31): the brachistochrone, determined by the differential equation, turns out to be a *cycloid*.   (The cycloid is described by a point in the circumference of a circle that rolls upon a straight line; in our case the straight line is the $x$-axis and the rolling proceeds upside down: the circle rolls under the $x$-axis.)

Let us observe, however, that we can see intuitively without resorting to formulas, that Snellius' law implies a differential equation.   In fact, this law determines the directions of the successive elements of the path represented in fig. 9.12, and this is precisely what a differential equation does.

Jean Bernoulli's solution of the problem of the brachistochrone, that we have discussed here, has a peculiar artistic quality. Looking at fig. 9.11 or fig. 9.12, we may see intuitively the key idea of the solution. If we can see this idea clearly, without effort, anticipating what it implies, we may notice that there is a real work of art before us.

The key idea of Jean Bernoulli's solution is, of course, reinterpretation. The geometric figure (fig. 9.11 or 9.12) is conceived successively in two different interpretations, is seen in two different "contexts": first in a mechanical context, then in an optical context. Does any discovery consist in an unexpected contact and subsequent interpenetration of two different contexts?

**5. Archimedes' discovery of the integral calculus.** It so happens that one of the greatest mathematical discoveries of all times was guided by physical intuition. I mean Archimedes' discovery of that branch of science that we call today the integral calculus. Archimedes found the area of the parabolic segment, the volume of the sphere and about a dozen similar results by a uniform method in which the idea of equilibrium plays an important role. As he says himself, he "investigated some problems in mathematics by means of mechanics."[2]

If we wish to understand Archimedes' work, we have to know something about the state of knowledge from which he started.

The geometry of the Greeks attained its peak in Archimedes' time; Eudoxus and Euclid were his predecessors, Apollonius his contemporary. We have to mention a few specific points that may have influenced Archimedes' discovery.

As Archimedes himself relates, Democritus found the volume of the cone; he stated that it is one-third of the volume of a cylinder with the same base and the same altitude. We know nothing about Democritus' method, but there seems to be some reason to suspect that he considered what we would call today a variable cross-section of the cone parallel to its base.[3]

Eudoxus was the first to prove Democritus' statement. In proving this and similar results, Eudoxus invented his "method of exhaustion" and set a standard of rigor for Greek mathematics.

We have to realize that the Greeks knew, in a certain sense, "coordinate geometry." They were used to handle loci in a plane by considering the distances of a moving point from two fixed axes of reference. If the sum of the squares of these distances is constant and the axes of reference are perpendicular to each other, the locus is a circle—this proposition belongs to coordinate geometry, but not yet to analytic geometry. Analytic geometry

---

[2] The *Method* of Archimedes, edited by Thomas L. Heath, Cambridge, 1912. Cf. p. 13. This booklet will be quoted as *Method* in the following footnotes. See also *Oeuvres complètes d'Archimède*, translated by P. Ver Eecke, pp. 474–519.

[3] Cf. *Method*, pp. 10–11.

begins in the moment when we express the relation mentioned in algebraic symbols as

$$x^2 + y^2 = a^2.$$

The Mechanics of the Greeks never attained the excellence of their Geometry, and started much later. If we take vague discussions by Aristotle and others for what they are worth, we can say that Mechanics as a science begins with Archimedes. He discovered, as everybody knows, the law of floating bodies. He also discovered the principle of the lever and the main properties of the center of gravity which we shall need in a moment.

Now we are prepared for discussing the most spectacular example of Archimedes' work; we wish to find, with his method, the *volume of the sphere*. Archimedes regards the sphere as generated by a revolving circle, and he regards the circle as a locus, characterized by a relation between the distances of a variable point from two fixed rectangular axes of reference. Written in modern notation, this relation is

$$x^2 + y^2 = 2ax,$$

the equation of a circle with radius $a$ that touches the $y$-axis at the origin. See fig. 9.13 which differs only slightly from Archimedes' original figure; the circle, revolving about the $x$-axis, generates a sphere. I think that the use of modern notation does not distort Archimedes' idea. On the contrary, it seems to me that this notation is suggestive. It suggests motives which may lead us to Archimedes' idea today and which are, perhaps, not too different from the motives that led Archimedes himself to his discovery.

In the equation of the circle there is the term $y^2$. Observe that $\pi y^2$ is the area of a *variable cross-section of the sphere*. Yet Democritus found the volume of the cone by examining the variation of its cross-section. This leads us to rewriting the equation of the circle in the form

$$\pi x^2 + \pi y^2 = \pi 2ax.$$

Now we can interpret $\pi x^2$ as the variable cross-section of a cone, generated by the rotation of the line $y = x$ about the $x$-axis, see fig. 9.13. This suggests to seek an analogous interpretation of the remaining term $\pi 2ax$. If we do not see such an interpretation, we may try to rewrite the equation in still other shapes, and so we have a chance to hit upon the form

(A)          $$2a(\pi y^2 + \pi x^2) = x\pi(2a)^2.$$

Much is concentrated in this equation (A). Looking at equation (A), noticing the various lengths and areas arising in it and suitably *disposing them in the figure*, we may witness the birth of a great idea; it will be born from the intimate union of formula (A) with fig. 9.13.

We notice the areas of three circular disks, $\pi y^2$, $\pi x^2$, and $\pi(2a)^2$. The three circles are the intersections of the same plane with three solids of revolution. The plane is perpendicular to the $x$-axis and at the distance $x$ from the origin $O$. The three solids of revolution are a sphere, a cone, and a cylinder. They are described by the three lines the equations of which are (A), $y = x$, and $y = 2a$, respectively, when the right hand portion of fig. 9.13 rotates about the $x$-axis. The cone and the cylinder have the same base and the same altitude. The radius of the common base and the common altitude have the same length $2a$. The vertex of the cone is at the origin $O$.

Archimedes treats differently the disks the areas of which appear on different sides of the equation (A). He leaves the disk with radius $2a$,

Fig. 9.13. The birth of the Integral Calculus.

cross-section of the cylinder, in its original position, at the distance $x$ from the origin. Yet he removes the disks with radii $y$ and $x$, cross-sections of the sphere and the cone, respectively, from their original position and transports them to the point $H$ of the $x$-axis with abscissa $-2a$. We let these disks with radii $y$ and $x$ hang with their center vertically under $H$, suspended by a string of negligible weight, see fig. 9.13. (This string is an addition, of negligible weight, to Archimedes' original figure.)

Let us regard the $x$-axis as a *lever*, a rigid bar of negligible weight, and the origin $O$ as its *fulcrum* or point of suspension. Equation (A) deals with *moments*. (A moment is the product of the weight and the arm of the lever.) Equation (A) expresses that the moment of the two disks on the left-hand side equals the moment of the one disk on the right-hand side and so, by the mechanical law discovered by Archimedes, the lever is in equilibrium.

As $x$ varies from 0 to $2a$, we obtain all cross-sections of the cylinder; these cross-sections *fill* the cylinder. To each cross-section of the cylinder there correspond two cross-sections suspended from the point $H$ and these cross-sections *fill* the sphere and the cone, respectively. *As their corresponding cross-sections, the sphere and the cone, hanging from H, are in equilibrium with the cylinder.* Therefore, by Archimedes' mechanical law, the moments must be equal. Let us call $V$ the volume of the sphere, let us recall the expression for the volume of the cone (due to Democritus) and also the volume of the cylinder and the obvious location of its center of gravity. Passing from the moments of the cross-sections to the moments of the corresponding solids, we are led from equation (A) to

(B) $$2a\left(V + \frac{\pi(2a)^2 2a}{3}\right) = a\pi(2a)^2 2a$$

which readily yields[4]
$$V = \frac{4\pi a^3}{3}.$$

Looking back at the foregoing, we see that the decisive step is that from (A) to (B), from the filling cross-sections to the full solids. Yet this step is only heuristically assumed, not logically justified. It is plausible, even very plausible, but not demonstrative. It is a guess, not a proof. And Archimedes, representing the great tradition of Greek mathematical rigor, knows this full well: "The fact at which we arrived is not actually demonstrated by the argument used; but the argument has given a sort of indication that the conclusion is true."[5] This guess, however, is a guess with a prospect. The idea goes much beyond the requirements of the problem at hand, and has an immensely greater scope. The passage from (A) to (B), from the cross-section to the whole solid is, in more modern language, the transition from the infinitesimal part to the total quantity, from the differential to the integral. This transition is a great beginning, and Archimedes, who was a great enough man to see himself in historical perspective, knew it full well: "I am persuaded that this method will be of no little service to mathematics. For I foresee that this method, once understood, will be used to discover other theorems which have not yet occurred to me, by other mathematicians, now living or yet unborn."[6]

### EXAMPLES AND COMMENTS ON CHAPTER IX.

**1.** Given in a plane a point $P$ and two intersecting lines $l$ and $m$, none of which passes through $P$. Let $Y$ be a variable point on $l$ and $Z$ a variable

---

[4] I presented this derivation several times in my classes and once I received a compliment I am proud of. After my usual "Are there any questions?" at the end of the derivation, a boy asked: "Who paid Archimedes for this research?" I must confess that I was not prompt enough to answer: "In those days such research was sponsored only by Urania, the Muse of Science."

[5] *Method*, p. 17.

[6] *Method*, p. 14.

point on $m$. Determine $Y$ and $Z$ so that the perimeter of $\triangle PYZ$ be a minimum.

Give two solutions, one by physical considerations, the other by geometry.

**2.** Three circles in a plane, exterior to each other, are given in position. Find the triangle with minimum perimeter that has one vertex on each circle. Give two different physical interpretations.

**3.** *Triangle with minimum perimeter inscribed in a given triangle.* Given $\triangle ABC$. Find three points $X$, $Y$, and $Z$ on the sides $BC$, $CA$, and $AB$ of the triangle, respectively, such that the perimeter of $\triangle XYZ$ is a minimum.

Give two different physical interpretations.

**4.** Generalize ex. 3.

**5.** Criticize the solution of ex. 1. Does it apply to all cases?

**6.** Criticize the solution of ex. 3. Does it apply to all cases?

**7.** Give a rigorous solution of ex. 3 for acute triangles. [Partial variation, ex. 1, ex. 5.]

**8.** Criticize the solutions of sect. 1 (4) and sect. 2 (2) for the problem of the traffic center. Do they apply to all cases?

**9.** *Traffic center of four points in space.* Given a tetrahedron with vertices at the points $A$, $B$, $C$, and $D$. Assume that there is a point $X$ inside the tetrahedron such that the sum of its distances from the four vertices

$$AX + BX + CX + DX$$

is a minimum. Show that the angles $\angle AXB$ and $\angle CXD$ are equal and are bisected by the same straight line; point out other pairs of angles similarly related. [Do you know a related problem? An analogous problem? Could you use its result, or the method of its solution?]

**10.** *Traffic center of four points in a plane.* Consider the extreme case of ex. 9 in which the points $A$, $B$, $C$, and $D$, in the same plane, are the four vertices of a convex quadrilateral $ABCD$. Do the statements of ex. 9 remain valid in this extreme case?

**11.** *Traffic network for four points.* Let $A$, $B$, $C$, and $D$ be four fixed points, and $X$ and $Y$ two variable points, in a plane. If the minimum of the sum of five distances $AX + BX + XY + YC + YD$ is so attained that all six points $A$, $B$, $C$, $D$, $X$, and $Y$ are distinct, the three lines $XA$, $XB$, and $XY$ are equally inclined to each other and so are the three lines $YC$, $YD$, and $YX$.

**12.** *Unfold and straighten.* There is still another useful interpretation of fig. 9.3. Draw $l$, $A^*X$, and $XB$ on a sheet of transparent paper, then fold the sheet along the line $l$: you obtain fig. 9.1 (with $A^*$ instead of $A$). Imagine the fig. 9.1 originally drawn in this sophisticated manner on a folded transparent sheet. In order to find the position of $X$ that renders $AX + XB$ a minimum, unfold the sheet, draw a straight line from $A$ (or rather $A^*$ in fig. 9.3) to $B$, and then fold the sheet back.

**13. Billiards.** On a rectangular billiard table there is a ball at the point $P$. It is required to drive the ball in such a direction that after four successive reflections on the four sides of the rectangle the ball should return to its original position $P$. [Fig. 9.14.]

**14. Geophysical exploration.** At the point $E$ of the horizontal surface of the earth an explosion takes place. The sound of this explosion is propagated in the interior of the earth and reflected by an oblique plane layer $OR$ which includes the angle $\alpha$ with the earth's surface. The sound coming from $E$ can attain a listening post $L$ at another point of the earth's surface in $n$

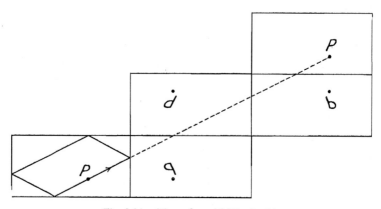

Fig. 9.14. The reflected billiard table.

different ways. (One of the $n$ paths is constructed in fig. 9.15 by the method of ex. 12.) Being given $n$ (observed by suitable apparatus) give limits between which $\alpha$ is included.

**15.** Given, in space, a straight line $l$ and two points, $A$ and $B$, not on $l$. On the line $l$ find a point $X$ such that the sum of its distances from the two given points $AX + XB$, be a minimum.

Do you know a related problem? A more special problem? Could you use its result, or the method of its solution?

**16.** Solve ex. 15 by using a tangent level surface.

**17.** Solve ex. 15 by paper-folding.

**18.** Solve ex. 15 by mechanical interpretation. Is the solution consistent with ex. 16 and 17?

**19.** Given three skew straight lines in space, $a$, $b$, and $c$. Show that the triangle with one vertex on each given line and minimum perimeter has the following property: the line joining its vertex on the line $a$ to the center of its inscribed circle is perpendicular to $a$.

**20.** Consider the particular case of ex. 19 in which the three skew lines are three edges of a cube. Where are the vertices of the desired triangle?

Where is the center of its inscribed circle? What is its perimeter, if the volume of the cube is $8a^3$?

**21.** Given three skew straight lines in space, $a$, $b$, and $c$. Let $X$ vary along $a$, $Y$ along $b$, $Z$ along $c$, and $T$ freely in space. Find the minimum of $XT + YT + ZT$.

**22.** Specialize ex. 21 as ex. 20 specializes ex. 19.

**23.** *Shortest lines on a polyhedral surface.* The end walls of a rectangular room are squares; the room is 20 feet long, 8 feet wide, and 8 feet high. A spider is on one of the end walls, 7 feet above the floor and midway between the side walls. The spider perceives a fly on the opposite wall 1 foot above

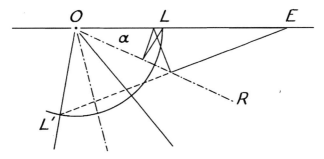

Fig. 9.15. Reflections underground.

the floor and also midway between the side walls. Show that the spider has *less* than 28 feet to travel along the walls, or the ceiling, or the floor, to attain the fly. [Ex. 17.]

**24.** *Shortest lines (geodesics) on a curved surface.* We regard a curved surface as the limit of a polyhedron. As the polyhedron approaches the curved surface, the number of its faces tends to ∞, the longest diagonal of any face tends to 0, and the faces tend to become tangential to the surface.

On a polyhedral surface, the shortest line between two points is a polygon. It may be a plane polygon all points of which lie in one plane, or it may be a skew polygon the points of which are not contained in one plane. (Both cases can be illustrated by the solution of ex. 23, the first case by (1), the second case by (2) and (3).)

On a curved surface, a shortest line is called a "geodesic" because shortest lines play a role in geodesy, the study of the earth's surface. A geodesic may be a plane curve, fully contained in one plane, or it may be a "space curve" ("tortuous" curve) the points of which cannot be contained in one plane. At any rate, a geodesic must have some intrinsic geometric relation to the surface on which it is a shortest line. What is this relation?

(1) We consider the polygonal line $ABC \ldots L$. Even if $ABC \ldots L$ is a skew polygon, two consecutive segments of it, as $HI$ and $IJ$, lie in the same

plane. If *ABC . . . L* is the shortest line on a polyhedral surface between its endpoints *A* and *L*, each of its intermediate vertices *B*, *C*, *D*, . . . , *H*, *I*, *J*, . . . *K* lies on an edge of the polyhedron. The plane that contains the segments *HI* and *IJ* contains also the bisector of $\angle HIJ$, and this bisector is perpendicular to the edge of the polyhedron that passes through *I*; see ex. 16 or ex. 18.

We consider a curve. Even if the curve is tortuous, an infinitesimal (very short) arc of it can be regarded as a plane (almost plane) arc. The plane of the infinitesimal arc is the *osculating plane* at its midpoint. The osculating plane is analogous to the plane in which two successive segments of a skew polygon lie. If the curve is a geodesic, that is, a shortest line on a surface, analogy suggests that *the osculating plane of a geodesic at a point passes through the normal to the surface at that point.*

(2) A geodesic can be interpreted physically as a rubber band stretched along a smooth (frictionless) surface. Let us examine the equilibrium of a small portion of the rubber band. The forces acting on this portion are two tensions of equal amount acting tangentially at the two endpoints of the short arc, and the reaction of the frictionless surface acting normally to it. The reaction of the surface, compounded into a resultant force, and the two tensions at the endpoints are in equilibrium. Therefore, these three forces are parallel to the same plane. Yet two "neighboring tangents" determine the osculating plane which, therefore, contains the normal to the surface.

(3) Each arc of a geodesic is a geodesic. In fact, if in a curve there is a portion that is not the shortest between its endpoints and, therefore, can be replaced by a shorter arc between the same endpoints, the whole curve cannot be a shortest line. Hence it is natural to expect that a geodesic possesses some distinctive property in each of its points. The property suggested by two very different heuristic considerations, (1) and (2), is a property of this kind.

(4) Look out for examples to test the heuristically obtained result. What are the shortest lines on a sphere? Do they have the property suggested? Do other lines on the spherical surface have the same property?

**25.** A material point moves without friction on a smooth rigid surface. No exterior forces (such as gravitation) act on the point (except, of course, the reaction of the surface). Give reasons why the point should be expected to describe a geodesic.

**26.** *A construction by paper-folding.* Find a polygon inscribed in a circle if the sides are given in magnitude and in succession.

Let $a_1$, $a_2$, $a_3$, . . . $a_n$ denote the given lengths. The side of length $a_1$ is followed by the side of length $a_2$, this one by the side of length $a_3$, and so on; the side of length $a_n$ is followed by the side of length $a_1$. It is understood that any of the lengths $a_1$, $a_2$, . . . $a_n$ is less than the sum of the remaining $n - 1$ lengths.

There is a beautiful solution by paper-folding. Draw $a_1, a_2, \ldots a_n$ on cardboard as successive chords in a sufficiently large circle so that two consecutive chords have a common endpoint. Draw the radii from these endpoints to the center of the circle. Cut out the polygon bounded by the $n$ chords and the two extreme radii, fold the cardboard along the $n - 1$ other radii, and paste together the two radii along which the cardboard was cut. You obtain so an open polyhedral surface; it consists of $n$ rigid isosceles triangles, is bounded by $n$ free edges, of lengths $a_1, a_2, \ldots a_n$, respectively, and has $n$ dihedral angles which can still be varied. (We suppose that $n > 3$.)

What can you do to this polyhedral surface to solve the proposed problem?

**27.** *The die is cast.* The mass in the interior of a heavy rigid convex polyhedron need not be uniformly distributed. In fact, we can imagine a suitable heterogeneous distribution of mass the center of gravity of which coincides with an arbitrarily assigned interior point of the polyhedron. Thrown on the horizontal floor, the polyhedron will come to rest on one of its faces. This yields a mechanical argument for the following geometrical proposition.

Given any convex polyhedron $P$ and any point $C$ in the interior of $P$, we can find a face $F$ of $P$ with the following property: the foot of the perpendicular drawn from $C$ to the plane of $F$ is an *interior* point of $F$.

Find a geometrical proof for this proposition. (Observe that the face $F$ may, but need not, be uniquely determined by the property stated.)

**28.** *The Deluge.* There are three kinds of remarkable points on a contour map: peaks, passes (or saddle points with a horizontal tangent plane), and "deeps." (On fig. 8.7 $P$ is a peak, $S$ a pass.) A "deep" is the deepest point in the bottom of a valley from which the water finds no outlet. A deep is an "inverted" peak: on the contour-map, regard any level line of elevation $h$ as if it had the elevation $-h$. Then the map is "inverted"; it becomes the map of a landscape under the sea, the peaks become deeps, the deeps become peaks, but the passes remain passes. There is a remarkable connection between these three kinds of points.

*Suppose that there are $P$ peaks, $D$ deeps, and $S$ passes on an island. Then*

$$P + D = S + 1.$$

In order to derive this theorem intuitively, we imagine that a persistent rain causes the lake around the island to rise till finally the whole island is submerged. We may admit that all $P$ peaks are equally high, and that all $D$ deeps are on, or under, the level of the lake. In fact, we can imagine the peaks raised and the deeps depressed without changing their number. As it starts raining, the water gathers in the deeps; we have, counting the lake,

$D + 1$ sheets of water     and     1 island

at the beginning.  Just before the island is engulfed, only the peaks show above the water and so we have

1 sheet of water    and    $P$ islands

at the end.  How did the transition take place?

Let us imagine that, at any time, the several sheets of water are at the same elevation.  If there is no pass precisely at this elevation, the water can rise a little more without changing the number of the sheets of water, or the number of the islands.  When, however, the rising water just attains a pass, the least subsequent rise of its level will either unite two formerly

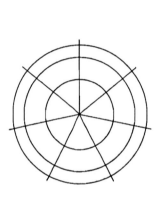

Fig. 9.16.  Neighborhood of a peak.

Fig. 9.17.  Neighborhood of a pass.

separated sheets of water or isolate a piece of land.  Therefore, each pass either decreases the number of the sheets of water by one unit, or increases the number of islands by one unit.  Looking at the total change, we obtain

$$(D + 1 - 1) + (P - 1) = S$$

which is the desired theorem.

(a)  Suppose now that there are $P$ peaks, $D$ deeps, and $S$ passes on the whole globe (some of them are under water) and show that

$$P + D = S + 2.$$

(b)  The last relation reminds us of Euler's theorem (see sect. 3.1–3.7 and ex. 3.1–3.9).  Could you use Euler's theorem to construct a geometrical proof for the result just obtained by an intuitive argument?  [Figs. 9.16 and 9.17 show important pieces of a more complete map in which not only some level lines are indicated, but also some "lines of steepest descent" which are perpendicular to the level lines.  These two kinds of lines subdivide the globe's surface into triangles and quadrilaterals.  Cf. ex. 3.2.]

(c)  Are there any remarks on the method?

**29. *Not so deep as a well.*** In order to find $d$, the depth of a well, you drop a stone into the well and measure the time $t$ between the moment of dropping the stone and the moment when you hear the stone striking the water.

(a) Given $g$, the gravitational acceleration, and $c$, the velocity of sound, express $d$ in terms of $g$, $c$, and $t$. (Neglect the resistance of the air.)

(b) If the well is not too deep, even the final velocity of the stone will be a small fraction of the velocity of sound and so we may expect that much the greater part of the measured time $t$ is taken up by the fall of the stone. Hence we should expect that

$$d = gt^2/2 - \text{correction}$$

where the correction is relatively small when $t$ is small.

In order to examine this guess, expand the expression obtained as answer to (a) in powers of $t$ and retain the first two non-vanishing terms.

(c) What would you regard as typical in this example?

**30. *A useful extreme case.*** An ellipse revolving about its major axis describes a so-called prolate spheroid, or egg-shaped ellipsoid of revolution. The foci of the rotating ellipse do not rotate: they are on the axis of revolution and are also called the foci of the prolate spheroid. We could make an elliptic mirror by covering the inner, concave side of the surface with polished metal; all light coming from one focus is reflected into the other focus by such an elliptic mirror; cf. sect. 1 (3). Elliptic mirrors are very seldom used in practice, but there is a limiting case which is very important in astronomy. What happens if one of the foci of the ellipsoid is fixed and the other tends to infinity?

**31.** Solve the differential equation of the brachistochrone found in sect. 4.

**32. *The Calculus of Variations*** is concerned with problems on the maxima and minima of quantities which depend on the shape and size of a variable curve. Such is the problem of the brachistochrone solved in sect. 4 by optical interpretation. The problem of geodesics, or shortest lines on a curved surface, discussed in ex. 24, also belongs to the Calculus of Variations, and the "isoperimetric problem" that will be treated in the next chapter belongs there too. Physical considerations, which can solve various problems on maxima and minima as we have seen, can also solve some problems of the Calculus of Variations. We sketch an example.

*Find the curve with given length and given endpoints that has the center of gravity of minimum elevation.* It is assumed that the density of heavy matter is constant along the curve which we regard as a uniform cord or chain. When the center of gravity of the chain attains its lowest possible position, the chain is in *equilibrium*. Now we can investigate the equilibrium of the chain in examining the forces acting on it, its weight and its tension. This investigation leads to a differential equation that determines the desired curve, the *catenary*. We do not go into detail. We just wish to remark that the solution sketched has the same basic idea as the mechanical solutions considered in sect. 2.

**33.** *From the equilibrium of the cross-sections to the equilibrium of the solids.* Archimedes did not state explicitly the general principle of his method, but he applied it to several examples, computing volumes, areas and centers of gravity, and the variety of these applications makes the principle perfectly clear. Let us apply the variant of Archimedes' method that has been presented in sect. 5 to some of his examples.

Prove Proposition 7 of the *Method*: Any segment of the sphere has to the cone with the same base and height the ratio that the sum of the radius of the sphere and the height of the complementary segment has to the height of the complementary segment.

**34.** Prove Proposition 6 of the *Method*: The center of gravity of a hemisphere is on its axis and divides this axis so that the portion adjacent to the vertex of the hemisphere has to the remaining portion the ratio of 5 to 3.

**35.** Prove Proposition 9 of the *Method*: The center of gravity of any segment of a sphere is on its axis and divides this axis so that the portion adjacent to the vertex has to the remaining part the ratio that the sum of the axis of the segment and four times the axis of the complementary segment has to the sum of the axis of the segment and double the axis of the complementary segment.

**36.** Prove Proposition 4 of the *Method*: Any segment of a paraboloid of revolution cut off by a plane at right angles to the axis has the ratio of 3 to 2 to the cone that has the same base and the same height as the segment.

**37.** Prove Proposition 5 of the *Method*: The center of gravity of a segment of a paraboloid of revolution cut off by a plane at right angles to the axis is on the axis and divides it so that the portion adjacent to the vertex is double the remaining portion.

**38.** *Archimedes' Method in retrospect.* What was in Archimedes' mind as he discovered his method, we shall never know and we can only vaguely guess. Yet we can set up a clear and fairly short list of such mathematical rules (well known today but unformulated in Archimedes' time) as we need to solve, with contemporary methods, the problems that Archimedes solved with his method. We need:

(1) Two general rules of the integral calculus:

$$\int cf(x)dx = c \int f(x)dx, \qquad \int [f(x) + g(x)]dx = \int f(x)dx + \int g(x)dx;$$

$c$ is a constant, $f(x)$ and $g(x)$ are functions.

(2) The value of four integrals:

$$\int x^n dx = x^{n+1}/(n + 1) \text{ for } n = 0, 1, 2, 3.$$

(3) The geometric interpretation of two integrals:

$$\int Q(x)dx, \qquad \int xQ(x)dx.$$

$Q(x)$ denotes a length in plane geometry and an area in solid geometry; it denotes in both cases the variable cross-section of a figure determined by a plane perpendicular to the $x$-axis. The first integral expresses an area or a volume, the second integral the moment of a uniformly filled area or volume, according as we consider a problem of plane or solid geometry.

Archimedes did not formulate these rules, although we cannot help thinking that he possessed them, in some form or other. He refrained even from formulating in general terms the underlying process, the passage from the variable cross-section to the area or volume, from the integrand to the integral as we would say today. He described this process in particular cases, he applied it to an admirable variety of cases, he doubtless knew it intimately, but he regarded it as merely heuristic, and he thought this a good enough reason for refraining from stating it generally.

Quote simple geometric facts which can yield intuitively the values of the four integrals listed under (2).[7]

---

[7] For other remarks on Archimedes' discovery cf. B. L. van der Waerden, *Elemente der Mathematik*, vol. 8, 1953, p. 121-129, and vol. 9, 1954, p. 1-9.

# X

# THE ISOPERIMETRIC PROBLEM

*The circle is the first, the most simple, and the most perfect figure.*—PROCLUS[1]
*Lo cerchio è perfettissima figura.*—DANTE[2]

**1. Descartes' inductive reasons.** In Descartes' unfinished work *Regulae ad Directionem Ingenii* (or *Rules for the Direction of the Mind*, which, by the way, must be regarded as one of the classical works on the logic of discovery) we find the following curious passage:[3] "In order to show by enumeration that the perimeter of a circle is less than that of any other figure of the same area, we do not need a complete survey of all the possible figures, but it suffices to prove this for a few particular figures whence we can conclude the same thing, by induction, for all the other figures."

In order to understand the meaning of the passage let us actually perform what Descartes suggests. We compare the circle to a few other figures, triangles, rectangles, and circular sectors. We take two triangles, the equilateral and the isosceles right triangle (with angles 60°, 60°, 60° and 90°, 45°, 45°, respectively). The shape of a rectangle is characterized by the ratio of its width to its height; we choose the ratios 1 : 1 (square), 2 : 1, 3 : 1, and 3 : 2. The shape of a sector of the circle is determined by the angle at the center; we choose the angles 180°, 90°, and 60° (semicircle, quadrant, and sextant). We assume that all these figures have the same area, let us say, 1 square inch. Then we compute the length of the perimeter of each figure in inches. The numbers obtained are collected in the following table; the order of the figures is so chosen that the perimeters increase as we read them down.

---

[1] Commentary on the first book of Euclid's *Elements*; on Definitions XV and XVI.

[2] *Convivio II*, XIII, 26.

[3] *Oeuvres de Descartes*, edited by Adam and Tannery, vol. 10, 1908, p. 390. The passage is unessentially altered; the property of the circle under consideration is stated here in a different form.

Table I.   Perimeters of Figures of Equal Area

| Circle | 3.55 |
|---|---|
| Square | 4.00 |
| Quadrant | 4.03 |
| Rectangle 3 : 2 | 4.08 |
| Semicircle | 4.10 |
| Sextant | 4.21 |
| Rectangle 2 : 1 | 4.24 |
| Equilateral triangle | 4.56 |
| Rectangle 3 : 1 | 4.64 |
| Isosceles right triangle | 4.84 |

Of the ten figures listed, which are all of the same area, the circle, listed at the top, has the shortest perimeter.   Can we conclude hence by induction, as Descartes seems to suggest, that the circle has the shortest perimeter not only among the ten figures listed but among all possible figures? By no means.   But it cannot be denied that our relatively short list suggests very strongly the general theorem.   So strongly, indeed, that if we added one or two more figures to the list, the suggestion could not be made much stronger.

I am inclined to believe that Descartes, in writing the passage quoted, thought of this last, more subtle point.   He intended to say, I think, that prolonging the list would not have much influence on our belief.

**2. Latent reasons.**   "Of all plane figures of equal area, the circle has the minimum perimeter."   Let us call this statement, supported by Table I, the *isoperimetric theorem*.[4]   Table I, constructed according to Descartes' suggestion, yields a fairly convincing inductive argument in favor of the isoperimetric theorem.   Yet why does the argument appear convincing?

Let us imagine a somewhat similar situation.   We choose ten trees of ten different familiar kinds.   We measure the specific weight of the wood of each tree, and pick out the tree the wood of which has the least specific weight.   Would it be reasonable to believe merely on the basis of these observations that the kind of tree that has the lightest wood among the ten kinds examined has also the lightest wood among all existing kinds of trees? To believe this would not be reasonable, but silly.

What is the difference from the case of the circle?   We are *prejudiced* in favor of the circle.   The circle is the most perfect figure; we readily believe that, along with its other perfections, the circle has the shortest perimeter for a given area.   The inductive argument suggested by Descartes appears so convincing because it corroborates a conjecture plausible from the start.

"The circle is the most perfect figure" is a traditional phrase.   We find it in the writings of Dante (1265–1321), of Proclus (410–485), and of still earlier writers.   The meaning of the sentence is not clear, but there may be something more behind it than mere tradition.

---

[4] An explanation of the name and equivalent forms will be given later (sect. 8).

**3. Physical reasons.** "Of all solids of equal volume, the sphere has the minimum surface." We call this statement the "isoperimetric theorem in space."

We are inclined to believe the isoperimetric theorem in space, as in the plane, without any mathematical demonstration. We are prejudiced in favor of the sphere, perhaps even more than in favor of the circle. In fact, nature itself seems to be prejudiced in favor of the sphere. Raindrops, soap bubbles, the sun, the moon, our globe, the planets are spherical, or nearly spherical. With a little knowledge of the physics of surface tension, we could learn the isoperimetric theorem from a soap bubble.

Yet even if we are ignorant of serious physics, we can be led to the isoperimetric theorem by quite primitive considerations. We can learn it from a cat. I think you have seen what a cat does when he prepares himself for sleeping through a cold night: he pulls in his legs, curls up, and, in short, makes his body as spherical as possible. He does so, obviously, to keep warm, to minimize the heat escaping through the surface of his body. The cat, who has no intention of decreasing his volume, tries to decrease his surface. He solves the problem of a body with given volume and minimum surface in making himself as spherical as possible. He seems to have some knowledge of the isoperimetric theorem.

The physics underlying this consideration is extremely crude.[5] Still, the consideration is convincing and even valuable as a sort of provisional support for the isoperimetric theorem. The elusive reasons speaking in favor of the sphere or the circle, hinted above (sect. 2), start condensing. Are they reasons of physical analogy?

**4. Lord Rayleigh's inductive reasons.** A little more than two hundred years after the death of Descartes, the physicist Lord Rayleigh investigated the tones of membranes. The parchment stretched over a drum is a "membrane" (or, rather, a reasonable approximation to the mathematical idea of a membrane) provided that it is very carefully made and stretched so that it is uniform throughout. Drums are usually circular in shape but, after all, we could make drums of an elliptical, or polygonal, or any other shape. A drum of any form can produce different tones of which usually the deepest tone, called the principal tone, is much the strongest. Lord Rayleigh compared the principal tones of membranes of different shapes, but of equal area and subject to the same physical conditions. He constructed the following Table II, very similar to our Table I in sect. 1. This Table II lists the same shapes as Table I, but in somewhat different order, and gives for each shape the pitch (the frequency) of the principal tone.[6]

---

[5] A better advised cat should not make the surface of his body a minimum, but its thermal conductance or, which amounts to the same, its electrostatic capacity. Yet, by a theorem of Poincaré, this different problem of minimum has the same solution, the sphere. See G. Pólya, *American Mathematical Monthly*, v. 54, 1947, p. 201–206.

[6] Lord Rayleigh, *The Theory of Sound*, 2nd ed., vol. 1, p. 345.

Table II.  Principal Frequencies of Membranes of Equal Area

| Circle | 4.261 |
|---|---|
| Square | 4.443 |
| Quadrant | 4.551 |
| Sextant | 4.616 |
| Rectangle 3 : 2 | 4.624 |
| Equilateral triangle | 4.774 |
| Semicircle | 4.803 |
| Rectangle 2 : 1 | 4.967 |
| Isosceles right triangle | 4.967 |
| Rectangle 3 : 1 | 5.736 |

Of the ten membranes listed, which are all of the same area, the circular membrane, listed at the top, has the deepest principal tone.  Can we conclude hence by induction that the circle has the lowest principal tone of *all* shapes?

Of course, we cannot; induction is never conclusive.  Yet the suggestion is very strong, still stronger than in the foregoing case.  We know (and Lord Rayleigh and his contemporaries also knew) that of all figures with a given area the circle has the minimum perimeter, and that this theorem can be demonstrated mathematically.  With this geometrical minimum property of the circle in our mind, we are inclined to believe that the circle has also the physical minimum property suggested by Table II.  Our judgement is influenced by analogy, and analogy has a deep influence.

The comparison of Tables I and II is highly instructive.  It yields various other suggestions which we do not attempt to discuss now.

**5. Deriving consequences.**  We have surveyed various grounds in favor of the isoperimetric theorem which are, of course, insufficient to prove it but sufficient to make it a reasonable conjecture.  A physicist, examining a conjecture in his science, derives consequences from it.  These consequences may or may not agree with the facts and the physicist devises experiments to find out which is the case.  A mathematician, examining a conjecture in his science, may follow a similar course.  He derives consequences from his conjecture.  These consequences may or may not be true and the mathematician tries to find out which is the case.

Let us follow this course in examining the isoperimetric theorem which we state now in the following form: *Of all plane figures of equal perimeter, the circle has the maximum area.*  This statement differs from that given above (sect. 2) and not merely verbally.  Yet it can be shown that the two statements are equivalent.  We postpone the proof (see sect. 8) and hasten to examine consequences.

(1) Dido, the fugitive daughter of a Tyrian king, arrived after many adventures at the coast of Africa where she became later the founder of Carthage and its first legendary queen.  Dido started by purchasing from

the natives a piece of land on the seashore "not larger than what an oxhide can surround." She cut the oxhide into fine narrow strips of which she made a very long string. And then Dido faced a geometric problem: what shape of land should she surround with her string of given length in order to obtain the maximum area?

In the interior of the continent the answer would be a circle, of course, but on the seashore the problem is different. Let us solve it, in assuming that the seashore is a straight line. In fig. 10.1 the arc $XYZ$ has a given length. We are required to make a maximum the area between this arc and the straight line $XZ$ (which lies on a given infinite straight line, but can be lengthened or shortened at pleasure).

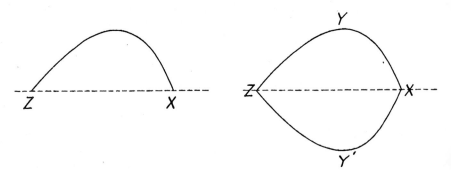

Fig. 10.1.  Dido's problem.          Fig. 10.2.  Solution by mirror image

To solve this problem we regard the given infinite straight line (the seashore) as a mirror; see fig. 10.2. The line $XYZ$ and its mirror image $XY'Z$ form jointly a *closed* curve $XYZY'$ of given length surrounding an area which is exactly double the one that we have to maximize. This area is a maximum when the closed curve is a circle of which the given infinite straight line (the seashore) is an axis of symmetry. Therefore, the solution of Dido's problem is a semicircle with center on the seashore.

(2) Jakob Steiner derived a host of interesting consequences from the isoperimetric theorem. Let us discuss one of his arguments which is especially striking. Inscribe in a given circle a polygon (fig. 10.3). Regard the segments of the circle (shaded in fig. 10.3) cut off by the sides of the inscribed polygon as rigid (cut out of cardboard). Imagine these rigid segments of the circle connected by flexible joints at the vertices of the inscribed polygon. Deform this articulated system by changing the angles at the joints. After deformation (see fig. 10.4) you obtain a new curve which is not a circle, but consists of successive circular arcs and has the same perimeter as the given circle. Therefore, by the isoperimetric theorem, the area of the new curve must be less than the area of the given circle. Yet the circular segments are rigid (of cardboard), their areas unchanged, and so the deformed polygon must take the blame for lessening the area:

*The area of a polygon inscribed in a circle is greater than the area of any other polygon with the same sides.* (The sides are the same in length and in order of succession.)

This consequence is elegant, but as yet unproved, insofar as we have not proved yet the isoperimetric theorem itself.

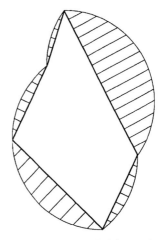

Fig. 10.3.  An inscribed polygon.

Fig. 10.4.  Flexible joints and cardboard segments.

(3) Let us combine Dido's problem with Steiner's method. Inscribe in a given semicircle a polygonal line; see fig. 10.5. Regard the segments cut off by the stretches of the polygonal line from the semicircle (shaded in fig. 10.5) as rigid (of cardboard). Place flexible joints at the vertices of

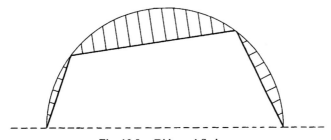

Fig. 10.5.  Dido and Steiner.

the polygonal line, vary the angles, and let the two endpoints shift along the line of the diameter, which you regard as given. You obtain so a new curve (fig. 10.6) consisting of circular arcs of the same total length as the semicircle, but including with the given infinite line less area than the semicircle, by virtue of the theorem that we have discussed under (1). Yet the circular segments are rigid (of cardboard) and so the deformed polygon is responsible for the lessening of the area. Hence the theorem: *The sides of a polygon are given, except one, in length and succession. The area becomes a maximum when the polygon is inscribed in a semicircle the diameter of which is the side originally not given.*

**6. Verifying consequences.** A physicist, having derived various consequences from his conjecture, looks for one that can be conveniently tested by experiments. If the experiments clearly contradict a consequence derived from it, the conjecture itself is exploded. If the experiments verify its consequences, the conjecture gains in authority, becomes more credible.

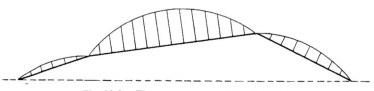

Fig. 10.6. The segments are of cardboard.

The mathematician may follow a similar course. He looks for accessible consequences of his conjecture which he could prove or disprove. A consequence disproved disproves the conjecture itself. A consequence proved renders the conjecture more credible and may hint a line along which the conjecture itself could be proved.

How about our own case? We have derived several consequences of the isoperimetric theorem; which one is the most accessible?

(1) Some of the consequences derived from the isoperimetric theorem in the foregoing section are, in fact, concerned with elementary problems on maxima. Is there any consequence that we could verify? Let us survey the various cases indicated by figs. 10.3–10.6. Which case is the simplest? The complexity of a polygon increases with the number of its sides. Therefore, the simplest polygon of all is the triangle; of course, we like the triangle best because we know the most about it. Now, the problem of figs. 10.3 and 10.4 makes no sense for triangles or, we may say, it is vacuous in the case of a triangle: a triangle with given sides is determined, rigid. There is no transition for a triangle like that from fig. 10.3 to fig. 10.4. Yet the transition from fig. 10.5 to 10.6 is perfectly possible for triangles. This may be the simplest consequence that we have derived so far from the isoperimetric theorem: let us examine it.

The simplest particular case of the result derived in sect. 5 (3) answers the following problem: *Given two sides of a triangle, find the maximum of the area;* see fig. 10.7. Sect. 5 (3) gives this answer: The area is a maximum when the triangle is inscribed in a semicircle, the diameter of which is the side originally not given. This means, however, that the area is a maximum when the two given sides include a right angle which is obvious (ex. 8.7).

We have succeeded in verifying a first consequence of the isoperimetric theorem. Such success naturally raises our spirits. What is behind the fact just verified? Could we generalize it? Could we verify some other consequence?

(2) In generalizing the problem discussed under (1), we arrive at the following: *All successive sides of a polygon are given in magnitude, except one. Find the maximum area.*

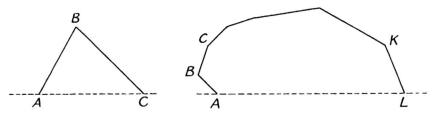

Fig. 10.7. The finger with one joint.          Fig. 10.8. The super-finger.

We introduce suitable notation and draw fig. 10.8. The lengths $AB$, $BC, \ldots KL$ are given; the length $LA$ is not given. We can imagine the broken line $ABC \ldots F \ldots KL$ as a sort of "super-finger"; the "bones" $AB, BC, \ldots KL$ are of invariable length, the angles at the joints $B, C, \ldots$ $F \ldots K$ variable. We are required to make the area $ABC \ldots KLA$ a maximum.

As in some problems that we have considered some time ago (sect. 8.4, 8.5), the characteristic difficulty seems to be that there are many variables (the angles at $B, C, \ldots F, \ldots$ and $K$). Yet we have just discussed, under (1), the extreme special case of the problem where there is just one variable angle (just one joint; fig. 10.7). It is natural to hope that we can use this special case as a stepping stone to the solution of the general problem.

Fig. 10.9. Only one joint is flexible.

In fact, let us take the problem as almost solved. Let us imagine that we have obtained already the desired values of all the angles except one. In fig. 10.9 we regard the angle at $F$ as variable, but all the other angles, at $B, C, \ldots K$ as fixed; the joints $B, C, \ldots K$ are rigid, only $F$ is flexible. We join $A$ and $L$ to $F$. The lengths $AF$ and $LF$ are invariable. The whole polygon $ABC \ldots F \ldots KLA$ is decomposed now into three parts, two of which are rigid (of cardboard) and only the third variable. The polygons $ABC \ldots FA$ and $LK \ldots FL$ are rigid. The triangle $AFL$ has two given sides, $FA$ and $FL$, and a variable angle at $F$. The area of this triangle, and with it the area of the whole polygon $ABC \ldots F \ldots KLA$, becomes a maximum when $\angle AFL$ is a right angle, as we have said a moment ago, under (1), in discussing fig. 10.7.

This reasoning obviously applies just as well to the other joints, that is, to the angles at $B$, $C$, ... and $K$ (fig. 10.8), and so we see: *the area of the polygon $ABC \ldots KLA$ cannot be a maximum unless the side originally not given, $AL$, subtends a right angle at each of the vertices not belonging to it, at $B$, $C$, ... $F$, ... $K$.* If there is a maximum area, it must be attained in the situation just described. We may take for granted that there is a maximum area and, remembering a little elementary geometry, describe the situation in other terms, as follows: *the maximum of the area is attained if, and only if, the polygon is inscribed in a semicircle, the diameter of which is the side originally not given.*

We have obtained here exactly the same result as in sect. 5 (3), but we did not use the isoperimetric theorem here and we did there.

(3) We have verified first, under (1), a very special consequence of the isoperimetric theorem, then, under (2), a much broader consequence. We have gathered now, perhaps, enough momentum to attack another broad consequence, derived above, in sect. 5(2).

We compare two polygons $ABC \ldots KL$ and $A'B'C' \ldots K'L'$; see fig. 10.10. The corresponding sides are equal $AB = A'B'$, $BC = B'C'$, ...$KL = K'L'$, $LA = L'A'$, but some of the angles are different; $ABC \ldots KL$ is inscribed in a circle, but $A'B'C' \ldots K'L'$ is not.

We join a vertex $J$ of $ABC \ldots KL$ to the center of the circumscribed circle and draw the diameter $JM$. If, by chance, the point $M$ coincides with a vertex of $ABC \ldots KL$, our task is greatly simplified (we could use then the result under (2) immediately). If not, $M$ lies on the circle between two adjacent vertices of the inscribed polygon, say, $A$ and $B$. Join $MA$, $MB$, consider $\triangle AMB$ (shaded in fig. 10.10) and construct over the base $A'B'$ the $\triangle A'M'B'$ (also shaded) congruent to $\triangle AMB$. Finally, join $J'M'$.

The polygon $AMBC \ldots KL$ is divided into two parts by the line $JM$ (see fig. 10.10; the corresponding polygon is correspondingly divided by $J'M'$). Apply to both parts the theorem proved under (2). The area of the polygon $MBC \ldots J$, inscribed in a semicircle, is not less than the area of $M'B'C' \ldots J'$; in fact, the corresponding sides are all equal, except that $MJ$, which forms the diameter of the semicircle, may differ from $M'J'$. For the same reason the area of $MALK \ldots J$ is not less than that of $M'A'L'K' \ldots J'$. By adding, we obtain that

$$\text{area } AMBC \ldots KL > \text{area } A'M'B'C' \ldots K'L'.$$

Yet

$$\triangle AMB \cong \triangle A'M'B'.$$

By subtracting, we obtain that

$$\text{area } ABC \ldots KL > \text{area } A'B'C' \ldots K'L'.$$

*The area of a polygon inscribed in a circle is greater than the area of any other polygon with the same sides.*
We have obtained here exactly the same result as in sect. 5 (2), but we did not use the isoperimetric theorem here and we did there.
(The first inequality, between the areas of the extended polygons, contains the sign $>$ although a conscientious reader may have expected the sign $\geqq$. Let us append the discussion of this somewhat more subtle point. I

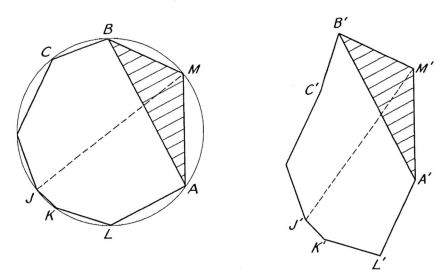

Fig. 10.10. One polygon is inscribed, the other is not.

say that the polygon $A'M'B'C' \ldots K'L'$ is not inscriptible in a circle; otherwise, $A'B'C' \ldots K'L'$ would also be inscriptible, which it is not. I say that the polygons $M'B'C' \ldots J'$ and $M'A'L'K' \ldots J'$ are not both inscriptible in a semicircle with diameter $M'J'$; otherwise, the whole polygon $A'M'B'C' \ldots K'L'$ would be inscriptible in a circle, which it is not. Hence the words "not less," used twice in the derivation of the inequality in question, can be replaced at least once by "greater.")[7]

**7. Very close.** The consequences that we have succeeded in verifying render the isoperimetric theorem highly plausible. Yet there is more. We may have the feeling that these consequences "contain a lot," that we are "very close" to the final solution, to the complete proof.

---

[7] The theorems and demonstrations of this section are due to Lhuilier; see footnote 5 of Chapter VIII.

(1) *Find the polygon with a given number of sides and a given perimeter that has the maximum area.*

If there is such a polygon, it must be *inscribed in a circle.* This much we can conclude immediately from our last remark, sect. 6(3).

On the other hand, take the problem as almost solved. Assume that you know already the correct position of all vertices except one, say, $X$. The $n - 1$ other vertices, say, $U, \ldots W, Y$ and $Z$, are already fixed. The whole polygon $U \ldots WXYZ$ consists of two parts: the polygon $U \ldots WYZ$ with $n - 1$ already fixed vertices, which is independent of $X$, and $\triangle WXY$, which depends on $X$. Of this triangle, $\triangle WXY$, you know the base $WY$ and the sum of the two other sides $WX + XY$; in fact, the remaining $n - 2$ sides of the polygon are supposed to be known, and you actually know the sum of all $n$ sides. The area of $\triangle WXY$ must be a maximum. Yet, as it is almost immediate, the $\triangle WXY$ with known base and perimeter attains its maximum area when it is isosceles (ex. 8.8). That is, $WX = XY$, two adjacent sides of the required polygon are equal. Therefore (by the symmetry of the conditions and the pattern of partial variation) any two adjacent sides are equal. All sides are equal: the desired polygon is *equilateral.*

The desired polygon, which is inscribed in a circle and also equilateral, is necessarily regular: *Of all polygons with a given number of sides and a given perimeter, the regular polygon has the largest area.*

(2) *Two regular polygons, one with n sides and the other with n + 1 sides, have the same perimeter. Which one has the larger area?*

The regular polygon with $n + 1$ sides has a larger area than any irregular polygon with $n + 1$ sides and the same perimeter, as we have just seen, under (1). Yet the regular polygon with $n$ sides, each equal to $a$, say, can be regarded as an irregular polygon with $n + 1$ sides: $n - 1$ sides are of length $a$, two sides of length $a/2$, and there is one angle equal to 180°. (Regard the midpoint of one side of the polygon, conceived in the usual way, as a vertex, and then you arrive at the present less-usual conception.) Therefore, *the regular polygon with n + 1 sides has a larger area than the regular polygon with n sides and the same perimeter.*

(3) *A circle and a regular polygon have the same perimeter. Which one has the larger area?*

Let us realize what the foregoing result, under (2), means. Let us take $n = 3, 4, \ldots$ and restate the result in each particular case. In passing from an equilateral triangle to a square with the same perimeter, we find the area increased. In passing from a square to a regular pentagon with the same perimeter, we again find the area increased. And so on, passing from one regular figure to the next, from pentagon to hexagon, from hexagon to heptagon, from $n$ to $n + 1$, we see that the area increases at each step as the perimeter remains unchanged. Ultimately, in the limit, we reach the circle. Its perimeter is still the same, but its area is obviously superior to the area of any regular polygon in the infinite sequence of which it is the limit. *The area of the circle is larger than that of any regular polygon with the same perimeter.*

(4) *A circle and an arbitrary polygon have the same perimeter. Which one has the larger area?*
The circle. This follows immediately from the foregoing (1) and (3).

(5) *A circle and an arbitrary curve have the same perimeter. Which one has the larger area?*
The circle. This follows from the foregoing (4), since any curve is the limit of polygons. We have proved the isoperimetric theorem!

**8. Three forms of the Isoperimetric Theorem.** In the foregoing (sect. 6 and 7) we have proved the following statement of the isoperimetric theorem:

I. *Of all plane figures of equal perimeter, the circle has the maximum area.*
In sect. 2, however, we discussed another statement.

II. *Of all plane figures of equal area the circle has the minimum perimeter.*
These two statements are different, and different not merely in wording. They need some further clarification.

(1) Two curves are called "isoperimetric" if their perimeters are equal. "Of all isoperimetric plane curves the circle has the largest area"—this is the traditional wording of statement I, which explains the name "isoperimetric theorem."

(2) We may call the two statements of the theorem (I and II) "conjugate statements" (see sect. 8.6). We shall show that these two conjugate statements are equivalent to each other by showing that they are both equivalent to the same third.

(3) Let $A$ denote the area and $L$ the length of the perimeter of a given curve. Let us assume that the given curve and a circle with radius $r$ are isoperimetric: $L = 2\pi r$. Then the first form (statement I) of the isoperimetric theorem asserts that

$$A \leq \pi r^2.$$

Substituting for $r$ its expression in terms of $L$, $r = L/2\pi$, we easily transform the inequality into

$$\frac{4\pi A}{L^2} \leq 1.$$

We call this inequality the *isoperimetric inequality* and the quotient on the left hand side the *isoperimetric quotient*. This quotient depends only on the shape of the curve and is independent of its size. In fact, if, without changing the form, we enlarge the linear dimensions of the curve in the ratio 1 : 2, the perimeter becomes $2L$ and the area $4A$, but the quotient $A/L^2$ remains unchanged, and the same is true of $4\pi A/L^2$ and of enlargements in any ratio. Some authors call $A/L^2$ the isoperimetric quotient; we have introduced the

.

factor $4\pi$ to make our isoperimetric quotient equal to 1 in the case of the circle. With this terminology, we can say:

III. *Of all plane figures, the circle has the highest isoperimetric quotient.*[8]

This is the third form of the isoperimetric theorem.

(4) We have arrived at the third form of the theorem in coming from figures with equal perimeter. Let us now start from statement III and pass to figures with equal area. Let us assume that a curve with area $A$ and perimeter $L$ has the same area as a circle with radius $r$. That is, $A = \pi r^2$. Substituting for $A$ this expression, we easily transform the isoperimetric inequality into $L \geq 2\pi r$. That is, the perimeter of the figure is greater than that of a circle with equal area. We arrive so at the second conjugate form of the theorem, at statement II.

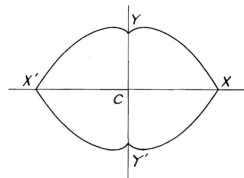

Fig. 10.11. Dido's problem, complicated by a cape.

Fig. 10.12. Reflection solves it sometimes.

(5) We could, of course, proceed along the same line of argument in the opposite direction and, passing through III, derive I from II. And so we can satisfy ourselves that all three forms are equivalent.

**9. Applications and questions.** If Dido bargained with the natives in the neighborhood of a cape, her problem was, perhaps, more similar to the following than to that discussed in sect. 5 (1).

*Given an angle* (the infinite part of a plane between two rays drawn from the same initial point). *Find the maximum area cut off from it by a line of given length.*

In fig. 10.11 the vertex of the given angle is called $C$ (cape). The arbitrary line connecting the points $X$ and $Y$ is supposed to have the given length $l$. We are required to make a maximum the three-cornered area between this curve and the seashore. We may shift the endpoints $X$ and $Y$ of the curve and modify its shape, but cannot change its length $l$.

The problem is not too easy, but is one of those problems which a particular choice of the data renders more accessible. If the angle at $C$ is a right angle,

---

[8] Abbreviating "isoperimetric quotient" as I.Q., we could say that the circle has the highest I.Q.

we may take the mirror image of the figure first with respect to one side of the angle, then with respect to the other. We obtain so a new figure, fig. 10.12, and a new problem. The line $XY$, quadruplicated by reflections, yields a new *closed* line of given length, $4l$. The area to be maximized, quadruplicated by reflections, yields a new area to be maximized, *completely surrounded* by the new given curve. By the isoperimetric theorem, the solution of the new problem is a circle. This circle has two given axes of symmetry, $XX'$ and $YY'$, and so its center is at the intersection of these axes, at the point $C$. Therefore, the solution of the original problem (Dido's problem) is a *quadrant*: a quarter of a circle with center at the vertex of the given angle.

We naturally recall here the solution of sect. 5 (1) based on fig. 10.2 and observe that it is closely analogous to the present solution. It is easy to see that there are an infinity of further cases in which this kind of solution works. If the given angle at $C$ is $360°/2n = 180°/n$, we can transform, by repeated reflections, the curve $XY$ with given length $l$ into a new closed curve with length $2nl$ and the proposed problem into a new problem, the solution of which is a circle, by virtue of the isoperimetric theorem. The cases treated in sect. 5 (1) and in the present section are just the first two cases in this infinite sequence, corresponding to $n = 1$ and $n = 2$.

That is, if the angle at $C$ is of a special kind ($180°/n$ with integral $n$) the solution of our problem (fig. 10.11) is a circular arc with center at $C$. It is natural to expect that this form of the solution is independent of the magnitude of the angle (at least as long as it does not exceed $180°$). That is, we conjecture that the solution of the problem of fig. 10.11 is the arc of a circle with center at $C$, whether the angle at $C$ is, or is not, of the special kind $180°/n$. This conjecture is an inductive conjecture, supported by the evidence of an infinity of particular cases, $n = 1, 2, 3, \ldots$ . Is this conjecture true?

The foregoing application of the isoperimetric theorem and the attached question may make us anticipate many similar applications and questions. Our derivation of the theorem raises further questions; its analogues in solid geometry and mathematical physics suggest still other questions. The isoperimetric theorem, deeply rooted in our experience and intuition, so easy to conjecture, but not so easy to prove, is an inexhaustible source of inspiration.

## EXAMPLES AND COMMENTS ON CHAPTER X

### First Part

**1.** *Looking back.* In the foregoing (sect. 6–8) we have proved the isoperimetric theorem—have we? Let us check the argument step by step.

There seems to be no objection against the simple result of sect. 6 (1). Yet, in solving the problem of sect. 6 (2), we assumed the existence of the maximum without proof; and we did the same in sect. 7 (1). Do these unproved assumptions invalidate the result?

**2.** *Could you derive some part of the result differently?* Verify the simplest non-trivial particular case of the result found in sect. 5 (2) directly. That is, prove independently of sect. 6 (3) that the area of a quadrilateral inscribed in a circle is greater than the area of any other quadrilateral with the same sides. [Ex. 8.41.]

**3.** *Restate with more detail* the argument of sect. 7 (2): construct a polygon with $n + 1$ sides that has the same perimeter as, but a greater area than, the regular polygon with $n$ sides.

**4.** Prove independently of sect. 7 (3) that a circle has a larger area than any regular polygon with the same perimeter.

**5.** Prove, more generally, that a circle has a larger area than any circum-scribable polygon with the same perimeter.

**6.** Restate with more detail the argument of sect. 7 (5). Does it prove the statement I of sect. 8? Is there any objection?

**7.** *Can you use the method for some other problem?* Use the method of sect. 8 to prove that the following two statements are equivalent:

"Of all boxes with a given surface area the cube has the maximum volume."

"Of all boxes with a given volume the cube has the minimum surface area."

**8.** *Sharper form of the Isoperimetric Theorem.* Compare the statements I, II, and III of sect. 8 with the following.

I'. *The area of a circle is larger than that of any other plane curve with the same perimeter.*

II'. *The perimeter of a circle is shorter than that of any other plane curve with the same area.*

III'. *If A is the area of a plane curve and L the length of its perimeter, then*

$$\frac{4\pi A}{L^2} \leqq 1$$

*and equality is attained if, and only if, the curve is a circle.*

Show that I', II', and III' are equivalent to each other. Have we proved I'?

**9.** Given a curve $C$ with perimeter $L$ and area $A$; $C$ is not a circle. Construct a curve $C'$ with the same perimeter $L$, but with an area $A'$ larger than $A$.

This problem is important (why?) but not too easy. If you cannot solve it in full generality, solve it in significant special cases; put pertinent questions that could bring you nearer to its general solution; try to restate it; try to approach it in one way or the other.

**10.** Given a quadrilateral $C$ with a re-entrant angle, the perimeter $L$ and the area $A$. Construct a triangle $C'$ with the same perimeter $L$, but with an area $A'$ larger than $A$.

**11.** Generalize ex. 10.

**12.** The information "$C$ is not a circle" is "purely negative." Could you characterize $C$ more "positively" in some manner that would give you a foothold for tackling ex. 9?

[Any three points on any curve are on the same circle, or on a straight line. What about four points?]

**13.** Given a curve $C$ with perimeter $L$ and area $A$; there are four points $P$, $Q$, $R$, and $S$ on $C$ which are not on the same circle, nor on the same straight line. Construct a curve $C'$ with the same perimeter $L$, but with an area $A'$ larger than $A$. [Ex. 2.]

**14.** Compare the following two questions.

We consider curves with a given perimeter. If $C$ is such a curve, but not the circle, we can construct another curve $C'$ with a greater area. (In fact, this has been done in exs. 10–13. The condition that $C$ is not a circle is essential; our construction fails to increase the area of the circle.) Can we conclude hence that the circle has the greatest area?

We consider positive integers. If $n$ is such an integer, but not 1, we can construct another integer $n'$ greater than $n$. (In fact, set $n' = n^2$. The condition $n > 1$ is essential; our construction fails for $n = 1$ as $1^2 = 1$.) Can we conclude hence that 1 is the greatest integer?

Point out the difference if there is any.

**15.** Prove the statement $I'$ of ex. 8.

### Second Part

**16.** *The stick and the string.* Given a stick and a string, each end of the stick attached to the corresponding end of the string (which, of course, must be longer than the stick). Surround with this contraption the largest possible area.

Lay down the stick. Its endpoints $A$ and $B$ determine its position completely. Yet the string can take infinitely many shapes, forming an arbitrary curve with given length that begins in $A$ and ends in $B$; see fig. 10.13. One of the possible shapes of the string is a circular arc which includes a segment of a circle with the stick. Complete the circle by adding another segment (shaded in fig. 10.14, I) and add the same segment to the figure included by the stick and an arbitrary position of the string (fig. 10.14, II). The circle I of fig. 10.14 has a larger area than any other curve with the same perimeter, and II of fig. 10.14 *is* such a curve. Subtracting the same (shaded) segment from I and II, we find the result: the area surrounded by the stick and the string is a *maximum when the string forms a circular arc*.

This result remains valid if we add to the variable area of fig. 10.13 any invariable area along its invariable straight boundary line. This remark is often useful.

State the conjugate result. That is, formulate the fact that has the same relation to theorem II of sect. 8 as the fact found in the foregoing has to theorem I of sect. 8.

**17.** Given an angle (the infinite part of a plane between two rays drawn from the same initial point) and two points, one on each side of the angle. Find the maximum area cut off from the angle by a line of given length

*B*                                        *A*

Fig. 10.13.    Stick and string.

connecting the two given points.    (In fig. 10.11 the points $\overset{\text{.}}{X}$ and $Y$ are given.)

**18.** Given an angle, with an opening less than 180°, and a point on one of its sides.    Find the maximum area cut off from the angle by a line of given length that begins at the given point.    (In fig. 10.11 the point $X$ is given, but $Y$ is variable.)

**19.** Given an angle with an opening less than 180°.    Find the maximum area cut off from the angle by a line of given length.    (In fig. 10.11, the points $X$ and $Y$ are variable.    A conjecture was stated in sect. 9.)

**20.** Given an angle with an opening less than 180°.    Find the maximum area cut off from the angle by a straight line of given length.

**21.** *Two sticks and two strings.*    We have two sticks, $AB$ and $CD$.    A first string is attached to the last point $B$ of the first stick at one end and to the first point $C$ of the second stick at the other end.    Another string connects similarly $D$ and $A$.    Surround with this contraption the largest possible area.

**22.** Generalize.

**23.** Specialize, and obtain by specializing an elementary theorem that played an important role in the text.

**24.** Given a circle in space. Find the surface with given area bounded by the given circle that includes the maximum volume with the disk rimmed by the given circle. [Do you know an analogous problem?]

**25.** *Dido's problem in solid geometry.* Given a trihedral angle (one of the eight infinite parts into which space is divided by three planes intersecting in one point). Find the maximum volume cut off from the trihedral angle by a surface of given area.

This problem is too difficult. You are only asked to pick out a more accessible special case.

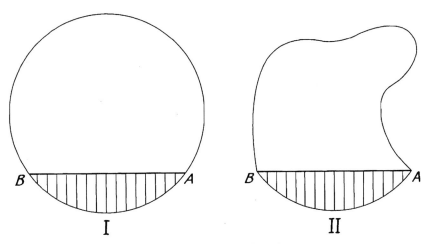

Fig. 10.14. The principle of circular arc.

**26.** Find a problem analogous to ex. 25 of which you can foresee the result. [Generalize, specialize, pass to the limit, . . . .]

**27.** *Bisectors of a plane region.* We consider a plane region surrounded by a curve. An arc that joins two points of the surrounding curve is called a *bisector* of the region if it divides the region into two parts of equal area.

Show that any two bisectors of the same region have at least one common point.

**28.** Compare two bisectors of a square. One is a straight line parallel to one of the sides that passes through the center of the square. The other is one-quarter of a circle the center of which is a vertex. Which of the two is shorter?

**29.** Find the shortest straight bisector of an equilateral triangle.

**30.** Find the shortest bisector of an equilateral triangle.

**31.** Show that the shortest bisectors of a circle are its diameters.

**32.** Find the shortest bisector of an ellipse.

**33.** Try to formulate general theorems covering ex. 28–32.

**34.** *Bisectors of a closed surface.*[9]   A not self-intersecting closed curve on a closed surface is called a *bisector* of the surface if it divides the surface into two parts (open surfaces) of equal area.

Show that any two bisectors of the same surface have at least one common point.

**35.** A shortest bisector of the surface of a polyhedron consists of pieces each of which is either a straight line or a circular arc.

**36.** A shortest bisector of the surface of a regular solid is a regular polygon. Find its shape and location and the number of solutions for each of the five regular solids. (You may experiment with a model of the solid and a rubber band.)

**37.** Show that the shortest bisectors of a spherical surface are the great circles.

**38.** Try to find a generalization of ex. 37 that covers also a substantial part of ex. 36. [Ex. 9.23, 9.24.]

**39.** Given a sphere $S$ with radius $a$. We call a *diaphragm* of $S$ that part of a spherical surface intersecting $S$ that is within $S$. Prove:

(1) All diaphragms passing through the center of $S$ have the same area.

(2) No diaphragm bisecting the volume of $S$ has an area less than $\pi a^2$.

The last statement, and the analogous cases discussed, suggest a conjecture. State it. [Ex. 31, 37.]

**40.** *A figure of many perfections.* We consider a plane region surrounded by a curve. We wish to survey some of the many theorems analogous to the isoperimetric theorem: Of all regions with a given area, the circle has the minimum perimeter.

We met already with a theorem of this kind. In sect. 4, we considered some inductive evidence for the statement: Of all membranes with a given area, the circular membrane emits the deepest principal tone.

Let us now regard the region as a homogeneous plate of uniform thickness. We consider the moment of inertia of this plate about an axis perpendicular to it that passes through its center of gravity. This moment of inertia, which we call the "polar moment of inertia," depends, other things being equal, on the size and shape of the plate. Of all plates with a given area, the circular plate has the minimum polar moment of inertia.

This plate, if it is a conductor of electricity, can also receive an electric charge, proportional to its electrostatic capacity. Also the capacity depends

---

[9] We consider here only closed surfaces of the "topological type" of the sphere and exclude, for instance, the (doughnut shaped) torus.

on the size and shape of the plate. Of all plates with a given area, the circular plate has the minimum capacity.

Now let the region be a cross-section of a homogeneous elastic beam. If we twist such a beam about its axis, we may observe that it resists the twisting. This resistance, or "torsional rigidity," of the beam depends, other things being equal, on the size and shape of the cross-section. Of all cross-sections with a given area, the circular cross-section has the maximum torsional rigidity.[10]

Why is the circle the solution of so many and so different problems on minima and maxima? What is the "reason"? Is the "perfect symmetry" of the circle the "true reason"? Such vague questions may be stimulating and fruitful, provided that you do not merely indulge in vague talk and speculation, but try seriously to get down to something more precise or more concrete.

**41. An analogous case.** Do you see the analogy between the isoperimetric theorem and the theorem of the means? (See sect. 8.6.)

The length of a closed curve depends in the same manner on each point, or on each element, of the curve. Also the area of the region surrounded by the curve depends in the same manner on each point, or element, of the curve. We seek the maximum of the area when the length is given. As both quantities concerned are of such a nature that no point of the curve plays a favored role in their definition, we need not be surprised that the solution is the only closed curve that contains each of its points in the same way and any two elements of which are superposable: the circle.

The sum $x_1 + x_2 + \ldots + x_n$ is a symmetric function of the variables $x_1, x_2, \ldots x_n$; that is, it depends in the same manner on each variable. Also the product $x_1 x_2 \ldots x_n$ depends in the same manner on each variable. We seek the maximum of the product when the sum is given. As both quantities concerned are symmetric in the $n$ variables, we need not be surprised that the solution requires $x_1 = x_2 = \ldots = x_n$.

Besides the area and the length there are other quantities depending on the size and shape of a closed curve which "depend in the same manner on each element of the curve"; we listed several such quantities in ex. 40. We seek the maximum of a quantity of this kind when another quantity of the same kind is given. Is the solution, if there is one, necessarily the circle?

Let us turn to the simpler analogous case for a plausible answer. Let us consider two symmetric functions, $f(x_1, x_2, \ldots x_n)$ and $g(x_1, x_2, \ldots x_n)$, of $n$ variables and let us seek the extrema of $f(x_1, x_2, \ldots x_n)$ when we are given that $g(x_1, x_2, \ldots x_n) = 1$. There are cases in which there is no maximum, other cases in which there is no minimum, and still other cases in which neither a maximum nor a minimum exists. The condition

---

[10] For proofs of the theorems indicated and for similar theorems, see G. Pólya and G. Szegö, *Isoperimetric Inequalities in Mathematical Physics*, Princeton University Press, 1951.

$x_1 = x_2 = \ldots = x_n$ plays an important rôle,[11] yet it need not be satisfied when a maximum or a minimum is attained. There is, however, a simple fact. If

$$x_1 = a_1, \quad x_2 = a_2, \quad x_3 = a_3, \ldots x_n = a_n$$

is a solution, also

$$x_1 = a_2, \quad x_2 = a_1, \quad x_3 = a_3, \ldots x_n = a_n$$

is a solution, by the symmetry of the functions $f$ and $g$. Therefore, if $a_1 \neq a_2$, there are are least two different solutions. *If there is a unique solution* (that is, if the extremum is attained, and attained for just one system of values of $x_1, x_2, \ldots x_n$) *the solution requires* $x_1 = x_2 = \ldots = x_n$.

"Comparaison n'est pas raison," say the French. Of course, such comparisons as the preceding cannot yield a binding reason, only a heuristic indication. Yet we are quite pleased sometimes to receive such an indication. Take as an illustration

$$f(x_1, x_2, \ldots x_n) = (x_1 + x_2 + \ldots + x_n)^2,$$

$$g(x_1, x_2, \ldots x_n) = (x_1^2 + x_2^2 + \ldots + x_n^2)/n$$

and find the extrema of $f$ under the condition $g = 1$ considering (1) all real values of $x_1, x_2, \ldots x_n$ and (2) only non-negative real values of these variables.

**42.** *The regular solids.* Find the polyhedron with a given number $n$ of faces and with a given surface-area that has the maximum volume.

This very difficult problem is suggested by the analogous problem of sect. 7 (1) which also suggests a conjecture: if there is a regular solid with $n$ faces, it yields the maximum volume. However plausible this conjecture may seem, it turned out wrong in two cases out of five. In fact, the conjecture is

correct for   $n = 4, 6, 12,$

incorrect for $n = 8, 20.$

What is the difference? Try to observe some simple geometrical property that distinguishes between the two kinds of regular solids.

**43.** *Inductive reasons.* Let $V$ denote the volume of a solid and $S$ the area of its surface. By analogy, sect. 8 (3) suggests to define

$$\frac{36\pi V^2}{S^3}$$

as the isoperimetric quotient in solid geometry. By analogy, we may conjecture that the sphere has the highest isoperimetric quotient. Table III supports this conjecture inductively.

---

[11] G. H. Hardy, J. E. Littlewood, and G. Pólya, *Inequalities.* See pp. 109–10, and the theorems there quoted.

Check some of the figures given in Table III and add new material. In particular, try to find a solid with an isoperimetric quotient higher than that of the regular icosahedron.

Table III. The Isoperimetric Quotient $36\pi V^2/S^3$

| Sphere | 1.0000 |
|---|---|
| Icosahedron | 0.8288 |
| Best double cone | 0.7698 |
| Dodecahedron | 0.7547 |
| Best prism | 0.6667 |
| Octahedron | 0.6045 |
| Cube | 0.5236 |
| Best cone | 0.5000 |
| Tetrahedron | 0.3023 |

For "best" double cone, prism, and cone see ex. 8.38, 8.35, and 8.52, respectively.

# FURTHER KINDS OF PLAUSIBLE REASONS

*The most simple relations are the most common, and this is the foundation*
*upon which induction rests.*—LAPLACE[1]

**1. Conjectures and conjectures.** All our foregoing discussions dealt with the rôle of conjectures in mathematical research. Our examples gave us an opportunity to familiarize ourselves with two kinds of plausible arguments speaking for or against a proposed conjecture: we discussed inductive arguments, from the verification of consequences, and arguments from analogy. Are there other kinds of useful plausible arguments for or against a conjecture? The examples of the present chapter aim at clarifying this question.

We should also realize that there are conjectures of various kinds: great and small, original and routine conjectures. There are conjectures which played a spectacular rôle in the history of science, but also the solution of the most modest mathematical problem may need some correspondingly modest conjecture or guess. We begin with examples from the classroom and then proceed to others which are of historical importance.

**2. Judging by a related case.** Working at a problem, we often try to guess. Of course, we would like to guess the whole solution. If, however, we do not succeed in this, we are quite satisfied if we can guess this or that feature of the solution. At least, we should like to know whether our problem is "reasonable." We ask ourselves: Is our problem reasonable? *Is it possible to satisfy the condition? Is the condition sufficient to determine the unknown? Or is it insufficient? Or redundant? Or contradictory?*[2]

Such questions come naturally and are particularly useful at an early stage of our work when they need not a final answer but just a provisional

---

[1] *Essai philosophique sur les probabilités;* see *Oeuvres complètes de Laplace*, vol. 7, p. CXXXIX.
[2] *How to Solve It*, p. 111.

answer, a guess, and there are cases in which we can guess the answer quite reasonably and with very little trouble.

   As an illustration we consider an elementary problem in solid geometry. The axis of a cylinder passes through the center of a sphere. The surface of the cylinder intersects the surface of the sphere and divides the solid sphere into two portions: the "perforated sphere" and the "plug". The first portion is outside the cylinder, the second inside. See fig. 11.1, which should be rotated about the vertical line *AB*. *Given r, the radius of the sphere, and h, the height of the cylindrical hole, find the volume of the perforated sphere.*

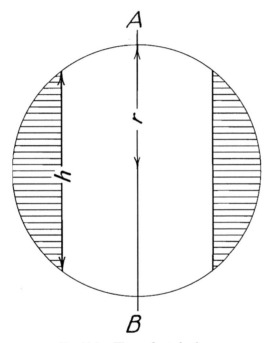

Fig. 11.1.   The perforated sphere.

   In familiarizing ourselves with the proposed problem, we arrive quite naturally at the usual questions: *Are the data sufficient to determine the unknown? Or are they insufficient? Or redundant?* The data $r$ and $h$ seem to be just enough. In fact, $r$ determines the size of the sphere and $h$ the size of the cylindrical hole. Knowing $r$ and $h$, we can determine the perforated sphere in shape and size and we also need $r$ and $h$ to determine it so.

   Yet, computing the required volume, we find that it is equal to $\pi h^3/6$; see ex. 5. This result looks extremely paradoxical. We have convinced ourselves that we need both $r$ and $h$ to determine the shape and size of the perforated sphere and now it turns out that we do not need $r$ to determine its volume; this sounds quite incredible.

Yet, there is no contradiction.   If $h$ remains constant and $r$ increases, the perforated sphere changes considerably in shape: it becomes wider (which tends to increase the volume) but its outer surface becomes flatter (which tends to decrease the volume).   Only we did not foresee (and it appears rather unlikely *a priori*) that these two tendencies balance exactly and the volume remains unchanged.

In order to understand both the present particular case and the underlying general idea, we need a distinction.   We should distinguish clearly between two related, but different, problems.   Being given $r$ and $h$, we may be required to determine

(a) the volume and

(b) the shape and size

of the perforated sphere.   Our original problem was (a).   We have seen intuitively that the data $r$ and $h$ are both necessary and sufficient to solve (b).   It follows hence that these data are also sufficient to solve (a), but not that they are necessary to solve (a); in fact, they are not.

In answering the question "Are the data necessary?" we *judged by a related case*, we *substituted* (b) for (a), we *neglected the distinction between the original problem* (a) *and the modified problem* (b).   From the heuristic viewpoint, such neglect is defensible.   We needed only a provisional, but quick, answer. Moreover, such a difference is *usually* negligible: the data which are necessary to determine the shape and size are *usually* also necessary to determine the volume.   We became involved into a paradox by forgetting that our conclusion was only heuristic, or by believing in some confused way that the unusual can never happen.   And, in our example, the unusual did happen.

*Judging a proposed problem by a modified problem* is a defensible, reasonable heuristic procedure.   We should not forget, however, that the conclusion at which we arrive by such a procedure is only provisional, not final; only plausible, but by no means certain to be true.

**3. Judging by the general case.**   The following problem can be suitably discussed in a class of Algebra for beginners.

The testament of a father of three sons contains the following dispositions. "The part of my eldest son shall be the average of the parts of the two others and three thousand dollars.   The part of my second son shall be exactly the average of the parts of the two others.   The part of my youngest son shall be the average of the parts of the two others less three thousand dollars." *What are the three parts?*

*Is the condition sufficient to determine the unknowns?*   There is quite a good reason to say yes.   In fact, there are three unknowns, say, $x$, $y$, and $z$, the parts of the eldest, the second, and the youngest son, respectively.   Each of the three sentences quoted from the testament can be translated into an equation.   Now, *in general, a system of three equations with three unknowns determines the unknowns*.   Thus we are quite reasonably led to think that the condition of the proposed problem is sufficient to determine the unknowns.

In writing down, however, the three equations, we obtain the following system:

$$x = \frac{y + z}{2} + 3000$$

$$y = \frac{x + z}{2}$$

$$z = \frac{x + y}{2} - 3000.$$

Adding these three equations we obtain

$$x + y + z = x + y + z.$$

or

$$0 = 0.$$

Therefore, any equation of the system is a consequence of the other two equations. Our system contains only two *independent* equations and, therefore, in fact, it is *not* sufficient to determine the unknowns.

The problem is essentially modified if the testament contains also the following sentence: "I divide my whole fortune of 15,000 dollars among my three sons." This system adds to the above system the equation

$$x + y + z = 15,000.$$

We have now a more comprehensive system of four equations. Yet *in general, a system of four equations with three unknowns is contradictory.* In fact, however, the present system is not contradictory, but just sufficient to determine the unknowns and yields

$$x = 7000, \quad y = 5000, \quad z = 3000.$$

The apparent contradictions of this not too deep example are not too difficult to disentangle, but a careful explanation may be useful.

It is not true that "a system of $n$ equations with $n$ unknowns determines the unknowns." In fact, we have just seen a counter-example with $n = 3$. What matters here, however, is not a mathematical theorem, but a heuristic statement, in fact, the following statement: "A system of $n$ equations with $n$ unknowns determines, *in general*, the unknowns." The term "in general" can be interpreted in various ways. What matters here is a somewhat vague and rough "practical" interpretation: a statement holds "in general" if it holds "in the great majority of such cases as are likely to occur naturally."

Treating a geometrical or physical problem algebraically, we try to express an intuitively submitted condition by our equations. We try to express a different clause of the condition by each equation and we try to cover the whole condition. If we succeed in collecting as many equations as we have unknowns, we hope that we shall be able to determine the unknowns.

Such hope is reasonable. Our equations "occurred naturally"; we may expect that we are "in the general case." Yet the example of the present section did not occur naturally; it was fabricated to show up the lack of absolute certainty in the heuristic statement. Therefore, this example does not invalidate at all the underlying heuristic principle.

We rely on something similar in everyday life. We are, quite reasonably, not too much afraid of things which are very unusual. Letters get lost and trains crash, yet I still send a letter and board a train without hesitation. After all, lost letters and train wrecks are extremely unusual; only to a very small percentage of letters or trains happens such an accident. Why should it happen just now? Similarly, $n$ equations with $n$ unknowns quite naturally obtained may be insufficient to determine the unknowns. Yet, in general, this does not happen; why should it happen just now?

We cannot live and we cannot solve problems without a modicum of optimism.

**4. Preferring the simpler conjecture.** "Simplex sigillum veri," or "Simplicity is the seal of truth," said the scholastics. Today, as humanity is older and richer with the considerable scientific experience of the intervening centuries, we should express ourselves more cautiously; we know that the truth can be immensely complex. Perhaps the scholastics did not mean that simplicity is a necessary attribute of truth; perhaps they intended to state a heuristic principle: "What is simple has a good chance to be true." It may be even better to say still less and to confine ourselves to the plain advice: "Try the simplest thing first."

This common sense advice includes (somewhat vaguely, it is true) the heuristic moves discussed in the foregoing. That the volume changes when the shape changes is not only the usual case, but also the simplest case. That a system of $n$ equations with $n$ unknowns determines the unknowns is not only the general case, but also the simplest case. It is reasonable to try the simplest case first. Even if we were obliged to return eventually to a closer examination of more complex possibilities, the previous examination of the simplest case may serve as a useful preparation.

Trying the simplest thing first is part of an attitude which is advantageous in face of problems little or great. Let us attempt to imagine (with sweeping simplification and, doubtless, with some distortion) Galileo's situation as he investigated the law of falling bodies. If we wished to count the era of modern science from a definite date, the date of this investigation of Galileo's could be considered as the most appropriate.

We should realize Galileo's position. He had a few forerunners, a few friends sharing his views, but was strenuously opposed by the dominating philosophical school, the Aristotelians. These Aristotelians asked, "Why do the bodies fall?" and were satisfied with some shallow, almost purely verbal explanation. Galileo asked, "How do the bodies fall?" and tried to find an answer from experiment, and a precise answer, expressible in numbers and mathematical concepts. This substitution of "How" for "Why," the search for an answer by experiment, and the search for a mathematical law condensing experimental facts are commonplace in modern science, but they were revolutionary innovations in Galileo's time.

A stone falling from a higher place hits the ground harder. A hammer falling from a higher point drives the stake deeper into the ground. The further the falling body gets from its starting point, the faster it moves—so much is clear from unsophisticated observation. *What is the simplest thing?* It seems simple enough to assume that the velocity of a falling body starting from rest is *proportional to the distance traveled.* "This principle appears very natural," says Galileo, "and corresponds to our experience with machines which operate by percussion." Still, Galileo rejected eventually the proportionality of the velocity to the distance as "not merely false, but impossible."[3]

Galileo's objections against the assumption that appeared so natural to him at first can be more clearly and strikingly formulated in the notation of the calculus. This is, of course, an anachronism; the calculus was invented after Galileo's time and, in part at least, under the impact of Galileo's discoveries. Still, let us use calculus. Let $t$ denote the time elapsed since the beginning of the fall and $x$ the distance traveled. Then the velocity is $dx/dt$ (one of Galileo's achievements was to formulate a clear concept of velocity). Let $g$ be an appropriate positive constant. Then that "simplest assumption," the proportionality of speed to the distance traveled, is expressed by the differential equation

$$(1) \qquad \frac{dx}{dt} = gx.$$

We have to add the initial condition

$$(2) \qquad x = 0 \quad \text{as} \quad t = 0.$$

From equations (1) and (2) it follows that

$$(3) \qquad \frac{dx}{dt} = 0 \quad \text{as} \quad t = 0;$$

this expresses that the falling body starts from rest.

We obtain, however, by integrating the differential equation (1) that

$$\int \frac{dx}{x} = \int g dt$$

$$\log x = gt + \log c$$

---

[3] See *Le Opere di Galileo Galilei,* edizione nazionale, vol. 8, p. 203, 373, 383.

where $c$ is some positive constant.  This yields

$$x = ce^{gt}, \qquad \frac{dx}{dt} = gce^{gt}.$$

We obtain hence, however, that

$$x = c > 0, \qquad \frac{dx}{dt} = gc > 0 \qquad \text{as } t = 0$$

in contradiction to (2) and (3): a motion satisfying the differential equation (1) *cannot start from rest*.  And so the assumption that appeared so "natural," just "the simplest thing," is, in fact, self-contradictory: "not merely false, but impossible" as Galileo expressed himself.

Yet what is the "next simplest thing"?  It may be to assume that the velocity of a falling body starting from rest is proportional to the time elapsed.  This is the well known law at which Galileo eventually arrived. It is expressed in modern notation by the equation

$$\frac{dx}{dt} = gt$$

and a motion satisfying this equation can certainly start from rest.

**5. Background.**  We cannot but admire Galileo's intellectual courage, his freedom from philosophical prejudice and mysticism.  Yet we must also admire Kepler's achievements; and Kepler, a contemporary of Galileo, was deeply involved in mysticism and the prejudices of his time.

It is difficult for us to realize Kepler's attitude.  The modern reader is amazed by a title as "A prodrome to cosmographic dissertations, containing the COSMIC MYSTERY, on the admirable proportion of celestial orbits and the genuine and proper causes of the number, magnitude, and periodic motions of the heavens, demonstrated by the five regular geometric solids."  The contents are still more amazing: astronomy mixed with theology, geometry scrambled with astrology.  Yet however extravagant some of the contents may appear, this first work of Kepler marks the beginning of his great astronomical discoveries and gives besides a lively and attractive picture of his personality.  His thirst for knowledge is admirable, although it is almost equalled by his hunger for mystery.

As the title of the work quite correctly says, Kepler set out to discover a cause or a reason for the number of the planets, for their distances from the sun, for the period of their revolutions.  He asks, in fact: Why are there just six planets?  Why are their orbits just so disposed?  These questions sound strange to us, but did not sound so to some of his contemporaries.[4]

---

[4] Kepler rejects Rhaeticus' explanation that there are six planets, since 6 is the first "perfect number."

One day he thought that he found the secret and he jotted down in his notebook: "The earth's orbit or sphere is the measure of all. Circumscribe about it a dodecahedron: the sphere surrounding it is Mars. Circumscribe about Mars a tetrahedron: the sphere surrounding it is Jupiter. Circumscribe about Jupiter a cube: the sphere surrounding it is Saturn. Now, inscribe in the Earth an icosahedron: the sphere contained in it is Venus. Inscribe in Venus an octahedron: the sphere contained in it is Mercury. Here you have the reason for the number of the planets."

That is, Kepler imagines 11 concentric surfaces, 6 spheres alternating with the 5 regular solids. The first and outermost surface is a sphere and each surface is surrounded by the preceding. Each sphere is associated with a planet: the radius of the sphere is the distance (mean distance) of the planet from the sun. Each regular solid is inscribed in the preceding, surrounding sphere and circumscribed about the following, surrounded sphere.

And Kepler adds: "I shall never succeed in finding words to express the delight of this discovery."

Kepler (in this respect a modern scientist) carefully compared his conjecture with the facts. He computed a table which is presented here in a slightly modernized form as Table I.

Table I.   Kepler's theory compared with the observations

| (1) Planets | (2) Copernicus' observation | (3) Kepler's theory | (4) Regular solids |
|---|---|---|---|
| Saturn | | | |
| | .635 | .577 | Cube |
| Jupiter | | | |
| | .333 | .333 | Tetrahedron |
| Mars | | | |
| | .757 | .795 | Dodecahedron |
| Earth | | | |
| | .794 | .795 | Icosahedron |
| Venus | | | |
| | .723 | .577 | Octahedron |
| Mercury | | | |

Column (1) lists the planets in order of decreasing distances from the sun; it contains six entries, one more than the following columns. Column (2) contains the ratio of the distances of two consecutive planets from the sun, according to Copernicus; each ratio is inserted between the lines in which the names of the respective planets are marked; the distance of the outer planet is the denominator. Column (4) lists the five regular solids in the order chosen by Kepler. Column (3) lists the ratio of the radii of the inscribed and circumscribed spheres for the corresponding regular solid.

The numbers on the same line should agree.    In fact, the agreement is good in two cases, and very bad in the remaining three.

Now Kepler (which reminds us in less glorious way of modern scientists) starts shifting his standpoint and modifying his original conjecture.    (The main modification is that he compares the distance of Mercury from the sun not to the radius of the sphere inscribed in the octahedron, but to the radius of the circle inscribed in the square in which a certain plane of symmetry intersects the octahedron.)    Yet he does not arrive at any startling agreement between conjecture and observation.    Still, he sticks to his idea.    The sphere is "the most perfect figure," and next to it the five regular solids, known to Plato, are the "noblest figures."    Kepler thinks for a moment that the countless crowd of fixed stars may have something to do with the undistinguished multitude of irregular solids.    And it seems "natural" to him that the sun and the planets, the most excellent things created, should be somehow related to Euclid's most excellent figures.    This could be the secret of the creation, the "Cosmic Mystery."

To modern eyes Kepler's conjecture may look preposterous.    We know many relations between observable facts and mathematical concepts, but these relations are of a quite different character.    No useful relation is known to us which would have any appreciable analogy to Kepler's conjecture.    We find it most strange that Kepler could believe that there is anything deep hidden behind the number of the planets and could ask such a question: Why are there just six planets?

We may be tempted to regard Kepler's conjecture as a queer aberration. Yet we should consider the possibility that some theories which we are respectfully debating today may be considered as queer aberrations in a not far away future, if they are not completely forgotten.    I think that Kepler's conjecture is highly instructive.    It shows with particular clarity a point that deserves to be borne in mind: the credence that we place in a conjecture is bound to depend on our whole *background*, on the whole *scientific atmosphere* of our time.

**6. Inexhaustible.**    The foregoing example brings into the foreground an important feature of plausible reasoning.    Let us try to describe it with some degree of generality.

We have a certain conjecture, say *A*.    That is, *A* is a clearly formulated, but not proved, proposition.    We suspect that *A* is true, but we do not actually know whether *A* is true or not.    Still we have some confidence in our conjecture *A*.    Such confidence may, but need not, have an articulate basis.    After prolonged and apparently unsuccessful work at some problem, there emerges quite suddenly a conjecture *A*.    This conjecture *A* may appear as the only possible escape from an entangled situation; it may appear as almost certain, although we could not tell why.

After a while, however, some more articulate reasons may occur to us that speak distinctly in favor of *A*, although they do not prove *A*: reasons from analogy, from induction, from related cases, from general experience, or from the inherent simplicity of *A* itself. Such reasons, without providing a strict demonstration, can make *A* very plausible. Yet it should be a warning to us that we trusted that conjecture without any of those more distinctly formulated arguments. And we perceived those arguments successively. There was a first clear point that we succeeded in detaching from an obscure background. Yet there was something more behind this point in the background since afterwards we succeeded in extracting another clear argument. And so there may be something more behind each clarified point. Perhaps that background is inexhaustible. *Perhaps our confidence in a conjecture is never based on clarified grounds alone*; such confidence may need somehow our *whole background* as a basis.

Still, plausible grounds are important, and clarified plausible grounds are particularly important. In dealing with the observable reality, we can never arrive at any demonstrative truth, we have always to rely on some plausible ground. In dealing with purely mathematical questions, we may arrive at a strict demonstration. Yet it may be very difficult to arrive at it, and the consideration of provisional, plausible grounds may give us temporary support and may lead us eventually to the discovery of the definitive demonstrative argument.

Heuristic reasons are important although they prove nothing. To clarify our heuristic reasons is also important although behind each reason so clarified there may be something more—some still obscure and still more important ground, perhaps.[5]

This suggests another remark: If in each concrete case we can clarify only a few of our plausible grounds, and in no concrete case exhaust them, how could we hope to describe exhaustively the kinds of plausible grounds in the abstract?

**7. Usual heuristic assumptions.** Two of our examples (sect. 2 and 3) bring up another point. Let us recall briefly one of the situations and touch upon a similar situation.

In working at some problem, you obtain from apparently different sources as many equations as you have unknowns. You ought to know that $n$ equations are not always sufficient to determine $n$ unknowns: the equations could be mutually dependent, or contradictory. Still, such a case is exceptional, and so it may be reasonable to hope that your equations will determine your unknowns. Therefore you go ahead, manipulate your equations, and see what follows from them. If there is contradiction or indetermination, it will show itself somehow. On the other hand, if you arrive at a neat result, you may feel more inclined to spend time and effort on a strict demonstration.

In solving another problem you are led to integrate an infinite series

---

[5] *How to Solve It*, p. 224.

term by term. You ought to know that such an operation is not always permissible, and could yield an incorrect result. Still, such a case is exceptional, and so it may be reasonable to hope that your series will behave. Therefore, it may be expedient to go ahead, see what follows from your formula not completely proved, and postpone worries about a complete proof.

We touched here upon two *usual heuristic assumptions*, one about systems of equations, the other about infinite series. In each branch of mathematics there are such assumptions, and one of the principal assets of the expert in that branch is to know the current assumptions and to know also how he can use them and how far he can trust them.

Of course, you should not trust any guess too far, neither usual heuristic assumptions nor your own conjectures. To believe without proof that your guess is true would be foolish. Yet to undertake some work in the hope that your guess *might* be true, may be reasonable. Guarded optimism is the reasonable attitude.

### EXAMPLES AND COMMENTS ON CHAPTER XI

**1.** Of a triangle, we are given the base $a$, the altitude $h$ perpendicular to $a$ and the angle $\alpha$ opposite to $a$. We should (a) construct the triangle, (b) compute its area. Are all the data necessary?

**2.** Of a trapezoid, we are given the altitude $h$ perpendicular to the two parallel sides, the middle line $m$ which is parallel to the two parallel sides and at the same distance from both, and the angles $\alpha$ and $\beta$ between one of the two parallel sides and the two remaining (oblique) sides. We should (a) construct the trapezoid, (b) compute its area. Are all the data necessary?

**3.** A zone is a portion of the surface of the sphere contained between two parallel planes. The altitude of the zone is the distance of the two planes. Given $r$ the radius of the sphere, $h$ the altitude of the zone, and $d$ the distance of that bounding plane from the center of the sphere which is nearer to the center, find the surface of the zone. Any remarks?

**4.** A first sphere has the radius $a$. A second sphere, with radius $b$, intersects the first sphere and passes through its center. Compute the area of that portion of the surface of the second sphere which lies inside the first sphere. Any remarks? Check the extreme cases.

**5.** Reconsider the example of sect. 2 and prove the solution.

**6.** A spherical segment is a portion of the sphere contained between two parallel planes. Its surface consists of three parts: of a zone of the sphere and of two circles, called the base and the top of the segment. We use the following notation:

$a$ is the radius of the base,

$b$ the radius of the top,

$h$ the altitude (the distance between the base and the top),

$M$ the area of the middle · cross-section (parallel to, and at the same distance from, the base and the top)
$V$ the volume of the segment.
Being given $a$, $b$, and $h$, find $Mh - V$.
Any remarks? Check some extreme cases.

**7.** The axis of a cone passes through the center of a sphere. The surface of the cone intersects the surface of the sphere in two circles and divides the solid sphere into two portions: the "conically perforated sphere" and the "plug" (see fig. 11.2 which should be rotated about the line $AB$); the plug is inside the cone. Let $r$ denote the radius of the sphere, $c$ the length

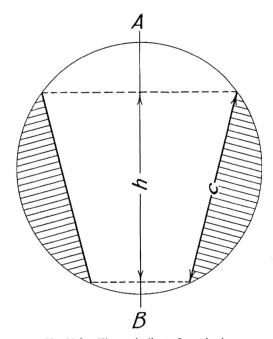

Fig. 11.2.   The conically perforated sphere.

of the chord that in rotating generates the conical hole and $h$ (the height of the perforated sphere) the projection of $c$ onto the axis of the cone. Given $r$, $c$, and $h$, find the volume of the conically perforated sphere. Any remarks?

**8.** The axis of a paraboloid of revolution passes through the center of a sphere and the two surfaces intersect in two circles. Compute the ring-shaped solid between the two surfaces (inside the sphere and outside the paraboloid) being given $r$ the radius of the sphere, $h$ the projection of the ring-shaped solid on the axis of the paraboloid, and $d$ the distance of the center of the sphere from the vertex of the paraboloid. (Rotate fig. 11.3 about $OX$.) Any remarks?

**9.** Of a trapezoid, given the lower base $a$, the upper base $b$, and the height $h$, perpendicular to both bases; $a > b$. The trapezoid, revolving about its lower base, describes a solid of revolution (a cylinder topped by two cones) of which find (a) the volume and (b) the surface area. Are the data sufficient to determine the unknown?

**10.** Ten numbers taken in a definite order, $u_1$, $u_2$, $u_3$, ... $u_{10}$, are so connected that, from the third onward, each of them is the sum of the two foregoing numbers:

$$u_n = u_{n-1} + u_{n-2}, \text{ for } n = 3, 4, \ldots 10.$$

Being given $u_7$, find the sum of all ten numbers $u_1 + u_2 + \ldots + u_{10}$. Are the data sufficient to determine the unknown?

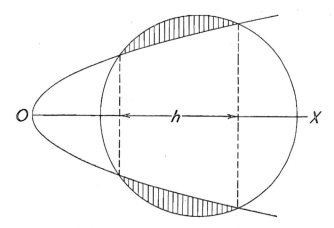

Fig. 11.3.   The parabolically perforated sphere.

**11.** Compute
$$\int_0^\infty \frac{dx}{(1 + x^2)(1 + x^\alpha)}.$$

Any remarks?   Check the cases $\alpha = 0$, $\alpha \to \infty$, $\alpha \to -\infty$.

**12.** Generalize ex. 11.   [Try the simplest thing first.]

**13.** Write one equation with one unknown that does not determine the unknown.

**14.** One equation may determine several unknowns if the nature of the unknowns is restricted by a suitable additional condition. For example, if $x$, $y$, and $z$ are real numbers they are completely determined by the equation

$$x^2 + y^2 + z^2 = 0.$$

Find all systems of positive integers $x$, $y$ satisfying the equation $x^2 + y^2 = 128$.

**15.** Find all systems of positive integers $x, y, z, w$ satisfying the equation $x^2 + y^2 + z^2 + w^2 = 64$.

**16.** *The general case.* Consider the system of three linear equations with three unknowns

$$a_1 x + b_1 y + c_1 z = d_1,$$
$$a_2 x + b_2 y + c_2 z = d_2,$$
$$a_3 x + b_3 y + c_3 z = d_3.$$

We assume that the 12 given numbers $a_1, b_1, c_1, d_1, a_2, \ldots d_3$ are real. The system is called *determinate* if there is just one solution (just one set $x, y, z$ of three numbers satisfying it), *indeterminate* if there are an infinity of solutions, and *inconsistent* if there is no solution. Seen from various standpoints, the case in which the system is determinate appears as the general, usual, normal, regular case and the other cases appear as exceptional, unusual, abnormal, irregular.

(a) Geometrically, we can interpret the set of three numbers $x, y, z$ as a point in a rectangular coordinate system and each equation as the set of points satisfying it, as a plane. (For this interpretation we have to assume, in fact, that on the left-hand side of each equation there is at least one non-vanishing coefficient, but let us assume this.) The system of three equations is determinate if the three planes have just one common point. When they have two common points, they have a straight line in common and so the system is indeterminate. When the three planes are parallel to the same straight line, but have no point common to all three, the system is inconsistent. If the three planes are in a "general position," if they are "chosen at random," they have just one point in common and the system is determinate.

(b) Algebraically, the system of three equations is determinate if, and only if, the determinant of the 9 coefficients on the left-hand sides does not vanish. Therefore, the system is determinate, unless a particular condition or *restriction* is imposed upon the coefficients, in form of an equation.

(c) We may interpret the set of nine (real) coefficients $(a_1, a_2, a_3, b_1, \ldots c_3)$ as a point in nine-dimensional space. The points corresponding to systems that are not determinate (indeterminate or inconsistent) satisfy an equation (the determinant $= 0$) and so they form a manifold of *lower dimension* (an eight-dimensional "hypersurface").

(d) It is *infinitely improbable* that a system of three linear equations with three unknowns given at random is not determinate. Cf. ex. 14.23.

**17.** For each of the five regular solids, consider the inscribed sphere and the circumscribed sphere and compute the ratio of the radii of these two spheres.

**18.** Column (3) of Table I would remain unchanged if we interchanged the cube and the octahedron or the dodecahedron and the icosahedron. This would leave Kepler's theory embarrassingly indeterminate. Yet

Kepler displays a singular ingenuity in detecting reasons why one of these five noble solids should be of higher nobility than, and take precedence over, another, as a baron takes precedence over a baronet. Find some simple geometrical property that distinguishes the three solids that Kepler placed around the Earth's orbit from the two that he placed in this orbit.

**19.** *No idea is really bad.* "Many a guess has turned out to be wrong but nevertheless useful in leading to a better one." "No idea is really bad, unless we are uncritical. What is really bad is to have no idea at all."[6] I use such sentences almost daily to comfort one or the other student who comes forward with some honest but naïve idea. These sentences apply both to trivial everyday situations and to scientific research. They apply most spectacularly to Kepler's case.

To Kepler himself, with his mind in that singular transition from the medieval to the modern standpoint, his idea of combining the six planets with the five regular solids appeared as brilliant. Yet I cannot imagine that Galileo, Kepler's contemporary, could have conceived such an idea. To a modern mind this idea must appear as pretty bad from the start, because it has so little relation to the rest of our knowledge about nature. Even if it had been in better agreement with the observations, Kepler's conjecture would be weakly supported, because it lacks the support of analogy with what is known otherwise.

Yet Kepler's guess which turned out to be wrong was most certainly useful in leading to a better one. It led Kepler to examine more closely the mean distances of the planets, their orbits, their times of revolution for which he hoped to find some similar "explanation," and so it led finally to Kepler's celebrated laws of planetary motion, to Newton, and to our whole modern scientific outlook.

**20.** *Some usual heuristic assumptions.* This subject would deserve a fuller treatment, yet we have to restrict ourselves to a very short list and sketchy comments. We must be careful to interpret the words "in general" in a "practical," necessarily somewhat vague, sense.

"If in a system of equations there are as many equations as unknowns, the unknowns are determined, *in general.*"

If, in a problem, there are as many "conditions" as available parameters, it is reasonable to start out with the tentative assumption that the problem has a solution. For instance, a quadratic form of $n$ variables has $n(n + 1)/2$ coefficients, and an orthogonal substitution in $n$ variables depends on $n(n - 1)/2$ parameters. Therefore, it is pretty plausible from the outset that, by a suitable orthogonal substitution, any quadratic form of $n$ variables can be reduced to the expression

$$\lambda_1 y_1^2 + \lambda_2 y_2^2 + \ldots + \lambda_n y_n^2 ;$$

---

[6] *How to Solve It*, pp. 207–8.

$y_1, y_2, \ldots y_n$ are the new variables introduced by the substitution, and $\lambda_1, \lambda_2, \ldots \lambda_n$ suitable parameters. In fact, this expression depends on $n$ parameters and

$$n(n + 1)/2 = n(n - 1)/2 + n.$$

This remark, coming after a proof of the proposition in the particular cases $n = 2$ and $n = 3$, and an explanation of the geometric meaning of these cases, may create a pretty strong presumption in favor of the general case.

"Two limit operations are, *in general*, commutative."

If one of the limit operations is the summation of an infinite series and the other is integration, we have the case mentioned in sect. 7.[7]

"What is true up to the limit, is true at the limit, *in general*."[8]

Being given that $a_n > 0$ and $\lim\limits_{n \to \infty} a_n = a$, we *cannot* conclude that $a > 0$; merely $a \geq 0$ is true. We consider a curve as the limit of an inscribed polygon and a surface as the limit of an inscribed polyhedron. Computing the length of the curve as the limit of the length of an inscribed polygon yields the correct result, yet computing the area of the surface as the limit of the area of an inscribed polyhedron may yield an incorrect result.[9] Although it can easily mislead us, the heuristic principle stated is most fertile in inspiring suggestions. See, for instance, ex. 9.24.

"Regard an unknown function, *at first*, as monotonic."

We followed something similar to this advice in sect. 2 as we assumed that with the change of the shape of a body its volume will change, too, and we were misled. Nevertheless, the principle stated is often useful. We may have to prove an inequality of the form

$$\int_a^b f(x)dx < \int_a^b g(x)dx$$

where $a < b$. We may begin by trying to prove more, namely that

$$f(x) < g(x).$$

This boils down to the initial assumption that the function with derivative $g(x) - f(x)$ is monotonic. (The problem is to compare the values of this function for $x = a$ and $x = b$.) The principle stated is contained in the more general heuristic principle "try the simplest thing first."

"*In general*, a function can be expanded in a power series, the very first term of which yields an acceptable approximation and the more terms we take, the better the approximation becomes."

Without the well-understood restriction "in general" this statement would be monumentally false. Nevertheless, physicists, engineers, and other

---

[7] See G. H. Hardy, *A Course of Pure Mathematics*, 7th ed., p. 493–496.
[8] Cf. William Whewell, *The Philosophy of the Inductive Sciences*, new ed., vol. I, p. 146.
[9] See H. A. Schwarz, *Gesammelte Mathematische Abhandlungen*, vol. 2, pp. 309–311.

scientists who apply the calculus to their science seem to be particularly fond of it. It includes another principle even more sweeping than the one that we have stated previously: "Regard an unknown function, *at first*, as linear." In fact, if we have the expansion

$$f(x) = a_0 + a_1 x + a_2 x^2 + a_n x^3 + \ldots,$$

we may take approximately

$$f(x) \sim a_0 + a_1 x.$$

(Observe that Galileo, who did not know calculus, had already a strong preference for the linear function; see sect. 4.) The present principle underlies the importance often attributed to the initial term of the relative error; see sect. 5.2. The principle was often useful in suggesting some idea close to the truth, yet it may easily suggest something very far from the truth.

In fact, a physicist (or an engineer, or a biologist) may be led to believe that a physical quantity $y$ depends so on another physical quantity $x$ that there is a differential equation

$$\frac{dy}{dx} = f(y).$$

Now, the integration involved by this equation may be too difficult, or the form of the function $f(y)$ may be unknown. In both cases the physicist expands the function $f(y)$ in powers of $y$ and he may regard the differential equations hence following as successive approximations:

$$\frac{dy}{dx} = a_0,$$

$$\frac{dy}{dx} = a_0 + a_1 y,$$

$$\frac{dy}{dx} = a_0 + a_1 y + a_2 y^2.$$

Yet the curves satisfying these three equations are of very different nature and the approximation may turn out totally misleading. Fortunately, the physicists rely more on careful judgment than on careful mathematics and so they obtained good results by similar procedures even in cases in which the mathematical fallacy was less obvious and, therefore, more dangerous than in our example.

**21.** *Optimism rewarded.* The quantities $a, b, c, d, e, f, g$, and $h$ are given. We investigate whether the system of four equations for the four unknowns $x, y, u$, and $v$

$$(S) \qquad \begin{aligned} ax + by + cv + du &= 0, \\ ex + fy + gv + hu &= 0, \\ hx + gy + fv + eu &= 0, \\ dx + cy + bv + au &= 0 \end{aligned}$$

admits any solution different from the trivial solution $x = y = u = v = 0$. This system $(S)$ has, as we know, a non-trivial solution if, and only if, its determinant vanishes, but we wish to avoid the direct computation of this determinant with four rows. The peculiar symmetry of the system $(S)$ may suggest to set

$$u = x, \qquad v = y.$$

Then the first equation of the system $(S)$ coincides with the fourth, and the second with the third, so that the system of four equations reduces to a system of only two distinct equations

$$(a + d)x + (b + c)y = 0,$$
$$(e + h)x + (f + g)y = 0.$$

This system admits a non-trivial solution if, and only if, its determinant vanishes.

Yet we can also reduce the system $(S)$ by setting

$$u = -x, \qquad v = -y.$$

Again, we obtain only two distinct equations

$$(a - d)x + (b - c)y = 0,$$
$$(e - h)x + (f - g)y = 0.$$

The vanishing of the determinant of either system of two equations involves the vanishing of the determinant of the system $(S)$. Hence we may suspect (if we are optimistic enough) that this latter determinant with four rows is the *product* of the two other determinants, each with two rows.

(a) Prove this, and generalize the result to determinants with $n$ rows.

(b) In which respect have we been optimistic?

**22.** Take the coordinate system as in sect. 9.4. The $x$-axis is horizontal and the $y$-axis points downward. Join the origin to the point $(a,b)$

(1) by a straight line

(2) by a circular arc with center on the x-axis.

A material point starting from rest at the origin attains the point $(a,b)$ in time $T_1$ or $T_2$ according as it slides down (without friction) following the path (1) or the path (2). Galileo suggested (as reported in sect. 9.4) that $T_1 > T_2$. After some work this inequality turns out to be equivalent to the following:

$$\int_0^h [x(1 - x)]^{-3/4}\, dx < 4h^{1/4}(1 - h)^{-1/4}$$

if we set

$$a^2(a^2 + b^2)^{-1} = h.$$

We could try to prove the inequality by expanding both sides in powers of $h$. What would be the simplest (or "most optimistic") possibility?

**23.** *Numerical computation and the engineer.*   The layman is inclined to think that the scientist's numerical computations are infallible, but dull.   Actually, the scientist's numerical computation may be exciting adventure, but unreliable.   Ancient astronomers tried, and modern engineers try, to obtain numerical results about imperfectly known phenomena with imperfectly known mathematical tools.   It is hardly surprising that such attempts may fail; it is more surprising that they often succeed.   Here is a typical example.   (The technical details, which are suppressed here, will be published elsewhere.)

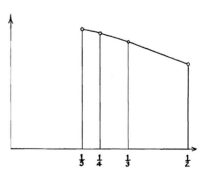

Fig. 11.4.  A trial: the abscissa is $n$.    Fig. 11.5.  Another trial: the abscissa is $1/n$.

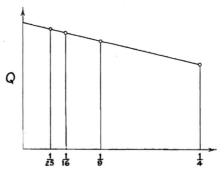

Fig. 11.6.  The abscissa is $1/n^2$: success!

An engineer wishes to compute a certain physical quantity $Q$ connected with a square of side 1.   (In fact, $Q$ is the torsional rigidity of a beam with square cross-section, but the reader need not know this—in fact, he need not even know what torsional rigidity is.)   An exact solution runs into mathematical difficulties, and so the engineer, as engineers often do, resorts to approximations.   Following a certain method of approximation, he divides the given square into equal "elements," that is, $n^2$ smaller squares each of the area $1/n^2$.   (In approximating a double integral, we also divide the given area into elements in this way.)   It can be reasonably expected that the approximate value tends to the true value as $n$ tends to infinity.

In fact, however, as $n$ increases, the difficulty of the computation also increases, and so rapidly that it soon becomes unmanageable. The engineer considers only the cases $n = 2, 3, 4, 5$ and obtains the corresponding approximate values for $Q$:

$$0.0937 \qquad 0.1185 \qquad 0.1279 \qquad 0.1324$$

Let us not forget that these numbers correspond to the values

$$1/4 \qquad 1/9 \qquad 1/16 \qquad 1/25$$

of a small square used in computing, respectively.

The engineer graphs these results. He decides to plot the approximate values obtained for $Q$ as ordinates, but he is hesitant about the choice of the abscissa. He tries first $n$ as abscissa, then $1/n$, and finally $1/n^2$ (which is the numerical value of the area of the small square used in the approximation): see figs. 11.4, 11.5, and 11.6, respectively. The last choice is the best; the four points in fig. 11.6 are *nearly on the same straight line*. Observing this, the engineer produces the line till it intersects the vertical axis and regards the ordinate of the intersection as a "good" approximation to $Q$.

(a) Why? What is the underlying idea?

(b) Check fig. 11.6 numerically: join each point to the next by a straight line and compute the three slopes.

(c) Choose the two most reliable points in fig. 11.6, use the straight line passing through them in the engineer's construction, compute the resulting approximation to $Q$, and compare it with 0.1406, the true value of $Q$.

# FINAL REMARK

The reader who went through the foregoing chapters and did some of the foregoing problems had a good opportunity to acquaint himself with some aspects of plausible reasoning. To form a general idea of the nature of plausible reasoning is the aim of the remaining five chapters of this work, collected in Vol. II. This aim deserves, I believe, considerable theoretical interest, but it may have also some practical value: we may perform a concrete task better if we understand more of the underlying abstract idea.

The formulation of certain patterns of plausible reasoning is the principal object of Vol. II. Yet these patterns will be extracted from, and discussed in close contact with, concrete examples. Therefore Vol. II will add several mathematical examples to those treated in the present Vol. I and will treat them in the same manner.

# SOLUTIONS

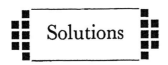

# Solutions

**1.** The primes ending in 1.

**2.** [Stanford 1948]

$$(n^2 + 1) + (n^2 + 2) + \ldots + (n + 1)^2 = n^3 + (n + 1)^3$$

The terms on the left-hand side are in arithmetic progression.

**3.** $1 + 3 + \ldots + (2n - 1) = n^2$.

**4.** 1, 9, 36, 100, ... are squares. See *How to Solve It*, p. 104.

**5.** [Stanford 1949] $\left(\dfrac{n + 1}{2}\right)^2$ or $\left(\dfrac{n + 1}{2}\right)^2 - \dfrac{1}{4}$, according as $n$ is odd or even. A uniform law for both cases: the integer nearest to $(n + 1)^2/4$.

**6.** First question: Yes. Second question: No; 33 is not a prime.

**7.** Not for you, if you have some experience with primes [ex. 1, 6, 9]. In fact, (1) can be proved (as particular case of a theorem of Kaluza, *Mathematische Zeitschrift*, vol. 28 (1928) p. 160–170) and (2) disproved: the next coefficient (of $x^7$) is $-3447 = -3 \cdot 3 \cdot 383$. The "formal computation" has a clear meaning. Setting

$$\left(\sum_0^\infty n! x^n\right)^{-1} = \sum_0^\infty u_n x^n,$$

we define $u_0 = 1$ and $u_1, u_2, u_3, \ldots$ recurrently by the equations

$$0! u_n + 1! u_{n-1} + 2! u_{n-2} + \ldots + (n - 1)! u_1 + n! u_0 = 0$$

for $n = 1, 2, 3, \ldots$ .

**8.** On the basis of the observed data it is quite reasonable to suspect that $A_n$ is positive and increases with $n$. Yet this conjecture is totally mistaken. By more advanced tools (integral calculus, or theory of analytic functions of a complex variable) we can prove that, for large $n$, the value of $A_n$ is approximately $(-1)^{n-1} (n - 1)! (\log n)^{-2}$.

**10.** In the case $2n = 60$ we have to make 9 or 7 trials ($p = 3, 5, 7, 11, 13, 17, 19, 23, 29$ or $p' = 31, 37, 41, 43, 47, 53, 59$) according as we follow the first or the second procedure. It is likely that for higher values of $n$ there will be still a greater difference between the number of trials, in favor of the second procedure.

No solution: **9, 11, 12, 13, 14.**

## SOLUTIONS, CHAPTER II

**1.** I think that C or D is the "right generalization" and B "overshoots the mark." B is too general to give any specific suggestion. You may prefer C or D; the choice depends on your background. Yet both C and D suggest to begin with the linear equations and lead eventually to the following plan: express two unknowns in terms of the third unknown from the first two (linear!) equations and, by substituting these expressions in the last equation, obtain a quadratic equation for the third unknown. (In the present case A, can you express *any* two unknowns from the first two equations?) There are two solutions:

$$(x,y,z) = (1, -2, 2), \qquad (29/13, -2/13, 2).$$

**2.** Rotated 180° about its axis, the pyramid coincides with itself. The right generalization of this pyramid is a solid having an *axis of symmetry* of this kind and the *simplest* solution is a plane passing through the axis and the given point. (There are an infinity of other solutions; by continuity, we can prescribe a straight line through which the bisecting plane should pass.) Note that a regular pyramid with pentagonal base does *not* admit a comparably simple solution. Compare *How to Solve It*, pp. 98–99.

**3.** A is a special case of B if we allow in B that *P* may coincide with *O*, yet the two problems are equivalent: the planes required in A and B are parallel to one another, and so the solution of each problem involves that of the other.

The more general problem B is more accessible, provided that $P \neq O$: choose $Q$ and $R$, on the two other lines, so that $OP = OQ = OR$. The plane passing through $P$, $Q$, and $R$ satisfies the condition of the problem. Therefore, if A is proposed, there is advantage in passing to the more general B.

**4.** A is a special case of B (for $p = 1$), yet the two problems are equivalent: the substitution $x = yp^{1/2}$ reduces B to A.

The more general problem B is more accessible: differentiate the easy integral

$$\int_{-\infty}^{\infty} (p + x^2)^{-1} dx$$

twice with respect to the parameter $p$. Therefore, if A is proposed, there is advantage in passing to the more general B.

Observe the parallel situation in ex. 3.

**6.** The extreme special case in which one of the circles degenerates into a point is more accessilbe and we can reduce to it the general case.   In fact, a common outer tangent of two circles remains parallel to itself when both radii decrease by the same amount, and a common inner tangent remains parallel to itself when one of the radii increases, and the other decreases by the same amount.   In both cases, we can reduce one of the circles to a point, without changing the direction of the common tangent.

**8.** The special case in which one of the sides of the angle at the circumference passes through the center of the circle is "leading".   From two such special angles, we can combine the general angle at the circumference by addition or subtraction.   (This is the gist of the classical proof; Euclid III 20.)   For a striking example of a "leading" special case see *How to Solve It*, pp. 166–170.

**12.** If two straight lines in a plane are cut by three parallel lines, the corresponding segments are proportional.   This helps to prove the more difficult analogous theorem in solid geometry; see Euclid XI 17.

**13.** The diagonals of a parallelogram intersect in their common midpoint.

**14.** The sum of any two sides of a triangle is greater than the third side. The simpler of the two analogous theorems (Euclid I 20) is used in the proof of the more difficult (Euclid XI 20).

**15.** Parallelepiped, rectangular parallelepiped (box), cube, bisecting plane of a dihedral angle.   *The bisecting planes of the six dihedral angles of a tetrahedron meet in one point which is the center of the sphere inscribed in the tetrahedron.*

**16.** Prism, right prism, sphere.   *The volume of a sphere is equal to the volume of a pyramid the base of which has the same area as the surface of the sphere and the altitude of which is the radius.*

**17.** Let us call a pyramid an *isosceles pyramid* if all edges starting from its apex are equal.   All lateral faces of an isosceles pyramid are isosceles triangles.   *The base of an isosceles pyramid is inscribed in a circle and the altitude of the isosceles pyramid passes through the center of this circle.* Cf. ex. 9.26.

**22.** Yes.   Interchanging $x$ and $-x$ we do not change $x^2$ or the product that represents $(\sin x)/x$ according to $E$.

**23.** Prediction: from $E$ follows

$$\left(1 - \frac{x^2}{4\pi^2}\right)\left(1 - \frac{x^2}{9\pi^2}\right)\left(1 - \frac{x^2}{16\pi^2}\right) \cdots = -\frac{\pi^2}{x(x+\pi)}\frac{\sin x - \sin \pi}{x - \pi}$$

and so, for $x \to \pi$, by the definition of the derivative,

$$\left(1 - \frac{1}{4}\right)\left(1 - \frac{1}{9}\right)\left(1 - \frac{1}{16}\right) \cdots = -\frac{\pi^2}{\pi \cdot 2\pi}\cos \pi = \frac{1}{2}.$$

Verification:

$$\left(1 - \frac{1}{4}\right)\left(1 - \frac{1}{9}\right) \cdots \left(1 - \frac{1}{n^2}\right) = \frac{1 \cdot 3}{2 \cdot 2} \cdot \frac{2 \cdot 4}{3 \cdot 3} \cdot \frac{3 \cdot 5}{4 \cdot 4} \cdots \frac{(n-1)(n+1)}{n \cdot n}$$

$$= \frac{1}{2} \frac{n+1}{n} \to \frac{1}{2}.$$

**24.** $1/6$. As ex. 23, or special case $k = 2$ of ex. 25.

**25.** Prediction: if $k$ is a positive integer

$$\prod_{n=1}^{k-1}\left(1 - \frac{k^2}{n^2}\right) \prod_{n=k+1}^{\infty}\left(1 - \frac{k^2}{n^2}\right) = \lim_{x \to k\pi} \frac{k^2 \pi^2}{x(x + k\pi)} \frac{\sin x}{k\pi - x}$$

$$= (1/2)\,(-\cos k\pi) = (-1)^{k-1}/2.$$

Verification: for $N \geq k + 1$

$$\prod_{n=1}^{k-1} \frac{(n - k)\,(n + k)}{n\ n} \prod_{n=k+1}^{N} \frac{(n - k)\,(n + k)}{n\ n}$$

$$= \frac{(-1)^{k-1}\,(k - 1)!\,(N - k)! \cdot (N + k)!/(k!2k)}{(N!/k)^2}$$

$$= \frac{(-1)^{k-1}}{2} \frac{(N - k)!\,(N + k)!}{N!\ N!}$$

$$= \frac{(-1)^{k-1}}{2} \frac{(N + 1)\,(N + 2) \ldots (N + k)}{N\ \ (N - 1) \ldots (N - k + 1)} \to \frac{(-1)^{k-1}}{2}$$

as $N$ tends to $\infty$.

**26.** $\pi/4$, the area of the circle with diameter 1. From $E$, for $x = \pi/2$

$$\frac{2}{\pi} = \left(1 - \frac{1}{4}\right)\left(1 - \frac{1}{16}\right)\left(1 - \frac{1}{36}\right)\left(1 - \frac{1}{64}\right)\left(1 - \frac{1}{100}\right) \cdots$$

$$= \frac{1 \cdot 3}{2 \cdot 2} \cdot \frac{3 \cdot 5}{4 \cdot 4} \cdot \frac{5 \cdot 7}{6 \cdot 6} \cdot \frac{7 \cdot 9}{8 \cdot 8} \cdot \frac{9 \cdot 11}{10 \cdot 10} \cdots .$$

This formula, due to Wallis (1616–1703), was well-known to Euler. There is another way of stating Wallis' formula:

$$\frac{1}{\pi} = \lim_{n \to \infty} \left(\frac{1 \cdot 3 \cdot 5 \ldots (2n - 1)}{2 \cdot 4 \cdot 6 \ldots \ \ 2n}\right)^2 n.$$

**27.** $x = \pi z$ in ex. 21 and definition of an infinite product.

**28.** Yes. From ex. 27

$$\frac{\sin \pi(z + 1)}{\pi} = \lim_{n \to \infty} \frac{z + n + 1}{z - n} \cdot \frac{(z + n) \ldots (z + 1)\,z(z - 1) \ldots (z - n)}{(-1)^n\,(n!)^2}.$$

**29.** From ex. 27 and ex. 26

$\cos \pi z = \sin \pi(-z + 1/2)$

$$= \pi \lim \frac{(-z+n+\frac{1}{2}) \ldots (-z+\frac{3}{2})(-z+\frac{1}{2})(-z-\frac{1}{2}) \ldots (-z+\frac{1}{2}-n)}{(-1)^n(n!)^2}$$

$$= \lim \frac{(2n-1-2z) \ldots (3-2z)(1-2z)(1+2z)(3+2z) \ldots (2n-1+2z)}{(2n-1) \quad \ldots \quad 3 \quad \cdot \quad 1 \quad \cdot \quad 1 \quad \cdot \quad 3 \quad \ldots (2n-1)}$$

$$\cdot \lim \frac{-z+n+\frac{1}{2}}{n} \cdot \pi \lim \left(\frac{1 \cdot 3 \cdot 5 \ldots (2n-1)}{2 \cdot 4 \cdot 6 \ldots \quad 2n}\right)^2 n$$

$$= \left(1 - \frac{4z^2}{1}\right)\left(1 - \frac{4z^2}{9}\right)\left(1 - \frac{4z^2}{25}\right) \ldots .$$

**30.** Yes. From $E$ and ex. 29

$$\frac{2 \sin \pi z/2}{\pi z} \cdot \cos \pi z/2 = \left(1 - \frac{z^2}{4}\right)\left(1 - \frac{z^2}{16}\right)\left(1 - \frac{z^2}{36}\right) \ldots$$

$$\cdot \left(1 - \frac{z^2}{1}\right)\left(1 - \frac{z^2}{9}\right)\left(1 - \frac{z^2}{25}\right) \ldots$$

$$= \left(1 - \frac{z^2}{1}\right)\left(1 - \frac{z^2}{4}\right)\left(1 - \frac{z^2}{9}\right)\left(1 - \frac{z^2}{16}\right) \ldots$$

$$= \frac{\sin \pi z}{\pi z}.$$

**31.** Prediction: for $x = \pi$ ex. 29 yields $\cos \pi = -1$.

Verification: the product of the first $n$ factors

$$\frac{-1 \cdot 3}{1 \cdot 1} \cdot \frac{1 \cdot 5}{3 \cdot 3} \cdot \frac{3 \cdot 7}{5 \cdot 5} \cdot \frac{5 \cdot 9}{7 \cdot 7} \ldots \frac{(2n-3)(2n+1)}{(2n-1)(2n-1)} = -\frac{2n+1}{2n-1} \to -1.$$

**32.** Prediction: ex. 29 yields $\cos 2\pi = 1$.

Verification: as ex. 31, or ex. 31 and ex. 35.

**33.** Prediction: for $x = n\pi$ ($n = 1, 2, 3, \ldots$) ex. 29 yields

$$\left(1 - \frac{4n^2}{1}\right)\left(1 - \frac{4n^2}{9}\right)\left(1 - \frac{4n^2}{25}\right) \ldots = \cos \pi n = (-1)^n.$$

Verification: from $\cos 0 = 1$ and ex. 35, or directly as ex. 31.

**34.** Yes. As ex. 22.

**35.** Yes. By result, or method, of ex. 28.

**36.** $1 - \sin x = 1 - \cos\left(\frac{\pi}{2} - x\right) = 2\sin^2\left(\frac{\pi}{4} - \frac{x}{2}\right)$

$$= \left(\frac{\sin \pi(1 - 2z)/4}{\sin \pi/4}\right)^2 ;$$

we set $x = \pi z$. By ex. 27

$\dfrac{\sin \pi(1 - 2z)/4}{\sin \pi/4}$

$$= \lim \frac{n+(1-2z)/4}{n+1/4} \cdots \frac{1+(1-2z)/4}{1+1/4} \frac{(1-2z)/4}{1/4} \frac{-1+(1-2z)/4}{-1+1/4} \cdots$$

$$\frac{-n+(1-2z)/4}{-n+1/4}$$

$$= \lim \frac{4n+1-2z}{4n+1} \cdots \frac{5-2z}{5} \frac{1-2z}{1} \frac{3+2z}{3} \cdots \frac{4n-1+2z}{4n-1}$$

$$= \left(1 - \frac{2z}{1}\right)\left(1 + \frac{2z}{3}\right)\left(1 - \frac{2z}{5}\right) \cdots \left(1 + \frac{2z}{4n-1}\right)\left(1 - \frac{2z}{4n+1}\right) \cdots .$$

**37.** By passing to the logarithms and differentiating in ex. 21 or ex. 27. The precise meaning of the right hand side is

$$\lim_{n \to \infty} \left(\frac{1}{x + n\pi} + \cdots + \frac{1}{x + \pi} + \frac{1}{x} + \frac{1}{x - \pi} + \cdots + \frac{1}{x - n\pi}\right).$$

**38.** By ex. 37

$$\cot x = \frac{1}{x} + \sum_{n=1}^{\infty} \left(\frac{1}{x + n\pi} + \frac{1}{x - n\pi}\right)$$

$$= \frac{1}{x} + \sum_{n=1}^{\infty} \frac{2x}{x^2 - n^2\pi^2}$$

$$= \frac{1}{x} - 2x \sum_{n=1}^{\infty} \left(\frac{1}{n^2\pi^2} + \frac{x^2}{n^4\pi^4} + \frac{x^4}{n^6\pi^6} + \cdots\right).$$

Let us set

$$y = \cot x = \frac{1}{x} + a_1 x + a_2 x^3 + a_3 x^5 + \cdots .$$

Then, expressing the coefficient of $x^{2n-1}$, we find

$$S_{2n} = 1 + \frac{1}{2^{2n}} + \frac{1}{3^{2n}} + \frac{1}{4^{2n}} + \frac{1}{5^{2n}} + \cdots = -\frac{a_n \pi^{2n}}{2}$$

for $n = 1, 2, 3, \ldots$ . In order to find the coefficients $a_1, a_2, a_3, \ldots$ , we use the differential equation

$$y' + y^2 = -1.$$

Substituting for $y$ and $y'$ their expansions and comparing the coefficients of like powers of $x$, we obtain relations between the coefficients $a_1$, $a_2$, $a_3$, $\ldots$ which we can most conveniently survey in the array

| | $x^{-2}$ | $1$ | $x^2$ | $x^4$ | $x^6$ | $\ldots$ |
|---|---|---|---|---|---|---|
| $y'$ | $-1$ | $a_1$ | $3a_2$ | $5a_3$ | $7a_4$ | $\ldots$ |
| | $1$ | $2a_1$ | $2a_2$ | $2a_3$ | $2a_4$ | $\ldots$ |
| | | | $a_1^2$ | $2a_1a_2$ | $2a_1a_3$ | $\ldots$ |
| $y^2$ | | | | $a_2^2$ | | $\ldots$ |
| | | | | | | $\ldots$ |
| | $0$ | $-1$ | $0$ | $0$ | $0$ | $\ldots$ |

Cf. ex. 5.1.　We obtain so the relations

$$3a_1 = -1, \quad 5a_2 + a_1^2 = 0, \quad 7a_3 + 2a_1a_2 = 0, \ldots$$

and hence successively

$$S_{2n} = -\frac{a_n \pi^{2n}}{2} = \frac{\pi^2}{6}, \frac{\pi^4}{90}, \frac{\pi^6}{945}, \frac{\pi^8}{9450}, \ldots$$

for $\qquad n = 1, 2, 3, 4, \ldots$.

**39.** Method of ex. 37 and 38 applied to result of ex. 36.　We set now

$$y = \cot\left(\frac{\pi}{4} - \frac{x}{2}\right) = b_1 + b_2 x + b_3 x^2 + b_4 x^3 + \ldots.$$

Then

$$T_n = 1 + \frac{(-1)^n}{3^n} + \frac{1}{5^n} + \frac{(-1)^n}{7^n} + \frac{1}{9^n} + \frac{(-1)^n}{11^n} + \ldots = \frac{b_n \pi^n}{2^{n+1}}.$$

Now $y$ satisfies the differential equation

$$2y' = 1 + y^2$$

which (observe that $b_1 = 1$) yields the array

| | $1$ | $x$ | $x^2$ | $x^3$ | $x^4$ | $\ldots$ |
|---|---|---|---|---|---|---|
| | $1$ | | | | | |
| | $1$ | $2b_2$ | $2b_3$ | $2b_4$ | $2b_5$ | $\ldots$ |
| | | | $b_2^2$ | $2b_2b_3$ | $2b_2b_4$ | $\ldots$ |
| $y^2$ | | | | | $b_3^2$ | $\ldots$ |
| | | | | | | $\ldots$ |
| $2y'$ | $2b_2$ | $4b_3$ | $6b_4$ | $8b_5$ | $10b_6$ | $\ldots$ |

Hence we obtain first the relations $2b_2 = 2$, $4b_3 = 2b_2$, $6b_4 = 2b_3 + b_2^2$, $\ldots$ and then the values for

$$n = 1, 2, 3, 4, 5, 6, \ldots$$

$$T_n = \frac{\pi}{4}, \frac{\pi^2}{8}, \frac{\pi^3}{32}, \frac{\pi^4}{96}, \frac{5\pi^5}{1536}, \frac{\pi^6}{960}, \ldots$$

**40.** Generally

$$S_{2n}\left(1 - \frac{1}{2^{2n}}\right) = 1 + \frac{1}{2^{2n}} + \frac{1}{3^{2n}} + \frac{1}{4^{2n}} + \frac{1}{5^{2n}} + \cdots$$

$$- \frac{1}{2^{2n}} \qquad\qquad - \frac{1}{4^{2n}} \qquad\qquad - \cdots$$

$$= T_{2n}.$$

This can be used to check the numerical work in ex. 38 and ex. 39:

$$\frac{1}{6} \cdot \frac{3}{4} = \frac{1}{8}, \quad \frac{1}{90} \cdot \frac{15}{16} = \frac{1}{96}, \quad \frac{1}{945} \cdot \frac{63}{64} = \frac{1}{960}.$$

**41.**

$$\int_0^1 (1 - x^2)^{-1/2} \arcsin x \cdot dx$$

$$= \int_0^1 (1 - x^2)^{-1/2} x \, dx + \frac{1}{2}\frac{1}{3} \int_0^1 (1 - x^2)^{-1/2} x^3 \, dx + \cdots$$

$$= 1 + \frac{1}{2}\frac{1}{3}\frac{2}{3} + \frac{1}{2}\frac{3}{4}\frac{1}{5}\frac{2 \cdot 4}{3 \cdot 5} + \frac{1}{2}\frac{3}{4}\frac{5}{6}\frac{1}{7}\frac{2 \cdot 4 \cdot 6}{3 \cdot 5 \cdot 7} + \cdots$$

$$= 1 + \frac{1}{3^2} + \frac{1}{5^2} + \frac{1}{7^2} + \cdots .$$

Now evaluate the integral we started from ($= (\pi/2)^2/2$) and use ex. 40. Cf. Euler, *Opera Omnia*, ser. 1, vol. 14, p. 178–181.

**42.**

$$\int_0^1 (1 - x^2)^{-1/2} (\arcsin x)^2 dx$$

$$= \int_0^1 (1 - x^2)^{-1/2} x^2 dx + \frac{2}{3}\frac{1}{2} \int_0^1 (1 - x^2)^{-1/2} x^4 dx + \cdots$$

$$= \frac{1}{2}\frac{\pi}{2} + \frac{2}{3}\frac{1}{2} \cdot \frac{1}{2}\frac{3}{4}\frac{\pi}{2} + \frac{2}{3}\frac{4}{5}\frac{1}{3} \cdot \frac{1}{2}\frac{3}{4}\frac{5}{6}\frac{\pi}{2} + \cdots$$

$$= \frac{\pi}{4}\left(1 + \frac{1}{2^2} + \frac{1}{3^2} + \frac{1}{4^2} + \cdots\right).$$

Now evaluate the integral we started from ($= (\pi/2)^3/3$). The expansion of (arcsin $x$)² that we have used will be derived in ex. 5.1.    Cf. Euler, *Opera Omnia*, ser. 1, vol. 14, p. 181–184.

**43.** (a)  $\displaystyle\sum_{n=1}^{\infty} \frac{x^n}{n^2} = \int_0^x \sum_{n=1}^{\infty} \frac{t^{n-1}}{n}\, dt = -\int_0^x t^{-1} \log(1-t)\, dt;$

integrate by parts and then introduce as new variable of integration $s = 1 - t$.

(b)  $x = 1/2$ which renders the greater of the two values, $x$ and $1 - x$, as small as possible.

**44.** If $P_n(x) = 0$, we have

$$\left(1 + \frac{ix}{n}\right)^n = \left(1 - \frac{ix}{n}\right)^n,$$

$$1 + \frac{ix}{n} = e^{2\pi ki/n}\left(1 - \frac{ix}{n}\right),$$

$$x = \frac{n}{i}\,\frac{e^{\pi ki/n} - e^{-\pi ki/n}}{e^{\pi ki/n} + e^{-\pi ki/n}} = n \tan\frac{k\pi}{n}$$

where we take $k = 0, 1, 2, \ldots n - 1$ if $n$ is odd.

**45.** If $n$ is odd, we can take in the expression of the roots, see ex. 44,

$$k = 0,\ \pm 1,\ \pm 2,\ \ldots \pm (n-1)/2.$$

Therefore,

$$\frac{P_n(x)}{x} = \prod_{k=1}^{(n-1)/2}\left(1 - \frac{x^2}{n^2 \tan^2(k\pi/n)}\right).$$

Observe that, for fixed $k$,

$$\lim_{n \to \infty} n \tan(k\pi/n) = k\pi.$$

Only a relatively small step is needed to carry us from the point now attained to a proof that is acceptable according to modern standards.    A somewhat different arrangement of Euler's argument due to Cauchy seems to have served as a model to Abel as he, led by analogy, discovered the representation of the elliptic functions by infinite products.    Cf. A. Cauchy, *Oeuvres complètes*, ser. 2, vol. 3, p. 462–465, and N. H. Abel, *Oeuvres complètes*, vol. 1, p. 335–343.

**46.** The sum of a finite number of terms is the same in whatever order the terms are taken.    The mistake was to extend this statement uncritically to an infinite number of terms, that is, to assume that the sum of an *infinite series* is the same in whatever order the terms are taken.    The assumed statement is false; our example shows that it is false.    The protection against such a mistake is to go back to the definitions of the terms used and to rely only on rigorous proofs based on these definitions.    Thus, the sum of an infinite series is, by definition, the *limit* of a certain sequence (of the sequence of the "partial sums") and interchanging an infinity of terms, as we did, we change essentially the defining sequence. (Under a certain *restrictive condition* a rearrangement of the terms of an infinite series does not change the sum; see Hardy, *Pure Mathematics*, p. 346–347, 374, 378–379. Yet this condition is not satisfied in the present case.)

No solution: **5, 7, 9, 10, 11, 18, 19, 20, 21.**

**SOLUTIONS, CHAPTER III**

**1.** Yes: $F = 2n$, $V = n + 2$, $E = 3n$.

**2.** (1) Yes: $F = m(p + 1)$, $V = pm + 2$, $E = m(p + 1) + pm$. (2) $p = 1$, $m = 4$.

**3.** (1) Exclude for a moment the tetrahedron; the remaining six polyhedra form three pairs. The two polyhedra in the same pair, as cube and octahedron, are so connected that they have the same $E$, but the $F$ of one equals the $V$ of the other. The tetrahedron remains alone, but it is connected with itself in this peculiar way. (2) Take the cube. Take any two neighboring faces of the cube and join their centers by a straight line. The 12 straight lines so obtained form the edges of a regular octahedron. This octahedron is inscribed in the cube, its 6 vertices lie in the centers of the 6 faces of the cube. Reciprocally, the centers of the 8 faces of the regular octahedron are the 8 vertices of a cube inscribed in the octahedron. A similar reciprocal relation holds between the polyhedra of the same pair also in the other cases. (Use cardboard models for the dodecahedron and the icosahedron.) The tetrahedron has this peculiar relation to itself: the centers of its 4 faces are the vertices of an inscribed tetrahedron. (3) The passage from one polyhedron of a pair to the other preserves Euler's formula.

**4.** By $E$ *red* boundary lines, the sphere is divided into $F$ countries; there are $V$ points that belong to the boundary of more than two countries. Choose in each country a point, the "capital" of the country. Connect the capitals of any two neighboring countries by a "road" so that each road crosses just one boundary line and different roads do not cross each other; draw these roads in *blue*. There are precisely $E$ blue lines (roads); they divide the sphere into $F'$ countries with $V'$ points belonging to the boundary of three or more of these countries. Satisfy yourself that $V' = F$ and $F' = V$. The relation between the red and blue subdivisions of the sphere is reciprocal, the passage from one to the other preserves Euler's formula.

**5.** Euler's formula will hold after "roofing" (sect. 4) if, and only if, it did hold before roofing. Yet by roofing all nontriangular faces of a given polyhedron we obtain another polyhedron with triangular faces only.

**6.** Analogous to ex. 5: "truncating" introduces vertices with three edges as "roofing" introduces triangular faces. We could also reduce the present case to ex. 5 by using ex. 4.

**7.** (1) $N_0 = V$, $N_1 = E$, $N_2 = F - 1$. The subscripts 0, 1, 2 indicate the respective dimensionality, see sect. 7. (2) $N_0 - N_1 + N_2 = 1$.

**8.** (1) Set $l + m = c_1$, $lm = c_2$. Then

$$N_0 = (l + 1)(m + 1) \qquad = 1 + c_1 + c_2,$$

$$N_1 = (l + 1)m + (m + 1)l \qquad = c_1 + 2c_2,$$

$$N_2 = lm \qquad = c_2.$$

(2) Yes, $N_0 - N_1 + N_2 = 1$, although this simple subdivision of a rectangle cannot be generated exactly in the manner described in ex. 7.

**9.** $N_2 \, 180° = (N_0 - 3) \, 360° + 180°$. In trying to come closer to our goal which is the equation (2) in the solution of ex. 7, we transform this successively into

$$2N_0 - N_2 - 5 = 0,$$

$$2N_0 - 3N_2 + 2N_2 - 3 = 2.$$

By counting the edges in two different ways, we obtain

$$3N_2 = 2N_1 - 3.$$

The last two equations yield

$$N_0 - N_1 + N_2 = 1,$$

which proves Euler's formula, in view of ex. 7 (2).

**10.** (1) Let $l + m + n = c_1$, $lm + ln + mn = c_2$ and $lmn = c_3$. Then

$$N_0 = (l + 1)(m + 1)(n + 1) = 1 + c_1 + c_2 + c_3,$$

$$N_1 = l(m + 1)(n + 1) + m(l + 1)(n + 1) + n(l + 1)(m + 1)$$

$$= c_1 + 2c_2 + 3c_3,$$

$$N_2 = (l + 1)mn + (m + 1)ln + (n + 1)lm = c_2 + 3c_3,$$

$$N_3 = lmn = c_3.$$

(2) Yes, $N_0 - N_1 + N_2 - N_3 = 1$.

**11.** We dealt with the case $n = 3$ in sect. 16. In dealing with this case we did not use any simplifying circumstance that would be specific to the particular case $n = 3$. Therefore, this particular case may well "represent" the general case (in the sense of ex. 2.10) as hinted already in sect. 17. The reader should repeat the discussion of sect. 16, saying $n$ for 3, $n + 1$ for 4, $P_n$ for 7, and $P_{n+1}$ for 11, with a little caution. See also ex. 12.

**12.** Follow the suggestions of sect. 17 and the analogy of ex. 11. Given $n$ planes in general position. They dissect the space into $S_n$ parts. Adjoin one more plane; it is intersected by the foregoing $n$ planes in $n$ straight lines which, *being in a general position*, determine on it $P_n$ regions. Each such plane region acts as a "diaphragm"; it divides an old compartment of space (one of those $S_n$ compartments) into two new compartments, makes one old compartment disappear and two new compartments appear and adds so finally a unit to the previous number $S_n$ of compartments. Hence the relation that we desired to prove.

**13.** See the third column of the table in sect. 14.

**14.** The second column of the table in sect. 14 agrees with

$$S_n = 1 + n + \frac{n(n-1)}{1 \cdot 2} + \frac{n(n-1)(n-2)}{1 \cdot 2 \cdot 3}$$

$$= \binom{n}{0} + \binom{n}{1} + \binom{n}{2} + \binom{n}{3};$$

we used the usual notation for binomial coefficients.

**15.** Finite 3, infinite 8.

**16.** Let $P_n^\infty$ denote the number of those among the $P_n$ parts defined in ex. 11 which are infinite. By observation, for

$$n = 1, 2, 3$$

$$P_n^\infty = 2, 4, 6.$$

Guess: $P_n^\infty = 2n$. Proof: Take a point in one of the finite parts and imagine an ever-increasing circle with this point as center. When this circle becomes very large, the $P_n - P_n^\infty$ finite parts practically coincide with its center. Now, $n$ different lines through the center of the circle intersect the periphery in $2n$ points and divide it into $2n$ parts. Hence, in fact, $P_n^\infty = 2n$.

$$P_n - P_n^\infty = 1 - n + \frac{n(n-1)}{2}.$$

For instance, the answer to ex. 15 is

$$1 - 4 + 6 = 3.$$

**17.** Same as ex. 18, by analogy to the solution of ex. 16.

**18.** Same as ex. 19.

**19.** See ex. 20.

**20.** We consider $n$ circles in the plane any two of which intersect in general position. We call $S_n^\infty$ the number of parts into which these circles dissect the plan, in view of ex. 17, 18, and 19. In analogy with sect. 16, notice that the number of parts into which a circle is divided by $n$ circles intersecting it is

$2n$ (general position assumed). Observe (think of all three interpretations of $S_n^\infty$):

$$n = 1 \quad 2 \quad 3 \quad 4$$

$$2n = 2 \quad 4 \quad 6 \quad 8$$

$$S_n^\infty = 2 \quad 4 \quad 8 \quad 14.$$

Guess: $S_{n+1}^\infty = S_n^\infty + 2n$. Proof: as ex. 11, 12. For instance,

$$S_5^\infty = S_4^\infty + 8 = 14 + 8 = 22;$$

This is the solution of ex. 17, 18, and 19. Further guesses:

$$S_n^\infty = 2 \binom{n}{0} + 2 \binom{n}{2},$$

$$S_n - S_n^\infty = - \binom{n}{0} + \binom{n}{1} - \binom{n}{2} + \binom{n}{3}.$$

**21.** See ex. 22–30.

**22.** Wrong: $F = 1$, $V = E = 0$, $1 + 0 \neq 0 + 2$.

**23.** Wrong: $F = 2$, $V = 0$, $E = 1$, $2 + 0 \neq 1 + 2$.

**24.** Wrong: $F = 3$, $V = 0$, $E = 2$, $3 + 0 \neq 2 + 2$.

**25.** Right: $F = 3$, $V = 2$, $E = 3$, $3 + 2 = 3 + 2$.

**26.** Wrong: $F = p + 1$, $V = 0$, $E = p$, $(p + 1) + 0 \neq p + 2$;  see ex. 22, 23, 24 for the cases $p = 0, 1, 2$, respectively. Observe that, in the present case, the solution of ex. 2 (1) becomes inapplicable.

**27.** The case $m = 3$, $p = 0$ is right, see ex. 25, and so is, more generally, the case $m \geq 3$: $F = m$, $V = 2$, $E = m$, $m + 2 = m + 2$. The case $m = 0$, $p = 0$ is wrong, see ex. 22. The remaining two cases *can* be so interpreted that they appear right. (1) $m = 1$, $p = 0$: one country with an interior barrier that has two endpoints, $F = 1$, $V = 2$, $E = 1$, $1 + 2 = 1 + 2$. (2) $m = 2$, $p = 0$: two countries separated by two arcs and two corners, $F = 2$, $V = 2$, $E = 2$, $2 + 2 = 2 + 2$. The more obvious interpretation given in ex. 23 yields "wrong." With the present interpretation the solution of ex. 2 (1) remains applicable to the case $m > 0$, $p = 0$.

**28.** $m \geq 3$, $p \geq 1$. The proof uses the fact that, in any convex polyhedron, at least three edges surround a face and at least three edges meet in a vertex.

**29.** Ex. 22–28 suggest two conditions: (1) A country counted in $F$, as a face of a convex polyhedron, should be of the "type of a circular region"; a full sphere is not of this type, neither is an annulus of this type. (2) A line counted in $E$, as an edge of a convex polyhedron, should terminate in corners;

the full periphery of a circle does not terminate so (it does not terminate at all). Ex. 22 fails to satisfy (1), ex. 23 fails to satisfy (2), ex. 24 fails to satisfy either, ex. 25 or, more generally, the case $m > 0, p = 0$, interpreted as in the solution of ex. 27, satisfies both (1) and (2).

**30.** (1) Take the case $(3, 2)$ of ex. 2 (1), cf. ex. 26, but erase on each meridian the arc between the two parallel circles: $F = 7$, $V = 8$, $E = 12$, $7 + 8 \neq 12 + 2$, there is a spherical zone among the countries and therefore conflict with condition (1), but not with condition (2), of ex. 29. (2) $F = 1$, $V = 1$, $E = 0$ (one country, encompassing the whole globe, except a mathematical point at the north pole); right, $1 + 1 = 0 + 2$, no conflict with (1) or (2) of ex. 29. Etc.

**32.** $3F_3 + 4F_4 + 5F_5 + \ldots = 3V_3 + 4V_4 + 5V_5 + \ldots = 2E$.

**33.** $4\pi$, $12\pi$, $8\pi$, $36\pi$, $20\pi$, respectively.

**34.** $\Sigma\alpha = \pi F_3 + 2\pi F_4 + 3\pi F_5 + \ldots$.

**35.** By ex. 34, 32, 31

$$\Sigma\alpha = \pi\Sigma(n - 2)F_n = 2\pi(E - F).$$

**36.** A convex spherical polygon with $n$ sides can be dissected into $n - 2$ spherical triangles. Therefore,

$$A = \alpha_1 + \alpha_2 + \ldots + \alpha_n - (n - 2)\pi$$
$$= 2\pi - (\pi - \alpha_1) - (\pi - \alpha_2) - \ldots - (\pi - \alpha_n)$$
$$= 2\pi - a_1' - a_2' - \ldots - a_n'$$
$$= 2\pi - P'.$$

**37.** The faces of the polyhedron passing through one of the vertices include an interior solid angle; its supplement is called by Descartes the exterior solid angle. Describe a sphere with radius 1 around the vertex as center, but keep only that sector of the sphere that is contained in the exterior solid angle; the sectors so generated at the several vertices of the polyhedron form, when shifted together, a full sphere as the circular sectors in the analogous plane figure (fig. 3.7) form, when shifted together, a full circle. We regard as the measure of a solid angle the area of the corresponding spherical polygon: the joint measure of all the exterior solid angles of the polyhedron is, in fact, $4\pi$.

**38.** Let $P_1$, $P_2$, $\ldots P_V$ denote the perimeters of the spherical polygons that correspond to the $V$ interior solid angles of the polyhedron. Then, by ex. 36 and 37,

$$\Sigma\alpha = P_1 + P_2 + \ldots + P_V$$
$$= 2\pi - A_1' + 2\pi - A_2' + \ldots + 2\pi - A_V'$$
$$= 2\pi V - 4\pi.$$

**39.** By ex. 35 and 38

$$2\pi(E - F) = \Sigma\alpha = 2\pi(V - 2).$$

**40.** By ex. 31, 32

$$3F = 3F_3 + 3F_4 + 3F_5 + \ldots$$

$$\leqq 3F_3 + 4F_4 + 5F_5 + \ldots = 2E$$

which yields the first of the six proposed inequalities. The case of equality is attained when $F = F_3$, that is, when all faces are triangles. Eliminating first $E$ and then $F$ from Euler's theorem and the inequality just proved, we obtain the remaining two inequalities in the first row; they go over into equations if, and only if, all faces are triangles. Interchanging the roles of $F$ and $V$, as suggested by ex. 3 and 4, we obtain the three proposed inequalities in the second row; they go over into equations if, and only if, all vertices of the polyhedron are three-edged. Some of the inequalities proved are given in Descartes' notes.

**41.** From Euler's theorem

$$6F - 2E = 12 + 2(2E - 3V)$$

and hence, by ex. 31, 32, and 40,

$$3F_3 + 2F_4 + F_5 = 12 + 2(2E - 3V) + F_7 + 2F_8 + \ldots$$

$$3F_3 + 2F_4 + F_5 \geqq 12$$

and so *any convex polyhedron must have some faces with less than six sides.*

No solution: **31.**

**SOLUTIONS, CHAPTER IV**

**1.** $R_2(25) = 12$, see sect. 2; $S_3(11) = 3$.

**2.** $R_2(n)$ denotes the number of the lattice points in a plane that lie on the periphery of a circle with radius $\sqrt{n}$ and center at the origin. (Take the case $n = 25$, ex. 1, and draw the circle.) $R_3(n)$ is the number of lattice points in space on the surface of the sphere with radius $\sqrt{n}$ and center at the origin.

**3.** If $p$ is an odd prime, $R_2(p^2) = 12$ or 4 according as $p$ divided by 4 leaves the remainder 1 or 3.

**4.** The comparison of the tables suggests: if $p$ is an odd prime, either both $p$ and $p^2$ are expressible as a sum of two squares, or neither $p$ nor $p^2$ is so expressible. A more precise conjecture is also somewhat supported by our observations: if $p$ is an odd prime, $R_2(p) = 8$ or 0, according as $p$ divided by 4 leaves the remainder 1 or 3.

**5.** If $p = x^2 + y^2$, it follows that

$$p^2 = x^4 + 2x^2y^2 + y^4 = (x^2 - y^2)^2 + (2xy)^2.$$

That is, if $R_2(p) > 0$ also $R_2(p^2) > 0$. This is only one half of our less precise, and only a small part of our more precise, conjecture. (If we know that $R_2(p^2) > 0$, a conclusion concerning $R_2(p)$ is definitely less obvious.) Still, it seems reasonable that such a partial verification greatly strengthens our confidence in the less precise conjecture, and strengthens somewhat our confidence in the more precise conjecture too.

**6.** $R_3(p) = 0$ for $n = 7, 15, 23$, and $28$, and for no other $n$ up to 30; see Table II on pp. 74–75.

**7.** The respective contributions to $S_4(n)$ are: (1) 24, (2) 12, (3) 6, (4) 4. (5) 1.

**8.** First, refer to the cases distinguished in ex. 7. If $S_4(4u)$ is odd, the case (5) necessarily arises, and so

$$4u = a^2 + a^2 + a^2 + a^2,$$
$$u = a^2.$$

Second, to any divisor $d$ of $u$ corresponds the divisor $u/d$ and these two divisors are different unless $u = d^2$. Therefore, the number of the divisors of $u$ is odd or even, according as $u$ is or is not a square, and the same holds for the sum of these divisors, since each divisor of $u$ is odd, as $u$ itself. We conjectured in sect. 6 that $S_4(4u)$ and the sum of the divisors of $u$ are equal; we proved now that these two numbers leave the same remainder when divided by 2. Having proved a part of our conjecture, we have, of course, more faith in it.

**9.**

(1) $24 \times 2^4 = 8 \times 48$        (6) $24 \times 2^3 = 8 \times 24$

(2) $12 \times 2^4 = 8 \times 24$        (7) $12 \times 2^3 = 8 \times 12$

(3) $6 \times 2^4 = 8 \times 12$        (8) $4 \times 2^3 = 8 \times 4$

(4) $4 \times 2^4 = 8 \times 8$        (9) $12 \times 2^2 = 8 \times 6$

(5) $1 \times 2^4 = 8 \times 2$        (10) $6 \times 2^2 = 8 \times 3$

(11) $4 \times 2 = 8 \times 1$.

**10.** See Table II, p. 74. Check at least a few entries. It follows from ex. 9 that $R_4(n)$ is divisible by 8.

**11.** Trying to notice at least fragmentary regularities (as we did in sect. 6), you may be led to grouping some more conspicuous cases as follows:

| (1) | 2 | 3 | 5 | 7 | 11 | 13 | 17 | 19 | 23 | 29 |
|---|---|---|---|---|---|---|---|---|---|---|
|  | 3 | 4 | 6 | 8 | 12 | 14 | 18 | 20 | 24 | 30 |
| (2) |  | 2 |  | 4 |  | 8 |  | 16 |  |  |
|  |  | 3 |  | 3 |  | 3 |  | 3 |  |  |
| (3) | 4 | 8 | 12 | 16 | 20 | 24 | 28 |  |  |  |
|  | 3 | 3 | 12 | 3 | 18 | 12 | 24. |  |  |  |

In (1), (2), and (3) the first line gives $n$, the second line $R_4(n)$.

**12.** Done in solution of ex. 11: (1) primes, (2) powers of 2, (3) numbers divisible by 4.

**13.** By the analogy of sect. 6 and a little observation, the law is relatively easy to discover when $n$ is not divisible by 4. Therefore, we concentrate on the case (3) in the solution of ex. 11.

$$
\begin{array}{lccccccc}
n & = & 4 & 8 & 12 & 16 & 20 & 24 & 28 \\
n/4 & = & 1 & 2 & 3 & 4 & 5 & 6 & 7 \\
R_4(n)/8 & = & 3 & \mathbf{3} & 12 & 3 & 18 & \mathbf{12} & 24.
\end{array}
$$

A number in heavy print in the third line is the *sum of all divisors* of the corresponding number in the second line—and, therefore, the sum of *some* divisors of the corresponding number in the first line, in which we are really interested. This observation leads to another trial:

$$
\begin{array}{lcccc}
n & = & 4 & 8 & 12 & 16 \\
R_4(n)/8 & = & 1+2 & 1+2 & 1+2+3+6 & 1+2 \\
n & = & 20 & 24 & 28 \\
R_4(n)/8 & = & 1+2+5+10 & 1+2+3+6 & 1+2+7+14.
\end{array}
$$

Which divisors are added together? Which divisors are omitted?

**14.** $R_4(n)$, the number of representations of $n$ as a sum of four squares, equals 8 times the sum of those divisors of $n$ which are not divisible by 4. (If $n$ itself is not divisible by 4, none of its divisors is, and hence the rule is simpler in this more frequent case.)

**15.** Correspondingly to the columns of Table II:

| | | | |
|---|---|---|---|
| 31 | $25 + 4 + 1 + 1$ | $12 \times 16$ | $32 = 31 + 1$ |
| | $9 + 9 + 9 + 4$ | $4 \times 16$ | |
| 32 | $16 + 16$ | $6 \times 4$ | $3 = 2 + 1$ |
| 33 | $25 + 4 + 4$ | $12 \times 8$ | $48 = 33 + 11 + 3 + 1$ |
| | $16 + 16 + 1$ | $12 \times 8$ | |
| | $16 + 9 + 4 + 4$ | $12 \times 16$ | |

**16.** $5 = 1 + 1 + 1 + 1 + 1 = 4 + 1$

$$R_8(5) = \binom{8}{5} \; 2^5 + 8 \cdot 7 \cdot 2^2 = 2016 = 16 \times 126.$$

$$40 = 25 + 9 + 1 + 1 + 1 + 1 + 1 + 1$$
$$40 = 9 + 9 + 9 + 9 + 1 + 1 + 1 + 1$$

$$S_8(40) = 8 \cdot 7 + \binom{8}{4} = 126.$$

**17.** Ex. 16. Table III has been actually constructed by a method less laborious than that of ex. 16; cf. ex. 6.17 and 6.23.

**18.** Within the limits of Table III, both $R_8(n)$ and $S_8(8n)$ increase steadily with $n$ whereas $R_4(n)$ and $S_4(4(2n-1))$ oscillate irregularly.

**19.** The analogy with $R_4(n)$ and $S_4(4(2n-1))$ points to divisors. One fragmentary regularity is easy to observe: if $n$ is odd, $R_8(n)/16$ and $S_8(8n)$ are exactly equal; if $n$ is even, they are different, although the difference is relatively small in most cases.

**20.** Odd and even already in ex. 19. Powers of 2:

| $n$ | 1 | 2 | 4 | 8 | 16 |
|---|---|---|---|---|---|
| $S(8n)$ | 1 | 8 | 64 | 512 | 4096. |

Also the second line consists of powers of 2:

| $n$ | $2^0$ | $2^1$ | $2^2$ | $2^3$ | $2^4$ |
|---|---|---|---|---|---|
| $S(8n)$ | $2^0$ | $2^3$ | $2^6$ | $2^9$ | $2^{12}$. |

What is the law of the exponents?

**21.** If $n$ is a power of 2, $S(8n) = n^3$. This (and the smooth increase of $R_8(n)$ and $S_8(8n)$) leads to constructing the following table.

| $n$ | $R_8(n)/16 - n^3$ | $S_8(8n) - n^3$ |
|---|---|---|
| 1 | 0 | 0 |
| 2 | $-1$ | 0 |
| 3 | 1 | 1 |
| 4 | 7 | 0 |
| 5 | 1 | 1 |
| 6 | $-20$ | 8 |
| 7 | 1 | 1 |
| 8 | 71 | 0 |
| 9 | 28 | 28 |
| 10 | $-118$ | 8 |
| 11 | 1 | 1 |
| 12 | 260 | 64 |
| 13 | 1 | 1 |
| 14 | $-336$ | 8 |
| 15 | 153 | 153 |
| 16 | 583 | 0 |
| 17 | 1 | 1 |
| 18 | $-533$ | 224 |
| 19 | 1 | 1 |
| 20 | 946 | 64 |

In the column concerned with $R_8(n)$, the $+$ and $-$ signs are regularly distributed.

**22.** Cubes of divisors!

| $n$ | $R_8(n)/16 = S_8(8n)$ |
|---|---|
| 1 | $1^3$ |
| 3 | $3^3 + 1^3$ |
| 5 | $5^3 + 1^3$ |
| 7 | $7^3 + 1^3$ |
| 9 | $9^3 + 3^3 + 1^3$ |
| 11 | $11^3 + 1^3$ |
| 13 | $13^3 + 1^3$ |
| 15 | $15^3 + 5^3 + 3^3 + 1^3$ |
| 17 | $17^3 + 1^3$ |
| 19 | $19^3 + 1^3$ |

| $n$ | $R_8(n)/16$ | $S_8(8n)$ |
|---|---|---|
| 2 | $2^3 - 1^3$ | $2^3$ |
| 4 | $4^3 + 2^3 - 1^3$ | $4^3$ |
| 6 | $6^3 - 3^3 + 2^3 - 1^3$ | $6^3 + 2^3$ |
| 8 | $8^3 + 4^3 + 2^3 - 1^3$ | $8^3$ |
| 10 | $10^3 - 5^3 + 2^3 - 1^3$ | $10^3 + 2^3$ |
| 12 | $12^3 + 6^3 + 4^3 - 3^3 + 2^3 - 1^3$ | $12^3 + 4^3$ |
| 14 | $14^3 - 7^3 + 2^3 - 1^3$ | $14^3 + 2^3$ |
| 16 | $16^3 + 8^3 + 4^3 + 2^3 - 1^3$ | $16^3$ |
| 18 | $18^3 - 9^3 + 6^3 - 3^3 + 2^3 - 1^3$ | $18^3 + 6^3 + 2^3$ |
| 20 | $20^3 + 10^3 - 5^3 + 4^3 + 2^3 - 1^3$ | $20^3 + 4^3$ |

**23.** (1) $(-1)^{n-1}R_8(n)/16$ equals the sum of the cubes of all odd divisors of $n$, less the sum of the cubes of all even divisors of $n$. (2) $S_8(8n)$ equals the sum of the cubes of those divisors of $n$ whose co-divisors are odd. (If $d$ is a divisor of $n$, we call $n/d$ the co-divisor of $d$.) See ex. 6.24 on the history of these theorems and references.

**24.** Construct the table

| 0 | 3 | 6 | 9 | 12 |
|---|---|---|---|---|
| 5 | 8 | 11 | 14 | |
| 10 | 13 | | | |

which should be imagined as extending without limit to the right and downwards and shows that the only positive integers not expressible in the proposed form are 1, 2, 4, 7.

**25.** Case $a = 3$, $b = 5$ in ex. 24; $a$ and $b$ are co-prime. Last integer non-expressible $ab - a - b = (a - 1)(b - 1) - 1$. This is incomparably easier to prove than the laws concerned with sums of squares.

**26.** (1) is generally true. (2) is not generally true, but the first exception is $n = 341$. (See G. H. Hardy and E. M. Wright, *An introduction to the theory of numbers*, Oxford, 1938, p. 69, 72.)

## SOLUTIONS, CHAPTER V

**1.** [Cf. Putnam 1948]

(a) $x + \dfrac{2}{3} x^3 + \dfrac{2}{3}\dfrac{4}{5} x^5 + \cdots + \dfrac{2}{3}\dfrac{4}{5}\dfrac{6}{7} \cdots \dfrac{2n}{2n + 1} x^{2n+1} + \cdots$ .

(b) Having verified the differential equation, put

$$y = a_0 x + a_1 x^3 + a_2 x^5 + \cdots + a_n x^{2n+1} + \cdots .$$

To compare the coefficients of like powers you may use the array

| | 1 | $x^2$ | $x^4$ | $\ldots$ | $x^{2n}$ | |
|---|---|---|---|---|---|---|
| | | | | | | |
| $y'$ | $a_0$ | $3a_1$ | $5a_2$ | $\ldots$ | $(2n + 1)a_n$ | $\ldots$ |
| $-x^2 y'$ | | $-a_0$ | $-3a_1$ | $\ldots$ | $-(2n - 1)a_{n-1}$ | $\ldots$ |
| $-xy$ | | $-a_0$ | $-a_1$ | $\ldots$ | $-a_{n-1}$ | $\ldots$ |
| | | | | | | |
| | 1 | 0 | 0 | $\ldots$ | 0 | $\ldots$ |

which yields $a_0 = 1$ and $(2n + 1)a_n = 2na_{n-1}$ for $n \geq 1$.

**2.** [Cf. Putnam 1950]

(a) $y = \dfrac{x}{1} + \dfrac{x^3}{1 \cdot 3} + \dfrac{x^5}{1 \cdot 3 \cdot 5} + \cdots + \dfrac{x^{2n-1}}{1 \cdot 3 \cdot 5 \ldots (2n - 1)} + \cdots$ .

(b) This expansion satisfies

$$y' = 1 + \dfrac{x^2}{1} + \dfrac{x^4}{1 \cdot 3} + \dfrac{x^6}{1 \cdot 3 \cdot 5} + \cdots ,$$

$$y' = 1 + xy.$$

The given product $y$ satisfies the same differential equation. Both the expansion and the product vanish when $x = 0$. Hence, they are identical.

**3.** The relations between the coefficients $a_n$ derived from

$$\frac{1}{1+x} + \frac{4a_1x}{(1+x)^3} + \frac{16a_2x^2}{(1+x)^5} + \cdots$$

$$= 1 + a_1x^2 + a_2x^4 + \cdots = f(x)$$

are exhibited in the array (see ex. 1)

| 1 | $-1$ | 1 | $-1$ | 1 |
|---|---|---|---|---|
| $4a_1$ | $-4a_1 \cdot 3$ | $4a_1 \cdot 6$ | $-4a_1 \cdot 10$ | |
| | $16a_2$ | $-16a_2 \cdot 5$ | $16a_2 \cdot 15$ | |
| | | $64a_3$ | $-64a_3 \cdot 7$ | |
| | | | $256a_4$ | |
| 1 | 0 | $a_1$ | 0 | $a_2$ |

They yield

$$f(x) = 1 + \left(\frac{1}{2}\right)^2 x^2 + \left(\frac{1}{2}\frac{3}{4}\right)^2 x^4 + \left(\frac{1}{2}\frac{3}{4}\frac{5}{6}\right)^2 x^6 + \left(\frac{1}{2}\frac{3}{4}\frac{5}{6}\frac{7}{8}\right)^2 x^8 + \cdots$$

This example is of historical interest.   See Gauss, *Werke*, vol. 3, p. 365–369.

**4.** Study the arrangement of the following array (see ex. 1, 3):

| $f(x)^3$ | $a_0^3$ | $3a_1a_0^2$ | $3a_2a_0^2$ | $3a_3a_0^2$ | $3a_4a_0^2$ |
|---|---|---|---|---|---|
| | | | $3a_1^2a_0$ | $6a_2a_1a_0$ | $6a_3a_1a_0$ |
| | | | $a_1^3$ | $3a_2^2a_0$ |
| | | | | $3a_2a_1^2$ |
| $3f(x)f(x^2)$ | $3a_0^2$ | $3a_1a_0$ | $3a_2a_0$ | $3a_3a_0$ | $3a_4a_0$ |
| | | | $3a_0a_1$ | $3a_1a_1$ | $3a_2a_1$ |
| | | | | | $3a_0a_2$ |
| $2f(x^3)$ | $2a_0$ | | $2a_1$ | | |
| | $6a_1$ | $6a_2$ | $6a_3$ | $6a_4$ | $6a_5$ |

Starting from $a_0 = 1$, we obtain recursively $a_1$, $a_2$, $a_3$, $a_4$, and $a_5 = 8$. See G. Pólya, *Zeitschrift für Kristallographie*, vol. (A) 93 (1936) p. 415–443, and *Acta Mathematica*, vol. 68 (1937) p. 145–252.

**5.** From comparing the expansions in sect. 1.

**6.** (a) $\varepsilon^2/15$.   (b) $+\infty$.   In both extreme cases, the error is positive, the approximate value larger than the true value.

**7.** (a) $\varepsilon^2/15$.   (b) $1/3$.   In both extreme cases, the approximate value is larger than the true value.

**8.** $4\pi(a^2 + b^2 + c^2)/3$.   There is some reason to suspect that this approximation yields values in excess of the true values.   See G. Pólya, *Publicaciones del Instituto di Matematica*, Rosario, vol. 5 (1943).

**9.** In passing from the integral to the series, use the binomial expansion and the integral formulas ex. 2.42.

$$P = 2\pi a \left[ 1 - \frac{1}{2} \sum_{1}^{\infty} \frac{1}{2} \frac{3}{4} \cdots \frac{2n-1}{2n} \frac{\varepsilon^{2n}}{2n-1} \right],$$

$$P' = 2\pi a \left[ 1 - \sum_{1}^{\infty} \frac{3}{4} \frac{7}{8} \cdots \frac{4n-1}{4n} \frac{\varepsilon^{2n}}{4n-1} \right].$$

**10.** Use the solution of ex. 9 and put $\dfrac{1}{2} \dfrac{3}{4} \dfrac{5}{6} \cdots \dfrac{2n-1}{2n} = g_n$. Then $g_1 > g_n$ for $n \geq 2$ and, for $\varepsilon > 0$,

$$E - P = 2\pi a \sum_{2}^{\infty} (g_1 g_n - g_n^2) \, \varepsilon^{2n}/(2n - 1) > 0.$$

**11.** The initial term of the relative error of $P''$ is

$$- \left[ \alpha + 3(1 - \alpha) \right] \varepsilon^4/64 + \cdots$$

and so it is of order 4 unless $\alpha = 3/2$ and $P'' = P + (P - P')/2$.

**12.** $(P'' - E)/E = 3 \cdot 2^{-14}\varepsilon^8 + \cdots$ when $\varepsilon$ small

$$= (3\pi - 8)/8 = .1781 \text{ when } \varepsilon = 1$$

$$= .00019 \text{ when } \varepsilon = 4/5.$$

Hence the conjecture $P'' > E$.   See G. Peano, *Applicazioni geometriche del calcolo infinitesimale*, p. 231–236.

**13.**

$$e^p = \lim_{n \to \infty} \left( 1 + \frac{p}{n} \right)^n.$$

Therefore, the desired conclusion is equivalent to

$$\limsup_{n \to \infty} \left( \frac{n(a_1 + a_{n+p})}{(n+p)a_n} \right)^n \geq 1.$$

The opposite assumption implies

$$\frac{n(a_1 + a_{n+p})}{(n+p)a_n} < 1$$

for $n > N$, $N$ fixed, or, which is the same,

$$\frac{a_{n+p}}{n+p} - \frac{a_n}{n} < -\frac{a_1}{n+p}.$$

Consider the values $n = (m - 1)p$:

$$\frac{a_{mp}}{mp} - \frac{a_{(m-1)p}}{(m-1)p} < - \frac{a_1}{mp},$$

$$\frac{a_{(m-1)p}}{(m-1)p} - \frac{a_{(m-2)p}}{(m-2)p} < - \frac{a_1}{(m-1)p},$$

. . . . . . . . . . . . . . . . .

As in sect. 5, we conclude that, with a suitable constant $C$,

$$\frac{a_{mp}}{m} < C - a_1 \left(1 + \frac{1}{2} + \cdots + \frac{1}{m}\right)$$

and this leads for $m \to \infty$ to a contradiction to the hypothesis $a_n > 0$.

**14.** The example $a_n = n^c$ of sect. 4 suggests

$$a_1 = 1,$$
$$a_n = n \log n \qquad \text{for } n = 2, 3, 4, \ldots .$$

With this choice

$$\left(\frac{a_1 + a_{n+p}}{a_n}\right)^n = \left\{\frac{1 + (n + p)\,[\log n + \log\,(1 + (p/n))]}{n \log n}\right\}^n$$

$$= \left\{\frac{(n + p)\log n + 1 + (n + p)\left[\dfrac{p}{n} - \dfrac{p^2}{2n^2} + \cdots\right]}{n \log n}\right\}^n$$

$$= \left\{1 + \frac{p + \alpha_n}{n}\right\}^n \to e^p$$

since $\alpha_n \to 0$.

**15.** The mantissas in question are the 900 ordinates of the slowly rising curve $y = \log x - 2$ that correspond to the abscissas $x = 100, 101, \ldots, 999$; log denotes the common logarithm. Table I specifies how many among these 900 points on the curve fall in certain horizontal strips of width $1/10$. Let us consider the points at which the curve enters and leaves such a strip. If $x_n$ is the abscissa of such a point

$$\log x_n - 2 = n/10,$$
$$x_n = 100 \cdot 10^{n/10},$$

where $n = 0, 1, 2, \ldots, 10$. The number of integers in any interval is approximately equal to the length of the interval: the difference is less than one unit. Therefore, the number of the mantissas in question with first figure $n$ is $x_{n+1} - x_n$, with an error less than 1. Now,

$$x_{n+1} - x_n = 100(10^{1/10} - 1)10^{n/10}$$

is the $n$th term of a geometric progression with ratio

$$10^{1/10} = 1.25893.$$

Predict and observe the analogous phenomenon in a six-place table of common logarithms.

**16.** The periodical repetition can (and should) be regarded as a kind of symmetry; but it is present in all cases, and so we shall not mention it again. The following kinds of symmetry play a role in our classification.

(1) Center of symmetry. Notation: $c$, $c'$.

(2) Line of symmetry. Notation: $h$ if the line is horizontal, $v$ or $v'$ if it is vertical.

(3) Gliding symmetry: the frieze shifted horizontally *and* reflected in the central horizontal line *simultaneously*, coincides with itself (in friezes 5, 7, *a*, *b*). Notation: $g$.

The following types of symmetry are represented in fig. 5.2. (The dash ' is used to distinguish two elements of symmetry of the same kind, as $c$ and $c'$, or $v$ and $v'$, when their situation in the figure is essentially different.)

1, d: no symmetry (except periodicity)

2, g: $c$, $c'$, $c$, $c'$, $\ldots$

3, f: $v$, $v'$, $v$, $v'$, $\ldots$

4, e: $h$

5, a: $g$

6, c: $h$; $(v, c)$, $(v', c')$, $(v, c)$, $(v', c')$, $\ldots$

7, b: $g$; $v$, $c$, $v$, $c$, $\ldots$ .

All possible kinds of symmetry are represented in fig. 5.2, as you may convince yourself inductively.

**17.** Three different kinds of symmetry are represented: 1 is of the same type as 2, 3 as 4. Try to find all types. Cf. G. Pólya, *Zeitschrift für Kristallographie*, vol. 60 (1924) p. 278–282, P. Niggli, *ibid.*, p. 283–298, and H. Weyl, *Symmetry*, Princeton, 1952.

**18.** Disregard certain details, depending on the style of the print. Then (1) vertical line of symmetry, (2) horizontal line of symmetry, (3) center of symmetry, (4) all three preceding kinds of symmetry jointly, (5) no symmetry. The same for the five curves representing the five equations in rectangular coordinates. Some variant of this problem can be used to enliven a class of analytic geometry.

## SOLUTIONS, CHAPTER VI

**2.** $x(1 - x)^{-2}$. Particular case of ex. 3, with $f(x) = (1 - x)^{-1}$; obtain it also by combining ex. 4 and 5.

**3.** $xf'(x) = \sum_{n=0}^{\infty} na_n x^n$.

**4.** $xf(x) = \sum\limits_{n=1}^{\infty} a_{n-1}x^n.$

**5.** $(1 - x)^{-1}f(x) = \sum\limits_{n=0}^{\infty} (a_0 + a_1 + \ldots + a_n)x^n;$  particular case of ex. 6.

**6.** $f(x)g(x) = \sum\limits_{n=0}^{\infty} (a_0b_n + a_1b_{n-1} + \ldots + a_nb_0)x^n.$

**7.** $D_3 = 1, D_4 = 2, D_5 = 5, D_6 = 14.$  For $D_6$, refer to fig. 6.1; there are 2 different dissections of type I, 6 of type II, and 6 of type III.

**8.** The recursion formula is verified for $n = 6$:

$$14 = 1 \times 5 + 1 \times 2 + 2 \times 1 + 5 \times 1.$$

Choose a certain side as the base of the polygon (the horizontal side in fig. 6.2) and start the dissection by drawing the second and third sides of the triangle $\Delta$ whose first side is the base. Having chosen $\Delta$, you still have to dissect a polygon with $k$ sides to the left of $\Delta$ and another polygon with $n + 1 - k$ sides to the right; both polygons jointly yield $D_k D_{n+1-k}$ possibilities. Choose $k = 2, 3, 4, \ldots n - 1$; of course, you have to interpret suitably the case $k = 2$.

**9.** By ex. 4 and 6, the recursion formula for $D_n^-$ yields

$$xg(x) = D_2x^3 + [g(x)]^2.$$

Choose the solution of this quadratic equation the expansion of which begins with $x^2$:

$$g(x) = (x/2)[1 - (1 - 4x)^{1/2}].$$

Expanding and using the notation for binomial coefficients, you find:

$$D_n = -\frac{1}{2}\binom{1/2}{n-1}(-4)^{n-1}.$$

**10.** Better so

$$\sum_{u=-\infty}^{\infty} x^{u^2} \sum_{v=-\infty}^{\infty} x^{v^2} \sum_{w=-\infty}^{\infty} x^{w^2} = \sum_{-\infty}^{\infty}\sum_{-\infty}^{\infty}\sum_{-\infty}^{\infty} x^{u^2+v^2+w^2}$$

where $u, v,$ and $w$ range over all integers (from $-\infty$ to $+\infty$) independently, so that the triple sum is extended over all the lattice points of space (see ex. 4.2). To see that this *is* the generating function of $R_3(n)$, you just collect those terms in which the exponent $u^2 + v^2 + w^2$ has the same value $n$.

**11.**
$$\sum_{n=0}^{\infty} R_k(n)x^n = \left[\sum_{n=0}^{\infty} R_1(n)x^n\right]^k.$$

**12.**
$$\sum_{n=1}^{\infty} S_k(n)x^n = \left[\sum_{n=1}^{\infty} x^{(2n-1)^2}\right]^k.$$

**13.** Let $I, J, K$, and $L$ denote certain power series all coefficients of which are integers. Then the generating functions of $R_1(n)$, $R_2(n)$, $R_4(n)$, and $R_8(n)$ are of the form

$$1 + 2I,$$

$$(1 + 2I)^2 = 1 + 4J,$$

$$(1 + 4J)^2 = 1 + 8K,$$

$$(1 + 8K)^2 = 1 + 16L,$$

respectively.

**14.**     $x + x^9 + x^{25} + x^{49} + x^{81} + \cdots$

$$= x(1 + x^8 + x^{24} + x^{48} + x^{80} + \cdots) = xP$$

where $P$ denotes a power series in which the coefficient of $x^n$ vanishes when $n$ is not divisible by 8. The generating functions of

$$S_1(n), \quad S_2(n), \quad S_4(n), \quad S_8(n)$$

are

$$xP, \quad x^2P^2, \quad x^4P^4, \quad x^8P^8,$$

respectively.

**15.** From ex. 6 and 11

$$G^{k+l} = G^k G^l$$

where $G$ stands for the generating function of $R_1(n)$.

**16.** Analogous to ex. 15, from ex. 6 and 12.

**17.** Use ex. 15 and 16 with $k = l = 4$. The actual computation was performed by this method, with occasional checks from other sources, as ex. 4.16 and ex. 23.

**18.** (1) From ex. 14 and ex. 16 follows

$$S_4(4)S_4(8n - 4) + S_4(12)S_4(8n - 12) + \cdots + S_4(8n - 4)S_4(4) = S_8(8n).$$

It was conjectured in sect. 4.6 that $S_4(4(2n - 1)) = \sigma(2n - 1)$ and in ex. 4.23 that $S_8(8u) = \sigma_3(u)$ if $u$ is an odd number.

(2) $\sigma(1)\sigma(9) + \sigma(3)\sigma(7) + \sigma(5)\sigma(5) + \sigma(7)\sigma(3) + \sigma(9)\sigma(1)$

$$= 2(1 \times 13 + 4 \times 8) + 6 \times 6$$

$$= 126 = 5^3 + 1^3 = \sigma_3(5).$$

(3) It seems reasonable that such a verification increases our confidence in both conjectures, to some degree.

**19.**     $\sum \left[ \dfrac{u-1}{2} - 5\, \dfrac{k(k+1)}{2} \right] s_{u-k(k+1)} = 0$

for $u = 1, 3, 5, \ldots$; the summation is extended over all non-negative integers $k$ satisfying the inequality

$$0 \leqq k(k+1) < u.$$

**20.**   $\sigma(3) = 4\sigma(1)$

$2\sigma(5) = 3\sigma(3)$

$3\sigma(7) = 2\sigma(5) + 12\sigma(1)$

$4\sigma(9) = \sigma(7) + 11\sigma(3).$

The last is true, since

$$4 \times 13 = 8 + 11 \times 4.$$

**21.** The recursion formula has been proved for $S_4(4(2n - 1))$ in ex. 19. This recursion formula means, in fact, an infinite system of relations which determine $S_4(4(2n - 1))$ unambiguously if $S_4(4)$ is given. Now, we know that

$$S_4(4) = \sigma(1) = 1.$$

If $\sigma(2n - 1)$ satisfies the same system of recursive relations as $S_4(4(2n - 1))$,

$$S_4(4(2n - 1)) = \sigma(2n - 1)$$

for $n = 1, 2, 3, \ldots$ because the system is unambiguous. If, conversely, the last equation holds, $\sigma(2n - 1)$ satisfies those recursive relations.

**22.** Assume that

$$G = a_0 + a_1 x + a_2 x^2 + a_3 x^3 + \cdots ,$$

$$H = u_0 + u_1 x + u_2 x^2 + u_3 x^3 + \cdots ,$$

$$G^k = H.$$

It follows, as in ex. 19, that

$$GxH' - kxG'H = 0.$$

Equate to 0 the coefficient of $x^n$:

$$\sum_{m=0}^{n} [n - (k + 1)m] a_m u_{n-m} = 0.$$

Consider $a_0, a_1, a_2, \ldots$ as given. From the last equation you can express $u_n$ in terms of $u_{n-1}, u_{n-2}, \ldots u_1, u_0$ provided that $a_0 \neq 0$. Observe that $u_0 = a_0^k$.

**23.** Apply ex. 22 to the case $k = 8$,

$$G = 1 + 2x + 2x^4 + 2x^9 + 2x^{16} + \cdots .$$

By ex. 11 the result of ex. 22 yields

$$nR_8(n) = 2(9 - n)R_8(n - 1) + 2(36 - n)R_8(n - 4)$$

$$+ 2(81 - n)R_8(n - 9) + \cdots .$$

Set $R_8(n)/16 = r_n$. Then $r_0 = 1/16$ and we find $r_1, r_2, r_3, \ldots r_{10}$ successively from

$$r_1 = 16r_0$$
$$2r_2 = 14r_1$$
$$3r_3 = 12r_2$$
$$4r_4 = 10r_3 + 64r_0$$
$$5r_5 = 8r_4 + 62r_1$$
$$6r_6 = 6r_5 + 60r_2$$
$$7r_7 = 4r_6 + 58r_3$$
$$8r_8 = 2r_7 + 56r_4$$
$$9r_9 = 54r_5 + 144r_0$$
$$10r_{10} = -2r_9 + 52r_6 + 142r_1$$

In using these formulas numerically, we have an important check: the right hand side of the equation that yields $r_n$ must be divisible by $n$.

The same method yields a recursion formula for $R_k(n)$, $k$ being any given integer $\geq 2$, and also for $S_k(n)$.

**25.** Call $s$ the infinite product. Computing $-xd \log s/dx$ and using No. 10 of Euler's memoir, you obtain

$$k \, \Sigma \sigma(l) x^l = \frac{\Sigma n a_n x^n}{1 - \Sigma a_m x^m};$$

the limits for all three sums are 1 and $\infty$. Multiply with the denominator of the right hand side and focus the coefficient of $x^n$.

Euler's case is $k = 1$. Also the case $k = 3$ yields a relatively simple result (see the work of Hardy and Wright, quoted in ex. 24, p. 282 and 283, theorems 353 and 357). In the other cases we do not know enough about the law of $a_n$.

No solution: **1, 24.**

### SOLUTIONS, CHAPTER VII

**1.** [Stanford 1950]

$$1 - 4 + 9 - 16 + \ldots + (-1)^{n-1} n^2 = (-1)^{n-1} \frac{n(n+1)}{2}.$$

The step from $n$ to $n + 1$ requires to verify that

$$(-1)^n (n+1)^2 = (-1)^n \frac{(n+1)(n+2)}{2} - (-1)^{n-1} \frac{n(n+1)}{2}.$$

**2.** To prove

$$P_n = \binom{n}{0} + \binom{n}{1} + \binom{n}{2},$$

$$S_n = \binom{n}{0} + \binom{n}{1} + \binom{n}{2} + \binom{n}{3},$$

$$S_n^\infty = 2\binom{n}{0} + 2\binom{n}{2} \qquad \bullet$$

we use ex. 3.11, ex. 3.12 combined with the expression of $P_n$, and ex. 3.20, respectively. Then, supposing the above expressions, we are required to verify

$$P_{n+1} - P_n = \binom{n}{0} + \binom{n}{1}$$

$$S_{n+1} - S_n = \binom{n}{0} + \binom{n}{1} + \binom{n}{2}$$

$$S_{n+1}^\infty - S_n^\infty = 2\binom{n}{1}.$$

All three follow from the well known fact (the basic relation in the Pascal triangle) that

$$\binom{n+1}{k+1} - \binom{n}{k+1} = \binom{n}{k}.$$

**3.** *How to Solve It*, p. 103–110.

**4.** $\dfrac{3}{4}, \dfrac{2}{3} = \dfrac{4}{6}, \dfrac{5}{8}, \dfrac{3}{5} = \dfrac{6}{10}, \cdots \dfrac{n+1}{2n}.$

The step from $n$ to $n+1$ requires to verify that

$$1 - \frac{1}{(n+1)^2} = \frac{n+2}{2n+2}\frac{2n}{n+1}.$$

Cf. ex. 2.23.

**5.** $-\dfrac{3}{1}, -\dfrac{5}{3}, -\dfrac{7}{5}, -\dfrac{9}{7}, \cdots -\dfrac{2n+1}{2n-1}.$

The step from $n$ to $n+1$ requires to verify that

$$1 - \frac{4}{(2n+1)^2} = \frac{2n+3}{2n+1}\frac{2n-1}{2n+1}.$$

Cf. ex. 2.31.

**6.** The general case is, in fact, equivalent to the limiting case

$$\frac{x}{1-x} = \frac{x}{1+x} + \frac{2x^2}{1+x^2} + \frac{4x^4}{1+x^4} + \frac{8x^8}{1+x^8} + \cdots$$

from which the particular case proposed is derived as follows: substitute $x^{16}$ for $x$ and multiply by 16, obtaining

$$\frac{16x^{16}}{1-x^{16}} = \frac{16x^{16}}{1+x^{16}} + \frac{32x^{32}}{1+x^{32}} + \cdots ;$$

then subtract from the original equation. If we set $2^{n+1} = m$, the step from $n$ to $n+1$ requires

$$\frac{mx^m}{1+x^m} = -\frac{2mx^{2m}}{1-x^{2m}} + \frac{mx^m}{1-x^m}.$$

**7.** To prove

$$1 + 3 + 5 + 7 + \ldots + 2n - 1 = n^2.$$

The step from $n$ to $n+1$ requires to verify that

$$2n + 1 = (n+1)^2 - n^2.$$

**8.** The $n$th term in the fourth row of the table is

$$(1+2) + (4+5) + \ldots + (3n - 5 + 3n - 4) + 3n - 2$$
$$= 3 + 9 + \ldots + [6(n-1) - 3] + 3n - 2$$
$$= 6\frac{n(n-1)}{2} - 3(n-1) + 3n - 2 = 3n^2 - 3n + 1.$$

The step from $n - 1$ to $n$ requires, in fact,

$$n^3 - (n-1)^3 = 3n^2 - 3n + 1.$$

**9.** After $n^2$, $n^3$ and $n^4$, the generalization concerned with $n^k$ is obvious. The simple case of $n^2$ was known since antiquity; Alfred Moessner discovered the rest quite recently by empirical induction, and Oskar Perron proved it by mathematical induction. See *Sitzungsberichte der Bayerischen Akademie der Wissenschaften*, Math.-naturwissenschaftliche Klasse, 1951, p. 29–43.

**10.** For $k = 1$ the theorem reduces to the obvious identity

$$1 - n = -(n-1).$$

The step from $k$ to $k+1$ requires to verify that

$$(-1)^{k+1}\binom{n}{k+1} = (-1)^{k+1}\binom{n-1}{k+1} - (-1)^k\binom{n-1}{k}$$

which is the basic relation in the Pascal triangle, already encountered in ex. 2.

**11.** [Stanford 1946]. Call the required number of pairings of $2n$ players $P_n$. The $n$th player can be matched with any one of the other $2n - 1$ players. Once his antagonist is chosen

$$2n - 2 = 2(n - 1)$$

players remain who can be paired in $P_{n-1}$ ways. Hence

$$P_n = (2n - 1)P_{n-1}.$$

**12.** Call $A_n$ the statement proposed to prove, concerned with $f_n(x)$. Instead of $A_n$ we shall prove $A'_n$.

$A'_n$. The function $f_n(x)$ is a quotient *the denominator of which is* $(1 - x)^{n+1}$ and the numerator a polynomial *of degree n* the constant term of which is 0 and the other coefficients positive integers.

Observe that $A'_n$ asserts more than $A_n$; the points in which $A'_n$ goes beyond $A_n$ are emphasized by italics. Assuming $A'_n$, set

$$(1 - x)^{n+1} f_n(x) = P_n(x) = a_1 x + a_2 x^2 + \ldots + a_n x^n$$

where $a_1, a_2, \ldots a_n$ are supposed to be positive integers. From the recursive definition we derive the recursive formula

$$P_{n+1}(x) = x[(1 - x)P'_n(x) + (n + 1)P_n(x)]$$

and this shows that the coefficients of $x$, $x^2$, $x^3$, $\ldots x^n$ and $x^{n+1}$ in $P_{n+1}(x)$ are $a_1$, $na_1 + 2a_2$, $(n - 1) a_2 + 3a_3$, $\ldots 2a_{n-1} + na_n$, $a_n$, respectively, which makes the assertion obvious.

**13.** (1) The sum of all the coefficients of $P_n(x)$ is $n!$. In fact, this sum is $P_n(1)$ and the recursive formula yields

$$P_{n+1}(1) = (n + 1)P_n(1).$$

(2) $P_n(x)/x$ is a reciprocal polynomial or

$$P_n(1/x)x^{n+1} = P_n(x).$$

In fact, assume that $a_1 = a_n$, $a_2 = a_{n-1}$, $\ldots$ ; the corresponding relations for the coefficients of $P_{n+1}(x)$ follow from their expression given at the end of the solution of ex. 12.

**16.** $Q_1 = 1$,  $Q_2 = 3$,  $Q_3 = 45$,  $Q_4 = 4725$

$Q_2/Q_1 = 3$,  $Q_3/Q_2 = 15$,  $Q_4/Q_3 = 105$

suggest

$$Q_n = 1^n 3^{n-1} 5^{n-2} \ldots (2n - 3)^2 (2n - 1)^1.$$

In fact, the definition yields

$$\frac{Q_{n+1}}{Q_{n-1}} = \frac{Q_n Q_{n+1}}{Q_{n-1} Q_n} = \frac{(2n)!(2n + 1)!}{(n!2^n)^2} = [1 \cdot 3 \cdot 5 \ldots (2n - 1)]^2(2n + 1)$$

and hence you prove the general law by inference from $n - 1$ to $n + 1$.

Observe that

$$\frac{2n!}{n!2^n} = \frac{1 \cdot 2 \cdot 3 \cdot 4 \cdot 5 \cdot 6 \dots (2n-1)2n}{2 \quad \cdot \quad 4 \quad \cdot \quad 6 \dots \quad \quad \cdot \quad 2n}.$$

**17.** The reasoning that carries us from 3 to 4 applies to the passage from $n$ to $n+1$ with one exception: it breaks down, as it must, in the passage from 1 to 2.

**18.** From $n = 3$ to $n + 1 = 4$: consider the lines $a$, $b$, $c$, and $d$. Consider first the case that there are two different lines among these four lines, for example $b$ and $c$. Then the point of intersection of $b$ and $c$ is uniquely determined and must also lie on $a$ (because, allegedly, the statement holds for $n = 3$) and also on $d$ (for the same reason). Therefore, the statement holds for $n + 1 = 4$. If, however, no two among the four given lines are different, the statement is obvious. This reasoning breaks down, as it must, in the passage from 2 to 3.

No solution: **14, 15.**

### SOLUTIONS, CHAPTER VIII

**1.** (1) Straight line, (2) perpendicular, (3) common perpendicular, (4) segments of line through given point and center (Euclid III, 7, 8), (5) segment of perpendicular through center, no maximum, (6) segments of line joining centers. The case in which the minimum distance is 0 has been consistently discarded as obvious, although it may be important.

**2.** (1) straight line, (2) perpendicular, (3) common perpendicular, (4) perpendicular, (5) common perpendicular, (6) see sect. 4, (7) segments of line joining point to center, (8) segment of perpendicular through center, no maximum, (9) segment of perpendicular through center, no maximum, (10) segments of line joining centers. The case in which the minimum distance is 0 has been discarded.

**3.** (1) concentric circles, (2) parallel lines, (3) concentric circles.

**4.** (1) concentric spheres, (2) parallel planes, (3) coaxial cylinders, (4) concentric spheres.

**5.** (2) See sect. 3. Others similar.

**6.** (6) There is just one cylinder that has the first given line as axis and the second given line as tangent. The point of contact is an endpoint of the shortest distance. Others similar.

**7.** Call one of the given sides the base. Keep the base in a fixed position, let the other side rotate about its fixed endpoint and call its other endpoint $X$. The prescribed path of $X$ is a circle, the level lines are parallel to the base, the triangle with maximum area is a right triangle (which is obvious).

**8.** Call the given side the base, keep it in a fixed position, call the opposite vertex $X$, and let $X$ vary. The prescribed path of $X$ is an ellipse, the level lines are parallel to the base, the triangle with maximum area is isosceles.

**9.** The level lines are straight lines $x + y =$ const., the prescribed path is (one branch of) an equilateral hyperbola with equation $xy = A$, where $A$ is the given area.    It is clear by symmetry that there is a point of contact with $x = y$.

**10.** Consider expanding concentric circles with the given point as center. It seems intuitive that there is a first circle hitting the given line; its radius is the shortest distance.    This is certainly so in the cases ex. 1 (2) and (4).

**11.** Crossing means passing from one side of the level line to the other, and on one side $f$ takes higher, on the other lower, values than at the point of crossing.

**12.** You may, but you need not.    The highest point may be the peak $P$ (you may wish to see the view) or the pass $S$ (you may cross it hiking from one valley to the other) or the initial point of your path, or its final point, or an angular point of it.

**13.** (1) The level line for $180°$ is the segment $AB$, the level line for $0°$ consists of the straight line passing through $A$ and $B$, except the segment $AB$. Any other level line consists of two circular arcs both with endpoints $A$ and $B$, and symmetrical to each other with respect to the straight line passing through $A$ and $B$.    (2) If two level lines are different, one lies inside the other; $\angle AXB$ takes a higher value on the inner, a lower value on the outer, level line.    With suitable interpretation, this applies also to $0°$.

**14.** The minimum is attained at the point where the line $l$ crosses the line through $A$ and $B$, that is, a level line.    This does not contradict the principle laid down in ex. 11; on both sides of this particular level line $\angle AXB$ takes higher values than on it.

**15.** Notation as in sect. 6.    Keep $c$ constant for a moment.    Then, since $V = abc/3$ is given, also $ab$ is constant, and

$$S = 2ab + 2(a + b)c$$

is a minimum, when $2(a + b)$, the perimeter of a rectangle with given area $ab$, is a minimum.    This happens, when $a = b$.    Now, change your standpoint and keep another edge constant.

**16.** Keep one of the sides constant.    Then you have the case of ex. 8, and the two other sides must be equal (the triangle is isosceles).    Any two sides are accessible to this argument, and the triangle should be equilateral.

**17.** Keep the plane of the base and the opposite vertex fixed, but let vary the base which is a triangle inscribed in a given circle.    The altitude is constant; the area of the base (and with it the volume) becomes a maximum when the base is equilateral, by sect. 4 (2).    We can choose any face as base, and so each face must be equilateral and, therefore, the tetrahedron regular when the maximum of the volume is attained.

**18.** Regard the triangle between $a$ and $b$ as the base. Without changing the corresponding altitude, change the base into a right triangle; this change increases the area of the base (ex. 7) and, therefore, the volume. You could treat now another pair of sides similarly, yet it is better to make $c$ perpendicular both to $a$ and to $b$ right away.

**19.** Fixing the endpoint on the cylinder, you find, by ex. 2 (7), that the shortest distance is perpendicular to the sphere. Fixing the endpoint on the sphere, you may convince yourself that the shortest distance is also perpendicular to the cylinder. Therefore, it should be perpendicular to both. This can be shown also directly.

**20.** The procedure of ex. 19 shows that the shortest distance should be perpendicular to both cylinders. In fact, it falls in the same line as the shortest distance between the axes of the cylinders; see sect. 4 (1).

**21.** The procedure of ex. 19 and an analogue of ex. 10 in space.

**22.** By hypothesis

$$f(X,Y,Z, \ldots ) \leqq f(A,B,C, \ldots )$$

for all admissible values of $X, Y, Z, \ldots$ . Therefore, in particular

$$f(X,B,C, \ldots ) \leqq f(A,B,C, \ldots )$$
$$f(X,Y,C, \ldots ) \leqq f(A,B,C, \ldots )$$

and so on; $X, Y, Z, \ldots$ may be variable numbers, lengths, angles, points, . . . .

**24.** Either $x_1 = y_1 = z_1$ (exceptional case) or, for $n \geqq 1$, of the three values $x_n, y_n, z_n$ just two are different. Call $d_n$ the absolute value of the difference; for example

$$d_1 = \left| x_1 - z_1 \right|, \qquad d_2 = \left| y_2 - x_2 \right|.$$

By definition

$$\pm d_2 = x_2 - y_2 = \frac{z_1 + x_1}{2} - y_1 = \frac{z_1 + x_1}{2} - z_1$$

$$= \frac{x_1 - z_1}{2} = \pm \frac{d_1}{2}$$

or $d_2 = d_1/2$. In the same way

$$d_n = d_{n-1}/2 = d_{n-2}/4 = \ldots = d_1/2^{n-1}$$

and so

$$\left| x_n - \frac{l}{3} \right| = \left| x_n - \frac{x_n + y_n + z_n}{3} \right| = \frac{\left| x_n - y_n + x_n - z_n \right|}{3}$$

$$\leqq 2d_n/3 \to 0.$$

**25.** We consider $n$ positive numbers $x_1, x_2, \ldots x_n$ with given arithmetic mean $A$,

$$x_1 + x_2 + \ldots + x_n = nA.$$

If $x_1, x_2, \ldots x_n$ are not all equal, one of them, say $x_1$, is the smallest and another, say $x_2$, is the largest. (The choice of the subscripts is a harmless simplification, just a matter of convenient notation. We did *not* make the unwarranted assumption that *only* $x_1$ takes the smallest value.) Then

$$x_1 < A < x_2.$$

Let us put now

$$x_1' = A, \qquad x_2' = x_1 + x_2 - A, \qquad x_3' = x_3, \ldots x_n' = x_n.$$

Then

$$x_1 + x_2 + \ldots + x_n = x_1' + x_2' + \ldots + x_n'$$

and

$$x_1' x_2' - x_1 x_2 = Ax_1 + Ax_2 - A^2 - x_1 x_2 = (A - x_1)(x_2 - A) > 0.$$

Therefore

$$x_1 x_2 x_3 \ldots x_n < x_1' x_2' x_3' \ldots x_n'.$$

If $x_1', x_2', \ldots x_n'$ are not all equal, we repeat the process obtaining another set of $n$ numbers $x_1'', x_2'', \ldots x_n''$ such that

$$x_1' + x_2' + \ldots + x_n' = x_1'' + x_2'' + \ldots + x_n''$$

$$x_1' x_2' x_3' \ldots x_n' < x_1'' x_2'' x_3'' \ldots x_n''.$$

The set $x_1', x_2', \ldots x_n'$ has at least one term equal to $A$, the set $x_1'', x_2'', \ldots x_n''$ has at least two terms equal to $A$. At the latest, the set $x_1^{(n-1)}, x_2^{(n-1)}, \ldots$ $x_n^{(n-1)}$ will contain $n-1$ and, therefore, $n$ terms equal to $A$, and so

$$x_1 x_2 \ldots x_n < x_1^{(n-1)} x_2^{(n-1)} \ldots x_n^{(n-1)}$$

$$= A^n = \left( \frac{x_1 + x_2 + \ldots + x_n}{n} \right)^n.$$

**26.** Connecting the common initial point of the perpendiculars $x, y, z$ with the three vertices of the triangle, we subdivide the latter into three smaller triangles. Expressing that the sum of the areas of these three parts is equal to the area of the whole, we obtain $x + y + z = l$. The equation $x = $ const. is represented by a line parallel to the base of the equilateral triangle, the equation $y = z$ by the altitude. The first segment of the broken line in fig. 8.9 is parallel to the base and ends on the altitude. The first step of ex. 25 is represented by a segment parallel to the base and ends on the line with equation $y = l/3$, which is parallel to another side and passes through the center. The second step is represented by a segment along the line $y = l/3$ and ends at the center.

**27.** For the solution modelled after ex. 25 see Rademacher-Toeplitz, pp. 11–14, 114–117.

**28.** Partial variation.

**29.** At the point where the extremum is attained, the equations

$$\frac{\partial f}{\partial x} + \lambda \frac{\partial g}{\partial x} = 0, \qquad \frac{\partial f}{\partial y} + \lambda \frac{\partial g}{\partial y} = 0$$

hold with a suitable value of $\lambda$. This condition is derived under the assumption that $\partial g/\partial x$ and $\partial g/\partial y$ do not both vanish. Under the further assumption that $\partial f/\partial x$ and $\partial f/\partial y$ do not both vanish, the equations with $\lambda$ express that the curve $g = 0$ (the prescribed path) is tangent to the curve $f = $ const. (a level line) that passes through the point of extremum, at this point.

**30.** At a peak, or at a pass, $\partial f/\partial x = \partial f/\partial y = 0$. At an angular point of the prescribed path, $\partial g/\partial x$ and $\partial g/\partial y$ (if they exist) are both $= 0$. An extremum at the initial (or final) point of the prescribed path does not fall at all under the analytic condition quoted in ex. 29 which is concerned with an extremum relative to *all* points $(x,y)$ in a certain neighborhood, satisfying $g(x,y) = 0$.

**31.** The condition is

$$\frac{\partial f}{\partial x} + \lambda \frac{\partial g}{\partial x} = \frac{\partial f}{\partial y} + \lambda \frac{\partial g}{\partial y} = \frac{\partial f}{\partial z} + \lambda \frac{\partial g}{\partial z} = 0.$$

It assumes that the three partial derivatives of $g$ are not all 0. Under the further assumption that the three partial derivatives of $f$ are not all 0, the three equations express that the surface $g = 0$ and the surface $f = $ const. passing through the point of extremum are tangent to each other at that point.

**32.** The condition consists of three equations of which the first, relative to the $x$-axis, is

$$\frac{\partial f}{\partial x} + \lambda \frac{\partial g}{\partial x} + \mu \frac{\partial h}{\partial x} = 0.$$

It assumes that three determinants of which the first is

$$\frac{\partial g}{\partial y} \frac{\partial h}{\partial z} - \frac{\partial g}{\partial z} \frac{\partial h}{\partial y}$$

do not all vanish. Under the further assumption that the three partial derivatives of $f$ are not all 0, the three equations express that the curve which is the intersection of the two surfaces $g = 0$ and $h = 0$, is tangent to the surface $f = $ const., passing through the point of extremum, at that point.

**34.** The desired conclusion is: the cube alone attains the maximum. Therefore, when the inequality becomes an equality, the cube should appear; that is, we should have $x = y$, or (looking at the areas) $x^2 = xy$. Yet, the inequality used (without success) becomes an equality when $2x^2 = 4xy$: we could have predicted hence that it will fail.

With, or without, peeking at sect. 6, we divide $S$ into three pairs of opposite faces

$$S = 2x^2 + 2xy + 2xy$$

and apply the theorem of the means:

$$(S/3)^3 \geq 2x^2 \cdot 2xy \cdot 2xy = 8x^4y^2 = 8V^2.$$

The equality is attained if, and only if, $2x^2 = 2xy$, or $x = y$; that is, only for the cube.

Draw the moral: foreseeing the case of equality may guide your choice, may yield a cue.

**35.** Let $V$, $S$, $x$, and $y$ stand for volume, surface, radius, and altitude of the cylinder, respectively, so that

$$V = \pi x^2 y, \qquad S = 2\pi x^2 + 2\pi xy.$$

The desired conclusion, $y = 2x$, guides our choice: with

$$S = 2\pi x^2 + \pi xy + \pi xy$$

the theorem of the means yields

$$(S/3)^3 \geq 2\pi x^2 \cdot \pi xy \cdot \pi xy = 2\pi^3 x^4 y^2 = 2\pi V^2,$$

with equality just for $y = 2x$.

**36.** Ex. 34 is a particular case, ex. 35 a limiting case. Let $V$, $S$, $y$, and $x$ stand for volume, surface, altitude of the prism, and for the length of a certain side of its base, respectively. Let $a$ and $l$ denote the area and the perimeter, respectively, of a polygon similar to the base in which the side corresponding to the side of length $x$ of the base is of length 1. Then

$$V = ax^2 y, \qquad S = 2ax^2 + lxy.$$

In ex. 34 and 35, the maximum of $S$ is attained, when the area of the base (now $ax^2$) is $S/6$. Expecting that this holds also in the present general case, we have a cue; we set

$$S = 2ax^2 + lxy/2 + lxy/2$$

and obtain, using the theorem of the means,

$$(S/3)^3 \geq 2ax^2 \cdot (lxy)^2/4 = [l^2/(2a)]V^2$$

with equality if $ax^2 = lxy/4 = S/6$.

**37.** Let $V$, $S$, $x$, and $y$ stand for the volume of the double pyramid, its surface, a side of its base, and the altitude of one of the constituent pyramids, respectively. Then

$$V = 2x^2y/3, \qquad S = 8x[(x/2)^2 + y^2]^{1/2}/2.$$

In the case of the regular octahedron, the double altitude of a constituent pyramid equals the diameter of the base, or

$$2y = 2^{1/2}x, \quad \text{or} \quad 2y^2 = x^2.$$

Having obtained this cue, we set

$$S^2 = 4x^2(x^2 + 2y^2 + 2y^2),$$

$$(S^2/3)^3 \geq 4^3x^6x^2 2y^2 2y^2 = 4^4x^8y^4 = (6V)^4.$$

Equality occurs only if $x^2 = 2y^2$. Note that in this case $S = 3^{1/2}2x^2$.

**38.** Let $V$, $S$, $x$, and $y$ denote the volume of the double cone, its surface, the radius of its base, and the altitude of one of the constituent cones, respectively. Then

$$V = 2\pi x^2y/3, \qquad S = 2 \cdot 2\pi x(x^2 + y^2)^{1/2}/2.$$

Consider the right triangle with legs $x, y$ and hypotenuse $(x^2 + y^2)^{1/2}$. If the projection of $x$ on the hypotenuse is $1/3$ of the latter (as we hope that it will be in the case of the minimum),

$$x^2 = (x^2 + y^2)/3$$

or $2x^2 = y^2$. Having obtained this cue, we set

$$S^2 = 2\pi^2x^2(2x^2 + y^2 + y^2),$$

$$(S^2/3)^3 \geq 8\pi^6x^6 \cdot 2x^2y^2y^2 = \pi^2(3V)^4.$$

Equality occurs only if $2x^2 = y^2$. Note that in this case $S = 3^{1/2}2 \cdot \pi x^2$.

**39.** Let $V$, $S$, and $y$ denote the volume of the double pyramid, its surface, and the altitude of one of the constituent pyramids, respectively. Let $x$, $a$, and $l$ be connected with the base of the double pyramid in the same way as in the solution of ex. 36 with the base of the prism. Let $p$ stand for the radius of the circle inscribed in the base. Then

$$V = 2ax^2y/3,$$

$$ax^2 = lxp/2,$$

$$S = 2lx(p^2 + y^2)^{1/2}/2 = (4a^2x^4 + l^2x^2y^2)^{1/2}.$$

In ex. 37 and 38, $S$ is a minimum when $S = 3^{1/2}2ax^2$ which yields

$$l^2x^2y^2 = 8a^2x^4.$$

Noticing this cue, we set

$$S^2 = 4a^2x^4 + l^2x^2y^2/2 + l^2x^2y^2/2,$$

$$(S^2/3)^3 \geqq 4a^2x^4 \cdot (l^2x^2y^2)^2/4 = (l^4/a^2)(3V/2)^4.$$

Equality occurs if, and only if, the base $ax^2 = S/(3^{1/2}2)$.

**40.** There is a plausible conjecture: the equilateral triangle has the minimum perimeter for a given area, or the maximum area for a given perimeter. Let $a$, $b$, $c$, $A$, and $L = 2p$ stand for the three sides of the triangle, its area, and the length of its perimeter, respectively. By Heron's formula,

$$A^2 = p(p - a)\,(p - b)\,(p - c).$$

The use of the theorem of the means is strongly suggested: $A$ should not be too large, when $p$ is given; the right hand side is a product. How should we apply the theorem? There is a cue: *if the triangle is equilateral,* $a = b = c$, or $p - a = p - b = p - c$. Therefore, we proceed as follows:

$$A^2/p = (p - a)\,(p - b)\,(p - c)$$

$$\leqq \left(\frac{p - a + p - b + p - c}{3}\right)^3$$

$$= (p/3)^3.$$

That is, $A^2 \leqq L^4/(2^43^3)$, and there is equality only in the case of the equilateral triangle. Cf. ex. 16.

**41.** There is a plausible conjecture: the square.

Let $a$ and $b$ include the angle $\phi$, $c$ and $d$ the angle $\psi$, and $\phi + \psi = \varepsilon$. We obtain

$$2A = ab \sin \phi + cd \sin \psi.$$

Expressing in two different ways the diagonal of the square that separates $\phi$ and $\psi$, we obtain

$$a^2 + b^2 - 2ab \cos \phi = c^2 + d^2 - 2cd \cos \psi.$$

We have now three relations to eliminate $\phi$ and $\psi$. Adding

$$(a^2 + b^2 - c^2 - d^2)^2 = 4a^2b^2 \cos^2 \phi + 4c^2d^2 \cos^2 \psi - 8abcd \cos \phi \cos \psi$$

$$16A^2 = 4a^2b^2 \sin^2 \phi + 4c^2d^2 \sin^2 \psi + 8abcd \sin \phi \sin \psi$$

we obtain

$$16A^2 + (a^2 + b^2 - c^2 - d^2)^2 = 4a^2b^2 + 4c^2d^2 - 8abcd \cos \varepsilon$$

$$= 4(ab + cd)^2 - 16abcd\,(\cos \varepsilon/2)^2.$$

Finally, noticing differences of squares and setting

$$a + b + c + d = 2p = L,$$

we find

$$A^2 = (p - a)(p - b)(p - c)(p - d) - abcd\,(\cos \varepsilon/2)^2.$$

In the probable case of equality (the square) the sides are equal, and so are the quantities $p - a, p - b, p - c, p - d$. With this cue, we obtain

$$A^2 \leqq (p - a)(p - b)(p - c)(p - d)$$

$$\leqq \left(\frac{p - a + p - b + p - c + p - d}{4}\right)^4$$

$$= (p/2)^4 = (L/4)^4.$$

In order that both inequalities encountered should become equalities, we must have $\varepsilon = 180°$ and $a = b = c = d$.

**42.** The prism is much more accessible than the two other solids which we shall tackle, after careful preparation, in ex. 46 and 47. Let $L$ denote the perimeter of the base and $h$ the altitude of the prism. Any lateral face is a parallelogram; its base is a side of the base of the solid, and its altitude $\geqq h$. Therefore, the lateral surface of the prism is $\geqq Lh$, and equality is attained if, and only if, all lateral faces are perpendicular to the base and so the prism a right prism.

**43.** Let $x_j, y_j$ be the coordinates of $P_j$, for $j = 0, 1, 2, \ldots n$, and put

$$x_j = x_{j-1} + u_j, \qquad y_j = y_{j-1} + v_j$$

for $j = 1, 2, \ldots n$. Then the left-hand side of the desired inequality is the length of the broken line $P_0P_1P_2 \ldots P_n$ and the right-hand side the length of the straight line $P_0P_n$, which is the shortest distance between its endpoints.

**44.** In the case $n = 2$, we examine (the notation is slightly changed) the assertion

$$(u^2 + v^2)^{1/2} + (U^2 + V^2)^{1/2} \geqq [(u + U)^2 + (v + V)^2]^{1/2}.$$

We transform it into equivalent forms by squaring and other algebraic manipulations:

$$(u^2 + v^2)^{1/2}(U^2 + V^2)^{1/2} \geqq uU + vV,$$

$$u^2V^2 + v^2U^2 \geqq 2uvUV,$$

$$(uV - vU)^2 \geqq 0.$$

In its last form, the assertion is obviously true. Equality is attained if, and only if,

$$u : v = U : V.$$

We handle the case $n = 3$, by applying repeatedly the case $n = 2$:

$$(u_1^2 + v_1^2)^{1/2} + (u_2^2 + v_2^2)^{1/2} + (u_3^2 + v_3^2)^{1/2}$$

$$\geqq [(u_1 + u_2)^2 + (v_1 + v_2)^2]^{1/2} + (u_3^2 + v_3^2)^{1/2}$$

$$\geqq [(u_1 + u_2 + u_3)^2 + (v_1 + v_2 + v_3)^2]^{1/2}.$$

And so on, for $n = 4, 5, \ldots$ . In fact, we use mathematical induction.

**45.** Let $h$ be the altitude and let the base be divided by the foot of the altitude into two segments, of lengths $p$ and $q$, respectively. We have to prove that

$$(p^2 + h^2)^{1/2} + (q^2 + h^2)^{1/2} \geqq 2\left[\left(\frac{p+q}{2}\right)^2 + h^2\right]^{1/2}$$

$$= [(p + q)^2 + (h + h)^2]^{1/2}$$

which is a case of ex. 43. For equality, we must have

$$p : h = q : h,$$

or $p = q$, that is, an isosceles triangle.

**46.** Let $h$ be the altitude of $P$, $a_1, a_2, \ldots a_n$ the sides of the base of $P$, and $p_1, p_2, \ldots p_n$ the perpendiculars from the foot of the altitude on the respective sides. Let $\Sigma$ denote a summation with $j$ ranging from $j = 1$ to $j = n$. Then

$$A = \Sigma a_j p_j / 2$$

$$S = A + \Sigma a_j (p_j^2 + h^2)^{1/2}/2.$$

These expressions become simpler for the right pyramid $P_0$, since all perpendiculars from the foot of the altitude on the sides have a common value $p_0$. Therefore,

$$A_0 = L_0 p_0 / 2$$

$$S_0 = A_0 + L_0 (p_0^2 + h^2)^{1/2}/2$$

$$= A_0 + (4A_0^2 + h^2 L_0^2)^{1/2}/2;$$

$P$ and $P_0$ have the same altitude $3V/A = 3V_0/A_0 = h$. Using ex. 43 and our assumptions, we obtain

$$2(S - A) = \Sigma[(a_j p_j)^2 + (a_j h)^2]^{1/2}$$

$$\geqq [(\Sigma a_j p_j)^2 + (\Sigma a_j h)^2]^{1/2}$$

$$= [4A^2 + h^2 L^2]^{1/2}$$

$$\geqq [4A^2 + h^2 L_0^2]^{1/2}$$

$$= 2(S_0 - A).$$

Therefore, $S \geqq S_0$. For equality, both inequalities encountered must become equalities and so two conditions must be satisfied. First,

$$p_1 : h = p_2 : h = \ldots = p_n : h,$$

that is, $P$ is a right pyramid. Second, $L = L_0$.

**47.** We take two steps: (1) We change the base of $D$ into that of $D_0$, and both pyramids, of which $D$ consists, into right pyramids, leaving, however, their altitudes unchanged. We obtain so a double pyramid $D'$ which is not necessarily a right double pyramid. (Its two constituent pyramids are right pyramids, but perhaps of different altitudes.) (2) We change $D'$ into $D_0$. Step (1) can only diminish the surface, by ex. 46. The altitudes of the two constituent pyramids of $D'$, of lengths $h_1$ and $h_2$, fall in the same straight line. Let $p_0$ denote the radius of the circle inscribed in the base of $D_0$. Then the surface of $D'$ is

$$S = [(p_0^2 + h_1^2)^{1/2} + (p_0^2 + h_2^2)^{1/2}]L_0/2$$

$$\geqq 2\left[p_0^2 + \left(\frac{h_1 + h_2}{2}\right)^2\right]^{1/2}\frac{L_0}{2} = S_0$$

by ex. 45.

**48.** Leaving the volume $V$ unchanged all the time, we take three steps: (1) Leaving the base unchanged, in shape and size, we transform the given prism into a right prism. (2) Leaving its area $A$ unchanged, we transform the base into a square. (3) We transform the right prism with square base into a cube. Steps (1) and (3) can only diminish the surface $S$, by exs. 42 and 34, respectively. Step (2) leaves the altitude $h = V/A$ unchanged, and can only diminish $L$, the perimeter of the base, by ex. 41; but $S = 2A + Lh$. Unless the prism is a cube from the outset, at least one of the three steps actually diminishes $S$. The weaker theorem 34 served as a stepping stone.

**49.** It follows from ex. 47, 41, and 37 as the foregoing ex. 48 follows from ex. 42, 41, and 34. Yet we can combine with advantage the two steps corresponding to (1) and (2) of ex. 48 into one step, thanks to the sharp formulation of ex. 47.

**50.** We start from any pyramid with triangular base (any tetrahedron, not necessarily regular). We transform it into a right pyramid, leaving unchanged the volume $V$ and the area of the base $A$, but changing (if necessary) the base into an equilateral triangle. This diminishes the perimeter of the base $L$, by ex. 40, and, therefore, the surface $S$, by ex. 46. The lateral faces of the new pyramid are isosceles triangles. Unless they happen to be equilateral, we take one of them as base, and repeat the process, diminishing again $S$. By the principle of partial variation (ex. 22), if there is at all a tetrahedron with a given $V$ and minimum $S$, it must be the regular tetrahedron.

**51.** See ex. 53.

**52.** See ex. 53.

**53.** Let $V$, $S$, and $y$ denote the volume, the surface, and the altitude of the pyramid, and let $x$, $a$, and $l$ be connected with the base of the pyramid in the same way as in the solution of ex. 36 with the base of the prism. Let $p$ stand for the radius of the circle inscribed in the base. Then

$$V = ax^2y/3$$

$$ax^2 = lxp/2$$

$$S = ax^2 + lx(p^2 + y^2)^{1/2}/2$$

$$= ax^2 + (4a^2x^4 + l^2x^2y^2)^{1/2}/2.$$

Trying to introduce expressions which depend on the shape, but not on the size, we are led to consider

$$\frac{S}{ax^2} = 1 + \left[1 + \left(\frac{ly}{2ax}\right)^2\right]^{1/2} = 1 + (1 + t)^{1/2}$$

(we introduced an abbreviation, setting $[ly/(2ax)]^2 = t$) and

$$\frac{S^3}{(3V)^2} = \frac{l^2}{4a} \frac{[1 + (1 + t)^{1/2}]^3}{t}.$$

As $V$ is given, and $S$ should be a minimum, the left-hand side should be a minimum. Therefore, the right-hand side should be a minimum. Yet the shape is given and so $l^2/a$ is given. Therefore, all that remains to do is to find the value of $t$ that makes the right hand side a minimum: this value is *independent of the shape*. It fits all special shapes equally, for example, the shapes mentioned in ex. 51 and 52. Yet there is a special shape, for which we know the result: if the base is an equilateral triangle, the best ratio $S : ax^2$, or total surface to base, is 4 : 1, by ex. 50. This remains true for all shapes, since $S/(ax^2)$ depends only on $t$, and yields

$$1 + (1 + t)^{1/2} = 4, \qquad t = 8.$$

**54.** The reader should copy the following table, substituting for the number of each problem a suitable figure.

|     | (1) | (2) | (3) | (4) | (5) | (6) | (7) |
| --- | --- | --- | --- | --- | --- | --- | --- |
| (a) | 34  | 35  | 36  | 42  | $u$ | 48  | $x$ |
| (b) | 51  | 52  | 53  | 46  | 50  | $v$ | $y$ |
| (c) | 37  | 38  | 39  | 47  | $w$ | 49  | $z$ |
| (d) | —   | —   | —   | —   | 40  | 41  | $n$ |

The rows: (a) prisms, (b) pyramids, (c) double pyramids, (d) polygons (relevant only for the last three columns).

The columns:  (1) right with square base, (2) right with circle as base, (3) right with base of given shape, (4) transition from oblique to right, (5) arbitrary triangular base, (6) arbitrary quadrilateral base, (7) arbitrary polygonal base with a given number $n$ of sides.

Analogy may suggest the theorems which can be expected to fill the gaps marked by the letters $u$, $v$, $w$, $x$, $y$, $z$, and $n$.  Here are some:

($u$) The prism that, of all triangular prisms with a given volume, has the minimum surface, has the following properties: its base is an equilateral triangle, the area of its base is $1/6$ of the total area of its surface, it is circumscribed about a sphere which touches each face at the center of the face.

($y$) The pyramid that, of all pyramids with $n$-sided polygonal base and a given volume, has the minimum surface, has the following properties: its base is a regular polygon, the area of its base is $1/4$ of the total area of its surface, it is circumscribed about a sphere which touches each face at the centroid of the face.

On the basis of the foregoing we can easily prove ($u$), ($v$), and ($w$), but ($x$), ($y$), and ($z$) depend on ($n$), which we shall discuss later; see sect. 10.7 (1).

**55.** (1) Using the *method* and the notation of sect. 6, we have

$$V = abc, \qquad S_5 = ab + 2ac + 2bc.$$

$$(S_5/3)^3 \geq ab \cdot 2ac \cdot 2bc = 4V^2$$

with equality if, and only if,

$$ab = 2ac = 2bc$$

or $a = b = 2c$: the box is one half of a cube.

(2) Using the *result* of sect. 6, we regard the plane of the face not counted in $S_5$ as a mirror.  The box together with its mirror image forms a new box of volume $2V$ the *whole* surface of which is given, $= 2S_5$.  By sect. 6, the new (double) box must be a cube when the maximum of the volume is attained.

**56.** Following ex. 55, regard the plane of the missing face as a mirror. Maximum for the triangular pyramid which is one half of a cube, halved by a diagonal plane.  Apply ex. 48, with an additional remark.

**57.** [Putnam 1950]  Following ex. 55 and 56, regard the planes of both missing faces as mirrors.  Maximum for the triangular pyramid which is one quarter of a cube; the cube is divided into four congruent fragments by two planes of symmetry, one a diagonal plane, the other perpendicular to the first and parallel to two faces.

**58.** Let $A$, $L$, $r$, and $s$ stand for the area, the perimeter, the radius, and the arc of the sector, respectively.  Then

$$A = rs/2, \qquad L = 2r + s$$

$$(L/2)^2 \geq 2r \cdot s = 4A;$$

we use the theorem of the means. Equality is attained when $s = 2r$ and the angle of the sector equals two radians.

**59.** Let $u$, $v$, and $w$ denote the sides of the triangle, $A$ the area, $\gamma$ the given angle opposite $w$. Then

$$2A = uv \sin \gamma.$$

(1)    $$[(u + v)/2]^2 \geq uv = 2A/\sin \gamma,$$

by the theorem of the means. Equality is attained when $u = v$, that is, the triangle is isosceles.

(2)    $$w^2 = u^2 + v^2 - 2uv \cos \gamma$$

$$= u^2 + v^2 - 4A \cot \gamma$$

$$(u^2 + v^2)/2 \geq uv = 2A/\sin \gamma.$$

The equality is attained, and so $w$ a minimum, when $u^2 = v^2$ and the triangle is isosceles.

(3) As both $u + v$ and $w$, also $u + v + w$ is a minimum, when the triangle is isosceles.

**60.** Use the notation of ex. 59. The given point lies on the side $w$. Draw from the given point parallels to $u$ and $v$, terminating on $v$ and $u$, and call them $a$ and $b$, respectively; $a$ and $b$ are given (they are, in fact, oblique coordinates). From similar triangles

$$\frac{v - b}{a} = \frac{b}{u - a} \quad \text{or} \quad \frac{a}{u} + \frac{b}{v} = 1$$

$$\frac{1}{4} = \left( \left[ \frac{a}{u} + \frac{b}{v} \right] / 2 \right)^2 \geq \frac{ab}{uv} = \frac{ab \sin \gamma}{2A}$$

$$A \geq 2ab \sin \gamma.$$

There is equality if, and only if,

$$\frac{a}{u} = \frac{b}{v} = \frac{1}{2}, \qquad u = 2a, \qquad v = 2b$$

and the given point is the midpoint of $w$.

**61.** Use the notation of sect. 6 and the theorem of the means.

(1)    $$V = abc \leq [(a + b + c)/3]^3 = [E/12]^3$$

(2)    $$S = 2ab + 2ac + 2bc$$

$$\leq a^2 + b^2 + a^2 + c^2 + b^2 + c^2$$

$$= 2(a + b + c)^2 - 4(ab + ac + bc)$$

that is,

$$3S \leq 2(E/4)^2.$$

In both cases, equality is attained only for $a = b = c$, that is, for the cube.

**62.** Use the notation of sect. 6 and the theorem of the means. The length is $c$, the girth $2(a + b)$, and

$$V = (2a \cdot 2b \cdot c)/4 \leqq [(2a + 2b + c)/3]^3/4 \leqq l^3/108.$$

Equality is attained only for $2a = 2b = c = l/3$.

**63.** Use the notation in the solution of ex. 35. Then

$$d^2 = (2x)^2 + (y/2)^2 = 2(x^2 + x^2 + y^2/8)$$

and, therefore, by the theorem of the means

$$V^2 = \pi^2 x^4 y^2 = 8\pi^2 x^2 \cdot x^2 \cdot y^2/8 \leqq 8\pi^2 (d^2/6)^3$$

with equality only if

$$x^2 = y^2/8 = d^2/6.$$

For the historical background cf. O. Toeplitz, *Die Entwicklung der Infinitesimalrechnung*, p. 78–79.

No solution: **23, 33.**

SOLUTIONS, CHAPTER IX

**1.** (1) Imagine two mirrors perpendicular to the plane of the drawing, the one through $l$ and the other through $m$. A person at $P$ looks at $m$ and sees himself from the side: the light coming from $P$ returns to $P$ after a first reflection in $l$ and a second in $m$. The light, choosing the shortest path, describes the desired $\triangle PYZ$ with minimum perimeter; the sides of $\triangle PYZ$ include equal angles with $l$ and $m$ at the points $Y$ and $Z$, respectively. (2) Let $P'$ and $P''$ be mirror images of $P$ with respect to $l$ and $m$, respectively. The straight line joining $P'$ and $P''$ intersects $l$ and $m$ in the required $Y$ and $Z$, respectively, and its length is that of the desired minimum perimeter. (By the idea of fig. 9.3, applied twice.)

**2.** (1) Light, after three successive reflections on three circular mirrors, returns to its source from the opposite direction. (2) A closed rubber band connects three rigid rings. Both interpretations suggest that the two sides of the required triangle that meet in a vertex on a given circle include equal angles with the radius.

**3.** Roundtrip of light or closed rubber band, as in ex. 2; $XY$ and $XZ$ are equally inclined to $BC$, etc.

**4.** A polygon with $n$ sides and minimum perimeter inscribed in a given polygon with $n$ sides has the following property: those two sides of the minimum polygon of which the common vertex lies on a certain side $s$ of the given polygon, are equally inclined to $s$. See, however, ex. 6 and 13.

**5.** Call $A$ the point of intersection of $l$ and $m$. Take a point $B$ on $m$ and a point $C$ on $l$ so that $\angle BAC$ (less than 180°) contains the point $P$ in its interior. Then, by reflection,

$$\angle P''AB = \angle BAP, \qquad \angle PAC = \angle CAP'$$

and hence

$$\angle P''AP' = 2\angle BAC.$$

The solution fails when $\angle P''AP' \geqq 180°$ or, which is the same, when the given $\angle BAC \geqq 90°$.

**6.** The solution cannot apply when the given triangle has an angle $\geqq 90°$, see ex. 5. The solution of ex. 4 is, of course, *a fortiori* liable to exception.

**7.** Fix for the moment $X$ in the position $P$ on the side $BC$. Then the solution (2) of ex. 1 applies (since $\angle BAC$ is acute, see ex. 5); the minimum perimeter is $P'P''$. Now the length $P'P''$ depends on $P$; it remains to find the minimum of $P'P''$. (As $P'P''$ itself was obtained as a minimum, we seek a minimum of the minima or a "minimum minimorum.") Now, by reflection, $P''A = PA = P'A$. Therefore, $\triangle P''AP'$ is isosceles; its angle at $A$ is independent of $P$ (see ex. 5) and so its *shape is independent of P*. Therefore, $P'P''$ becomes a minimum when $P'A = PA$ becomes a minimum, and this is visibly the case when $PA \perp BC$; cf. sect. 8.3. *The vertices of the triangle with minimum perimeter inscribed in a given acute triangle are the feet of the three altitudes of the given triangle.* Comparing this with the solution of ex. 3, we see that *the altitudes of an acute triangle bisect the respective angles of the inscribed triangle of which their feet are the vertices.* The latter result is, of course, quite elementary. The present solution is due to L. Fejér. Cf. Courant-Robbins, p. 346–353.

**8.** No. If $\triangle ABC$ has an angle $\geqq 120°$, the vertex of that angle is the traffic center. This is strongly suggested by the mechanical solution in sect. 2 (2).

**9.** [Putnam 1949] This is closely analogous to the simpler plane problem treated in sect. 1 (4), sect. 2 (2) and ex. 8. Which method should we adopt? (1) *Mechanical interpretation*, modelled on fig. 9.7. There are four pulleys, one at each of the four given points $A, B, C$, and $D$. Four strings are attached together at the point $X$; each string passes over one of the pulleys and carries a weight of one pound at its other endpoint. As in sect. 2 (2), a first consideration (of the potential energy) shows that the equilibrium position of this mechanical system corresponds to the proposed minimum problem. A second consideration deals with the forces acting on the point $X$. There are four such forces; they are equal in magnitude and in the direction of the four taut strings, going to $A, B, C$, and $D$, respectively. The resultant of the first two forces must counterbalance the resultant of the last two forces. Therefore, these resultants are in the same line which bisects both $\angle AXB$ and $\angle CXD$. The equality of these angles follows

from the congruence of the two parallelograms of forces (both are rhombi)
Similarly related pairs are: $\angle AXC$ and $\angle BXD$, $\angle AXD$ and $\angle BXC$
(2) *Partial variation and optical interpretation*, modelled on fig. 9.4. Keep
constant (for a moment) $CX + DX$, the sum of two distances. Then the
point $X$ has to vary on the surface of a prolate spheroid (ellipsoid of revolution) with foci at $C$ and $D$. We conceive this surface as a mirror. The
light starting from $A$, reflected at our spheroidal mirror and arriving at $B$
renders $AX + XB$ a minimum; along its path $\angle AXB$ is bisected by the
normal to the mirror at the point $X$. The same normal bisects $\angle CXD$ by
sect. 1 (3) or sect. 2 (1). Yet the equality of $\angle AXB$ and $\angle CXD$ is not so
easily obtained by this method: although both methods work equally well
in the simpler analogous case, they do not equally well apply to the present
theorem.

**10.** Yes. Wherever the point $X$ may be

$$AX + XC \geqq AC, \qquad BX + XD \geqq BD$$

since the straight path is the shortest between two points, and both
inequalities become equations if, and only if, $X$ is the point of intersection
of the diagonals $AC$ and $BD$: this is the traffic center. The statement of
ex. 9 remains fully correct in view of the fact that the normal to the plane of
the quadrilateral is a common bisector of $\angle AXC$ and $\angle BXD$, which are
both straight angles.

**11.** Follows by partial variation from the result of sect. 1 (4). Cf.
Courant-Robbins, p. 354–361.

**13.** Drive the ball parallel to a diagonal of the table. Fig. 9.14 applies
ex. 12 four times in succession. Imagine fig. 9.14 drawn on transparent
paper and fold it along the reflecting lines; then the several segments of the
straight line $PP$ just cover the rhombic path of the billiard ball. By the
way, we see here a case in which ex. 4 has an infinity of solutions.

**14.** By fig. 9.15 (which should be drawn on transparent paper)

$$n2\alpha < 180° < (n + 1)2\alpha,$$

$$90°/(n + 1) < \alpha < 90°/n.$$

Draw figures illustrating the cases $n = 1, 2, 3$. Consider the case $n = \infty$.

**15.** The particular case treated in sect. 1 and ex. 12 yields several
suggestions; see ex. 16, 17, and 18.

**16.** If $A$, $B$, and $AX + XB$ are given, the locus of $X$ is the surface of a
prolate spheroid (ellipsoid of revolution) with foci $A$ and $B$; such spheroids
are the level surfaces. The spheroid to which the given line $l$ is tangent
at the point $X$ yields the solution. The normal to this spheroid at the
point $X$ is *perpendicular to l*, and *bisects* $\angle AXB$ by a property of the ellipse
proved in sect. 1 (3) and sect. 2 (1).

**17.** Place a sheet of paper folded in two so that the crease coincides with $l$ and one half of the sheet (a half-plane) passes through $A$ and the other half through $B$. The desired shortest line is certainly described on this folded sheet. If the sheet is unfolded, the shortest line becomes straight. On the folded as on the unfolded sheet the lines $XA$ and $XB$ *include the same angle with $l$.*

**18.** Fasten one end of a rubber band of suitable length to $A$, pass the band over the rigid rod $l$ at $X$ and fasten the other end to $B$ so that the band is stretched: it forms so the shortest line required by ex. 15 (if the friction is negligible). Three forces act at the point $X$: two tensions equal in magnitude, one directed toward $A$ and the other toward $B$, and the reaction of the rod which is perpendicular to $l$ (since the friction is negligible). The parallelogram of forces is a rhombus and so a normal to $l$ bisects $\angle AXB$, as found in ex. 16. The reaction of the rod has no component parallel to the rod, and so the components of the tensions parallel to $l$ must be equal in amount (and opposite in direction). Therefore, $XA$ and $XB$ are equally inclined to $l$, as found in ex. 17. By the way, the equivalence of the results of ex. 16 and 17 can be shown by a little solid geometry. (Trihedral angles are congruent if they have three appropriate data in common.)

**19.** Closed rubber band around three knitting needles held rigidly. Partial variation and ex. 16, or ex. 18; the bisectors of the three angles of a triangle meet in the center of the inscribed circle. Ex. 3 is a limiting case.

**20.** Each vertex of the triangle is the midpoint of an edge of the cube. The triangle is equilateral; its center is the center of the cube; its perimeter is $3\sqrt{6}a$.

**21.** By partial variation, sect. 8.3, and sect. 1 (4) or 2 (2), $TX$, $TY$, and $TZ$ are perpendicular to $a$, $b$, and $c$, respectively, and equally inclined to each other (120°). We could call $T$ the "traffic center of three skew lines." The problem of sect. 1 (4) is an extreme case: $a$, $b$, and $c$ become parallel. There is an obvious generalization and there are some obvious analogous problems: the traffic center of three spheres, the traffic center of a point, a straight line and a plane, and so on.

**22.** The traffic center of three skew edges of a cube is, of course, the center of the cube. Represent clearly the rotation of the cube through 120° that interchanges the three given skew edges, and the situation of the triangle found in ex. 20.

**23.** In order to find the shortest line between two given points $A$ and $B$ on the surface of a polyhedron, imagine the polyhedral surface made of cardboard, of plane polygons hinged together and folded up suitably. Unfold the polyhedral surface in one plane (lay the cardboard flat on the desk): the shortest line required becomes the *straight line from $A$ to $B$.* Before unfolding it, however, we have to cut the polyhedral surface along suitable edges which the shortest line *does not cross*.  As we do not know in advance which faces and which edges the shortest line will cross, we have to examine all admissible combinations.  We turn now to the proposed problem, list the essential sequences of faces, and note after each sequence the square of the rectilinear distance from the spider to the fly along that sequence.

(1)  End wall, ceiling, end wall:

$$(1 + 20 + 7)^2 = 784;$$

(2)  End wall, ceiling, side wall, end wall:

$$(1 + 20 + 4)^2 + (4 + 7)^2 = 746;$$

(3)  End wall, ceiling, side wall, floor, end wall:

$$(1 + 20 + 1)^2 + (4 + 8 + 4)^2 = 740.$$

**24.** An arc of a great circle is a geodesic on the sphere.  A great circle is a plane curve; the plane in which the great circle lies is its osculating plane at all its points.  This plane passes through the center of the sphere, and therefore it contains all the normals to the sphere (all the radii) that pass through points of the great circle.  A small circle is not a geodesic; in fact, the plane of the small circle contains none of the normals to the sphere that pass through the points of the small circle.

**25.** By the conservation of energy, the magnitude of the velocity of the point is constant although, of course, the direction of the velocity varies. The difference of the velocity vectors at the two endpoints of a short arc of the trajectory is due to the normal reactions of the surface and is, therefore, almost normal to the surface.  This is the characteristic property of a geodesic; see ex. 24 (2).  Another version of the same argument: reinterpret the tensions along the rubber band of ex. 24 (2) as velocities along the trajectory; all vectors are of the same magnitude and the variation in the direction is due to normal reactions in both cases.

**26.** Push the $n$ free edges gently against a plane (your desk), forming a pyramid with $n$ isosceles lateral faces the base of which is the desired polygon. In fact, the base is inscribed in a circle the center of which is the foot of the altitude of the pyramid.  The radius of the circle is the third side of a right triangle of which the hypotenuse is the radius of the great circle described on the cardboard and the second side the altitude of the pyramid.

**27.** If the center of gravity is as close to the floor as possible, there is equilibrium.  As little mechanics as this is enough to suggest the desired solution: take a point $D$ on the surface of $P$ such that the distance $CD$ is a minimum.  An easy discussion shows that $D$ can neither be a vertex of $P$ nor lie on an edge of $P$, and that $CD$ is perpendicular to the face $F$ of $P$ on which $D$ lies.  See Pólya-Szegö, *Analysis*, vol. 2, p. 162, problem 1.

**28.** (a) Imagine the globe completely dried up, so that all peaks, passes, and deeps are exposed. Now cover just one deep with some water. The remaining part of the globe has $P$ peaks, $S$ passes, and $D - 1$ deeps, and can be regarded as an island. By the result proven in the text,

$$P + (D - 1) = S + 1.$$

(b) The level lines and the lines of steepest descent subdivide the globe into $F$ "countries"; this is the terminology of ex. 3.2. Take so many lines that each remarkable point, peak, deep, or pass becomes a vertex, as on figs. 9.16 and 9.17, and that no country has more than one remarkable point on its boundary.

We "distribute" each edge, or boundary line, equally between the two countries that it separates, giving $1/2$ of the edge to each country. Similarly, we distribute each vertex equally among the countries of which it is a vertex. In return, each country will contribute to the left-hand side of Euler's equation

$$V - E + F = 2;$$

it will contribute one unit to $F$ and a suitable fraction to $V$ and to $-E$. Let us compute this contribution for the various kinds of countries.

I. If there is no remarkable point on its boundary, the country is a quadrilateral, contained between two level lines and two lines of steepest descent. Its contribution to $V - E + F$ is

$$4 \times \frac{1}{4} - 4 \times \frac{1}{2} + 1 = 0.$$

II. If there is a peak or a deep on its boundary, the country is a triangle; see fig. 9.16. If the peak, or the deep, is a common vertex to $n$ countries, the contribution of each country to $V - E + F$ is

$$\left(2 \times \frac{1}{4} + \frac{1}{n}\right) - 3 \times \frac{1}{2} + 1 = \frac{1}{n}$$

and the joint contribution of all $n$ countries is $n \cdot 1/n = 1$.

III. If there is a pass on its boundary, the country is a quadrilateral; see fig. 9.17. Its contribution to $V - E + F$ is

$$\left(3 \times \frac{1}{4} + \frac{1}{8}\right) - 4 \times \frac{1}{2} + 1 = -\frac{1}{8}$$

and the joint contribution of the 8 countries of which the pass is the common vertex is $8 \cdot (-1/8) = -1$.

The grand total of all contributions is, by Euler's theorem,

$$P + D - S = 2.$$

(c) The proof using the idea of the "Deluge" is not properly an example of "physical mathematics": it uses ideas in touch with everyday experience, but not with any specific physical theory. The hint in part (b) of the question was misleading: it appeared to suggest that $P$, $D$, and $S$ are somehow analogous to $F$, $V$, and $E$, which is by no means the case. Still, it was a useful hint: it directed our attention to Euler's theorem. This is, however, quite natural: the ideas that guide us in the solution of problems are quite often mistaken but still useful.

**29.** (a) Let $t_1$ be the time of descent of the stone and $t_2$ the time of ascent of the sound. Then

$$t = t_1 + t_2, \qquad d = gt_1^2/2, \qquad d = ct_2.$$

Eliminating $t_1$ and $t_2$ and solving a quadratic equation, we find

$$d^{1/2} = -c(2g)^{-1/2} \pm [c^2(2g)^{-1} + ct]^{1/2}.$$

Since $t = 0$ should give $d = 0$, we have to choose the sign $+$. Then

$$d = c^2g^{-1} + ct - c^2g^{-1}[1 + 2gc^{-1}t]^{1/2}.$$

(b) $$d = gt^2/2 - g^2t^3/(2c) + \dots .$$

Neglecting the terms not written here, we can use the two terms retained as a suitable approximate formula.

(c) It is typical that we can foresee the principal term of the expansion and even the sign of the correction on the basis of physical considerations. Also the mathematical procedure used to obtain a suitable approximate formula is typical: we *expanded* (the expression for $d$) *in powers of a small quantity* (the time $t$). Cf. sect. 5.2.

**30.** The elliptic mirror becomes a parabolic mirror which collects all rays of light that fall on it parallel to its axis into its focus. Such a parabolic mirror is the most essential part of a reflecting telescope.

**31.** The equation is separable. We obtain

$$dx = \left(\frac{y}{c - y}\right)^{1/2} dy$$

by obvious transformations. We set

$$\left(\frac{y}{c - y}\right)^{1/2} = \tan \varphi,$$

introducing the auxiliary variable $\varphi$, and obtain

$$y = c \sin^2 \varphi, \qquad x = c(\varphi - (1/2) \sin 2\varphi).$$

We find $x$ by integration and have to choose the constant of integration so that the curve passes through the origin: $\varphi = 0$ implies $x = y = 0$. Setting $2\varphi = t$, $c = 2a$, we obtain the usual equations of the cycloid:

$$x = a(t - \sin t), \qquad y = a(1 - \cos t).$$

The cycloid passes through the point $A$, which is the origin. There is just one value of $a$ that makes the first branch of the cycloid (corresponding to $0 < t < 2\pi$) pass through the given point $B$. In order to see this, let $a$ vary from 0 to $\infty$; this "inflates" the cycloid which, sweeping over a quadrant of the plane, hits $B$ when inflated to the right size.

**33.** Let $a$ be the radius of the sphere (as in sect. 5), $h$ the height of the segment, $V$ its volume, and $C$ the volume of the cone with the same base as the segment and the same height $h$. The origin (the point $O$ in fig. 9.13) is the common vertex of the segment and the cone. From elementary geometry and the equation of the circle given in sect. 5

$$C = \frac{\pi(2ah - h^2)h}{3}.$$

Use fig. 9.13 but consider now only the cross-sections at the distance $x$ from 0 with $0 < x < h$. Passing from the equilibrium of the cross-sections expressed by equation (A) to the equilibrium of the solids (segment, cone—but not that with volume $C$—and cylinder) we find

(B)   $2a(V + \pi h^2 \cdot h/3) = (h/2)\pi(2a)^2 h.$

Hence

$$V = \frac{\pi h^2(3a - h)}{3} = \frac{a + (2a - h)}{2a - h} C;$$

$2a - h$ is the height of the complementary segment.

**34.** Write the equation of the circle considered in sect. 5 in the form

(A)   $2a\pi x^2 = x\pi y^2 + x\pi x^2.$

Only $\pi x^2$, the cross-section of a cone, hangs now from the point $H$ of fig. 9.13; the cross-section $\pi y^2$ of the sphere and the cross-section $\pi x^2$ of another cone (congruent with the first) remain in their original position (with abscissa $x$). Consider $0 < x < a$, pass to the equilibrium of the three solids, introduce $\bar{x}$, the abscissa of the center of gravity of the hemisphere, and remember the position of the center of gravity of a cone (its distance from the vertex is 3/4 of the altitude):

(B)   $2a \cdot \pi a^2 \cdot a/3 = \bar{x} \cdot 2\pi a^3/3 + (3a/4)\pi a^2 \cdot a/3,$

$$\bar{x} = 5a/8.$$

**35.** Keep the notation of ex. 33, but change that of ex. 34 in one respect: $\bar{x}$ denotes now the abscissa of the center of gravity of a segment with height $h$. Considering $0 < x < h$, pass from (A) of ex. 34 to

(B)   $2a \cdot \pi h^2 \cdot h/3 = \bar{x}V + (3h/4)\pi h^2 \cdot h/3$

which yields in view of the value of $V$ found in ex. 33

$$\frac{\bar{x}}{h - \bar{x}} = \frac{h + 4(2a - h)}{h + 2(2a - h)}.$$

**36.** Let $h$ denote the height and $V$ the volume of the segment. Write the usual equation of the parabola in the form

(A) $$2p \cdot \pi y^2 = x \cdot \pi(2p)^2.$$

Notice the cross-section $\pi y^2$ of the paraboloid and the cross section $\pi(2p)^2$ of a cylinder. Considering $0 < x < h$ and passing from the equilibrium of the cross-sections to that of the solids, we find

(B) $$2p \cdot V = (h/2)\pi(2p)^2 h,$$

$$V = \pi p h^2 = (3/2)\pi 2p h(h/3).$$

Notice that, by the equation of the parabola, $\pi 2ph$ is the base of the segment.

**37.** Keep the notation of ex. 36 and let $\bar{x}$ denote the abscissa of the center of gravity of the segment. Now write the equation of the parabola in the form

(A) $$x \cdot \pi y^2 = 2p \cdot \pi x^2.$$

Notice $\pi x^2$, the cross-section of a cone. Considering $0 < x < h$ and passing from the cross-sections to the solids, we find

$$\bar{x}V = 2p \cdot \pi h^2(h/3)$$

and hence, by ex. 36,

$$\bar{x} = 2h/3.$$

**38.** $n = 0$: volume of prism, area of parallelogram; $n = 1$: area of triangle, center of gravity of parallelogram or prism; $n = 2$: volume of cone or pyramid, center of gravity of triangle. $n = 3$: center of gravity of cone or pyramid.

Observe that the method of Archimedes as presented here in sect. 5 and ex. 33–38 would be suitable for a class of Analytic Geometry and could lend a new interest to this subject, which may so easily become dry and boring in the usual presentation. The propositions of the "Method" that we have not discussed can be similarly treated and could be similarly used.

No solution: **12, 32.**

### SOLUTIONS, CHAPTER X

**1.** No. The gap is not too bad: the existence of the maximum can be established with the help of the general theorem quoted in ch. VIII, footnote 3.

**2.** The explicit formula given in the solution of ex. 8.41 shows that $A^2 \leqq (p-a)(p-b)(p-c)(p-d)$, and equality is attained if, and only if, $\varepsilon = 180°$ in which case the quadrilateral is inscribed in a circle.

**3.** Let $A$, $B$, and $C$ be consecutive vertices of the regular polygon with $n$ sides, and $M$ the midpoint of the side $BC$. Replace $\triangle ABM$ by the isosceles triangle $\triangle AB'M$ (in which $AB' = B'M$) that has the same base $AM$ and the same perimeter and, therefore, a larger area; see ex. 8.8.

**4.** If we express both areas in terms of $r$, the radius of the circle, and $n$, the number of the sides of the polygon, it remains to prove the inequality:

$$\frac{\pi^2 r^2}{n \tan (\pi/n)} < \pi r^2.$$

It is more elegant to observe that a regular polygon is circumscribable about a circle: the desired result is a particular case of ex. 5.

**5.** A polygon with area $A$ and perimeter $L$ is circumscribed about a circle with radius $r$. Then, obviously, $\pi r^2 < A$. Lines drawn from the center of the circle to the vertices of the polygon divide it into triangles with the common altitude $r$; hence $A = Lr/2$. Combining both results obtained, we find

$$A = \frac{L^2 r^2}{4A} < \frac{L^2}{4\pi}.$$

Now, $L^2/(4\pi)$ is precisely the area of the circle that has the perimeter $L$.

**6.** Let $A$ denote the area and $L$ the perimeter of a given curve, and $r$ the radius of the circle with the same perimeter so that $L = 2\pi r$. Let $A_n$ denote the area and $L_n$ the perimeter of a polygon $P_n$ that tends to the given curve as $n \to \infty$. Then $A_n$ tends to $A$ and $L_n$ to $L$. Consider the polygon $P'_n$ that is similar to $P_n$ and has the perimeter $L$; the area of $P'_n$ is $A_n(L/L_n)^2$. Since $P'_n$ has the same perimeter as the circle with radius $r$, we conclude from sect. 7 (4) that

$$A_n(L/L_n)^2 < \pi r^2.$$

Passing to the limit, we find that

$$\lim_{n \to \infty} A_n(L/L_n)^2 = A \leqq \pi r^2.$$

This justifies statement I of sect. 8. Yet the text of sect. 7 (5) is objectionable: we definitely did *not* prove that $A < \pi r^2$, as that text appears to suggest. In fact, the relation expressed by $<$ can go over into that expressed by $\leqq$ as we pass to the limit.

**7.** Both statements are equivalent to the inequality

$$\frac{216 V^2}{S^3} \leqq 1;$$

$V$ denotes the volume and $S$ the area of the surface of the box. In sect. 8.6 we proved this inequality directly.

**8.** The equivalence of I′, II′, and III′ is shown by the same method as that of I, II, and III in sect. 8. Yet I′ is not equivalent to I. In fact, I′ explicitly denies the possibility left unsettled by I that a curve which is not a circle could have the same perimeter and also the same area as a circle.

The argument of sect. 7 (5) as amplified in ex. 6 proves I, but does not prove I': it proves $\leq$ which is enough for I, but not $<$ which would be needed for I'.

**9.** The solution of the proposed problem would add to I of sect. 8 what is still needed to obtain I' of ex. 8. So much for the importance; concerning the other points see ex. 10–13.

**10.** Call $C''$ the smallest triangle containing $C$, $L''$ its perimeter and $A''$ its area. Then, obviously, $L'' < L$ and $A'' > A$. Take as $C'$ the triangle similar to $C''$ with perimeter $L$; the area of $C'$ is $A' = A''(L/L'')^2 > A'' > A$.

**11.** If $C$ is any curve, but *not convex*, we consider first $C''$, the least convex curve containing $C$, and then $C'$, similar to $C''$, but having a perimeter equal to that of $C$. The whole argument of ex. 10, down to the final inequality, can be repeated in the more general situation.

**12.** Take two different points $P$ and $Q$ on the closed curve $C$. There must be on $C$ a third point $R$ that is not on the straight line through $P$ and $Q$, since $C$ cannot be wholly contained in a straight line. Consider the circle through $P$, $Q$, and $R$. If this circle does not coincide with $C$, there is a fourth point $S$ on $C$ which is not on the circle: the problem of ex. 9 is equivalent to that of ex. 13.

**13.** If $C$ is not convex, ex. 11 yields the desired construction. If $C$ is convex, $P$, $Q$, $R$, and $S$ are, in some order, the vertices of a convex quadrilateral. The region surrounded by $C$ consists of this quadrilateral and of four segments. Each segment is bounded by a side of the quadrilateral and by one of the four arcs into which $P$, $Q$, $R$, and $S$ divide $C$. Following Steiner's idea (see sect. 5 (2), figs. 10.3 and 10.4) we consider the four segments as rigid (of cardboard) and rigidly attached to the respective sides of the quadrilateral which we consider as articulated (with flexible joints at the four vertices). We adopt the notation of ex. 8.41. Then, by our main condition, $\varepsilon \neq 180°$. Let a slight motion of the articulated quadrilateral change $\varepsilon$ into $\varepsilon'$. We choose $\varepsilon'$ so close to $\varepsilon$ that the four arcs rigidly attached to the sides *still form a not self-intersecting* curve $C'$. Moreover, we choose $\varepsilon'$ so that

$$\left| \varepsilon' - 180° \right| < \left| \varepsilon - 180° \right|.$$

This implies that the area of $C'$ is larger than that of $C$, in virtue of the formula for $A^2$ given in the solution of ex. 8.41. Yet the $C'$, consisting of the same four arcs as $C$, has the same perimeter.

**14.** Both inferences have the same logical form. Yet the second inference, that leads to an obviously false result, must be incorrect. Therefore, also the first inference must be incorrect, although it aims at a result that might be true. The second inference is, in fact, an ingenious parody of the first, due to O. Perron.

The difference between the two cases must be some outside circumstance not mentioned in the proposed text. There is no greatest integer. Yet, among all isoperimetric curves there is one with the greatest area. This, however, we did not learn from ex. 10–13.

**15.** The curve $C$ is not a circle, but it has the same perimeter as a certain circle. The area of $C$ cannot be larger than that of the circle, by ex. 6. I say that the area of $C$ cannot be equal to that of the circle. Otherwise there would be, as we know from ex. 10–13, another curve $C'$ still with the same perimeter as the circle, but with a larger area, which is impossible in virtue of what we proved in ex. 6.

**16.** Given two points, $A$ and $B$ in fig. 10.13, joined by a straight line and a variable curve which include a region together. We consider the length of the curve and the area of the region. In the text we regarded the including length as given and sought the maximum of the included area. Here we regard the included area as given and seek the minimum of the including length. In both cases, the solution is the same: an arc of a circle. Even the proof is essentially the same. We may use fig. 10.14 here as there. Of course, there are obvious differences; the (unshaded) segment of the circle in fig. 10.14, I is constructed now from a given area, not from a given length, and we use now theorem II' of ex. 8, not theorem I'.

**17.** Use ex. 16: identify the points $X$ and $Y$ of fig. 10.11 with the points $A$ and $B$ of fig. 10.13, respectively, and add the invariable $\triangle XYC$ to fig. 10.13. There is maximum when the line of given length is an arc of circle.

**18.** In fig. 10.11 regard the line $CY$ as a mirror, let $X'$ be the mirror image of $X$, and apply ex. 17 to $\angle XCX'$ and the two given points $X$ and $X'$ on its two sides. There is maximum when the line of given length is an arc of circle perpendicular to $CY$ at the point $Y$.

**19.** Use partial variation. Regard $X$ as fixed: the solution is an arc of circle perpendicular to $CY$, by ex. 18. Regard $Y$ as fixed: the arc of circle is also perpendicular to $CX$. Finally, the solution is an arc of circle perpendicular both to $CX$ and to $CY$, and so its center is at $C$, as conjectured in sect. 9.

**20.** There is maximum when the straight line is perpendicular to the bisector of the angle. This would follow from symmetry, if we knew in advance that there is just one solution. The result follows without any such assumption from ex. 8.59 (2).

**21.** By the idea of fig. 10.14, there is maximum when the strings $BC$ and $DA$ are arcs of the *same* circle of which the sticks $AB$ and $CD$ are chords.

**22.** A closed line consisting of $2n$ pieces, $n$ sticks alternating with $n$ strings, surrounds a maximum area, when all the sticks are chords and all the strings are arcs of the same circle.

**23.** When all the strings of ex. 22 are of length 0, we obtain sect. 5 (2) and figs. 10.3 and 10.4.

**24.** Analogous to ex. 16: the rigid disk, the variable surface with given area and the circle that forms the rim of both correspond to the stick, the string, and the pair of points $A$ and $B$, respectively. The method of ex. 16 applies. (In fig. 10.14 rotate the circle I about its vertical diameter and do the same to the segment at the base of fig. II, but change its upper part in a more arbitrary manner.) Assuming the isoperimetric theorem in space, we obtain: the included volume is a maximum, when the surface with given area is a portion (a zone with one base) of a sphere.

**25.** Take the three planes perpendicular to each other and take for granted the isoperimetric theorem in space. Then the trihedral angle becomes an octant and you can use the analogue of fig. 10.12 in space. By successive reflections on the three planes, the surface cutting the octant becomes a *closed* surface; its area and the volume surrounded are eight times the given area and the volume cut off by the original surface, respectively. The closed surface with given area that surrounds the maximum volume is the sphere. Therefore, in our special case of the proposed problem there is maximum when the surface with given area is a portion (1/8) of a sphere with center at the vertex of the trihedral angle.

**26.** The configuration considered in the solution of ex. 25 is the special case $n = 2$ of the following general situation. There are $n + 1$ planes; $n$ planes pass through the same straight line and divide the space into $2n$ equal wedges (dihedral angles) and the last plane is perpendicular to the $n$ foregoing. These $n + 1$ planes divide the space into $4n$ equal trihedral angles to any one of which the method of repeated reflections, used in ex. 25, applies and yields the same result: *the volume cut off is a maximum when the surface of a given area is a portion of a sphere with center at the vertex of the trihedral angle.*

(There are three more configurations containing trihedral angles to which the method applies and yields the same result. These configurations are connected with the regular solids, the first with the tetrahedron, the second with the cube and the octahedron, and the third with the dodecahedron and the icosahedron. Their study requires more effort or more preliminary knowledge and so we just list them in the following table which starts with the simple configuration described above.

| Planes | Parts of space | Angles | | |
|--------|----------------|------|------|------|
| $n + 1$ | $4n$ | 90° | 90° | 180°/$n$ |
| 6 | 24 | 90° | 60° | 60° |
| 9 | 48 | 90° | 60° | 45° |
| 15 | 120 | 90° | 60° | 36° |

The "planes" are planes of symmetry, the "parts of space" trihedral angles, and the "angles" are included by the three planes bounding the trihedral angle.)

It is natural to conjecture that the result remains valid for *any* trihedral angle. This conjecture is supported inductively by the cases listed and also by analogy; the similarly obtained similar conjecture about angles in a plane (sect. 9) has been proved (ex. 19).

It is even natural to extend the conjecture to polyhedral angles and there we can find at least a limiting case accessible to verification. We call here "cone" the infinite part of space described by an acute plane angle rotating about one of its sides. We seek the surface with given area that cuts off the maximum volume from the cone. It can be proved that this surface is (1) a surface of revolution, (2) a portion of a sphere, and (3) that the center of this sphere is the vertex of the cone. We cannot go into detail here, but it should be observed that part (2) of the proof results in the same way from ex. 24 as the solution of ex. 17 results from ex. 16. The problem of ex. 25, raised by Steiner, still awaits a complete solution.

**27.** If a region with area $A$ had two bisectors without any common point, it would be divided by them into three sub-regions, two with area $A/2$ and a third with a non-vanishing area, which is obviously impossible.

**28.** The straight line is shorter: $1 < (\pi/2)^{1/2}$.

**29.** See ex. 30.

**30.** Assume that the endpoints of a given bisector lie on two different sides which meet in the vertex $O$, but none of the endpoints coincides with $O$. By suitable reflections (idea of fig. 10.12) we obtain six equal triangles one of which is the original triangle and six equal arcs one of which is the given bisector. The six triangles form a regular hexagon with center $O$. The six arcs form a closed curve which surrounds one-half of the area of the hexagon and especially the point $O$ in which three axes of symmetry of the curve meet. If the length of the bisector is a minimum, the closed curve must be a circle or a regular hexagon, according as all bisectors are admitted (the present ex. 30) or only straight bisectors are admitted (ex. 29); we have to use theorem II' of ex. 8 or the theorem conjugate to that of sect. 7 (1), respectively. The solution of ex. 30 is the sixth part of a circle with center in one of the vertices, the solution of ex. 29 is a line parallel to one of the sides; in each case there are three solutions. The given bisector may have some other situation (both endpoints on the same side, or at the same vertex, and so on) but the discussion of these situations corroborates the result obtained.

**31.** [Cf. Putnam 1946]   Let $O$ be the center of the circle.   If the straight line segment $PP'$ is bisected by $O$, we call the points $P$ and $P'$ *opposite* to each other. We call two curves opposite to each other if one of them consists of the points opposite to the points of the other. Let now $A$ and $B$ be the endpoints of bisector which we call shortly $AB$. Let the points $A'$, $B'$ and the arc $A'B'$ be opposite to $A$, $B$, and $AB$, respectively. Then $A'B'$ is a bisector. Let $P$ be a common point of $AB$ and $A'B'$ (ex. 27) and $P'$ opposite to $P$. Then also $P'$ is a common point of $AB$ and $A'B'$. Let $A$, $P$, $P'$, and $B$ follow each other in this order on $AB$ and let $PB'$ be the shorter (not longer) of the two arcs $PA$ and $PB'$. (This choice is possible; it is, in fact, just a matter of notation.) Consider the curve consisting of two pieces: the arc $B'P$ (of $A'B'$) and the arc $PB$ (of $AB$). This curve is (1) shorter (not longer) than $AB$, and (2) longer (not shorter) than the diameter $BB'$ which is the straight path from $B$ to $B'$. It follows from (1) and (2) that $AB$ is longer (not shorter) than the diameter $BB'$, and this is the theorem.

**32.** The minor axis. See ex. 33.

**33.** *The shortest bisector of any region is either a straight line or an arc of a circle.* See ex. 16. *If the region has a center of symmetry* (as the square, the circle and the ellipse have, but not the equilateral triangle) *the shortest bisector is a straight line.* The proof is almost the same as for the circle (ex. 31).

**34.** Practically the same as ex. 27.

**35.** Ex. 16.

**36.** In all five cases, the plane of the shortest bisector passes through the center of the circumscribed sphere.

Tetrahedron: square in a plane parallel to two opposite edges; 3 solutions.

Cube: square parallel to one of the faces; 3 solutions.

Octahedron: hexagon in a plane parallel to one of the faces; 4 solutions.

Dodecahedron: decagon in a plane parallel to one of the faces; 6 solutions.

Icosahedron: decagon in a plane perpendicular to an axis that joins two opposite vertices; 6 solutions.

The proof is greatly facilitated in the last four cases by a general remark; see ex. 38.

**37.** Let $O$ be the center of the sphere. Define opposite points and curves as in ex. 31. Let $b$ be a bisector. Then also $b'$, the curve opposite to $b$, is a bisector and $b$ and $b'$ have a common point $P$ (ex. 34). Also $P'$, the point opposite to $P$, is a common point. The points $P$ and $P'$ divide $b$ into two arcs none of which can be shorter than the shortest line joining $P$ and $P'$ which is one half of a great circle.

**38.** Four of the five regular solids (all except the tetrahedron) have a center of symmetry. *A closed surface which has a center of symmetry has a bisector which is a geodesic.* The proof is almost the same as for the sphere (ex. 37).

**39.** (See *Elemente der Mathematik*, vol. 4 (1949), p. 93 and vol. 5 (1950), p. 65, problem 65.) Call $d$ the distance of the rim of the diaphragm from its vertex. The area of the diaphragm is $\pi d^2$; this proposition is due to Archimedes. Cf. ex. 11.4.

(1) If the center of $S$ is the vertex of the diaphragm, $d = a$, $\pi d^2 = \pi a^2$.

(2) Let $l$ be the straight line that joins the center $C$ of the sphere $S$ and the center $C'$ of the other sphere of which the diaphragm is a portion. Let $A$ be the intersection of $l$ with $S$ that lies on the same side of $C$ as $C'$, and $D$ and $B$ the intersections of $l$ with the diaphragm and with the plane that passes through the rim of the diaphragm, respectively. If the diaphragm bisects the volume of $S$, the points $A$, $B$, $C$, and $D$ follow each other *in this order* along $l$. The point of $l$ nearest to the rim of the diaphragm is $B$, and $D$ is farther from this rim than $C$. Therefore, $d > a$, $\pi d^2 > \pi a^2$.

(3) Conjecture: No surface bisecting the volume of the sphere with radius $a$ has an area less than $\pi a^2$. The proof may be difficult.

**41.** (1) Maximum $f = n^2$, attained when

$$x_1 = x_2 = \ldots = x_n = 1 \text{ or } -1.$$

Minimum $f = 0$, attained for infinitely many different systems $x_1, \ldots x_n$ when $n \geqq 3$.

(2) Maximum same as before and unique. Minimum $f = n$, attained when

$$x_1 = n^{1/2}, \qquad x_2 = \ldots = x_n = 0$$

and in $n - 1$ similar cases.

**42.** The conjecture is correct for the regular solids with three-edged vertices, but it is incorrect if a vertex has more than three edges. See M. Goldberg, *Tôhoku Mathematical Journal*, vol. 40 (1935) p. 226–236.

No solution: **40, 43.**

### SOLUTIONS, CHAPTER XI

**1.** (a) yes (b) no: $\alpha$ is unnecessary, the area is $ah/2$.

**2.** (a) yes (b) no: $\alpha$ and $\beta$ are unnecessary, the area is $mh$.

**3.** $2\pi rh$, *independent* of $d$. Solution by the method of Archimedes or by integral calculus: from $x^2 + y^2 = r^2$, follows

$$y^2 \left(\frac{dy}{dx}\right)^2 + y^2 = r^2, \qquad \int_d^{d+h} 2\pi y \left[1 + \left(\frac{dy}{dx}\right)^2\right]^{1/2} dx = 2\pi rh.$$

**4.** Let $h$ be the altitude of the zone of which the area is required. From similar right triangles $h : a = a : 2b$, and so the area required is $2\pi bh = \pi a^2$, *independent* of $b$. The zone becomes a full sphere when $b = a/2$ and a circle when $b = \infty$. Cf. ex. 10.39.

**5.** Have you observed the analogy with ex. 1–4? The volume of the perforated sphere can be obtained elementarily, or by the use of analytic geometry and integral calculus as

$$\int_{-h/2}^{h/2} \pi y^2 dx - \pi y_1^2 h = \pi h^3/6;$$

$x^2 + y^2 = r^2$ and $y_1$ is the ordinate corresponding to $x = h/2$.

**6.** $\pi h^3/12$, *independent* of $a$ and $b$. Solution similar to that of ex. 5, connected with that of ex. 7. In the extreme case $a = b = 0$ the segment becomes a full sphere with diameter $h$. If $h$ is small, the difference between $Mh$ and $V$ is intuitively seen to be very small.

**7.** $\pi c^2 h/6$, *independent* of $r$. If $c = h$, the cone degenerates into a cylinder and we have the case of sect. 2 and ex. 5. Solution similar to that of ex. 8.

**8.** With $O$ as origin and $OX$ as $x$-axis, the equations of the circle and the parabola in fig. 11.3 are

$$(x - d)^2 + y^2 = r^2, \qquad 2px = y^2,$$

respectively. Let $x_1$ and $x_2$ denote the abscissas of the points of intersection of the two curves, $x_1 < x_2$. Then $x_2 - x_1 = h$ and the volume required is

$$\pi \int_{x_1}^{x_2} [r^2 - (x - d)^2 - 2px]\, dx = \pi \int_{x_1}^{x_2} (x_2 - x)(x - x_1)\, dx = \pi h^3/6,$$

*independent* of $r$ and $d$; substitute $x - x_1 = t$. We used the decomposition in factors of a polynomial of the second degree when the two roots and the coefficient of $x^2$ are known.

**9.** (a) yes; the volume is $\pi h^2 (a + 2b)/3$. (b) no.

**10.** Yes. As $u_1$ and $u_2$ can be arbitrarily given, there are an infinity of possible systems $u_1, u_2, \ldots u_{10}$ satisfying the recursive relation $u_n = u_{n-1} + u_{n-2}$. We examine two special systems:

$u_1', u_2', u_3', \ldots u_{10}'$ with $u_1' = 0$, $u_2' = 1$

$u_1'', u_2'', u_3'', \ldots u_{10}''$ with $u_1'' = 1$, $u_2'' = 1$.

We find

$$u_7' = 8, \; u_1' + u_2' + \ldots + u_{10}' = 88,$$

$$u_7'' = 13, \; u_1'' + u_2'' + \ldots + u_{10}'' = 143.$$

With a little luck, we may observe that

$$(*) \qquad u_1' + u_2' + \ldots + u_{10}' = 11u_7', \quad u_1'' + u_2'' + \ldots + u_{10}'' = 11u_7''$$

and then guess, and finally prove, that

$$(**) \qquad\qquad u_1 + u_2 + \ldots + u_{10} = 11u_7.$$

The proof is so: we verify directly that

$$(***) \qquad\qquad u_n = (u_2 - u_1)u_n' + u_1 u_n''$$

holds for $n = 1$ and $n = 2$ and we conclude hence, using the recursive relation, that it holds also for $n = 3, 4, 5, \ldots 10$. Adding the two observed equations (*) after having multiplied the first by $u_2 - u_1$ and the second by $u_1$, we conclude from (***) the desired (**). Main idea of the proof: the general solution $u_n$ of our recursive relation (more aptly called a linear homogeneous difference equation of the second order) is a linear combination of two independent particular solutions $u_n'$ and $u_n''$ (as the general integral of a linear homogeneous differential equation of the second order is a linear combination of two particular integrals).

**II.**
$$\int_0^\infty \frac{1}{1 + x^\alpha} \frac{dx}{1 + x^2} = \int_0^\infty \frac{x^{-\alpha}}{x^{-\alpha} + 1} \frac{1}{x^{-1} + x} \frac{dx}{x}$$

$$= \int_0^\infty \frac{x^\alpha}{1 + x^\alpha} \frac{dx}{1 + x^2}$$

$$= \frac{1}{2} \int_0^\infty \frac{1 + x^\alpha}{1 + x^\alpha} \cdot \frac{dx}{1 + x^2} = \frac{\pi}{4}$$

*independent* of $\alpha$. In passing from the second form to the third we introduced $x^{-1}$ as new variable of integration. For $\alpha = 0, \infty, -\infty$ the given integral reduces to

$$\int_0^\infty \frac{1}{2} \frac{dx}{1 + x^2}, \qquad \int_0^1 \frac{dx}{1 + x^2}, \qquad \int_1^\infty \frac{dx}{1 + x^2},$$

respectively. These cases could suggest the above solution.

**12.** The most obvious fact of this kind is:

$$\int_{-\infty}^{\infty} f(u)\,du = 0 \quad \text{if} \quad f(-u) = -f(u).$$

Set $u = \log x, \quad f(\log x) = F(x)$:

$$\int_{0}^{\infty} F(x)x^{-1}dx = 0 \quad \text{if} \quad F(x^{-1}) = -F(x).$$

This suggests the following generalization:

$$\int_{0}^{\infty} g(x)[1 + h(x)]x^{-1}dx = \int_{0}^{\infty} g(x)x^{-1}dx \text{ if } g(x^{-1}) = g(x),\ h(x^{-1}) = -h(x).$$

Ex. 11 is the particular case:

$$g(x) = \frac{x}{2(1 + x^2)}, \qquad h(x) = \frac{1 - x^\alpha}{1 + x^\alpha}.$$

**13.** $0x = 0$, or $x^2 - 4 = (x - 2)(x + 2)$, etc.

**14.** $x = y = 8$; it is enough to try $x = 8, 9, 10, 11$.

**15.** $x = y = z = w = 4$, by trials.

**17.** [Cf. Stanford 1948]  The planes of symmetry of a regular solid pass through its center and divide a sphere with the same center into spherical triangles.  The three radii through the three vertices of this spherical triangle pass through a vertex, the center of a face and the midpoint of an edge, respectively.  The corresponding angles of the spherical triangle are $\pi/v$, $\pi/f$, and $\pi/2$.  Let us call $c$ the side (hypotenuse) of the spherical triangle opposite the angle $\pi/2$.  The ratio of the radius of the inscribed sphere to that of the circumscribed sphere is $\cos c$ and, by spherical trigonometry,

$$\cos c = \cot(\pi/f)\cot(\pi/v).$$

The numbers $f$ and $v$, and the resulting value of $\cos c$, are displayed in the following table for the Tetrahedron, Hexahedron (cube), Octahedron, Dodecahedron, and Icosahedron.

|  | T | H | O | D | I |
|---|---|---|---|---|---|
| $f =$ | 3 | 4 | 3 | 5 | 3 |
| $v =$ | 3 | 3 | 4 | 3 | 5 |
| $\cos c =$ | $\dfrac{1}{3}$ | $\dfrac{1}{\sqrt{3}}$ | | $\sqrt{\dfrac{5 + 2\sqrt{5}}{15}}$ | |

H. Weyl, *Symmetry*, Princeton, 1952, reproduces Kepler's original figure; see p. 76, fig. 46.

**18.** See ex. 10.42.

**21.** (a) We call a determinant with $n$ rows central-symmetric if its elements $a_{j,k}$ satisfy the condition

$$a_{j,k} = a_{n+1-j,n+1-k} \text{ for } j, k = 1, 2, 3, \ldots n.$$

*A central-symmetric determinant with n rows is the product of two determinants. Either both factors have n/2 rows, or one factor has $(n + 1)/2$ rows and the other $(n - 1)/2$ rows, according as n is even or odd. Examples:*

$$\begin{vmatrix} a & b \\ b & a \end{vmatrix} = (a + b)(a - b), \quad \begin{vmatrix} a & b & c \\ d & c & d \\ c & b & a \end{vmatrix} = \begin{vmatrix} a + c & b \\ 2d & c \end{vmatrix} (a - c),$$

$$\begin{vmatrix} a & b & c & d \\ e & f & g & h \\ h & g & f & e \\ d & c & b & a \end{vmatrix} = \begin{vmatrix} a + d & b + c \\ e + h & f + g \end{vmatrix} \begin{vmatrix} f - g & e - h \\ b - c & a - d \end{vmatrix}.$$

Proof: Put $n = 2m$ or $n = 2m + 1$, according as $n$ is even or odd. Add the last column to the first, then the column preceding the last to the second, and so on, till the first $m$ columns are changed. After that, subtract the first row from the last, then the second row from the row preceding the last, and so on, till the last $m$ rows are changed. These operations introduce either a rectangle $m \times (m + 1)$, or a square $m \times m$, consisting of vanishing elements in the south-west corner.

(b) The determinant with four rows could be divisible by both determinants with two rows, *without* being their product, namely, if these determinants with two rows had a common factor. Optimistically, we assumed that there is no such common factor: we tried the simplest assumption and succeeded.

**22.** Most optimistic: the coefficient of any power of $h$ on the left-hand side is less than, or equal to, the coefficient of the same power on the right-hand side. This is really the case: after division by $4h^{1/4}$, the constant term is 1 on both sides and, for $n \geq 1$, the coefficients of $h^n$ are

$$\frac{3}{4} \frac{7}{8} \frac{11}{12} \cdots \frac{4n-1}{4n} \frac{1}{4n+1}, \qquad \frac{1}{4} \frac{5}{8} \frac{9}{4} \cdots \frac{4n-3}{4n}$$

on the respective sides. Obviously

$$3 \cdot 7 \cdot 11 \cdots (4n - 1) < 5 \cdot 9 \cdots (4n - 3)(4n + 1).$$

**23.** (a) Call $P_n$ the proximate value obtained with the method in question when the square is subdivided into $n^2$ smaller squares. Assume that $P_n$ can be expanded in powers of $n^{-1}$:

$$P_n = Q_0 + Q_1 n^{-1} + Q_2 n^{-2} + \cdots .$$

("In general, a function can be expanded in a power series." Cf. ex. 20.) As $n \to \infty$, $P_n \to Q_0$, and we infer that $Q_0 = Q$. Now, the four points in fig. 11.6 are closer to a straight line than those in fig. 11.5. This circumstance *suggests* that $Q_1 = 0$ and the terms $n^{-3}$, $n^{-4}$, ... are negligible even for small $n$. This leads to

$$P_n \sim Q + Q_2 n^{-2}$$

which represents, if we take $n^{-2}$ as abscissa and $P_n$ as ordinate, a straight line (approximately). In some more or less similar cases it has been proved that the error of approximation is of the order $1/n^2$, and in the light of such analogy the guess appears less wild.

(b) The columns of the following table contain: (1) values of $n$, (2) ordinates, (3) differences of ordinates, (4) abscissas, (5) differences of abscissas, (6) slopes computed as the ratio of (3) to (5), except that in (5) and (6) the sign — is omitted.

| (1) | (2) | (3) | (4) | (5) | (6) |
|-----|-----|-----|-----|-----|-----|
| 2 | 0.0937 | | 0.2500 | | |
| | | 0.0248 | | 0.1389 | 0.1785 |
| 3 | 0.1185 | | 0.1111 | | |
| | | 0.0094 | | 0.0486 | 0.1934 |
| 4 | 0.1279 | | 0.0625 | | |
| | | 0.0045 | | 0.0225 | 0.2000 |
| 5 | 0.1324 | | 0.0400 | | |

(c) It is natural to regard $n = 5$ as the most reliable computation and $n = 4$ as the next best. If the points $(x_1, y_1)$ and $(x_2, y_2)$ lie on the straight line with equation $y = mx + b$, we easily find (from a system of two equations for $m$ and $b$) that

$$b = \frac{y_1/x_1 - y_2/x_2}{1/x_1 - 1/x_2},$$

which, in the present case yields

$$Q \sim \frac{25 \times 0.1324 - 16 \times 0.1279}{25 - 16} = 0.1404.$$

If you have expected anything better than that, you are too sanguine.

No solution: **16, 19, 20.**

# BIBLIOGRAPHY

## I. CLASSICS

EUCLID, *Elements.* The inexpensive shortened edition in Everyman's Library is sufficient here. "EUCLID III 7" refers to Proposition 7 of Book III of the Elements.

DESCARTES, *Oeuvres,* edited by Charles Adam and Paul Tannery. The work "Regulae ad Directionem Ingenii," vol. 10, pp. 359–469, is of especial interest.

EULER, *Opera Omnia,* edited by the "Societas scientiarum naturalium Helvetica."

LAPLACE, *Oeuvres complètes.* The "Introduction" of vol. 7, pp. V–CLIII, also separately printed (and better known) under the title "Essai philosophique sur les probabilités" is of especial interest.

## II. SOME BOOKS OF SIMILAR TENDENCY

R. COURANT and H. ROBBINS, *What is mathematics?*

H. RADEMACHER and O. TOEPLITZ, *Von Zahlen und Figuren.*

O. TOEPLITZ, *Die Entwicklung der Infinitesimalrechnung;* of especial interest.

## III. RELATED FORMER WORK OF THE AUTHOR

*Books*

1. *Aufgaben und Lehrsätze aus der Analysis,* 2 volumes, Berlin, 1925. Jointly with G. Szegö. Reprinted New York, 1945.

2. *How to Solve It,* Princeton, 1945. The 5th printing, 1948, is slightly enlarged.

3. Wahrscheinlichkeitsrechnung, Fehlerausgleichung, Statistik. From *Abderhalden's Handbuch der biologischen Arbeitsmethoden,* Abt. V, Teil 2, pp. 669–758.

*Papers*

1. Geometrische Darstellung einer Gedankenkette. *Schweizerische Pädagogische Zeitschrift,* 1919, 11 pp.

2. Wie sucht man die Lösung mathematischer Aufgaben? *Zeitschrift für mathematischen und naturwissenschaftlichen Unterricht,* v. 63, 1932, pp. 159–169.

3. Wie sucht man die Lösung mathematischer Aufgaben? *Acta Psychologica,* v. 4, 1938, pp. 113–170.

4. Heuristic reasoning and the theory of probability. *American Mathematical Monthly*, v. 48, 1941, pp. 450–465.

5. On Patterns of Plausible Inference. *Courant Anniversary Volume*, 1948, pp. 277–288.

6. Generalization, Specialization, Analogy. *American Mathematical Monthly*, v. 55, 1948, pp. 241–243.

7. Preliminary remarks on a logic of plausible inference. *Dialectica*, v. 3, 1949, pp. 28–35.

8. With, or without, motivation? *American Mathematical Monthly*, v. 56, 1949, pp. 684–691.

9. Let us teach guessing. *Etudes de Philosophie des Sciences, en hommage à Ferdinand Gonseth*, 1950, pp. 147–154. Editions du Griffon, Neuchatel, Switzerland.

10. On plausible reasoning. *Proceedings of the International Congress of Mathematicians*, 1950, v. 1, pp. 739–747.

## IV. PROBLEMS

Among the examples proposed for solution there are some taken from the *William Lowell Putnam Mathematical Competition* or the *Stanford University Competitive Examination in Mathematics.* This fact is indicated at the beginning of the solution with the year in which the problem was proposed as "Putnam 1948" or "Stanford 1946." The problems of the Putnam Examination are published yearly in the *American Mathematical Monthly* and most Stanford examinations have been published there too.

# PATTERNS OF
# PLAUSIBLE INFERENCE

VOLUME II
OF MATHEMATICS
AND PLAUSIBLE
REASONING

By G. POLYA

PRINCETON UNIVERSITY PRESS
PRINCETON, NEW JERSEY
1954

*Published,* 1954, *by Princeton University Press*
*London : Geoffrey Cumberlege, Oxford University Press*
*L.C. Card 53–6388*

COMPOSED BY THE PITMAN PRESS, BATH, ENGLAND
PRINTED IN THE UNITED STATES OF AMERICA

# PREFACE

Inductive reasoning is one of the many battlefields for conflicting philosophical opinions, and one that is still relatively lively today. The reader who went through Vol. I of this work has had a good opportunity to notice two things. First, inductive and analogical reasoning play a major rôle in mathematical discovery. Second, both inductive and analogical reasoning are particular cases of plausible reasoning. It seems to me more philosophical to consider the general idea of plausible reasoning instead of its isolated particular cases. The present Vol. II attempts to formulate certain patterns of plausible reasoning, to investigate their relation to the Calculus of Probability, and to examine in what sense they can be regarded as "rules" of plausible reasoning. Their relation to mathematical invention and instruction will also be briefly discussed.

The text of the present Vol. II does not often refer explicitly to Vol. I, and the reader can understand the main connections without looking up these references. Among the problems appended to the various chapters there are some that the reader cannot solve without referring to Vol. I, but on the whole one can read Vol. II in first approximation without having read Vol. I. Yet, of course, it is more natural to read Vol. II after Vol. I, the examples of which provide the investigation that lies ahead of us with experimental data and a richer background.

Such data and background are particularly desirable in view of the method that will be followed. I wish to investigate plausible reasoning in the manner of the naturalist: I collect observations, state conclusions, and emphasize the points in which my observations seem to support my conclusions. Yet I respect the judgment of the reader and I do not want to force or trick him into adopting my conclusions.

Of course, the views presented here have no pretension to be final. In fact, there are a few places where I feel clearly the need of some improvement, minor or major. Yet I believe that the main direction is right, and that the discussions, and especially the examples, of this work may elucidate the "double nature" and the "complementary aspects" of plausible and especially inductive reasoning, which appears sometimes as "objective" and sometimes as "subjective."

GEORGE POLYA

*Stanford University*
*May* 1953

# HINTS TO THE READER

THE section 2 of chapter VII is quoted as sect. 2 in chapter VII, but as sect. 7.2 in any other chapter. The subsection (3) of section 5 of chapter XIV is quoted as sect. 5 (3) in chapter XIV, but as sect. 14.5 (3) in any other chapter. We refer to example 26 of chapter XIV as ex. 26 in the same chapter, but as ex. 14.26 in any other chapter.

Some knowledge of elementary algebra and geometry may be enough to read substantial parts of the text. Thorough knowledge of elementary algebra and geometry and some knowledge of analytic geometry and calculus, including limits and infinite series, is sufficient for almost the whole text and the majority of the examples and comments. Yet more advanced knowledge is supposed in a few incidental remarks of the text, in some proposed problems, and in several comments. Usually some warning is given when more advanced knowledge is assumed.

The advanced reader who skips parts that appear to him too elementary may miss more than the less advanced reader who skips parts that appear to him too complex.

Some details of (not very difficult) demonstrations are often omitted without warning. Duly prepared for this eventuality, a reader with good critical habits need not spoil them.

Some of the problems proposed for solution are very easy, but a few are pretty hard. Hints that may facilitate the solution are enclosed in square brackets [ ]. The surrounding problems may provide hints. Especial attention should be paid to the introductory lines prefixed to the examples in some chapters, or prefixed to the First Part, or Second Part, of such examples.

The solutions are sometimes very short: they suppose that the reader has earnestly tried to solve the problem by his own means before looking at the printed solution.

A reader who spent serious effort on a problem may profit by it even if he does not succeed in solving it. For example, he may look at the solution, try to isolate what appears to him the key idea, put the book aside, and then try to work out the solution.

At some places, this book is lavish of figures or in giving small intermediate steps of a derivation. The aim is to render visible the *evolution* of a figure or a formula; see, for instance, fig. 16.1–16.5. Yet no book can have enough figures or formulas. A reader may want to read a passage "in first approximation" or more thoroughly. If he wants to read more thoroughly, he should have paper and pencil at hand: he should be prepared to write or draw any formula or figure given in, or only indicated by, the text. Doing so, he has a better chance to see the evolution of the figure or formula, to understand how the various details contribute to the final product, and to remember the whole thing.

# CONTENTS

# Volume  II

# Patterns of Plausible Inference

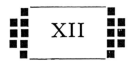

# XII

## SOME CONSPICUOUS PATTERNS

*I do not wish, at this stage, to examine the logical justification of this form of argumentation; for the present, I am considering it as a* practice, *which we can observe in the habits of men and animals.*—BERTRAND RUSSELL[1]

**1. Verification of a consequence.** In the first volume of this work on *Induction and Analogy in Mathematics* we found some opportunity to familiarize ourselves with the practice of plausible reasoning. In the present second volume we undertake to describe this practice in general terms. The examples of the first part have already indicated certain forms or *patterns* of plausible reasoning. In the present chapter we undertake to formulate some such patterns explicitly.[2]

We begin with a pattern of plausible inference which is of so general use that we could extract it from almost any example. Yet let us take an example which we have not yet discussed before.

The following conjecture is due to Euler:[3] *Any integer of the form $8n + 3$ is the sum of a square and of the double of a prime.* Euler could not prove this conjecture, and the difficulty of a proof appears perhaps even greater today than in Euler's time. Yet Euler verified his statement for all integers of the form $8n + 3$ under 200; for $n = 1, 2, \ldots 10$ see Table I.

Table I

$$11 = 1 + 2 \times 5$$
$$19 = 9 + 2 \times 5$$
$$27 = 1 + 2 \times 13$$
$$35 = 1 + 2 \times 17 = 9 + 2 \times 13 = 25 + 2 \times 5$$
$$43 = 9 + 2 \times 17$$
$$51 = 25 + 2 \times 13$$
$$59 = 1 + 2 \times 29 = 25 + 2 \times 17 = 49 + 2 \times 5$$
$$67 = 9 + 2 \times 29$$
$$75 = 1 + 2 \times 37 = 49 + 2 \times 13$$
$$83 = 1 + 2 \times 41 = 9 + 2 \times 37 = 25 + 2 \times 29 = 49 + 2 \times 17$$

---

[1] *Philosophy*, W. W. Norton & Co., 1927, p. 80.

[2] Parts of this chapter were used in my address "On plausible reasoning" printed in the *Proceedings of the International Congress of Mathematicians* 1950, vol. 1, p. 739–747.

[3] *Opera Omnia*, ser. 1, vol. 4, p. 120–124. In this context, Euler regards 1 as a prime; this is needed to account for the case $3 = 1 + 2 \times 1$.

Such empirical work can be easily carried further; no exception has been found in numbers under 1000.[4]  Does this prove Euler's conjecture?  By no means; even verification up to 1,000,000 would prove nothing.  Yet each verification renders the conjecture somewhat more credible, and we can see herein a general pattern.

Let $A$ denote some clearly formulated conjecture which is, at present, neither proved, nor disproved.  (For instance, $A$ may be Euler's conjecture that, for $n = 1, 2, 3, \ldots$ ,

$$8n + 3 = x^2 + 2p$$ .

where $x$ is an integer and $p$ a prime.)  Let $B$ denote some consequence of $A$; also $B$ should be clearly stated and neither proved, nor disproved.  (For instance, $B$ may be the first particular case of Euler's conjecture not listed in Table I which asserts that $91 = x^2 + 2p$.)  For the moment we do not know whether $A$ or $B$ is true.  We do know, however, that

$$A \text{ implies } B.$$

Now, we undertake to check $B$.  (A few trials suffice to find out whether the assertion about 91 is true or not.)  If it turned out that $B$ is false, we could conclude that $A$ also is false.  This is completely clear.  We have here a classical elementary pattern of reasoning, the "modus tollens" of the so-called hypothetical syllogism:

$$\frac{\begin{array}{c} A \text{ implies } B \\ B \text{ false} \end{array}}{A \text{ false}}$$

The horizontal line separating the two premises from the conclusion stands as usual for the word "therefore."  We have here *demonstrative inference* of a well-known type.

What happens if $B$ turns out to be true?  (Actually, $91 = 9 + 2 \times 41 = 81 + 2 \times 5$.)  There is no demonstrative conclusion: the verification of its consequence $B$ does not prove the conjecture $A$.  Yet such verification renders $A$ more credible.  (Euler's conjecture, verified in one more case, becomes somewhat more credible.)  We have here a pattern of *plausible inference*:

$$\frac{\begin{array}{c} A \text{ implies } B \\ B \text{ true} \end{array}}{A \text{ more credible}}$$

The horizontal line again stands for "therefore."  We shall call this pattern the *fundamental inductive pattern* or, somewhat shorter, the "inductive pattern."

---

[4] Communication of Professor D. H. Lehmer.

This inductive pattern says nothing surprising. On the contrary, it expresses a belief which no reasonable person seems to doubt: *The verification of a consequence renders a conjecture more credible.* With a little attention, we can observe countless reasonings in everyday life, in the law courts, in science, etc., which appear to conform to our pattern.

**2. Successive verification of several consequences.** In the present section, I use the phrase "discussion of a theorem" in the specific meaning "discussion, or survey, of some particular cases and some more immediate consequences of the theorem." I think that the discussion of the theorems presented is useful both in advanced and in elementary classes. Let us consider a very elementary example. Let us assume that you teach a class in solid geometry and that you have to derive the formula for the area of the lateral surface of the frustum of a cone. Of course, the cone is a right circular cone, and you are given the radius of the base $R$, the radius of the top $r$, and the altitude $h$. You go through the usual derivation and you arrive at the result:

*A. The area of the lateral surface of the frustum is*

$$\pi(R + r)\sqrt{(R - r)^2 + h^2}.$$

We call this theorem $A$ for future reference.

Now comes the discussion of the theorem $A$. You ask the class: *Can you check the result?* If there is no response, you give more explicit hints: Can you check the result by *applying* it? Can you check it by applying it to some *particular case you already know?* Eventually, with more or less collaboration from the part of your class, you get down to various known cases. If $R = r$, you obtain a first noteworthy particular case:

$B_1$. *The area of the lateral surface of a cylinder is $2\pi rh$.*

Of course, $h$ stands for the altitude of the cylinder and $r$ for the radius of its base. We call $B_1$ this consequence of $A$ for future reference. The consequence $B_1$ has been treated already in your class and so it serves as a confirmation of $A$.

You obtain another particular case of $A$ in setting $r = 0$ which yields:

$B_2$. *The area of the lateral surface of a cone is $\pi R\sqrt{R^2 + h^2}$.*

Here $h$ denotes the altitude of the cone and $R$ the radius of its base. Also this consequence $B_2$ of $A$ was known before and serves as a further confirmation of $A$.

There is a less obvious but interesting particular case corresponding to $h = 0$:

$B_3$. *The area of the annulus between two concentric circles with radii $R$ and $r$ is $\pi R^2 - \pi r^2$.*

This consequence $B_3$ of $A$ is known from plane geometry and yields still another confirmation of $A$.

The foregoing three particular cases, all known from previous study, support $A$ from three different sides; the three figures (cylinder, cone, and annulus, corresponding to $r = R$, $r = 0$, and $h = 0$, respectively) look quite different. You may mention also the very particular case $r = h = 0$.

$B_4$. *The area of a circle with radius R is $\pi R^2$.*

I have sometimes observed that a boy in the last row who seemed to sleep soundly toward the end of my careful derivation opened his eyes and showed some interest in the progress of the discussion. The derivation of the formula, apparently plain and easy, seemed abstruse and difficult to him. He was not convinced by the derivation. He is more convinced by the discussion: a formula that checks in so many and so different cases has a good chance to be correct, he thinks. And in thinking so he conforms to a pattern of plausible reasoning which is closely related to, but more sophisticated than, the fundamental inductive pattern:

$$A \text{ implies } B_{n+1}$$

$B_{n+1}$ is very different from the formerly verified consequences $B_1, B_2, \ldots B_n$ of $A$

$$B_{n+1} \text{ true}$$

$A$ much more credible

This pattern adds a qualification to the fundamental inductive pattern. Certainly the verification of any consequence strengthens our belief in a conjecture. Yet the verification of certain consequences strengthens our belief more and that of others strengthens it less. The pattern just given brings to our attention a circumstance which has a great influence on the strength of inductive evidence: the variety of the consequences tested. The verification of a new consequence counts more if the new consequence differs more from the formerly verified consequences.

Now let us look at the reverse of the medal. Take the example of the foregoing section 1. The successive cases in Table I in which Euler's conjecture is verified look very similar to each other—unless we notice some hidden clue, and it seems very difficult to notice such a clue. Therefore, sooner or later, we get tired of this monotonous sequence of verifications. Having verified a certain number of cases, we hesitate. Is it worth while to tackle the next case? The next case, if the result is negative, could explode the conjecture—but the next case is so similar in all known aspects to the cases already verified that we scarcely expect a negative result. The next case, if the result is positive, would increase our confidence in Euler's conjecture, but this increase in confidence would be so small that it is scarcely worth the trouble of testing that next case.

This consideration suggests the following pattern which is not essentially different from the pattern that we have just stated, but rather a complementary form of it:

$A$ implies $B_{n+1}$

$B_{n+1}$ is very similar to the formerly verified

consequences $B_1$, $B_2$, ... $B_n$ of $A$

$B_{n+1}$ true

_____

$A$ just a little more credible

*The verification of a new consequence counts more or less according as the new consequence differs more or less from the formerly verified consequences.*

**3. Verification of an improbable consequence.** In a little known short note[5] Euler considers, for positive values of the parameter $n$, the series

$$(1) \quad 1 - \frac{x^2}{n(n+1)} + \frac{x^4}{n(n+1)(n+2)(n+3)} - \frac{x^6}{n\ldots(n+5)} + \cdots$$

which converges for all values of $x$. He observes the sum of the series and its zeros for $n = 1, 2, 3, 4$.

$n = 1$: sum $\cos x$,      zeros $\pm \pi/2$, $\pm 3\pi/2$, $\pm 5\pi/2$, ...

$n = 2$: sum $(\sin x)/x$,      zeros $\pm \pi$, $\pm 2\pi$, $\pm 3\pi$, ...

$n = 3$: sum $2(1 - \cos x)/x^2$,      zeros $\pm 2\pi$, $\pm 4\pi$, $\pm 6\pi$, ...

$n = 4$: sum $6(x - \sin x)/x^3$,      no real zeros.

Euler observes a difference: in the first three cases all the zeros are real, in the last case none of the zeros is real. Euler notices a more subtle difference between the first two cases and the third case: for $n = 1$ and $n = 2$, the distance between two consecutive zeros is $\pi$ (provided that we disregard the zeros next to the origin in the case $n = 2$) but for $n = 3$ the distance between consecutive zeros is $2\pi$ (with a similar proviso). This leads him to a striking observation: in the case $n = 3$ all the zeros are double zeros. "Yet we know from Analysis," says Euler, "that two roots of an equation always coincide in the transition from real to imaginary roots. Thus we may understand why all the zeros suddenly become imaginary when we take for $n$ a value exceeding 3." On the basis of these observations he states a surprising conjecture: the function defined by the *series* (1) *has only real zeros, and an infinity of them, when* $0 < n \leq 3$, *but has no real zero at all when* $n > 3$. In this statement he regards $n$ as a continuously varying parameter.

In Euler's time questions about the reality of the zeros of transcendental equations were absolutely new, and we must confess that even today we possess no systematic method to decide such questions. (For instance, we

[5] *Opera Omnia*, ser. 1, vol. 16, sect. 1, p. 241–265.

cannot prove or disprove Riemann's famous hypothesis.)    Therefore, Euler's conjecture appears extremely bold.    I think that the courage and clearness with which he states his conjecture are admirable.

Yet Euler's admirable performance is understandable to a certain extent. Other experts perform similar feats in dealing with other subjects, and each of us performs something similar in everyday life.    In fact, Euler *guessed the whole from a few scattered details*.    Quite similarly, an archaeologist may reconstitute with reasonable certainty a whole inscription from a few scattered letters on a worn-out stone.    A paleontologist may describe reliably the whole animal after having examined a few of its petrified bones. When a person whom you know very well starts talking in a certain way, you may predict after a few words the whole story he is going to tell you. Quite similarly, Euler guessed the whole story, the whole mathematical situation, from a few clearly recognized points.

It is still remarkable that he guessed it from so few points, by considering just four cases, $n = 1, 2, 3, 4$.    We should not forget, however, that circumstantial evidence may be very strong.    A defendant is accused of having blown up the yacht of his girl friend's father, and the prosecution produces a receipt signed by the defendant acknowledging the purchase of such and such an amount of dynamite.    Such evidence strengthens the prosecution's case immensely.    Why?    Because the purchase of dynamite by an ordinary citizen is a very unusual event in itself, but such a purchase is completely understandable if the purchaser intends to blow up something or somebody. Please observe that this court case is very similar to the case $n = 3$ of Euler's series.    That all roots of an equation written at random turn out to be double roots is a very unusual event in itself.    Yet it is completely understandable that in the transition from two real roots to two imaginary roots a double root appears.    The case $n = 3$ is the strongest piece of circumstantial evidence produced by Euler and we can perceive herein a general pattern of plausible inference:

$$A \text{ implies } B$$
$$B \text{ very improbable in itself}$$
$$B \text{ true}$$

$$\overline{A \text{ very much more credible}}$$

Also this pattern appears as a modification or a sophistication of the fundamental inductive pattern (sect. 1).    Let us add, without specific illustration for the moment, the complementary pattern which explains the same idea from the reverse side:

$$A \text{ implies } B$$
$$B \text{ quite probable in itself}$$
$$B \text{ true}$$

$$\overline{A \text{ just a little more credible}}$$

*The verification of a consequence counts more or less according as the consequence is more or less improbable in itself.* The verification of the most surprising consequences is the most convincing.

By the way, Euler was right: 150 years later, his conjecture has been completely proved.[6]

**4. Inference from analogy.** At this stage it may be instructive to look back at the examples of the first volume on *Induction and Analogy*. We have formulated a few patterns of plausible inference in the foregoing sections of this chapter: how do those examples appear in the light of these patterns?

Let us reconsider two related examples (from sect. 10.1 and 10.4 of Volume I, respectively). One of these examples is connected with the isoperimetric theorem and Descartes, the other with a physical analogue of the isoperimetric theorem and Lord Rayleigh. We reproduce two tables from ch. X (called there Table I and Table II, here Table II and Table III, respectively) putting them side by side. Table II (as it is numbered in the present chapter) lists the perimeters of ten figures, each of which has the same area 1, and Table III lists the principal frequencies of the same ten figures (considered as vibrating membranes).

| Table II Perimeters of Figures of Equal Area | | Table III Principal Frequencies of Membranes of Equal Area | |
|---|---|---|---|
| Circle | 3.55 | Circle | 4.261 |
| Square | 4.00 | Square | 4.443 |
| Quadrant | 4.03 | Quadrant | 4.551 |
| Rectangle 3 : 2 | 4.08 | Sextant | 4.616 |
| Semicircle | 4.10 | Rectangle 3 : 2 | 4.624 |
| Sextant | 4.21 | Equilateral triangle | 4.774 |
| Rectangle 2 : 1 | 4.24 | Semicircle | 4.803 |
| Equilateral triangle | 4.56 | Rectangle 2 : 1 | 4.967 |
| Rectangle 3 : 1 | 4.64 | Isosceles right triangle | 4.967 |
| Isosceles right triangle | 4.84 | Rectangle 3 : 1 | 5.736 |

The perimeters in one table, the principal frequencies in the other, are increasingly ordered. Both tables start with the circle which has the shortest perimeter among the ten figures listed and also the lowest principal frequency, and this suggests two theorems:

*Of all plane figures with a given area the circle has the shortest perimeter.*

*Of all membranes with a given area the circle has the lowest principal frequency.*

---

[6] See the author's paper: Sopra una equazione transcendente trattata da Eulero, *Bolletino dell' Unione Matematica Italiana*, vol. 5, 1926, p. 64–68.

The first statement is the isoperimetric theorem, the second a celebrated conjecture of Lord Rayleigh. Our tables yield sound inductive evidence for both statements but, of course, no proof.

The situation has changed since we considered these tables in sect. 10.1 and 10.4. In the meantime we have seen a proof for the isoperimetric theorem (sect. 10.6–10.8, ex. 10.1–10.15). The geometrical minimum property of the circle, inductively supported by Table II, has been proved. It is natural to expect that the analogous physical minimum property of the circle, inductively supported by Table III, will also turn out to be true. In expecting this we follow an important pattern of plausible inference:

$$A \text{ analogous to } B$$
$$B \text{ true}$$
$$\overline{\phantom{AAAAAAAAAAAAAAAAA}}$$
$$A \text{ more credible}$$

*A conjecture becomes more credible when an analogous conjecture turns out to be true.* The application of this pattern to the situation discussed seems sensible. Yet there are further promising indications in this situation.

**5. Deepening the analogy.** The Tables II and III, side by side, seem to offer further suggestions. The ten figures considered do not appear in exactly the same sequence in both tables. There is something peculiar about this sequence. The arrangement in Table II appears not very different from that in Table III, but this is not the main point. The tables contain various kinds of figures: rectangles, triangles, sectors. How are the *figures of the same kind* arranged? How would a shorter table look listing only figures of one kind? The tables contain a few regular figures: the equilateral triangle, the square, and, let us not forget it, the circle. How are the regular figures arranged? Could we compare somehow figures of different kinds, for instance, triangles and sectors? Could we broaden the inductive basis by adding further figures to our tables? (In this we are much restricted. It is not difficult to compute areas and perimeters, but the principal frequency is difficult to handle and its explicit expression is known in very few cases only.) Eventually we obtain Table IV.

Table IV exhibits a remarkable parallelism between these two quantities depending on the shape of a variable plane figure: the perimeter and the principal frequency. (We should not forget that the area of the variable figure is fixed, $= 1$.) If we know the perimeter, we are by no means able to compute the principal frequency or vice versa. Yet, judging from Table IV, we should think that, in many simple cases, these two quantities *vary in the same direction.* Consider the two columns of numerical data in this table and pass from any row to the next row: if there is an increase in one of the columns, there is a corresponding increase in the other, and if there is a decrease in one of the columns, there is a corresponding decrease in the other.

Table IV

Perimeters and principal frequencies of figures of equal area

| Figure | Perimeter | Pr. frequency |
|---|---|---|
| Rectangles: | | |
| 1 : 1 (square) | 4.00 | 4.443 |
| 3 : 2 | 4.08 | 4.624 |
| 2 : 1 | 4.24 | 4.967 |
| 3 : 1 | 4.64 | 5.736 |
| Triangles: | | |
| 60° 60° 60° | 4.56 | 4.774 |
| 45° 45° 90° | 4.84 | 4.967 |
| 30° 60° 90° | 5.08 | 5.157 |
| Sectors: | | |
| 180° (semicircle) | 4.10 | 4.803 |
| 90° (quadrant) | 4.03 | 4.551 |
| 60° (sextant) | 4.21 | 4.616 |
| 45° | 4.44 | 4.755 |
| 36° | 4.68 | 4.916 |
| 30° | 4.93 | 5.084 |
| Regular figures: | | |
| circle | 3.55 | 4.261 |
| square | 4.00 | 4.443 |
| equilateral triangle | 4.56 | 4.774 |
| Triangles versus sectors: | | |
| tr. 60° 60° 60° | 4.56 | 4.774 |
| sector 60° | 4.21 | 4.616 |
| tr. 45° 45° 90° | 4.84 | 4.967 |
| sector 45° | 4.44 | 4.755 |
| tr. 30° 60° 90° | 5.08 | 5.157 |
| sector 30° | 4.93 | 5.084 |

Let us focus our attention on the rectangles. If the ratio of the length to the width increases from 1 to ∞, so that the shape varies from a square to an infinitely elongated rectangle, both the perimeter and the principal frequency seem to increase steadily. The square which, being a regular figure, is "nearest" to the circle among all quadrilaterals, has the minimum perimeter and also the minimum principal frequency. Of the three triangles listed, the equilateral triangle which, being a regular figure, is "nearest" to the circle among all triangles has the minimum perimeter and also the minimum principal frequency. The behavior of the sectors is more complex. As the angle of the sector varies from 180° to 0°, the perimeter first decreases, attains a minimum, and then increases; and the principal frequency varies in the same manner. Let us now look at the regular figures. The equilateral triangle has 3 axes of symmetry, the square has 4 such axes, and the circle an infinity.

As far as we can see from Table IV, both the perimeter and the principal frequency seem to decrease as the number of the axes of symmetry increases. In the last section of Table IV we matched each triangle against the sector whose angle is equal to the least angle of the triangle.　In all three cases, the sector turned out to be "more circular," having the shorter perimeter and the lower principal frequency.

What we definitely know about these regularities goes, of course, only as far as Table IV goes.　That these regularities hold beyond the limits of the experimental material collected is suggested and rendered plausible by Table IV, but is by no means proved.　And so Table IV led us to several new conjectures which are similar to Rayleigh's conjecture although, of course, of much more limited scope.

How does Table IV influence our confidence in Rayleigh's conjecture? Can we find in Table IV any reasonable ground for it that we did not notice before in discussing the Tables II and III?

We certainly can.　First of all, the Table IV contains a few more particular cases in which Rayleigh's conjecture is verified (the 30° 60° 90° triangle, the sectors with opening 45°, 36°, and 30°).　Yet there is more than that.　The analogy between the isoperimetric theorem and Rayleigh's conjecture has been considerably deepened; the facts listed in Table IV add several new aspects to this analogy.　Now it seems to be reasonable to consider a conclusion from analogy as becoming stronger if the analogy itself, on which the conclusion is based, becomes stronger.　And so Table IV considerably strengthens Rayleigh's case.

**6. Shaded analogical inference.**　Yet there is still something more. As we have observed, Table IV suggests several conjectures which are analogous to (but of more limited scope than) Rayleigh's conjecture.　Table IV suggests these conjectures and lends them some plausibility too.　Yet this circumstance quite reasonably raises somewhat the plausibility of Rayleigh's original conjecture.　If you think so too, you think according to the following pattern:

$A$ analogous to $B$

$B$ more credible

_____

$A$ somewhat more credible

*A conjecture becomes somewhat more credible when an analogous conjecture becomes more credible.*　This is a weakened or *shaded* form of the pattern formulated in sect. 4.

### EXAMPLES AND COMMENTS ON CHAPTER XII

**1.** Table I, exhibiting some inductive evidence for Euler's conjecture mentioned in sect. 1, is very similar to the table in sect. 1.3, or to Tables I,

II, and III in ch. IV, or to Euler's table given in support of his "Most Extraordinary Law of the Numbers Concerning the Sum of Their Divisors," see sect. 6.2. These tables resemble also two tables given in ch. III, one in sect. 3.1 (polyhedra), the other in sect. 3.12 (partitions of space). To which one of these two is the resemblance closer?

2. Euler, having verified his "Most Extraordinary Law" (cf. sect. 6.2) for $n = 1, 2, 3, 4, \ldots 20$, proceeds to verify it for $n = 101$ and $n = 301$. He had a good reason to examine 101 and 301 rather than 21 and 22 (which he clearly states at the outset of No. 7 of his memoir). Ignoring, or remembering only vaguely, the contents of Euler's Law, would you think that the verification of his two cases (101 and 301) has more probative value than would have the verification of the two next cases (21 and 22)?

3. Of a triangle, let $a$, $b$, and $c$ denote the sides, $2p = a + b + c$ the perimeter, $A$ the area.

Check Heron's formula

$$A^2 = p(p - a)\,(p - b)\,(p - c)$$

in as many ways as you can.

4. We consider a quadrilateral inscribed in a circle. Let $a$, $b$, $c$, and $d$ denote the sides, $2p = a + b + c + d$ the perimeter, $A$ the area.

It is asserted that

$$A^2 = (p - a)\,(p - b)\,(p - c)\,(p - d).$$

Check this assertion in as many ways as you can. Have you any comment?

5. Let $V$ denote the volume of a tetrahedron and

$$a\,,b\,,c\,,$$

$$e\,,f\,,g$$

the lengths of its six edges; the edges $a$, $b$, and $c$ end in the same vertex of the tetrahedron, $e$ is the edge opposite to $a$, $f$ to $b$, and $g$ to $c$. (Two edges of a tetrahedron are called opposite to each other if they have no vertex in common.) The edges $e$, $f$, and $g$ are the three sides of a face of the tetrahedron, opposite to the vertex in which $a$, $b$, and $c$ end. It is asserted that

$$144V^2 = 4a^2b^2c^2 + (b^2 + c^2 - e^2)\,(c^2 + a^2 - f^2)\,(a^2 + b^2 - g^2)$$

$$- a^2(b^2 + c^2 - e^2)^2 - b^2(c^2 + a^2 - f^2)^2 - c^2(a^2 + b^2 - g^2)^2.$$

Check this assertion in as many ways as you can. [Is the expression proposed for $V$ symmetric in the six edges?]

**6.** Set

$$a^n + b^n + c^n = s_n$$

for $n = 1, 2, 3, \ldots$ and define $p$, $q$, and $r$ by the identity in $x$

$$(x - a)(x - b)(x - c) = x^3 - px^2 + qx - r$$

so that

$$p = a + b + c,$$

$$q = ab + ac + bc,$$

$$r = abc.$$

(In the usual terminology, $p$, $q$, and $r$ are the "elementary symmetric functions" of $a$, $b$, and $c$, and $s_n$ a "sum of like powers.")  Obviously, $p = s_1$.  It is asserted that, for arbitrary values of $a$, $b$, and $c$,

$$q = \frac{2s_1^5 - 5s_1^2 s_3 + 3s_5}{5(s_1^3 - s_3)},$$

$$r = \frac{s_1^6 - 5s_1^3 s_3 - 5s_3^2 + 9s_1 s_5}{15(s_1^3 - s_3)}$$

provided that the denominator does not vanish.  Check these formulas in the particular case $a = 1$, $b = 2$, $c = 3$ and in three more cases displayed in the table:

| Case | $a$ | $b$ | $c$ |
|------|-----|-----|-----|
| (1)  | 1   | 2   | 3   |
| (2)  | 1   | 2   | −3  |
| (3)  | 1   | 2   | 0   |
| (4)  | 1   | 2   | −2  |

Devise further checks.  Especially, try to generalize the cases (2), (3), and (4).

**7.** Let $A$, $B_1$, $B_2$, $B_3$, and $B_4$ have the meaning given them in sect. 2.  Does the verification of $B_4$, coming after that of $B_1$, $B_2$, and $B_3$, supply additional inductive evidence for $A$?

**8.** Let us recall Euler's "Most Extraordinary Law" and the meaning of the abbreviations $T$, $C_1$, $C_2$, $C_3$, $\ldots$ $C_1^*$, $C_2^*$, $C_3^*$, $\ldots$ explained in sect. 6.3. Euler supported the theorem $T$, when he was not yet able to prove it, inductively, by verifying its consequences $C_1$, $C_2$, $C_3$, $\ldots C_{20}$.  (He went even further, perhaps.)  Then he discovered that also $C_1^*$, $C_2^*$, $C_3^*$, $\ldots$ are consequences of $T$, and verified $C_1^*$, $C_2^*$, $\ldots C_{20}^*$, $C_{101}^*$, $C_{301}^*$.  Thanks to these new verifications Euler's confidence was, presumably, much strengthened: but was it justifiably strengthened?  [Closer attention to detail is needed here than in ex. 2.]

**9.** We return to Euler's conjecture discussed in sect. 1; for the sake of brevity, we call it the "conjecture $E$." Let us note concisely the meaning of this abbreviation,

$$E: \qquad 8n + 3 = x^2 + 2p.$$

The idea that led Euler to his conjecture $E$ deserves mention. Euler devoted much of his work to those celebrated propositions of Number Theory that Fermat has stated without proof. One of these (we call it the "conjecture $F$") says that any integer is the sum of three trigonal numbers. Let us note concisely the meaning of this abbreviation,

$$F: \qquad n = \frac{x(x-1)}{2} + \frac{y(y-1)}{2} + \frac{z(z-1)}{2}.$$

Euler observed that *if* his conjecture $E$ were true, Fermat's conjecture $F$ would easily follow. That is, Euler satisfied himself that $E$ implies $F$. (For details, see the next ex. 10.) Bent on proving Fermat's conjecture $F$, Euler naturally desired that his conjecture $E$ should be true. Is this mere wishful thinking? I do not think so; the relations considered yield some weak but not unreasonable ground for believing Euler's conjecture $E$ according to the following scheme:

$$E \text{ implies } F$$
$$\underline{F \text{ credible} \qquad\qquad}$$
$$E \text{ somewhat credible}$$

Here is another pattern of plausible inference. The reader should compare it with the fundamental inductive pattern.

**10.** In proving that $E$ implies $F$ (in the notation of the foregoing ex. 9), Euler used a deeper result which he proved previously: a prime number of the form $4n + 1$ is a sum of two squares. (This was discussed inductively in ex. 4.4.) Taking this for granted, prove that actually $E$ implies $F$.

**11.** After having conceived his conjecture discussed in sect. 3, Euler tested it by computing the first zeros of his series for a few values of $n$. (By a "first zero" we mean a zero the absolute value of which is a minimum. If $x$ is a first zero of the series in question, also $- x$ is a zero, and $x$ and $- x$ are "first zeros." Therefore, $x$ is real if, and only if, $x^2$ is positive.) Of course, Euler had to compute these zeros approximately. A method (Daniel Bernoulli's method) which he frequently used for such a purpose yielded the following sequences of approximate values for the first zero $x$ in the cases $n = 1/2, 1/3, 1/4$.

| $n = 1/2$ | $n = 1/3$ | $n = 1/4$ |
|---|---|---|
| $4x^2 \sim 3.000$ | $9x^2 \sim 4.0000$ | $16x^2 \sim 5.0000$ |
| 3.281 | 4.2424 | 5.2232 |
| 3.291 | 4.2528 | 5.2302 |
| 3.304 | 4.2532 | 5.2304 |

In all three cases, the approximate values seem to tend to a positive limit regularly and rather rapidly. Euler takes this as a sign that the first zeros are real and sees herein a confirmation of his conjecture.

Let us realize the general scheme of Euler's heuristic conclusion. Let $A$ stand for his conjecture explained in sect. 3, concerning the reality of the zeros of his series. Let $B$ stand for the fact that for $n = 1/2$ the first zero is real. Obviously $A$ implies $B$. Now Euler did not prove $B$, he only made $B$ more credible. Therefore we have here the following pattern of plausible inference

$$A \text{ implies } B$$
$$B \text{ more credible}$$

----

$$A \text{ somewhat more credible}$$

The second premise is weaker than the second premise of the fundamental inductive pattern. The word "somewhat" is inserted to emphasize that also the conclusion is weaker than in the fundamental inductive pattern.

12. A modern mathematician can derive a more stringent heuristic conclusion from the numerical data of the foregoing ex. 11 than Euler himself derived. It can be shown that if Euler's series has only real zeros, the successive approximate values obtained by Daniel Bernoulli's method form necessarily an *increasing* sequence.[7] Let $A$ stand for the same conjecture as in the foregoing ex. 11, but let $B$ denote now another statement, namely the following: "For $n = 1/2$, the first four approximations obtained by Daniel Bernoulli's method form an increasing sequence, and the same holds for $n = 1/3$ and $n = 1/4$." Then both premises of the fundamental inductive pattern are known to hold:

$$A \text{ implies } B$$
$$B \text{ true}$$

and the resulting heuristic conclusion is stronger.

Two remarks may be added to the foregoing.

(1) Euler did not formulate the property just quoted of Daniel Bernoulli's method and certainly did not prove it. Yet there is a good chance that, in the possession of a vast experience with this method, he had some sort of inductive knowledge of it. So Euler, although he did not draw the modern mathematician's inference explicitly, possessed it in a less clarified form. And, presumably, he had in his rich mathematical background still other indications which he could not quite formulate and which we could not yet clarify today.

----

[7] See the author's paper "Remarks on power series," *Acta Scientiarum Mathematicarum*, v. 12B, 1950, p. 199–203.

(2) The numerical data quoted in ex. 11 led the author to suspect the general theorem proved l.c. This is a small but concrete example of the use of the inductive method in mathematical research.

**13.** In ch. IV we investigated inductively the sum of four odd squares; see sect. 4.3–4.6, Table I. Later we tackled the analogous problems involving four arbitrary squares and eight squares; see ex. 4.10–4.23 and Tables II and III. The former investigation certainly helped us to recognize the law in the latter cases. Should our confidence in the result of the latter investigation also be enhanced by the result of the former?

**14.** *Inductive conclusion from fruitless efforts.* Construct, by ruler and compasses, the side of a square equal in area to a circle of given radius. This is the strict formulation of the famous problem of the quadrature of the circle, conceived by the Greeks. It was not forgotten in the Middle Ages, although we cannot believe that many people then understood its strict formulation; Dante refers to it at the theological culmination of the *Divina Commedia*, toward the end of the concluding Canto. The problem was about two thousand years old as the French Academy resolved that manuscripts purporting to square the circle will not be examined. Was the Academy narrow-minded? I do not think so; after the fruitless efforts of thousands of people in thousands of years there was some ground to suspect that the problem is insoluble.

You are moved to give up a task that withstands your repeated efforts. You desist only after many and great efforts if you are stubborn or deeply concerned. You desist after a few cursory trials if you are easy going or not seriously concerned. Yet in any case there is a sort of inductive conclusion. The conjecture under consideration is:

*A.* It is impossible to do this task.

You observe:

*B.* Even I cannot do this task.

This, in itself, is *very* unlikely indeed. Yet certainly

$$A \text{ implies } B$$

and so your observation of *B* renders *A* more credible, by the fundamental inductive pattern.

The impossibility of squaring the circle, strictly formulated, was proved in 1882, by Lindemann, after the basic work of Hermite. There are other similar problems dating from the Greeks (the Trisection of an Angle and the Duplication of the Cube) that, after the accumulated evidence of fruitless efforts, have been ultimately proved insoluble. After fruitless efforts to construct a "perpetuum mobile" the physicists formulated the "principle of the impossibility of a perpetual motion," and this principle turned out remarkably fruitful.

# XIII

## FURTHER PATTERNS AND FIRST LINKS

*When we have intuitively understood some simple propositions . . . it is useful to go through them with a continuous, uninterrupted motion of thought, to meditate upon their mutual relations, and to conceive distinctly several of them, as many as possible, simultaneously. In this manner our knowledge will grow more certain, and the capacity of the mind will notably increase.—*
DESCARTES[1]

1. **Examining a consequence.** We consider a situation which frequently occurs in mathematical research. We wish to decide whether a clearly formulated mathematical proposition $A$ is true or not. We have, perhaps, some intuitive confidence in the truth of $A$, but that is not enough: we wish to prove $A$ or disprove it. We work at this problem, but without decisive success. After a while we notice a consequence $B$ of $A$. This $B$ is a clearly formulated mathematical proposition of which we know that it follows from $A$:

$$A \text{ implies } B.$$

Yet we do not know whether $B$ is true or not. Now it seems that $B$ is more accessible than $A$; for some reason or other we have the impression that we shall have better success with $B$ than we had with $A$. Therefore, we switch to examining $B$. We work to answer the question: is $B$ true or false? Finally we succeed in answering it. *How does this answer influence our confidence in $A$?*

That depends on the answer. If we find that $B$, this consequence of $A$, is false, we can infer with certainty that $A$ must also be false. Yet if we find that $B$ is true, there is no demonstrative inference: although its consequence $B$ turned out to be true, $A$ itself could be false. Yet there is a heuristic inference: since its consequence $B$ turned out to be true, $A$ itself seems to

---

[1] The eleventh of his Rules for the Direction of the Mind. See *Oeuvres*, edited by Adam and Tannery, vol. 10, 1908, p. 407.

deserve more confidence. According to the nature of our result concerning $B$, we follow a demonstrative or a heuristic pattern:

| Demonstrative | Heuristic |
|---|---|
| $A$ implies $B$ | $A$ implies $B$ |
| $B$ false | $B$ true |
| $A$ false | $A$ more credible |

We met these patterns already in sect. 12.1 where we called the heuristic pattern the fundamental inductive pattern. We shall meet with similar but different patterns in the following sections.

**2. Examining a possible ground.** We consider another situation that frequently occurs in mathematical research. We wish to decide whether the clearly formulated proposition $A$ is true or not, we wish to prove $A$ or disprove it. After some indecisive work we hit upon another proposition $B$ from which $A$ would follow. We do not know whether $B$ is true or not, but we have satisfied ourselves that

$$A \text{ is implied by } B.$$

Thus, if we could prove $B$, the desired $A$ would follow; $B$ is a possible ground for $A$. We may be tired of $A$, or $B$ may appear to us more promising than $A$; for some reason or other we switch to examining $B$. Our aim is now to prove or disprove $B$. Finally we succeed. How will our result concerning $B$ influence our confidence in $A$?

That depends on the nature of our result. If we find that $B$ is true, we can conclude that $A$ which is implied by $B$ (follows from $B$, is a consequence of $B$) is also true. Yet if we find that $B$ is false, there is no demonstrative conclusion: $A$ could still be true. But we have been obliged to discard a possible ground for $A$, we have one chance less to prove $A$, our hope to prove $A$ from $B$ has been wrecked: if there is any change at all in our confidence in $A$ in consequence of the disproof of $B$, it can only be a change for the worse. In short, according to the nature of our result concerning $B$, we follow a demonstrative or a heuristic pattern:

| Demonstrative | Heuristic |
|---|---|
| $A$ implied by $B$ | $A$ implied by $B$ |
| $B$ true | $B$ false |
| $A$ true | $A$ less credible |

Observe that the first premise is the same in both patterns. The second premises are diametrically opposite, and the conclusions are also opposite, although not quite as far apart.

The demonstrative inference follows a classical pattern, the "modus ponens" of the so-called hypothetical syllogism. The heuristic pattern is similar to, but different from, the fundamental inductive pattern, see sect. 1 or sect. 12.1. We can state the heuristic inference in words: *our confidence in a conjecture can only diminish when a possible ground for the conjecture is exploded.*

**3. Examining a conflicting conjecture.** We consider a situation which is not too usual in mathematical research but frequently occurs in the natural sciences. We examine two conflicting, incompatible conjectures $A$ and $B$. When we say that $A$ conflicts with $B$ or

$$A \text{ is incompatible with } B$$

we mean that the truth of one of the two propositions $A$ and $B$ necessarily implies the falsity of the other. Thus, $A$ may be true or not and $B$ may be true or not; we do not know which is the case except that we know that both cannot be true. They could both be false, however. A naturalist proposed the conjecture $A$ to explain some phenomenon, another naturalist proposed the conjecture $B$ to explain the same phenomenon differently. The explanations are incompatible; both naturalists cannot be right, but both could be wrong.

If one of the conjectures, say $B$, has been proved to be right, then the fate of the other is also definitely decided: $A$ must be wrong. If, however, $B$ has been disproved, the fate of $A$ is not yet definitely settled: also $A$ could be wrong. Yet, undeniably, by the disproof of a rival conjecture incompatible with it, $A$ can only gain. (The naturalist who invented $A$ would certainly think so.) And so we have again two patterns:

| *Demonstrative* | *Heuristic* |
|:---:|:---:|
| $A$ incompatible with $B$ | $A$ incompatible with $B$ |
| $B$ true | $B$ false |
| --- | --- |
| $A$ false | $A$ more credible |

*Our confidence in a conjecture can only increase when an incompatible rival conjecture is exploded.*

**4. Logical terms.** In the foregoing three sections we have seen three pairs of patterns. Each pair consists of a demonstrative pattern and of a heuristic pattern; the three demonstrative patterns are related to each other and the corresponding three heuristic patterns seem to be correspondingly related. The relations between the demonstrative patterns are clear relations of formal logic. In the next section we shall try to clarify the relations between the heuristic patterns. The present section prepares us for the next section by discussing a few simple terms of formal logic.[2]

---

[2] We treat here formal logic in the "old-fashioned" manner (using ordinary language and avoiding logical symbols as much as feasible) but with a few modern ideas. Some of the simpler logical symbols will be used incidentally later, especially in sect. 7.

(1) The term *proposition* may be taken in a more general meaning, but most of the time it will be sufficient and even advantageous to think of some clearly formulated mathematical proposition of which *for the moment we do not know whether it is true or not.* (A good example of a proposition for a more advanced reader is the celebrated "Riemann hypothesis": Riemann's $\xi$-function has only real zeros. We do not know, in spite of the efforts of many excellent mathematicians, whether this proposition is true or false.) We shall use capitals $A$, $B$, $C$, ... to denote propositions.

(2) The *negation* of the proposition $A$ is a proposition that is true if, and only if, $A$ is false. We let non-$A$ stand for the negation of $A$.

(3) The two statements "$A$ is false" and "non-$A$ is true" amount exactly to the same. We can substitute one for the other in any context without changing the import, the truth, or the falsity, of the whole text. Two statements which can be so substituted for each other are termed *equivalent.* Thus, the statement "$A$ is false" is equivalent to the statement "non-$A$ is true." It will be convenient to write this in the abbreviated form:

$$\text{"}A\text{ false" eq. "non-}A\text{ true."}$$

It is also correct to say that

$$\text{"}A\text{ true" eq. "non-}A\text{ false."}$$

$$\text{"non-}A\text{ true" eq. "}A\text{ false,"}$$

$$\text{"non-}A\text{ false" eq. "}A\text{ true."}$$

(4) We say that the two propositions $A$ and $B$ are *incompatible* with each other, if both cannot be true. The proposition $A$ can be true or false, $B$ can be true or false; if we consider $A$ and $B$ jointly, four different cases can arise:

| $A$ true, $B$ true | $A$ true, $B$ false |
|---|---|
| $A$ false, $B$ true | $A$ false, $B$ false. |

If we say that $A$ is incompatible with $B$, we mean that the first of these four cases (in the north-west corner) is excluded. Incompatibility is always mutual. Therefore,

$$\text{"}A\text{ incompatible with }B\text{" eq. "}B\text{ incompatible with }A\text{."}$$

(5) We say that $A$ implies $B$ (or $B$ is implied by $A$, or $B$ follows from $A$, or $B$ is a consequence of $A$, etc.) if $A$ and non-$B$ are incompatible. Thus the concept of implication is characterized by the following equivalence:

$$\text{"}A\text{ implies }B\text{" eq. "}A\text{ incompatible with non-}B\text{."}$$

To know that $A$ implies $B$ is important. For the moment we do not know whether $A$ is true or not and we are in the same state of ignorance concerning $B$. If, however, it should turn out some day that $A$ is true, we shall know right away that non-$B$ must be false and so $B$ must be true.

We know that

"$A$ incompatible with non-$B$" eq. "non-$B$ incompatible with $A$."

We know also that

"non-$B$ incompatible with $A$" eq. "non-$B$ implies non-$A$."

From the chain of the last three equivalences we conclude:

"$A$ implies $B$" eq. "non-$B$ implies non-$A$."

This last equivalence is quite important in itself and it will be essential in the following consideration.

(6) The few points of formal logic discussed in this section enable us already to clarify the relation between the demonstrative patterns encountered in the three foregoing sections.

Let us start from the demonstrative pattern formulated in sect. 1 (the "modus tollens"):

$$A \text{ implies } B$$
$$B \text{ false}$$
$$\overline{A \text{ false}}$$

It is understood that this pattern is generally applicable. Let us apply it in substituting non-$A$ for $A$ and non-$B$ for $B$. We obtain

$$\text{non-}A \text{ implies non-}B$$
$$\text{non-}B \text{ false}$$
$$\overline{\text{non-}A \text{ false}}$$

We have seen, however, in the foregoing that

"non-$A$ implies non-$B$" eq. "$B$ implies $A$"

"non-$B$ false" eq. "$B$ true"

"non-$A$ false" eq. "$A$ true"

Let us substitute for the premises and the conclusion of the last considered pattern the three corresponding equivalent statements here displayed. Then we obtain:

$$B \text{ implies } A$$
$$\underline{B \text{ true}}$$
$$A \text{ true}$$

which is the demonstrative pattern of sect. 2, the "modus ponens."

We leave to the reader to derive similarly the demonstrative pattern of sect. 3 from that of sect. 1.

**5. Logical links between patterns of plausible inference.** The discussion in the foregoing section was merely preparatory. We did not discuss those classical points of demonstrative logic for their own sake, but in order to prepare ourselves for the study of plausible inference. We considered the derivation of the "modus ponens" from the "modus tollens" not in some vain hope of presenting something new and surprising concerning such classical matters, but as a preparation for such questions as the following: Can we derive by pure formal logic the heuristic pattern of sect. 2 from the heuristic pattern of sect. 1?

No, obviously, we cannot. In fact, these patterns contain such statements as "$A$ becomes more credible" or "$A$ becomes less credible." Although everybody understands what this means, the consistent formal logician refuses to understand such statements, and he is even right. Pure formal logic has no place for such statements; it has no way to handle them.

We could, however, widen the domain of formal logic in an appropriate way. The main point seems to be to add the following equivalence to the classical contents of formal logic: "non-$A$ becomes more credible" is equivalent to "$A$ becomes less credible." In abbreviation

"non-$A$ more credible" eq. "$A$ less credible."

In admitting this, we obtain a useful tool. Now we can adapt the procedure of sect. 4 (6) to our present aim and proceed as follows. We start from the fundamental inductive pattern, given in sect. 1. We apply it to non-$A$ and non-$B$, instead of applying it to $A$ and $B$; that is, we substitute non-$A$ for $A$ and non-$B$ for $B$ in it. Then we apply three equivalences; two of these are purely logical (and discussed, by the way, in the foregoing sect. 4) and the third is the essential new equivalence introduced in the present section. By these steps we finally arrive at the heuristic pattern given in sect. 2.

It may be left to the reader to write down this derivation in detail and to derive the heuristic pattern of sect. 3 in the same manner from the fundamental inductive pattern of sect. 1.

**6. Shaded inference.** It seems to me that inductive reasoning in the mathematical domain is easier to study than in the physical domain. The reason is simple enough. In asking a mathematical question, you may hope to obtain a completely unambiguous answer, a perfectly sharp Yes or No. In addressing a question to Nature, you cannot hope to obtain an answer without some margin of uncertainty. You predict that a lunar eclipse will begin (the shadow will indent the disk of the moon) at such and such a time. Actually, you observe the beginning of the eclipse 4 minutes later than predicted. According to the standards of Greek astronomy your prediction would be amazingly correct, according to modern standards it is scandalously incorrect. A given discrepancy between prediction and observation can be interpreted as confirmation or refutation. Such interpretation depends on some kind of plausible reasoning the difficulties of which in "physical situations" begin a step earlier than in "mathematical situations." We shall try to reduce this distinction to its simplest expression.

Suppose that we are investigating a mathematical conjecture by examining its consequences. Let $A$ stand for the conjecture and $B$ for one of its consequences, so that $A$ implies $B$. We arrive at a final decision concerning $B$: we disprove $B$ or we prove it and, accordingly, we face one or the other of the following two situations:

$A$ implies $B$          $A$ implies $B$

$B$ false              $B$ true

We shall call these "mathematical situations." We have considered them repeatedly and we know what reasonable inference we can draw from each.

Suppose now that we are investigating a physical conjecture $A$ and that we test experimentally one of its consequences $B$. We cannot arrive at an absolute decision concerning $B$; our experiments may show, however, that $B$, or its contrary, is very hard to believe. Accordingly, we face one or the other of the following two situations:

$A$ implies $B$          $A$ implies $B$

$B$ scarcely credible     $B$ almost certain

We call these "physical situations." What inference is reasonable in these situations? (The empty space under the horizontal line that suggests the word "therefore" symbolizes the open question.)

In each of the four situations considered we have two data or *premises*. The first premise is the same in all four situations; all the difference between them hinges on the second premise. This second premise is on the level of pure formal logic in the "mathematical" situations, but on a much vaguer level in the "physical" situations. This difference seems to me essential; it may account for the additional difficulties of the physical situations.

Let us survey the four situations "with a continuous uninterrupted motion of thought," as Descartes liked to say (see the motto at the beginning of this chapter). Let us imagine that our confidence in $B$ changes gradually, varies "continuously." We imagine that $B$ becomes less credible, then still less credible, scarcely believable, and finally false. On the other hand, we imagine that $B$ becomes more credible, then still more credible, practically certain, and finally true. If the *strength of our conclusion varies continually in the same direction as the strength of our confidence in B*, there is little doubt what our conclusion should be since the two extreme cases ($B$ false, $B$ true) are clear.   In this manner we arrive at the following patterns:

$$A \text{ implies } B$$
$$B \text{ less credible}$$
$$\overline{A \text{ less credible}}$$

$$A \text{ implies } B$$
$$B \text{ more credible}$$
$$\overline{A \text{ somewhat more credible}}$$

The word "somewhat" in the second pattern is inserted to remind us that the conclusion is, of course, weaker than in the fundamental inductive pattern.   *Our confidence in a conjecture is influenced by our confidence in one of its consequences and varies in the same direction.*   We shall call these patterns *shaded*; the first is a shaded demonstrative pattern, the second is the shaded version of the fundamental inductive pattern.   The term "shaded" intends to indicate the weakening of the second premise: "less credible" instead of "false"; "more credible" instead of "true."   We have already used this term in this meaning in sect. 12.6.

We obtained the shaded patterns just introduced from their extreme cases, the "modus tollens" and the fundamental inductive pattern discussed in sect. 1, by weakening the second premise.   In the same manner we can obtain other shaded patterns from the patterns formulated in sects. 2 and 3. We state just one here (all are listed in the next section).   The heuristic pattern of sect. 3 yields the following shaded pattern:

$$A \text{ incompatible with } B$$
$$B \text{ less credible}$$
$$\overline{A \text{ somewhat more credible}}$$

**7. A table.**   In order to list concisely the patterns discussed in this chapter, it will be convenient to use some abbreviations.   We write

$$A \rightarrow B \quad \text{for} \quad A \text{ implies } B,$$

$$A \leftarrow B \quad \text{for} \quad A \text{ is implied by } B,$$

$$A \mid B \quad \text{for} \quad A \text{ incompatible with } B.$$

The symbols introduced are used by some authors writing on symbolic logic.[3]   In this notation, the two formulas

$$A \rightarrow B, \qquad B \leftarrow A$$

are exactly equivalent and so are the formulas

$$A \mid B, \qquad B \mid A.$$

---

[3] D. Hilbert and W. Ackerman, *Grundzuge der theoretischen Logik.*

We shall also abbreviate "credible" as "cr." and "somewhat" as "s." See Table I.

Table I

| | (1) Demonstrative | (2) Shaded Demonstrative | (3) Shaded Inductive | (4) Inductive |
|---|---|---|---|---|
| 1. Examining a consequence | $A \to B$ $B$ false | $A \to B$ $B$ less cr. | $A \to B$ $B$ more cr. | $A \to B$ $B$ true |
| | $A$ false | $A$ less cr. | $A$ s. more cr. | $A$ more cr. |
| 2. Examining a possible ground | $A \leftarrow B$ $B$ true | $A \leftarrow B$ $B$ more cr. | $A \leftarrow B$ $B$ less cr. | $A \leftarrow B$ $B$ false |
| | $A$ true | $A$ more cr. | $A$ s. less cr. | $A$ less cr. |
| 3. Examining a conflicting conjecture | $A \mid B$ $B$ true | $A \mid B$ $B$ more cr. | $A \mid B$ $B$ less cr. | $A \mid B$ $B$ false |
| | $A$ false | $A$ less cr. | $A$ s. more cr. | $A$ more cr. |

**8. Combination of simple patterns.** The following situation can easily arise in mathematical research. We investigate a theorem $A$. This theorem $A$ is clearly formulated, but we do not know and we wish to find out whether it is true or false. After a while we hit upon a possible ground: we see that $A$ can be derived from another theorem $H$

$A$ is implied by $H$

and so we try to prove $H$. We do not succeed in proving $H$, but we notice that one of its consequences $B$ is true. The situation is:

$A$ implied by $H$
$B$ implied by $H$
$B$ true
_____

Is there a reasonable inference concerning $A$ from these premises?

There is one, I think, and we can even obtain it by combining two of the patterns surveyed in the foregoing section. In fact, the fundamental inductive pattern yields:

$H$ implies $B$
$B$ true
_____
$H$ more credible

In obtaining this conclusion, we have used only two of our three premises. Let us combine the unused third premise with the conclusion just obtained:

one of our shaded patterns (in the intersection of the second row with the second column of Table I in sect. 7) yields:

$$A \text{ implied by } H$$
$$H \text{ more credible}$$
$$\overline{A \text{ more credible}}$$

The result is (as it should be) pretty obvious in itself: from the verification of the consequence $B$ of its possible ground $H$ some credit is reflected upon the proposition $A$ itself.

**9. On inference from analogy.** The situation discussed in the foregoing section can be construed as a link between the patterns discussed in this chapter and one of the most conspicuous forms of plausible inference: the conclusion from analogy.

I do not think that it is possible to explain the idea of analogy in completely definite terms of formal logic; at any rate, I have no ambition to explain it so. As we have discussed before, in sect. 2.4, analogy has to do with similarity *and* the intentions of the thinker. If you notice some similarity between two objects (or, preferably, between two systems of objects) and intend to reduce this similarity to definite concepts, you think analogically.

For example, you notice some similarity between two theorems $A$ and $B$; you observe some common points. Perhaps, you think, it will be possible some day to imagine a more comprehensive theorem $H$ that brings out all the essential common points and from which both $A$ and $B$ would naturally follow. If you think so you begin to think analogically.[4]

At any rate, let us consider the analogy of two theorems $A$ and $B$ as the intention to discover a common ground $H$ from which both $A$ and $B$ would follow:

$$A \text{ implied by } H, \qquad B \text{ implied by } H.$$

Let us not forget that we do not *have* $H$; we just *hope* that there is such an $H$.

Now we succeed in proving one of the two analogous theorems, say $B$. How does this event influence our confidence in the other theorem $A$? The situation has something in common with the situation considered in the foregoing sect. 8. There we reached a reasonable conclusion expressed by the compound pattern

$$A \text{ implied by } H$$
$$B \text{ implied by } H$$
$$B \text{ true}$$
$$\overline{A \text{ more credible}}$$

---

[4] Thus, the isoperimetric theorem and Rayleigh's conjecture, compared with each other in sect. 12.4, may suggest the idea of a common generalization.

There is, of course, the important difference that now we do not have $H$, we just hope for $H$. With this proviso, however, we can regard the two premises

$$A \text{ implied by } H$$
$$B \text{ implied by } H$$

as equivalent to one:

$$A \text{ analogous to } B.$$

In substituting this one premise for those two premises in the above compound pattern we arrive at the fundamental pattern of plausible inference first exhibited in sect. 12.4:

$$A \text{ analogous to } B$$
$$\underline{B \text{ true}}$$
$$A \text{ more credible}$$

**10. Qualified inference.** We come back again to the fundamental inductive pattern. It is the first pattern that we introduced and it is the most conspicuous form of plausible reasoning. It is concerned with the verification of a consequence of a conjecture and the resulting change in our opinion. It says something about the direction of this change; such a verification can only increase our confidence in the conjecture. It says nothing about the amount of the change; the increase of confidence can be great or small. Indeed, it can be tremendously great or ridiculously small.

The aim of the present section is to clarify the circumstances on which such an important difference depends. We begin by recalling one of our examples (sect. 12.3).

A defendant is accused of having blown up the yacht of his girl friend's father and the prosecution produces a receipt signed by the defendant acknowledging the purchase of such and such an amount of dynamite.

The evidence against the defendant appears very strong. Why does it appear so? Let us insist on the general features of the case. Two statements play an essential role.

$$A: \text{ the defendant blew up that yacht.}$$

$$B: \text{ the defendant acquired explosives.}$$

At the beginning of the proceedings, the court has to consider $A$ as a conjecture. The prosecution works to render $A$ more credible to the jurors, the defense works to render it less credible.

At the beginning of the proceedings also $B$ has to be considered as a conjecture. Later, after the introduction of that receipt in court (the authenticity of the signature was not challenged by the defense) $B$ has to be considered as a proven fact.

Certain relations between $A$ and $B$, however, should be clear from the beginning.

$A$ without $B$ is impossible.   If the defendant blew up the yacht, he had some explosives.   He had to acquire these explosives somehow: by purchase, stealing, gift, inheritance or otherwise.   That is

$$A \text{ implies } B.$$

$B$ without $A$ is not impossible, but must appear extremely unlikely from the outset.   To buy dynamite is very unusual anyhow for an ordinary citizen.   To buy dynamite without the intention of blowing up something or somebody would be nonsense.   It was easy to suspect that the defendant had quite strong emotional and financial grounds for blowing up that yacht. It was difficult to suspect any purpose for the purchase of dynamite, except blowing up the yacht.   And so, as we said, $B$ without $A$ appears extremely unlikely.

Let us nail down this important constituent of the situation: *the credibility of B, before the event, viewed under the assumption that A is not true.* We shall abbreviate this precise but long description as "the credibility of $B$ without $A$."   Thus, we can say:

$$B \text{ without } A \text{ is hardly credible.}$$

Now, we can see the essential premises and the whole pattern of the plausible inference that impressed us with its cogency:

$A$ implies $B$

$B$ without $A$ hardly credible

$B$ true

$A$ very much more credible

For better understanding let us imagine that that important constituent of the situation, the credibility of $B$ without $A$, changes gradually, varies continuously between its extreme cases.

$A$ implies $B$.   If, conversely, also $B$ implies $A$, so that $A$ and $B$ imply each other mutually, the credibility of $B$ without $A$ attains its minimum, is nil. In this case, if $B$ is true, also $A$ is true, so that the credibility of $A$ attains its maximum.

$A$ implies $B$.   That is, $B$ is certain when $A$ is true.   If the credibility of $B$ without $A$ approaches its maximum, $B$ is almost certain when $A$ is false. Therefore, $B$ is almost certain anyway.   When an event happens that looks almost certain in advance, we do not get much new information and so we cannot draw surprising consequences.   (The purchase of a loaf of bread, for instance, can hardly ever yield such a strong circumstantial evidence, as the purchase of dynamite.)

Let us assume, as in sect. 6, that the strength of the conclusion varies *continually in the same direction* when that influential factor, the credibility of $B$ without $A$, changes without change of direction. Then these two must vary in opposite directions and we arrive so at an important qualification of the strength of the conclusion in the fundamental inductive pattern:

$$A \text{ implies } B$$
$$\underline{B \text{ true}\phantom{xxxxx}}$$
$$A \text{ more credible}$$

*The strength of the conclusion increases when the credibility of B without A decreases.*
Let us put side by side the two extreme cases:

$$\left\{ \begin{array}{l} A \text{ implies } B \\ B \text{ without } A \text{ hardly credible} \end{array} \right. \qquad \left\{ \begin{array}{l} A \text{ implies } B \\ B \text{ almost certain anyway} \end{array} \right.$$

$$\underline{\phantom{xx}B \text{ true}\phantom{xxxxxxxxxx}} \qquad \underline{\phantom{xx}B \text{ true}\phantom{xxxxxxxxxx}}$$

$$A \text{ very much more credible} \qquad A \text{ very little more credible}$$

The first two premises are bracketed to express that the second is to be considered as a qualification of the first. That $A$ implies $B$ is the first premise in the fundamental inductive pattern. Here we qualify this premise; we add a modification which counts heavily in determining the strength of the conclusion. For the sake of comparison let us remember that in sect. 6 we modified the fundamental inductive pattern in another direction, in weakening its second premise.

**11. On successive verifications.** We have verified already $n$ consequences $B_1, B_2, \ldots, B_n$ of a certain conjecture $A$. Now we proceed to a next consequence $B_{n+1}$, we test it, and we find that also $B_{n+1}$ is true. What is the influence of this additional evidence on our confidence in $A$? Of course,

$$A \text{ implies } B_{n+1}$$
$$\underline{B_{n+1} \text{ is true}\phantom{xxx}}$$
$$A \text{ more credible}$$

Yet how strong is the conclusion? That depends on the credibility of $B_{n+1}$ without $A$ as we have seen in the foregoing section.

Now we may have had a good reason for believing in $B_{n+1}$ before it was verified even under the assumption that $A$ is not true. We have seen previously that $B_1, B_2, \ldots, B_n$ are true. If $B_{n+1}$ is very similar to $B_1, B_2, \ldots, B_n$, we may foresee *by analogy* that also $B_{n+1}$ will be true. If $B_{n+1}$ is very different from $B_1, B_2, \ldots, B_n$, it is not supported by such analogy and we may have very little reason to believe in $B_{n+1}$ without $A$. Therefore, *the strength of the additional confidence resulting from an additional verification increases when the analogy of the newly verified consequence with the previously verified consequences decreases.*

This expresses essentially the same thing as the complementary patterns formulated in sect. 12.2, but perhaps a little better. In fact, we may regard the explicit mention of analogy as an advantage.

**12. On rival conjectures.** If there are two different conjectures, $A$ and $B$, aimed at explaining the same phenomenon, we regard them as opposed to each other even if they are not proved to be logically incompatible. These conjectures $A$ and $B$ may or may not be incompatible, but one of them tends to render the other superfluous. This is enough opposition, and we regard $A$ and $B$ as *rival* conjectures.

There are cases in which we treat rival conjectures almost *as if* they were incompatible. For example, we have two rival conjectures $A$ and $B$ but, in spite of some effort, we cannot think of a third conjecture explaining the same phenomenon; then each of the two conjectures $A$ and $B$ is the "unique obvious rival" of the other. A short schematic illustration may clarify the meaning of the term.

Let us say that $A$ is the emission theory of light that goes back to Newton and that $B$ is the wave theory of light that originated with Huyghens. Let us also imagine that we discuss these matters in the time after Newton and Huyghens, but before Young and Fresnel when, in fact, much inconclusive discussion of these theories took place. Nobody showed, or pretended to show, that these two theories are logically incompatible, and still less that they are the only logically possible alternatives; but there were no other theories of light prominently in view, although the physicists had ample opportunity to invent such theories: each theory was the unique obvious rival of the other. And so any argument that seemed to speak against one of the two rival theories was readily interpreted as speaking for the other.

In general, the relation between rival conjectures is similar to the relation between rivals in any other kind of competition. If you compete for a prize, the weakening of the position of any of your rivals means some strengthening of your position. You do not gain much by a slight setback to one of your many obscure rivals. You gain more if such setback occurs to a dangerous rival. You gain still more if your most dangerous rival drops out of the race. If you have a unique obvious rival, any weakening or strengthening of his position influences your position appreciably. And something similar happens between competing conjectures. There is a pattern of plausible reasoning which we attempt to make somewhat more explicit in Table II.

Table II

| A incompatible with B B false | A incompatible with B B less credible |
|---|---|
| A more credible | A somewhat more credible |

| A rival of B B false | A rival of B B less credible |
|---|---|
| A a little more credible | A very little more credible |

The disposition of Table II is almost self-explanatory.　This Table contains four patterns arranged in two rows and two columns.　The first row contains two patterns already considered; see sect. 3, the end of sect. 6, and the last row of Table I.　In passing from the first row to the second row, we weaken the first premise; in fact we substitute for a clear relation of formal logic between $A$ and $B$ a somewhat diffuse relation which, however, makes some sense in practice.　This weakening of the first premise renders the conclusion correspondingly weaker, as the verbal expression attempts to convey.　In passing from the first column to the second column we weaken the second premise, which renders the conclusion correspondingly weaker.　The pattern in the southeast corner has no premise that would make sense in demonstrative logic and its conclusion is the weakest.

It is important to emphasize that the verbal expressions used are slightly misleading.　In fact, the specifications added to "credible" ("somewhat," "a little," "very little") should not convey any *absolute*, only a *relative*, degree of credibility.　They indicate only the change in strength as we pass from one row to the other, or from one column to the other.　Even the weakest of the four patterns may yield a weighty conclusion if the conviction that the conjecture $A$ has no other dangerous rival than $B$ is strong enough.　In fact, this pattern will play some role in the next chapter.

**13. On judicial proof.**　The reasoning by which a tribunal arrives at its decisions may be compared with the inductive reasoning by which the naturalist supports his generalizations.　Such comparisons have been already offered and debated by authorities on legal procedure.[5]　Let us begin the discussion of this interesting point by considering an example.

(1) The manager of a popular restaurant that is kept open to late hours returned to his suburban home, as usual, well after midnight.　As he left his car to open the door of his garage, he was held up and robbed by two masked individuals.　The police, searching the premises, found a dark grey rag in the front yard of the victim; the rag might have been used by one of the holdup men for covering his face.　The police questioned several persons in the nearby town.　One of the men questioned had an overcoat with a big hole in the lining, but otherwise in good condition.　The rag found in the front yard of the victim of the holdup was of the same material as the lining and fitted into the hole exactly.　The man with the overcoat was arrested and charged with participation in the holdup.

(2) Many of us may feel that such a charge was amply justified by the related circumstances.　But why?　What is the underlying idea?

The charge is not a statement of facts, but the expression of a suspicion, of a *conjecture*:

$A$. The man with the overcoat participated in the holdup.

---

[5] J. H. Wigmore, *The principles of judicial proof*, Boston, 1913; cf. p. 9–12, 15–17.

Such an official charge, however, should not be a gratuitous conjecture, but supported by relevant facts. The conjecture $A$ is supported by the fact $B$. The rag found in the front yard of the victim of the holdup is of the same material as, and fits precisely into the hole of, the lining of the overcoat of the accused.

Yet why do we regard $B$ as a justification for $A$? We should not forget that $A$ is just a conjecture: it can be true or false. If we wish to act fairly, we have to consider carefully both possibilities.

If $A$ is true, $B$ is readily understandable. We can easily imagine that a man in urgent need of a mask and not within reach of more suitable materials cuts out a piece from the lining of his overcoat. In a hurry to get away after the criminal act, such a man may lose his mask. Or, in a fright, he may even throw away his mask there and then instead of pocketing it and throwing it away at a safer place. In short, $B$ with $A$ looks readily credible.

If, however, $A$ is not true, $B$ appears inexplicable. If the man did not participate in a holdup or something, why should he spoil his perfectly good overcoat by cutting out a large piece from the lining? And why should that piece of lining arrive, of all places, upon the scene of a robbery by masked bandits? It could arrive there by mere coincidence, of course, but such a coincidence is hard to believe. In short, $B$ without $A$ is hardly credible.

And so we see that the conclusion that led to the charge against the man with the overcoat has the following pattern:

$$\frac{\left\{ \begin{array}{l} B \text{ with } A \text{ readily credible} \\ B \text{ without } A \text{ hardly credible} \end{array} \right.}{\begin{array}{c} B \text{ true} \\ \hline A \text{ more credible} \end{array}}$$

Yet this pattern of plausible reasoning is obviously related to another pattern of plausible reasoning that we have discussed before (in sect. 10):

$$\frac{\left\{ \begin{array}{l} A \text{ implies } B \\ B \text{ without } A \text{ hardly credible} \end{array} \right.}{\begin{array}{c} B \text{ true} \\ \hline A \text{ very much more credible} \end{array}}$$

The difference between the two patterns appears at the very beginning. The premise

$$B \text{ with } A \text{ readily credible}$$

is similar to, but weaker than, the premise

$$A \text{ implies } B$$

which, in fact, could also be worded as "$B$ with $A$ certain." Thus the former pattern (just discovered) appears as a "weakened form" of the latter pattern (introduced in sect. 10) and so eventually as a modification of the fundamental inductive pattern (formulated in sect. 12.1).

The case that led us to the formulation of the new pattern was fairly simple. Let us look at a more complex case.[6]

(3) At the time of the murder, Clarence B. Hiller, with his wife and four children, lived in a two-storey house in Chicago. The bedrooms of the family were on the second floor. At the head of the stairs leading to the second floor a gas light was kept burning at night. Shortly after 2 o'clock on Monday morning, Mrs. Hiller was awakened and noticed that this light was out. She awoke her husband and he went in his nightgown to the head of the stairway where he encountered an intruder. The men fought and in the struggle both fell to the foot of the stairway, where Hiller was shot twice; he died in a few moments. The shooting occurred about 2.25 a.m.

Just a little before the shooting one of Hiller's daughters had seen a man at the door of her bedroom, holding a match so that his face could not be seen. She was not frightened because her father used to get up at night and see if the children were all right. No one else in the family saw the intruder.

About three-quarters of a mile from Hiller's house there is a streetcar stop. Early in the morning on which the murder occurred, four police officers, who had gone off duty in the neighborhood shortly before, were sitting on a bench at this stop waiting for the streetcar. About 2.38 a.m. they saw a man approaching from a direction from which the bench could not be easily seen. The officers spoke to the man, but he continued walking with his right hand in his pocket. The officers stopped the man and searched him. There was a loaded revolver in the pocket; he was perspiring; fresh blood appeared at different places of his clothing; there was a slight wound on his left forearm, bleeding slightly. The officers (who did not know at this time of the murder) brought the man to the police station where he was examined. This man—we shall call him the defendant—was later charged with Mr. Hiller's murder.

The court had to examine and, after examination, to deny or uphold the accusation, that is, the following conjecture advanced by the prosecution:

*A.* The defendant shot and killed Mr. Hiller.

We survey the salient points of the evidence put forward in support of the conjecture *A.*

$B_1$. There was burned powder in two chambers of the cylinder of the revolver found on the defendant when he was arrested, and there was the smell of fresh smoke. In the judgment of the police officers the revolver had been fired twice within an hour before the arrest. The five cartridges with which the revolver was loaded bore exactly the same factory markings as

---

[6] Concerning the following case the reader should consult the findings of the Supreme Court of Illinois, almost fully reprinted in Wigmore, l.c. footnote 5, p. 83–88.

three undischarged cartridges found in the hallway of the Hiller house near the dead body.

$B_2$. The intruder entered the Hiller house through a rear window of the kitchen from which he first removed the screen. A person entering this window could support himself on the railing of the porch. On this railing which was freshly painted there was the imprint of four fingers of a left hand. Two employees of the identification bureau of the Chicago police testified that, in their judgment, the imprints on the railing were identical with the defendant's fingerprints.

$B_3$. Two experts not belonging to the Chicago police expressed the same opinion concerning the fingerprints. (One was an inspector of the Dominion Police at Ottawa, Canada; the other a former expert of the federal government in Washington, D.C.)

$B_4$. About 2.00 a.m., just before the shooting of Mr. Hiller, a prowler entered a house separated by a vacant lot from the Hiller house. Two women saw a man in the door of their bedroom with a lighted match over his head. Both women testified that this prowler was of the same size and build as the defendant. One of the women remembered also that the prowler wore a light-colored shirt and figured suspenders. Having inspected the shirt and the suspenders of the defendant exhibited in court, the witness testified that in her opinion the defendant was the same man that she saw on that night in the door.

$B_5$. The defendant gave a false name and a false address when he was arrested and he denied that he had ever been arrested before. In fact, he had been sentenced before on a charge of burglary, paroled, returned to the penitentiary for a violation of the parole, and released on parole a second time about six weeks before the night of the murder. He bought the revolver under a false name about two weeks after his second parole, pawned it, got it back, pawned it again and got it back a second time five hours before the shooting.

$B_6$. The defendant was unable to explain consistently the blood on his clothing, or the wound on his left forearm, or his whereabouts on the night of the shooting. Concerning his whereabouts he told two different stories, one after his arrest and another in court. The people whom he first asserted having visited that night denied that he called on them. The defendant then told the court that he visited a saloon, but no witness was found to corroborate this.

(4) All the facts, events, and circumstances related under the headings $B_1$, $B_2$, ... $B_6$ are readily understandable *if* the accusation $A$ is true. They all support $A$, but the weight of such support is not the same in all cases. Some of these facts would be explicable even if $A$ was not true. Some others, however, would appear as miraculous coincidences if $A$ was not true.

That the cartridges found in the revolver of the defendant are of the same make as the cartridges found next to the body of the victim proves little in itself, if this make is a usual make sold by all gunsmiths. Yet it proves a lot that just as many cartridges have been fired from this revolver as shots have been fired at the victim and within the same hour; it is difficult to explain away such a coincidence. The agreement of the fingerprints on the railing with the fingerprints of the defendant would be considered in itself as an almost decisive proof nowadays, but was not yet considered so at the time of the trial, in 1911. That the defendant lied about his name, address, and criminal past when he was arrested does not prove much; such a lie is understandable if the accusation $A$ is true, but also understandable if it is not true: the man would prefer to be let alone by the police anyway. Yet it weighs heavily that the defendant was not able to explain consistently his whereabouts on that fatal night. He must have known that the point is important, and his counsel certainly knew about the importance of an alibi. If the accusation $A$ was untrue and the defendant passed the night harmlessly or committing some minor crime, why did he not say so right away, or at least before it was too late?

All the details mentioned are readily understandable if the accusation $A$ is true. Yet the coincidence of so many details appears as inexplicable if the accusation $A$ is not true; it is extremely hard to believe that so many coincidences are due to mere chance. At any rate, the defense failed to propose a consistent explanation of the evidence submitted.

There was, of course, more evidence before the jury than related here, and there was one that no description can adequately render: the behavior of the defendant and the witnesses. The jury convicted the defendant of murder and the State Supreme Court upheld the conviction. We quote the last sentence of the opinion of the Chief Justice: "No one of these circumstances, considered alone, would be conclusive of his [the defendant's] guilt, but when all the facts and circumstances introduced in evidence are considered together, the jury were justified in believing that a verdict of guilty should follow as a logical sequence."[7]

(5) The statements $B_1, B_2, \ldots B_6$ listed under (3) fit pretty well the pattern introduced under (2). They fit even better another pattern which differs from it in just one point:

$$\begin{cases} B \text{ with } A \text{ readily credible} \\ B \text{ without } A \text{ less readily credible} \end{cases}$$

$$\underline{\hspace{2em} B \text{ true} \hspace{2em}}$$

$$A \text{ more credible}$$

Each of the statements $B_1, B_2, \ldots B_6$ can be meaningfully substituted for $B$ in this pattern in which, of course, $A$ has to be interpreted as the accusation. The statements $B_1, B_2, \ldots B_6$ are compound statements; they have

---

[7] Cf. Wigmore, *loc. cit.* 5, p. 88.

portions (some of which we have emphasized under (4)) each of which can be regarded in itself as relevant evidence: each such portion can also be meaningfully substituted for *B* in the above pattern. If we recall our discussion under (4) we may realize that, of course, the less easily credible *B* is without *A*, the stronger is the conclusion.

If we visualize how the successive witnesses unfolded the mass of evidence before the jury in the course of the proceedings, we may see more clearly the analogous roles of plausible reasoning in such proceedings and in a scientific inquiry in which several consequences of a conjecture are successively tested. (Cf. especially sect. 12.2.)

(6) The foregoing consideration clearly suggests a compound pattern of plausible reasoning that is related to the pattern stated under (5) just as the compound pattern introduced in sect. 12.2 is related to the fundamental inductive pattern of sect. 12.1. I do not enter upon this matter here; a reader more versed in the doctrine of legal evidence could take it up with more impressive examples and interpretations, but I will add one more illustration of the pattern.[8]

As Columbus and his companions sailed westward across an unknown ocean they were cheered whenever they saw birds. They regarded birds as a favorable sign, indicating the nearness of land. Although in this they were repeatedly disappointed, the underlying reasoning seems to me quite correct. Stated at length, this reasoning runs as follows:

{ When the ship is near the land, we often see birds.
{ When the ship is far from the land, we see birds less often.

Now we see birds.

Therefore, it becomes more credible that we are near the land.

This reasoning fits exactly the pattern formulated under (5): the presence of birds is regarded as circumstantial evidence for the nearness of land. Columbus' crew saw several birds on Thursday, the 11th of October, 1492, and the next day they discovered the first island of a New World.

The reader may notice that the pattern illustrated underlies much of our everyday reasoning.

### EXAMPLES AND COMMENTS ON CHAPTER XIII

### First Part

**1.** Following the method of sect. 4 (6), derive the demonstrative pattern mentioned in sect. 3 from the demonstrative pattern mentioned in sect. 1

---

[8] *How to Solve It*, pp. 212–221.

2. Supply the details of the demonstration sketched in sect. 5: derive the heuristic pattern of sect. 2 from the heuristic pattern of sect. 1.

3. Derive the heuristic pattern of sect. 3 from the heuristic pattern of sect. 1.

4. In a crossword-puzzle we have to find a word with 9 letters, and the clue is: "Disagreeable form of tiredness."[9]

The condition that the unknown word has to satisfy is ambiguously stated, of course. After a few unsuccessful attempts we may observe that "tiredness" has 9 letters, just as many as the unknown word, and this may lead us to the following conjecture:

A. The unknown word means "disagreeable" and is an *anagram* of TIREDNESS.

(Anagram means a rearrangement of the letters of the given word into a new word.) This conjecture A may appear quite likely. (In fact, "form of" may suggest "anagram of" in crossword jargon.) Tackling other unknown words of the puzzle, we find quite plausible solutions for two of them crossing the nine-letter word mentioned, for which we obtain two possible letters, placed as the following diagram indicates:

$$— \ — \ — \ — \ — \ — \ T \ — \ R.$$

We may regard this as evidence for our conjecture A.

(a) Why? Point out the appropriate pattern.

(b) Try to find the nine-letter word required. [In doing so, you have a natural opportunity to weigh the evidence for A. Here are a few helpful questions: Which letter is the most likely between T and R? How could the vowels E E I be placed? See also *How to Solve It*, pp. 147–149.]

5. Let us take up once more the court case already considered in sect. 12.3 and (more fully) in sect. 10. Let us consider again the charge (the *Factum Probandum*, the fact to prove, of the prosecution):

A. The defendant blew up the yacht.

Yet let us change our notation in another point: we consider here the statement

B. The defendant bought dynamite in such and such an amount, in such and such a shop, on such and such a day.

The change is that B denotes here not a general statement, but a specific fact. (Courts prefer, or should prefer, to deal with facts as distinctly specified as possible.) We again take B as proved. (Thus, B is a *Factum Probans*, a fact supporting the proof.)

Such change of notation as we have introduced cannot change the force of the argument. Yet what is now the pattern?

6. The defendants are a contractor and a public official. One was charged with giving, the other with accepting, a bribe. There was a specific charge:

---

[9] *The Manchester Guardian Weekly*, November 29, 1951.

the accusation asserted that the down payment for the new car of the official came from the contractor's pocket.   One of the witnesses for the prosecution was a motor car dealer; he testified that he received $875 on November 29 as down payment on the official's car.   Another witness was the manager of a local bank; he testified that $875 was withdrawn on November 27 (of the same year) from a usually inactive joint account of the contractor and his wife; the receipt was signed by the wife.   These facts were not challenged by the defense.

What would you regard as a strong point in this evidence?   Name the appropriate pattern.

**7.** The Blacks, the Whites, and the Greens live on the same street in Suburbia.   The Blacks and the Whites are neighbors and the Greens live just opposite.   One evening Mr. Black and Mrs. White had a long conversation over the fence.   It was rather dark, yet Mrs. Green did not fail to observe the conversation and jumped to the conclusion—you know which conclusion: that favorite conjecture of Mrs. Green.

Unfortunately, there is not much chance to stem the flood of gossip started by Mrs. Green.   If, however, taking a terrific risk, you wished to constitute yourself counsel for the defense of innocent people maligned by Mrs. Green, I can give you a fact: the Whites, who wished to move nearer to Mr. White's office for a long time, signed a lease for a house that belongs to Mr. Black's uncle, and this happened a few days after that conversation. Use this fact.

What is your defense and what is the pattern?

**8.** The prosecution tries to prove:

*A.* The defendant knew, and was capable of recognizing, the victim at the time of the crime.

The prosecution supports this by the undenied fact:

*C.* The defendant and the victim were both employed by the same firm for several months three years before the crime was committed.

Thus *A* is the Factum Probandum and *C* is advanced as Factum Probans. What is the pattern?   [The notation is devised to help you.   Is the size of the firm relevant?]

**9.** *On inductive research in mathematics and in the physical sciences.*   A difference between "mathematical situations" and "physical situations" that seems to be important from the standpoint of plausible reasoning has been pointed out in sect. 6.   There seem to be other differences of this kind, and one should be discussed here.

Coulomb discovered that the force between electrified bodies varies inversely as the square of the distance.   He supported this law of the inverse square by direct experiments with the torsionbalance.   Coulomb's experiments were delicate and the discrepancy between his theoretical and experimental numbers is considerable.  We cannot help suspecting that without the powerful analogy of Newton's law (the law of the inverse square in gravitational attractions), neither Coulomb himself nor his contemporaries would have considered his experiments with the torsion-balance as conclusive.

Cavendish discovered the law of the inverse square in electrical attractions and repulsions independently of Coulomb.  (Cavendish's researches were not published in his lifetime, and Coulomb's priority is incontestable.)  Yet Cavendish devised a more subtle experiment to support the law.  We need not discuss the details of his method,[10] only one feature of which is essential here: Cavendish considers the possibility that the intensity of the force is not $r^{-2}$ ($r$ is the distance of the electric charges) but, more generally, $r^{-\alpha}$ where $\alpha$ is some positive constant.  His experiment shows that $\alpha - 2$ cannot exceed in absolute value a certain numerical fraction.

Coulomb's experimental investigation is pretty similar to an inductive investigation in mathematics: he confronts particular consequences of a conjectural physical law with the observations, as a mathematician would confront particular consequences of a conjectural number-theoretic law with the observations.  Analogy may play an important rôle in the choice of the conjectural law, here and there.  Yet Cavendish's experimental investigation is of a different character; he does not consider only one conjectural law (the law $r^{-2}$) but several conjectural laws (the laws $r^{-\alpha}$).  These laws are different (different laws of electrical attraction correspond to different values of the parameter $\alpha$) but they are related, they belong to the same "family" of laws.  Cavendish confronts a whole family of laws with the observations and tries to pick out the law that agrees best with them.

This is the most characteristic difference between the two investigations: one aims at *one* conjecture, the other at a *family* of conjectures.  The first compares the observations with the consequences of one conjecture, the other compares them with the consequences of several conjectures simultaneously.  The first tries to judge on the basis of such comparison whether the proposed conjecture is acceptable or not, the other tries to find out the most acceptable (or least unacceptable) conjecture.  The first kind of inductive investigation is widely practised in mathematics and is not unusual in the physical sciences.  The second kind of inductive investigation is widely practised in the physical sciences, but we very seldom meet with it in mathematics.

In fact, the most typical kind of physical experiment aims at measuring some physical constant, at determining its value, as Cavendish's experiment aims at determining the value of the exponent $\alpha$.  In mathematics we could regard this or that investigation as an inductive research aiming at the

---

[10] Cf. J. C. Maxwell, *A treatise on electricity and magnetism*, 2nd ed. (1881), vol. 1, p. 76–82.

determination of some mathematical constant, but such investigations are quite exceptional.[11]

**10.** *Tentative general formulations.* Newton's often repeated "Hypotheses non fingo" is unilateral. It would be a mistake to interpret it as "Beware of conjectures": such advice, if followed, would ruin inductive investigation. Better advice is: Be quick in shaping conjectures; go slow in accepting them. Still better is Faraday's word: "The philosopher should be a man willing to listen to every suggestion, but determined to judge for himself." Of course, the philosopher that Faraday had in mind cultivates the experimental, not the traditional, philosophy.

The present study intends to be an "inductive" investigation of plausible reasoning. I present here, without inhibition, a few tentative generalizations. They apply to a few forms of plausible reasoning, but even their formulation should be carefully scrutinized before any attempt at a further extension.

(1) *Monotonicity.* The considerations of sect. 6 may suggest a rule: "The conclusion of a plausible inference varies monotonically, when one of its premises varies monotonically." This fits the case considered in sect. 6 and a few more cases some of which will be presently considered.

(2) *Continuity.* We shall need a term of demonstrative logic. We say that

$$A \text{ and } B \text{ are equivalent}$$

when $A$ and $B$ imply each other mutually, that is, when $A$ follows from $B$ and also $B$ follows from $A$. If $A$ and $B$ are equivalent, we may not know at present whether $A$ or $B$ is true, but we know that only two cases are possible: either both are true or both are false; $A$ and $B$ stand or fall together. A descriptive symbolic expression for the equivalence of $A$ and $B$ is:

$$A \rightleftarrows B.$$

The two arrows suggest that we can pass from the truth of any one of the two statements $A$ and $B$ to the truth of the other.

In sect. 10 we considered a suggestive connection between two statements $A$ and $B$. We considered the logical relation

$$A \text{ implies } B$$

along with the credibility of $B$ without $A$. Let us imagine that this credibility varies monotonically: $B$ becomes less and less credible without $A$. In the limit when $B$ becomes impossible without $A$, the truth of $B$ implies that of $A$. Yet we supposed that the truth of $A$ implies that of $B$ and so, in the limit, $A$ and $B$ imply each other mutually, become equivalent.

---

[11] By the way, Cavendish's experiment has an even broader scope: it tends to show that the law $r^{-2}$ is more acceptable than any other law $\varphi(r)$, without restricting the function $\varphi(r)$ to the form $r^{-\alpha}$; cf. Maxwell, *ibid.*, p. 76–82.

Let us observe now how the variation described affects our fundamental inductive pattern:

$$A \text{ implies } B$$
$$B \text{ true}$$

$$A \text{ more credible}$$

We assume that the second premise remains unchanged when the first premise varies as described. As $B$ becomes less and less credible without $A$, the conjecture $A$ becomes more and more credible by the verification of its consequence $B$. That is, the conclusion becomes stronger, gains weight. In the limit the conclusion becomes "$A$ is true" and so our pattern of plausible inference becomes in the limit the following (obvious) pattern of demonstrative inference

$$A \text{ and } B \text{ equivalent}$$
$$B \text{ true}$$

$$A \text{ true}$$

In short, our pattern of plausible inference has a "limiting form," which is a pattern of demonstrative inference. As the premises of the plausible inference "tend" to the corresponding premises of the limiting form, the plausible conclusion "approaches" its extreme limiting strength. Still shorter: there is a continuous transition from the heuristic pattern to a demonstrative pattern.

Most of this description fits a few more cases. Some of these are displayed in Table III. The symbols and abbreviations of Table I, explained in sect. 7, are used in Table III. Also the symbol $\leftrightarrows$, explained above, is used. Instead of non-$B$, defined in sect. 4, the shorter symbol $\bar{B}$ is used.

### Table III

| | $A \to B$ | $A \to B$ | $A \mid B$ | $A \mid B$ |
| | $B$ true | $B$ more cr. | $B$ false | $B$ less cr. |
| Approaching: | --- | --- | --- | --- |
| | $A$ more cr. | $A$ s. more cr. | $A$ more cr. | $A$ s. more cr. |
| | $A \rightleftarrows B$ | $A \leftrightarrows B$ | $A \rightleftarrows \bar{B}$ | $A \rightleftarrows \bar{B}$ |
| | $B$ true | $B$ more cr. | $B$ false | $B$ less cr. |
| Limiting: | --- | --- | --- | --- |
| | $A$ true | $A$ more cr. | $A$ true | $A$ more cr. |

(3) *Plausible from demonstrative?* Table III may yield another suggestion. The limiting patterns in this Table are much more obvious than the approaching patterns. Two of the limiting patterns are demonstrative, and the two others, although not purely demonstrative, are hardly questionable. The more debatable approaching patterns which are all patterns of plausible

inference, seem to arise from the corresponding limiting patterns by a uniform "weakening" process: stronger statements as

$$A \rightleftarrows B, \quad A \text{ true}, \quad A \text{ more cr.}$$

are systematically replaced by corresponding weaker statements as

$$A \rightarrow B, \quad A \text{ more cr.}, \quad A \text{ s. more cr.}$$

Could *all* forms of plausible inference be linked in some analogous way to forms of demonstrative, or almost demonstrative, inference?

## Second Part

Ex. 11 should be read first: it introduces (and excuses) what follows.

**11. *More personal, more complex.*** In the foregoing I did not discuss conjectures from my own published mathematical work. This is an omission since, after all, I cannot know any mathematician more intimately than myself. This omission could even appear suspicious to some readers. I do not think that such suspicion is justified. The reason for not discussing more complex specialized topics from my own research is not lack of frankness, but just the complexity and specialization of the topics; I thought it better to discuss simpler topics of more general interest.

The following ex. 12–19 suppose considerably more advanced knowledge than the bulk of the book. They are taken from my own research. I try to offer a representative sample. I include some conjectures that have appeared in print before, and others that have not. I include some conjectures from my "naïve" early work, when I had not begun to think explicitly about the subject of plausible reasoning, and I include conjectures from my later, less naïve work. Ex. 12 dates from my naïve days; it tells of the heuristic grounds that led me to a result. Ex. 13, 14, 15, and 16 deal with formerly published conjectures from my naïve years, ex. 17 with a formerly published conjecture from my less naïve years, and ex. 18 and 19 with conjectures that have not been printed before.[12]

---

[12] Papers by the author of this book are quoted without name in this footnote; "cf." introduces the page on which the conjecture is stated (sometimes in form of a question); a paper quoted with name brought the first proof of the conjecture discussed. *Ex.* 12: *Rendiconti, Circolo Matematico di Palermo*, vol. 34 (1912), p. 89–120. *Ex.* 13: *L'Intermédiaire des Mathématiciens*, vol. 21 (1914), p. 27, qu. 4340; G. Szego, *Math. Annalen*, vol. 76 (1915), p. 490–503. *Ex.* 14: *Math. Annalen*, vol. 77 (1916), p. 497–513; cf. p. 510; F. Carlson, *Math. Zeitschrift*, vol. 9 (1921), p. 1–13. *Ex.* 15: *L'Intermédiaire des Mathématiciens*, vol. 20 (1913), p. 145–146, qu. 4240; G. Szego, *Math. Zeitschrift*, vol. 13 (1922), p. 38. See also *Journal für die reine und angewandte Math.*, vol. 158 (1927), p. 6–18. *Ex.* 16: *Jahresbericht der Deutschen Math. Vereinigung*, vol. 28 (1919), p. 31–40; cf. p. 38. *Ex.* 17: *Proceedings of the National Academy of Sciences*, vol. 33 (1947), p. 218–221; cf. p. 219. *Ex.* 19: *Journal für die reine und angewandte Math.*, vol. 151 (1921), p. 1–31; see theorem I, p. 3.

I should add that even in my naïve years I was impressed and somewhat puzzled by the strength of the confidence with which some of my own conjectures inspired me, and I wondered what sort of reasons might underlie such confidence. The following sentences express fairly well my early views about the source of new conjectures.

"There is a straight line that joins two given points. A new theorem is often a generalization that joins two extreme particular cases, and is obtained by a sort of linear interpolation. There is a straight line with a given direction through a given point. A new theorem is often conceived in the fortunate moment when a general direction of inquiry meets with an appropriate particular case. A new theorem may also result from drawing a parallel."[13]

**12.** *There is a straight line that joins two given points.* The numbers $A_n^{(k)}$ disposed in the infinite square

$$
\begin{array}{cccccc}
A_0^{(1)} & A_1^{(1)} & A_2^{(1)} & A_3^{(1)} & \ldots & A_n^{(1)} \ldots \\
A_0^{(2)} & A_1^{(2)} & A_2^{(2)} & A_3^{(2)} & \ldots & A_n^{(2)} \ldots \\
\cdot & \cdot & \cdot & \cdot & & \cdot \\
A_0^{(k)} & A_1^{(k)} & A_2^{(k)} & A_3^{(k)} & \ldots & A_n^{(k)} \ldots \\
\cdot & \cdot & \cdot & \cdot & \cdot & \cdot \quad \cdot
\end{array}
$$

are connected with a function $f(x)$. (They are derived from the expansion of $f(x)$ in powers of $x$; both $A_n^{(k)}$ and $f(x)$ are real.) E. Laguerre discovered that $V(k)$, the number of changes of sign in the $k$th row $A_0^{(k)}$, $A_1^{(k)}$, $A_2^{(k)}$, $\ldots$ of the array, is connected with $R$, the number of roots of the equation $f(x) = 0$ between 0 and 1 (extremities excluded):

$$V(k) \geqq R,$$

and $V(k)$ can only decrease or remain unchanged as $k$ increases. Will the never increasing $V(k)$ ultimately attain $R$? Laguerre proposed this question, but left it unsolved. M. Fekete proved that $V(k)$ finally attains $R$ provided that $R = 0$, and this particular case suggested the conjecture that $V(k)$ always attains $R$. I was lucky enough to observe another connection between the numbers $A_n^{(k)}$ and the function $f(x)$: there are certain positive numbers $B_n^{(k)}$ (independent of the function $f(x)$) such that

$$\lim_{k \to \infty} A_n^{(k)}/B_n^{(k)} = f(0) \text{ for fixed } n,$$

$$\lim_{n \to \infty} A_n^{(k)}/B_n^{(k)} = f(1) \text{ for fixed } k.$$

That is, the vertical direction in the infinite square array is connected with $f(0)$, the horizontal direction with $f(1)$. These two extreme directions,

---

[13] Cf. G. Polya and G. Szego, *Analysis*, vol. 1, p. VI.

the vertical and the horizontal, may remind you of *two extreme points that demand to be joined*: what about the intermediate oblique directions? Just for a moment, take the conjecture that we aim at proving for granted: it implies some connection between the intermediate directions and the values that $f(x)$ takes when $x$ varies between 0 and 1. This suggests that $A_n^{(k)}/B_n^{(k)}$ may tend to $f(x)$ where $x$ is somehow connected with the limit of $n/k$. In fact, I found finally, and it was even easy to prove, that

$$\lim_{k,\,n \to \infty} A_n^{(k)}/B_n^{(k)} = f(x) \text{ if } \lim n/(n+k) = x.$$

This relation turned out the key to the solution of Laguerre's problem.

**13.** *There is a straight line with a given direction through a given point. Drawing a parallel.* In conjunction with the Fourier series of a positive function $f(x)$,

$$f(x) = a_0 + 2 \sum_1^\infty (a_n \cos nx + b_n \sin nx),$$

O. Toeplitz considered the equation in $\lambda$

$$(*) \quad \begin{vmatrix} a_0 - \lambda & a_1 - ib_1 & \cdots & a_{n-1} - ib_{n-1} \\ a_1 + ib_1 & a_0 - \lambda & \cdots & a_{n-2} - ib_{n-2} \\ & & \cdot & \\ a_{n-1} + ib_{n-1} & a_{n-2} + ib_{n-2} & \cdots & a_0 - \lambda \end{vmatrix} = 0.$$

His research revealed that the $n$ roots of this equation

$$\lambda_{n1}, \quad \lambda_{n2}, \quad \lambda_{n3}, \quad \cdots \quad \lambda_{nn}$$

"imitate" the $n$ equidistant values of the function $f(x)$

$$f\left(\frac{2\pi}{n}\right), \quad f\left(\frac{4\pi}{n}\right), \quad f\left(\frac{6\pi}{n}\right), \cdots f\left(\frac{2n\pi}{n}\right).$$

For example, the arithmetic mean of the $n$ roots

$$\frac{\lambda_{n1} + \lambda_{n2} + \cdots + \lambda_{nn}}{n} = a_0$$

corresponds to

$$\lim_{n \to \infty} \left[ f\left(\frac{2\pi}{n}\right) + f\left(\frac{4\pi}{n}\right) + \cdots + f\left(\frac{2n\pi}{n}\right) \right] \frac{1}{n} = \frac{1}{2\pi} \int_0^{2\pi} f(x)dx = a_0.$$

Let us try to *draw a parallel*: the geometric mean of the $n$ roots

$$[\lambda_{n1}\lambda_{n2} \cdots \lambda_{nn}]^{1/n} = D_n^{1/n}$$

where $D_n$ denotes the determinant with $n$ rows that we obtain in setting $\lambda = 0$ on the left-hand side of the equation $(*)$. This may correspond to

$$(**) \quad \lim_{n \to \infty} \left[ f\left(\frac{2\pi}{n}\right) f\left(\frac{4\pi}{n}\right) \cdots f\left(\frac{2n\pi}{n}\right) \right]^{1/n} = e^{\frac{1}{2\pi} \int_0^{2\pi} \log f(x)\, dx}$$

At this stage it is natural to look for a convenient particular case. For the particular function

$$f(x) = a_0 + 2a_1 \cos x + 2b_1 \sin x$$

there is no difficulty in computing $D_n$ and $\lim_{n \to \infty} D_n^{1/n}$, and this limit turns out to be equal to the value (**): *a general direction of inquiry met with an appropriate particular case,* and it would have been difficult not to state the conjecture: for any positive function $f(x)$

$$\lim_{n \to \infty} D_n^{1/n} = e^{\frac{1}{2\pi} \int_0^{2\pi} \log f(x)\, dx}$$

**14.** *The most obvious case may be the only possible case.* If the coefficients $a_0, a_1, a_2, \ldots a_n, \ldots$ of the power series

$$a_0 + a_1 z + a_2 z^2 + \cdots + a_n z^n + \cdots$$

are integers, and an infinity of these integers are different from 0, the series obviously diverges at the point $z = 1$ since its general term does not tend to 0. Therefore the radius of convergence of such a power series is $\leq 1$. The extreme value 1 of the radius of convergence may be attained. An obvious example is the series

$$1 + z + z^2 + \cdots + z^n + \cdots ;$$

it represents a function of extremely simple analytic character, the rational function $1/(1 - z)$. Another example is

$$z + z^2 + z^6 + \cdots + z^{n!} + \cdots ;$$

it represents a function of extremely complex analytic character, a non-continuable function. (This is the most familiar example of a series for which the circle of convergence is a singular line.) These two obvious examples are of opposite nature. Yet, surprisingly, any power series with integral coefficients and radius of convergence 1, the analytic nature of which could be ascertained, turned out to be similar to one or the other of these opposite examples: it represented either a rational function or a non-continuable function. Theorems by E. Borel and P. Fatou showed that extensive classes of functions of intermediate nature cannot be represented by such a series. As the author succeeded in proving similar theorems and in handling more examples, he unearthed mounting evidence for a conjecture: "The *most obvious* continuable analytic functions represented by a power series with integral coefficients and the radius of convergence 1, certain rational functions, are the *only such* functions." In other words: "If the radius of convergence of a power series with integral coefficients attains the extreme value 1, the function represented is necessarily of an extreme character: it is either extremely simple, a rational function, or extremely complicated, a non-continuable function."

**15.** *Setting the fashion. The power of words.* This example has various aspects which I shall try to discuss one after the other.

(1) E. Laguerre discovered several sequences of real numbers

$$\alpha_0, \; \alpha_1, \; \alpha_2, \; \ldots \; \alpha_n$$

with the following curious property: if the (otherwise arbitrary) equation of degree $n$

(I) $$a_0 + a_1 x + a_2 x^2 + \ldots + a_n x^n = 0$$

has real roots only, also the equation

(II) $$a_0 \alpha_0 + a_1 \alpha_1 x + a_2 \alpha_2 x^2 + \ldots + a_n \alpha_n x^n = 0$$

(into which the sequence transforms (I)) will have real roots only. Laguerre proposed, but left unsolved, the problem: to find a simple necessary and sufficient condition that characterizes this kind of sequences.

It is easy to find necessary conditions: we just apply a sequence of the desired kind to any equation that is known to have real roots only, and then we obtain a transformed equation all roots of which are necessarily real. For example, applying the sequence $\alpha_0, \; \alpha_1, \; \ldots \; \alpha_n$ to the equations

$$1 - x^2 = 0, \qquad x^2 - x^4 = 0, \qquad x^4 - x^6 = 0, \qquad \ldots$$

which, obviously, have real roots only, we easily find that $\alpha_0, \; \alpha_2, \; \alpha_4, \; \alpha_6, \; \ldots$ are necessarily of the *same sign*, all positive, or all negative (in the broad sense: 0 is not excluded). Applying the same sequence to the equations

$$x - x^3 = 0, \qquad x^3 - x^5 = 0, \qquad x^5 - x^7 = 0, \qquad \ldots$$

we find that also $\alpha_1, \; \alpha_3, \; \alpha_5, \; \alpha_7, \; \ldots$ are necessarily of the same sign. Bearing these remarks in mind and applying the sequence to the equation

(III) $$1 + \binom{n}{1} x + \binom{n}{2} x^2 + \ldots + x^n = 0$$

(all $n$ roots of which are equal to $-1$) we obtain the necessary condition that the roots of the equation

(IV) $$\alpha_0 + \binom{n}{1} \alpha_1 x + \binom{n}{2} \alpha_2 x^2 + \ldots + \alpha_n x^n = 0$$

are all real and of the same sign.

We obtained this last condition by accumulating several necessary conditions. Their accumulation should be rather strong: *is it strong enough to form a sufficient condition?* If it were so, the following curious proposition would hold: "If both equations (I) and (IV) have real roots only, and the roots of (IV) are all of the same sign, also the equation (II) has only real roots."

(2) This conjecture came rather early to my mind, but I could not trust it: it looked too strange. In the case $n = 2$, however, the conjecture was easily verified (the case $n = 1$ is completely trivial).

Yet, by chance, I came across a theorem proved by E. Malo: If the equation (I) has only real roots and the equation

$$\alpha_0 + \alpha_1 x + \alpha_2 x^2 + \ldots + \alpha_n x^n = 0$$

has only real roots of the same sign, the equation (II) must have only real roots. Malo's theorem was clearly analogous to that strange conjecture and made it look much less strange. Moreover, Malo's theorem was a consequence of that conjecture, as I could easily see, and so another broad and important consequence of the conjecture had been verified: the conjecture appeared much stronger.

(3) As it is easily seen, the conjecture can be restated as follows: "If the sequence of positive numbers $\alpha_0$, $\alpha_1$, ... $\alpha_n$ transforms the equation (III) into an equation with real roots only, it will transform an arbitrary equation with real roots only into an equation of the same nature." In other words, the equation (III) *sets the fashion*: its response to the sequence $\alpha_0$, $\alpha_1$, ... $\alpha_n$ is imitated by all equations whose roots are all real.

(4) Why does equation (III) set the fashion? Because all its roots coincide. This answer is the "right" answer in some sense. At any rate, as I found out later, equations (or functions) all roots (or zeros) of which coincide play an analogous role in several analogous problems: they set the fashion (they are "*tonangebend*").

Yet, as the conjecture in question was still a conjecture, I resorted to the following "explanation": All roots of the equation (III) are equal to $-1$. These roots are all crowded into one point of the real axis, they are as close together as possible. In such a situation they are, understandably, the most inclined to jump out of the real axis. Therefore, if the sequence $\alpha_0$, $\alpha_1$, ... $\alpha_n$ applied to the polynomial $(1 + x)^n$ with the most crowded roots does not succeed in driving out these roots of the real axis, it has still less chance to drive out the less crowded roots of other polynomials.

The logical value of this "explanation" is obviously nil, but that does not imply that its psychological value is also nil. I am convinced that this playful analogy was extremely important for me personally: it helped to keep the conjecture alive for years.

I should mention here that similar quaint verbal formulations have often been connected with my mathematical work. The sentence at the end of ex. 14 is a characteristic instance. Here are two more examples.

For more than two decades I was much interested in Fabry's well known gap-theorem on power series. There were two periods: a first "contemplative" period, and a second "active" period. In the active period I did some work connected with the theorem, and found various proofs, extensions, and analogues to it. In the contemplative period I did practically no work connected with the theorem, I just admired it, and recalled it from time to time in some curious, far fetched formulation as the following: "If it is infinitely improbable that in a power series a coefficient chosen at random be different from 0, then it is not only infinitely probable, but certain, that the power series is not continuable." Obviously, this sentence has neither logical, nor literary merit, but it served me well in keeping my interest alive.

The idea of a certain proof occurred to me fairly clearly, but for several days after that I did not attempt to work out the final shape of the proof. During these days I was obsessed by the word "transplantation." In fact, this word describes the decisive idea of the proof as precisely as it is possible for any single word to describe such a complex thing.

I made up, of course, various explanations for this "power of words" but, perhaps, it is better to wait with explanations till there are more examples.[14]

**16.** *This is too improbable to be a mere coincidence.* Let $f$ denote the number of prime factors of the integer $n$, and let us call $n$ "evenly factorized" or "oddly factorized" according as $f$ is even or odd. For example:

$$30 = 2 \times 3 \times 5 \quad \text{is oddly factorized}$$

$$60 = 2 \times 2 \times 3 \times 5 \quad \text{is evenly factorized.}$$

Prime numbers as 2, 3, 5, 7, 11, 13, 17, . . . are oddly factorized, squares as 4, 9, 16, . . . are evenly factorized, and the number 1 has to be regarded as evenly factorized since it has no prime factors and 0 is an even number. Among the twelve first numbers

| 1 | 2 | 3 | 4 | 5 | 6 | 7 | 8 | 9 | 10 | 11 | 12 |
|---|---|---|---|---|---|---|---|---|----|----|----|
| e | o | o | e | o | e | o | o | e | e  | o  | o  |

five are evenly factorized, and seven oddly.

If we look at the succession of e's and o's in the foregoing scheme, we can scarcely detect a simple rule. The two kinds of numbers appear to alternate irregularly, unpredictably, *at random.* The idea of chance occurs to us almost unavoidably: it is natural to think that we would obtain a similar sequence if, instead of taking the trouble to factorize a proposed integer, we just flipped a coin and wrote "e" or "o" according as head or tail turns up.

---

[14] Cf. J. Hadamard, The psychology of invention in the mathematical field, p. 84–85.

It is also natural to suspect that the coin is "fair," that heads and tails turn up about equally often, that the two kinds of integers, the evenly and the oddly factorized, are equally frequent.

Now it can be proved (the proof is difficult) that among the first $n$ integers about as many are evenly factorized as oddly factorized if $n$ is large. (The ratio tends to 1 as $n$ tends to $\infty$.)    This seems to corroborate our suspicions. Now you would more confidently expect that the evenly and oddly factorized numbers follow each other as a random succession of heads and tails. Expecting so, I started listing for each $n$ which kind of integers, the oddly or the evenly factorized, are in majority among the first $n$.    The first listings looked lopsided.    I was surprised and proceeded to larger values of $n$: it was still lopsided.    I was amazed and proceeded to still larger values of $n$, but it was still lopsided.    I got tired of numerical computation as I reached $n = 1500$ and had to admit that there is some observational evidence for the conjecture: "For $n \geq 2$, the evenly factorized integers are never in majority among the first $n$ integers."

You toss a coin 1500 times.    You count how many heads and how many tails you obtain in the first $n$ trials.    It can easily happen that you obtain less heads than tails. (Take "less" in the wide sense, as "not more.")    It could even happen that you obtain less heads than tails all the time from the second step on, for $n = 2, 3, 4, \ldots 1500$, but this cannot so easily happen.    *This is too improbable to be a mere coincidence*, we are tempted to say.    Yet the improbable did happen and has been observed as we played heads and tails with the factorized integers.    *There must be some reason.*    I did what I had to do: I took the conjecture for granted (tentatively, of course) and tried to derive consequences from it.    I was lucky enough to observe two points.    First, if the new conjecture were true, the truth of a far more important conjecture due to Riemann (on the $\zeta$-function) would necessarily follow.    Second, if a little more than the new conjecture were true (if the evenly factorized numbers were definitely in the minority from a certain $n$ upward) another important conjecture due to Gauss (on the classnumber of quadratic forms) would necessarily follow.    Both points seemed to speak in favor of the new conjecture.

The conjecture is no longer new, but its fate is still undecided.    A. E. Ingham derived consequences from it that may tend to render it less credible. On the other hand, D. H. Lehmer verified it by numerical computation up to $n = 600,000$.

**17.** *Perfecting the analogy.*    "Of all solids with a given volume, the sphere has the minimum surface."    This is the classical isoperimetric theorem in space, a remarkable physical analogue to which was discovered by H. Poincaré and strictly proved by G. Szego: "Of all solids with a given volume the sphere has the minimum electrostatic capacity."    It is natural to think that there should be further analogous theorems, and I was looking for such a theorem.    The field of force around a solid charged with electricity is similar to the field of flow around a solid moving with uniform velocity across an incompressible ideal fluid. (Both fields are sourceless and irrotational, and therefore satisfy the same partial differential equation.) The capacity of the solid in the electrostatic field corresponds roughly to the "virtual mass" of the solid in the hydrodynamic field. (The moving body stirs up the fluid and adds to its own kinetic energy that of the moving fluid; the "virtual mass" is a factor of this additional kinetic energy.) Both the capacity and the virtual mass are connected with the energy of the corresponding field. Yet there is a conspicuous difference: the capacity depends only on the shape and size of the solid, but the virtual mass depends also on the direction of the motion of the solid. In order to render the analogy more perfect, I had to construct a new concept: averaging the virtual mass over all the possible directions, we obtain the *average virtual mass*. And so the conjecture arose: "Of all solids with a given volume, the sphere has the minimum average virtual mass."

The ellipsoid is the only shape for which the virtual mass has been explicitly computed in all directions. In fact, it turned out that of all ellipsoids with a given volume the sphere has the minimum average virtual mass: the conjecture has been verified in an important particular case. I could support the conjecture also by analogy: I succeeded in proving the analogous hydrodynamical minimum property of the circle in two dimensions. So supported, the conjecture deserved to be publicly stated, at least I thought so.

The conjecture is neither proved nor disproved at this date, although G. Szegö and M. Schiffer have found interesting connected results which seem to support it.

**18.** *A new conjecture.* We consider the plane in which the rectangular coordinates are $x$ and $y$, and in this plane a region $R$ surrounded by a closed curve $C$. We seek a function $u = u(x, y)$ that satisfies one or the other of the two boundary conditions

(1) $\quad u = 0$,

(2) $\quad \dfrac{\partial u}{\partial n} = 0$

along the curve $C$ and the partial differential equation

(3) $\quad \dfrac{\partial^2 u}{\partial x^2} + \dfrac{\partial^2 u}{\partial y^2} + vu = 0$

within the region $R$. In (2) $n$ denotes the normal to the curve $C$; in (3) $v$ denotes a certain constant. Thus we have two different problems: in the first, we have to solve the differential equation (3) with the boundary condition (1); in the second with (2). Both problems are important in physics in connection with various phenomena of vibration. Both problems have the same trivial solution: $u = 0$ identically. Only for particular values of $\nu$ has one or the other problem a non-trivial, that is, not identically vanishing solution $u$: the problem with the boundary condition (1) for $\nu = \lambda_1, \lambda_2, \lambda_3, \ldots$ , the problem with the boundary condition (2) for $\nu = \mu_1, \mu_2, \mu_3, \ldots$ . It is

$$0 < \lambda_1 < \lambda_2 \leqq \lambda_3 \leqq \lambda_4 \leqq \ldots ,$$

$$0 = \mu_1 < \mu_2 \leqq \mu_3 \leqq \mu_4 \leqq \ldots .$$

These exceptional values $\lambda_1, \lambda_2, \lambda_3, \ldots$ and $\mu_1, \mu_2, \mu_3, \ldots$ are called the *eigenvalues* of the first and the second problem, respectively. The eigenvalues are connected with the frequencies of the characteristic vibrations in the corresponding physical phenomena.

I wish to state a new conjecture: *Let $A$ denote the area of the region $R$; then, for $n = 1, 2, 3, \ldots$*

$$\mu_n < 4\pi n A^{-1} < \lambda_n.$$

This is a challenging conjecture. As the shape of the region $R$ can arbitrarily vary and $n$ ranges over all the integers $1, 2, 3, \ldots$ , the conjecture covers an immense variety of particular cases of which not too much is known: the conjecture could be exploded by a numerical result concerning any of these particular cases. Yet the conjecture is not too badly supported.

(a) The conjecture is verified for $n = 1, 2, 3, \ldots$ when $R$ is a rectangle. This was, in fact, the particular case that suggested the conjecture.

(b) The conjecture is verified for $n = 1, 2$ when $R$ is of arbitrary shape. This particular case is very different from that mentioned under (a).

(c) The conjecture has been verified by numerical computation in a few particular cases in which the eigenvalues can be explicitly computed: for $n$ up to 25 and for special shapes of $R$ (the circle, a few circular sectors, a few triangles).

(d) It has been known for a long time that

$$\mu_1 \leqq \lambda_1, \quad \mu_2 \leqq \lambda_2, \quad \mu_3 \leqq \lambda_3, \ldots .$$

This agrees with the conjecture.

(e) It is also known that

$$\lim_{n \to \infty} \mu_n n^{-1} = \lim_{n \to \infty} \lambda_n n^{-1} = 4\pi A^{-1}$$

and this also agrees with the conjecture.

Of course, no amount of such verification can prove the conjecture or oblige anybody to believe in it to any degree. Yet such verifications can enhance the interest of the conjecture, spur us on to testing further consequences, and add a zest to scientific investigation which is the *raison d'être* of a conjecture.

Technical details about the conjecture stated (for some of which I am greatly indebted to Mr. Peter Szego) will be published elsewhere.

**19.** *Another new conjecture*: "If $F'(x)$ is an algebraic function and all the coefficients $a_1, a_2, a_3, \ldots$ of the series

$$F(x) = a_1 x + a_2 x^2 + a_3 x^3 + \cdots$$

are integers, also $F(x)$ is an algebraic function." In other words: "Unless the integral of an algebraic function is an algebraic function itself, it cannot be represented by a power series with integral coefficients."

For an illustration, consider the expansion

$$\arcsin 2x = 2 \sum_{0}^{\infty} \binom{2n}{n} \frac{x^{2n+1}}{2n+1}$$

The derivative of arc sin $2x$ is an algebraic function and all the coefficients in the expansion of this derivative are integers as we see from the above formula. Yet arc sin $2x$ is not an algebraic function. Therefore, if the proposed conjecture is true, there must be infinitely many coefficients in the above expansion that are not integers. This is easy to verify: if $2n + 1$ is a prime number, it is not a divisor of $2n!$ Thus the case examined supports the conjecture, which is also supported (and has been suggested) by the following fact: if we substitute "rational function" for "algebraic function" in the proposed conjectural statement, we change it into a true and proven statement. The conjecture is also supported by rather vague analogies, by the "atmosphere" surrounding the subject of power series with integral coefficients; see ex. 14.

There are many cases similar to that of arc sin $2x$ that could be easily tested but have not been examined yet. Without a thorough investigation of such accessible consequences such a conjecture should not be printed. I publish it here as an example of an undeveloped, rather imperfectly supported, conjecture.

**20.** *What is typical?* As far as I can see, there is nothing in the last examples (ex. 12–19) that would disagree with the general impression derived from the foregoing examples. In these last examples, as in the others, the conjecture in question was doubly supported: by some clear facts, and by a "general atmosphere." The former kind of support, by clear facts, seems to me well within the patterns outlined in this chapter and the foregoing: analogy and verified consequences are prominent, merely plausible consequences also play some rôle; support from two very different sides appears as vital. On the whole, ex. 12–19 appear to be typical.

None of the conjectures mentioned in ex. 13–17 has been refuted (to this date, at least). This could look atypical. I have not mentioned here, however, countless other conjectures of mine that have been refuted after a few minutes, or hours, or days; such shortlived conjectures are usually quickly forgotten. Of course, I published only conjectures that passed all the obvious tests I could think of and outlasted at least a few months' work; these are the sturdiest conjectures, with the best chances of survival. Typically, research consists in shaping many conjectures, exploding most of them, and establishing a few.

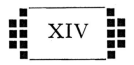

# XIV

## CHANCE, THE EVER-PRESENT RIVAL CONJECTURE

*. . . the probability that this coincidence is a mere work of chance is, therefore, considerably less than* $(1/2)^{60}$ *. . . Hence this coincidence must be produced by some cause, and a cause can be assigned which affords a perfect explanation of the observed facts.*—G. KIRCHHOFF[1]

**1. Random mass phenomena.** Everyday speech uses the words "probable," "likely," "plausible," and "credible" in meanings which are not sharply distinguished. Now, we single out the word "probable" and we shall learn to use this word in a specific meaning, as a technical term of a branch of science which is called the "Theory of Probability."[2]

This theory has a great variety of applications and aspects and, therefore, it can be conceived and introduced in various ways. Some authors regard it as a purely mathematical theory, others as a kind or branch of logic, and still others as a part of the study of nature. These various points of view may or may not be incompatible. We have to start by studying one of them, but we should not commit ourselves to any of them. We shall change our position somewhat in the next chapter, but in the present chapter we choose the viewpoint which is the most convenient in the great majority of applications and which the beginner can master most quickly. We regard here the theory of probability as a part of the study of nature, as the theory of certain observable phenomena, the *random mass phenomena.*[3] We can understand pretty clearly what this term means if we compare a few familiar examples of such phenomena.

---

[1] *Abhandlungen der k. Akademie der Wissenschaften,* Berlin, 1861, p. 78–80.

[2] In the foregoing, the words "probable" and "probability" have been sometimes used in a non-technical sense, but this will be carefully avoided in the present chapter and the next. The words "likely" and "likelihood" will be introduced as technical terms later in this chapter.

[3] In this essential point, and in several other points, the present exposition follows the views of Richard von Mises although it deviates from his definition of mathematical probability; cf. his book *Probability, Statistics, and Truth.*

(1) *Rainfall.* Rainfall is a mass phenomenon. It consists of a very grea  
number of single events, of the fall of a very great number of raindrops  
These raindrops, although very similar to each other, differ in variou  
respects: in size, in the place where they strike the ground, etc. There i  
something in the behavior of the raindrops that we properly describe a  
"random." In order to understand clearly the meaning of this term let u  
imagine an experiment.

Let us observe the first drops on the pavement as the rain starts falling  
We observe the pavement in the middle of some large public square  
sufficiently far from buildings or trees or anything that could obstruct th  
rain. We focus our attention on two stones which we call the "right-hanc  
stone" and the "left-hand stone." We observe the drops falling on thes  
stones and we note the order in which they strike. The first drop falls o  
the left-hand stone, the second drop on the right, the third again right, th  
fourth left, and so on, without apparent regularity as

$$L\ R\ R\ L\ L\ L\ R\ L\ R\ L\ R\ R\ L\ R\ R$$

(*R* for right, *L* for left). There is no regularity in this succession of the rain  
drops. In fact, having observed a certain number of drops, we canno  
reasonably predict which way the next drop will fall. We have notec  
above fifteen entries. Looking at them, can we predict what the sixteentl  
entry will be, *R* or *L*? Obviously, we cannot. On the other hand, there i  
some sort of regularity in the succession of the raindrops. In fact, we ca  
confidently predict that at the end of the rain the two stones will be equall  
wet. That is, the number of drops striking each stone will be very nearl  
proportional to the area of its free horizontal surface. Nobody doubts tha  
this is so, and the meteorologists certainly assume that this is so in con  
structing their rain-gauges. Yet there is something paradoxical. We ca  
foresee what will happen in the long run, but we cannot foresee the details  
The rainfall is a typical random mass phenomenon, *unpredictable in certai  
details, predictable in certain numerical proportions of the whole.*

(2) *Boys among the newborn.* In a hospital, the newborn babies are registerec  
in order as they are born. Boys and girls (*B* and *G*) follow each othe  
without apparent regularity as

$$G\ B\ B\ G\ B\ G\ B\ B\ G\ G\ B\ B\ B\ G\ G$$

Although we cannot predict the details of this random succession, we ca  
well predict an important feature of the final result obtained by summing u  
all such registrations in the United States during a year: the number of th  
boys will be greater than the number of the girls and, in fact, the ratio o  
these two numbers will be little different from the ratio 51.5 : 48.5. Th  
number of births in the United States is about 3 millions per year. W  
have here a random mass phenomenon of considerable dimensions.

(3) *A game of chance.* We toss a penny repeatedly, noting each time which side it shows, "heads" or "tails" ($H$ or $T$). We obtain so a succession without apparent regularity as

$$T\ H\ H\ H\ T\ H\ T\ H\ H\ T\ H\ T\ H\ T\ H\ T\ T$$

If we have the patience to toss the penny a few hundred times, a definite ratio of heads to tails emerges, which does not change much if we prolong our experiment still further.   If our penny is "unbiased," the ratio 50 : 50 of heads to tails should appear in the long run.   If the penny is biased, some other ratio will come into view.   At any rate we see again the characteristic features of a random mass phenomenon.   Constant proportions emerge in the long run, although the details are unpredictable.   There is a certain aggregate regularity, in spite of the irregularity of the individual happenings.

**2. The concept of probability.**   In the year 1943 the number of births in the United States, male, female, and total, was

$$1{,}506{,}959 \qquad 1{,}427{,}901 \qquad 2{,}934{,}860,$$

respectively.   We call

| | |
|---|---|
| 1,506,959 | the frequency of the male births, |
| 1,427,901 | the frequency of the female births. |

We call

$$\frac{1{,}506{,}959}{2{,}934{,}860} = 0.5135$$

the relative frequency of the male births and

$$\frac{1{,}427{,}901}{2{,}934{,}860} = 0.4865$$

the relative frequency of the female births.   In general, if an event of a certain kind occurs in $m$ cases out of $n$, we call $m$ the *frequency* of occurrence of that kind of event and $m/n$ its *relative frequency.*

Let us imagine that, throughout the whole year, the births are successively registered in the whole United States (as in the hospital that we have mentioned in the foregoing section).   If we look at the succession of male and female births, we have before us an extremely long series of almost three million entries beginning like

$$G\ B\ B\ G\ B\ G\ B\ B\ G\ G\ B\ B\ B\ G\ G.$$

As the mass phenomenon unfolds, we have, at each stage of the observation, a certain frequency of male births and also a certain relative frequency.   Let us note, after 1, 2, 3, . . . observations, the frequencies and the relative frequencies found up to that point:

| Observations | Event | Frequency of B | Relative frequency |
|---|---|---|---|
| 1 | G | 0 | 0/1 = 0.000 |
| 2 | B | 1 | 1/2 = 0.500 |
| 3 | B | 2 | 2/3 = 0.667 |
| 4 | G | 2 | 2/4 = 0.500 |
| 5 | B | 3 | 3/5 = 0.600 |
| 6 | G | 3 | 3/6 = 0.500 |
| 7 | B | 4 | 4/7 = 0.571 |
| 8 | B | 5 | 5/8 = 0.625 |
| 9 | G | 5 | 5/9 = 0.556 |
| 10 | G | 5 | 5/10 = 0.500 |
| 11 | B | 6 | 6/11 = 0.545 |
| 12 | B | 7 | 7/12 = 0.583 |
| 13 | B | 8 | 8/13 = 0.615 |
| 14 | G | 8 | 8/14 = 0.571 |
| 15 | G | 8 | 8/15 = 0.533 |

As far as we have tabulated it, the relative frequency oscillates pretty strongly (between the limits 0.000 and 0.667). Yet we have here only a very small number of observations. As we go further and further, the oscillations of the relative frequency will become less and less violent, and we can confidently expect that in the end it will oscillate very little about its final value 0.5135. *As the number of observations increases, the relative frequency appears to settle down to a stable final value, in spite of all the unpredictable irregularities of detail.* Such behavior, the emergence of a stable relative frequency in the long run, is typical of random mass phenomena.

An important aim of any theory of such phenomena must be to predict the final stable relative frequency or *long range relative frequency.* We have to consider the *theoretical value of long range relative frequency* and we shall call this theoretical value *probability.*

We wish to clarify this concept of probability. Naturally, we begin with the study of mass phenomena for which we can predict the long range relative frequency with some degree of reasonable confidence.

(1) *Balls in a bag.* A bag contains $p$ balls of various colors among which there are exactly $f$ white balls. We use this simple apparatus to produce a random mass phenomenon. We draw a ball, we look at its color and we write $W$ if the ball is white, but we write $D$ if it is of a different color. We put back the ball just drawn into the bag, we shuffle the balls in the bag, then

we draw again one and note the color of this second ball, $W$ or $D$.   In proceeding so, we obtain a random sequence similar to those considered in sect. 1:

$$W\ D\ D\ D\ W\ D\ D\ W\ W\ D\ D\ D\ W\ W\ D.$$

What is the long range relative frequency of the white balls?

Let us discuss the circumstances in which we can predict the desired frequency with reasonable confidence.   Let us assume that the balls are homogeneous and exactly spherical, made of the same material and having the same radius.   Their surfaces are equally smooth, and their different coloration influences only negligibly their mechanical behavior, if it has any influence at all.   The person who draws the balls is blindfolded or prevented in some other manner from seeing the balls.   The position of the balls in the bag varies from one drawing to the other, is unpredictable, beyond our control.   Yet the permanent circumstances are well under control: the balls are all the same shape, size, and weight; they are *undistinguishable* by the person who draws them.

Under such circumstances we see no reason why one ball should be preferred to another and we naturally expect that, in the long run, each ball will be drawn approximately *equally often*.   Let us say that we have the patience to make 10,000 drawings.   Then we should expect that each of the $p$ balls will appear about

$$\frac{10,000}{p}\ \text{times.}$$

There are $f$ white balls.   Therefore, in 10,000 drawings, we expect to get white

$$f\frac{10,000}{p} = 10,000\frac{f}{p}\ \text{times};$$

this is the expected frequency of the white balls.   To obtain the relative frequency, we have to divide the frequency by the number of observations, or drawings, that is, by 10,000.   And so we are led to the statement: the long range relative frequency, or *probability*, of the white balls is $f/p$.

The letters $f$ and $p$ are chosen to conform to the traditional mode of expression.   As we have to draw one of the $p$ balls, we have to choose one of $p$ *possible* cases.   We have good reasons (equal condition of the $p$ balls) not to prefer any of these $p$ possible cases to any other.   If we wish that a white ball should be drawn (for example, if we are betting on white), the $f$ white balls appear to us as *favorable* cases.   Hence we can describe the probability $f/p$ as the *ratio of the number of favorable cases to the number of possible cases*.

Pulling a ball from a bag, putting the ball back into the bag, shaking the bag, pulling another ball, and repeating this $n$ times seems to be a pretty silly occupation.   Do we waste our time in studying such a primitive game?

I do not think so.   The bag and the balls, handled in the described manner, generate a random mass phenomenon which is particularly simple and accessible.   Generalization naturally starts from the simplest, the most transparent particular case.   The science of dynamics was born when Galileo began studying the fall of heavy bodies.   The science of probability was born when Fermat and Pascal began studying games of chance which depend on casting a die, or drawing a card from a pack, or drawing a ball from a bag.   The fundamental concepts and laws of dynamics can be extracted from the simple phenomenon of falling bodies.   We use the bag and the balls to understand the fundamental concept of probability.

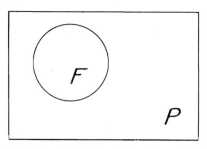

Fig. 14.1.   Probability defined by rainfall.

(2) *Rainfall.*   We return to the consideration of the random mass phenomenon from which we started in sect. 1.   The area of a horizontal surface is $P$ and the area of a certain portion of this surface is $F$; see fig. 14.1.   We observe the raindrops falling on this area $P$ and we are interested in the frequency of the raindrops falling on the subarea $F$.   We are inclined to predict without hesitation the long range relative frequency: the fraction of the total rain over the area that falls on the subarea will be very nearly $F/P$ if the rain consists of more than a few drops.   In other words, the probability that a raindrop striking the surface of area $P$ should strike the portion of area $F$ is $F/P$.   If we idealize the rainfall and consider a raindrop as a geometric point, we can also say: the probability that a point falling in the area $P$ should fall in the subarea $F$ is $F/P$.

In the last statement we consider each point of the area $P$ as a possible case and each point of the subarea $F$ as a favorable case.   The number of favorable cases as that of the possible cases is infinite, and it would not make sense to talk about the ratio of infinite numbers.   We can consider, however, the area of a surface as the *measure* of the points contained in the surface.   Using this term, we can describe the probability $F/P$ as the *ratio of the measure of favorable cases to the measure of possible cases.*

   **3. Using the bag and the balls.**   In deriving the fundamental principle of statics, Lagrange replaced an arbitrary system of forces by a suitable system of pulleys.   In the light of this Lagrangeian argument (the details of which are not needed here[4]) any case of equilibrium appears as a suitable combination of correctly balanced pulleys. The Calculus of Probability can be viewed in a similar manner; in fact, such a view is suggested by the early history of this science. Seen from this standpoint, any problem of probability appears comparable to a suitable problem about bags containing balls, and any random mass phenomenon appears as similar in certain essential respects to successive drawings of balls from a system of suitably combined bags. Let us illustrate this by a few simple examples.

(1) Instead of tossing a fair penny, we can draw a ball from a bag containing just two balls, one of which is marked with an $H$ and the other with a $T$ (heads and tails). Instead of casting an unbiased die, we can draw a ball from a bag containing exactly six balls, marked with 1, 2, 3, 4, 5, or 6 spots, respectively. Instead of drawing a card from a pack of cards, we can draw a ball from a bag containing 52 balls, suitably marked. It seems to be intuitively clear that substituting a bag with balls for pennies, dice, cards, and other similar contrivances in a suitable way, we do not change the odds in the usual games of chance. At least, we do not change the chances in that idealized version of these games in which the contrivances used (pennies, dice, etc.) are supposed to be perfectly symmetrical and, correspondingly, certain fundamental chances perfectly equal.

(2) Wishing to study the randomness in the distribution of boys and girls among the newborn, we may substitute for the actual mass phenomenon successive drawings from a bag containing 1,000 balls, 515 marked with $B$ and 485 marked with $G$. This substitution is, of course, theoretical and, as every theory is bound to be, it is tentative and approximate. Yet the point is that the bag and the balls enable us to formulate a theory.

(3) A meteorologist registers the succession of rainy and rainless days in a certain locality. His observations seem to show that, on the whole, each day tends to resemble the foregoing day: rainless days seem to follow rainless days more easily than rainy days and, similarly, rainy days seem to follow rainy days more easily than rainless days. Of course, a dependable regularity appears only in a long series of observations; the details are irregular, seem to be random.

The meteorologist may wish to express more clearly his impressions that we have just sketched. If he wishes to formulate a theory in terms of probability, he may consider three bags. Each bag contains the same number of balls, let us say 1,000 balls. Some of the balls are white, the others are black (white for rainless, black for rainy). Yet there are important differences between the bags. Each bag bears an inscription, easily visible to the person who draws the balls. One bag is inscribed "START," another "AFTER WHITE," and the third "AFTER BLACK." The ratio of balls of different color is different in different bags. In each bag the ratio of white balls to

---

4 See E. Mach, *Die Mechanik*, p. 59–62.

black balls approximates the observable ratio of rainless days to rainy days, but in different circumstances. In the bag "START" the ratio is that of rainless days to rainy days throughout the year, in the bag "AFTER WHITE" the ratio is that of rainless days to rainy days following a rainless day, and in the bag "AFTER BLACK" the ratio is that of rainless days to rainy days following a rainy day. Therefore, the bag "AFTER WHITE" contains *more* white balls than the bag "AFTER BLACK." The balls are drawn successively and each ball drawn, when its color has been noticed, is replaced into the bag from which it was drawn. The bag "START" is used but once, for the first ball. If the first ball is white, we use the bag "AFTER WHITE" for the second ball, but if the first ball is black, the second ball is drawn from the bag "AFTER BLACK." And so on, the color of the ball just drawn determines the bag from which the next ball should be drawn.

It is just a theory that the succession of white and black balls drawn under the described circumstances imitates the succession of rainless and rainy days with a reasonable approximation. Yet, on the face of it, this theory does not seem to be out of place. At any rate, this theory, or some similar theory, could deserve to be confronted with the observations.

(4) Take any English text (from Shakespeare, if you prefer) and replace each of the letters *a, e, i, o, u,* and *y* by *V* and each of the remaining twenty letters by *C.* (*V* means vowel and *C* means consonant.) You obtain a pattern as

$$C\,V\,C\,V\,V\,C\,C\,V\,C\,C\,V\,C\,V\,C\,C\,.$$

This irregular sequence is in some way opposite to that discussed in the foregoing subsection (3): each day tends to be like the foregoing day, but each letter tends to be unlike the foregoing letter. Still, we could imitate the succession of vowels and consonants by a succession of white and black balls drawn from three bags bearing the same inscriptions as before (in subsection (3)), yet the ratio of white balls to black balls should not be the same as before. To imitate realistically the succession of vowels and consonants the bag "AFTER WHITE" should contain *less* white balls than the bag "AFTER BLACK."

(5) There are two bags. The first bag contains *p* balls among which there are *f* white balls. The second bag contains *P* chips among which there are *F* white chips. Using both hands, I draw from both bags at the same time, a ball with the left hand and a chip with the right hand. What is the probability that both the ball and the chip turn out to be white?

We could, of course, repeat this primitive experiment sufficiently often, perhaps a thousand times, and so obtain an approximate value for the desired probability. Yet we can also try to guess it, and that is more interesting.

The result of the two simultaneous drawings is a "couple," consisting of a ball and a chip. There are *p* balls and *P* chips. As any ball can be coupled with any chip, there are *pP* possible couples; they are shown in fig. 14.2 where $p = 5, f = 2, P = 4, F = 3$. There is no reason to prefer any of the $p$ balls to any other ball, or any of the $P$ chips to any other chip. There seems to be *no reason to prefer any of the pP couples to any other couple.* In fact, in performing the experiment with the two bags, I am supposed to proceed blindly, at random, so that each hand draws independently of the other. "Let not thy left hand know what thy right hand doeth." It seems incredible that the chances of the ball that I draw with my left hand should be influenced by the chip that I draw with my right hand. Why should ball no. 1 be any more attracted by chip no. 1 than it is by chip no. 2?

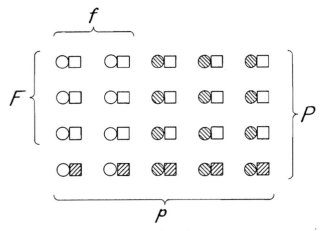

Fig. 14.2. Independent events.

And so we can imagine a bag, containing $pP$ mechanically indistinguishable objects (each object is a couple, a ball attached to a chip); one drawing from this one bag appears *equivalent* to the two simultaneous drawings from the two bags described at the outset. We have so $pP$ possible cases; it remains to find the number of favorable cases. A glance at fig. 14.2 shows that there are $fF$ couples consisting of a white ball and a white chip. And so we obtain the value of the desired probability: it is

$$\frac{fF}{pP} = \frac{f}{p} \cdot \frac{F}{P}$$

the *product* of two probabilities. In fact, $f/p$ is the probability of drawing a white ball from the first bag, and $F/P$ the probability of drawing a white chip from the second bag.

The essential point about the ball and the chip is that the drawing of one does not influence the chances of the other. In the usual terminology of the calculus of probability, such events are called *independent* of each other; the joint happening of both events is viewed as a *compound* event. The foregoing consideration motivates the rule: *The probability of a compound event is the product of the probabilities of the constituent events, provided that these constituent events are mutually independent.*

**4. The calculus of probability. Statistical hypotheses.** The theory of probability, as we see it, is a part of the study of nature, the theory of random mass phenomena.

The most striking achievement of the physical sciences is prediction. The astronomers predict with precision the eclipses of the sun and the moon, the position of the planets, and the return of comets which evade observation for several years. A great astronomer (Leverrier) succeeded even in predicting the position of a planet (Neptune), the very existence of which was not known before. The theory of probability predicts the frequencies in certain mass phenomena with some amount of success.

The astronomers base their predictions on former observations, on the laws of mechanics, the law of gravitation, and long difficult computations. Any branch of physical science bases its predictions on some theory or, we can say, on some conjecture, since no theory is certain and so every theory is a more or less reasonable, more or less well-supported, conjecture. In trying to predict the frequencies in a certain random mass phenomenon from the theory of probability we have to make some theoretical assumption about the phenomenon. Such an assumption, which has to be expressed in terms of probability concepts, is called a *statistical hypothesis.*

When we apply the theory of probability we have to compute probabilities (which are theoretical, approximate values of relative frequencies). When we try to find a probability, we have a problem to solve. The unknown of this problem is the desired probability. Yet, in order to determine this unknown, we need data and conditions in our problem. The data are usually probabilities and the conditions, on which the relation of the unknown probability to the given probabilities depends, constitute a statistical hypothesis.

As in the applications of the theory of probability the computation of probabilities plays a prominent role, this theory is usually called the *calculus of probability.* Thus, the aim of the calculus of probability is to compute new probabilities on the basis of given probabilities and given statistical hypotheses.

The reader who wishes to peruse the remaining part of this chapter must either know the elements of the calculus of probability, or he must take for granted certain results derived from these elements. Most of the time, the text will state the results without derivation; derivations will be given subsequently, in the First Part of the Exercises and Comments following this chapter, and in the corresponding Solutions. Yet even if the reader does not check the derivation of the results, he ought to have some insight into the underlying theoretical assumptions. We can make such assumptions intuitively understandable: we compare the random mass phenomenon that we examine to drawings from suitably filled bags under suitable conditions, as in the foregoing sect. 3.

The applications of the calculus of probability are of unending variety. The following sections of this chapter attempt to illustrate the principal types of applications by suitable elementary examples. Stress will be laid on the motivation of these applications, that is, on such preliminary considerations as make the choice of procedure plausible.

**5. Straightforward prediction of frequencies.** At the beginning of its history the calculus of probability was essentially a theory of certain games of chance. Yet the predictions of this theory were not tested experimentally on a large scale until modern times. We begin by discussing an experiment of this kind.

(1) W. F. R. Weldon cast 12 dice 26,306 times, noting each time how many of these 12 dice have shown more than four spots.[5] The results of his observations are listed in column (4) of Table I; column (1) shows the number of the dice among the 12 that have turned up five or six spots. Thus, in 26,306 trials it never happened that all twelve dice showed more than four spots. The most frequent case was that in which four out of the twelve dice showed five or six spots; this happened 6,114 times.

Table I

| (1) Nr. of 5 or 6 | (2) Excess I | (3) Predicted I | (4) Observed | (5) Predicted II | (6) Excess II |
|---|---|---|---|---|---|
| 0 | + 18 | 203 | 185 | 187 | + 2 |
| 1 | + 67 | 1216 | 1149 | 1146 | − 3 |
| 2 | + 80 | 3345 | 3265 | 3215 | − 50 |
| 3 | + 101 | 5576 | 5475 | 5465 | − 10 |
| 4 | + 159 | 6273 | 6114 | 6269 | + 155 |
| 5 | − 176 | 5018 | 5194 | 5115 | − 79 |
| 6 | − 140 | 2927 | 3067 | 3043 | − 24 |
| 7 | − 76 | 1255 | 1331 | 1330 | − 1 |
| 8 | − 11 | 392 | 403 | 424 | + 21 |
| 9 | − 18 | 87 | 105 | 96 | − 9 |
| 10 | − 1 | 13 | 14 | 15 | + 1 |
| 11 | − 3 | 1 | 4 | 1 | − 3 |
| 12 | 0 | 0 | 0 | 0 | 0 |
| Total | 0 | 26,306 | 26,306 | 26,306 | 0 |

How can the theory predict the observed numbers listed in column (4) of Table I? If we assume that the dice are "fair" and that the trials with different dice, or with the same die at different times, are independent of

⁵ *Philosophical Magazine*, ser. 5, vol. 50, 1900, p. 167–169; in a paper by Karl Pearson.

each other, we can compute the relevant probabilities.   Under our assumption (which is properly termed a "statistical hypothesis") the probability that exactly 4 dice out of 12 should show 5 or 6 spots is

$$P = 495 \left(\frac{1}{3}\right)^4 \left(\frac{2}{3}\right)^8 = \frac{126,720}{531,441}.$$

Now, by definition, the probability is the theoretical value of long range relative frequency.   If the event with probability $P$ shows itself $m$ times in $n$ trials, we expect that

$$\frac{m}{n} = P \text{ approximately}$$

or

$$m = Pn \text{ approximately}.$$

Therefore, we should expect that exactly 4 dice will show five or six spots out of the 12 dice cast in about

$$Pn = \frac{126,720}{531,441} 26,306 = 6,273$$

cases out of $n = 26,306$ trials.   (Observe that we can compute this number 6,273 before the trials start.)   Now, this predicted value 6,273 does not seem to be "very different" from the observed number 6,114, and so our first impression about the practical applicability of the theory of probability may be quite good.

The number 6,273 is listed in column (3) of Table I at the proper place, in the same row as the number 4 in column (1).   All the numbers in column (3) are similarly computed.   In order to compare more conveniently the predicted values in column (3) with the observed numbers in column (4), we list the differences (predicted less observed) in column (2).   With their meaning in mind, we survey the columns (2), (3), and (4).   Is the agreement between experience and theory satisfactory?   Are the observed numbers sufficiently close to the predicted values?

There is, obviously, some agreement between the columns (3) and (4). Both columns of numbers have the same general aspect: the maximum is attained at the same point (in the same row) and the numbers first increase to the maximum and then decrease steadily to 0 in very much the same fashion in both columns.   The deviation of the observed number from the predicted value appears relatively small in most cases; the agreement, at a first glance, looks quite good.   On the other hand, however, the number of trials, 26,306, appears pretty large.   Are the deviations sufficiently small in view of the large number of trials?

This seems to be the right question.   Yet we cannot answer it off-hand; we had better postpone it till we know a little more; see sect. 7 (3).   Yet without any special knowledge, just with a little common sense, we can draw quite a sharp conclusion from Table I. A physicist would easily notice the following point about the columns (3) and (4). The differences are listed in column (2). Some of these differences are positive, others negative. If these differences were randomly distributed, the + and − signs should be intermingled in some disorderly fashion. In fact, however, the + and − signs are sharply separated: the theoretical values are too large up to a certain point, and too small from that point onward. In such a case, the physicist speaks of a *systematic* deviation of the theory from the experiment, and he regards such a systematic deviation as a grave objection against the theory.

And so the agreement between the theory of probability and Weldon's observations, which looked quite good at first, begins to look much less good.

(2) Yet who is responsible for that systematic deviation? The theoretical values have been computed according to the rules of the calculus of probability on the basis of a certain assumption, a "statistical hypothesis." We need not blame the rules of the calculus; the fault may be with the statistical hypothesis. In fact, this statistical hypothesis has a weak point: we assumed that the dice used in the experiment were "fair." When gentlemen play a game of dice, they should assume that the dice are fair, but for a naturalist such an assumption is unwarranted.

In fact, let us look at the example of the physicist. Galileo discovered the law of falling bodies that we write today in the usual notation as an equation:

$$s = gt^2/2 \, ;$$

$s$ stands for space (distance), $t$ for time. More exactly, Galileo discovered the *form* of the dependence of $s$ on $t$: the distance is proportional to the square of the time $t$. Yet he made no theoretical prediction about the constant $g$ that enters into this proportionality; the suitable value of $g$ has to be found by experiments. In this respect, as in many other respects, natural science followed the example of Galileo; in countless cases the theory yielded the general form of a natural law, and the experiment had to determine the numerical values of the constants that enter into the mathematical expression of the law. And this procedure works in our example, too.

If a die is "fair," none of the six faces is preferable to the others, and so the probability for casting 5 or 6 spots is

$$\frac{2}{6} = \frac{1}{3}.$$

Even if the die is not fair, there is a certain probability $p$ for casting 5 or 6 spots; $p$ may be different from 1/3. (Yet not very different in an ordinary die, otherwise we would consider the die as "loaded.") We take $p$ as a constant that has to be determined by experiment. And now, we modify our original statistical hypothesis: we *assume* that all twelve dice used have the *same probability* $p$ for showing 5 or 6 spots. (This is a simple assumption but, of course, pretty arbitrary. We cannot believe that it is exactly true; we can only hope that it is not very far from the truth. There is virtually no chance that the dice are exactly equal, but they may be only slightly different.) We keep unchanged the other part of our former statistical hypothesis (different dice and different trials are considered as independent).

On the basis of this new statistical hypothesis we can again assign theoretical values corresponding to the observations listed in column (4) of Table I. For example, the theoretical value corresponding to the observed value 6,114 is

$$495 \, p^4 \, (1 - p)^8 \, 26,306;$$

it depends on $p$, and also the theoretical values corresponding to the other numbers in column (4) depend on $p$.

It remains to determine $p$ from the experiments that we are examining. We cannot hope to determine $p$ from experiments exactly, only in some reasonable approximation. If we change our standpoint for a moment and consider the casting of a single die as a trial,

$$12 \times 26,306 = 315,672$$

trials have been performed; this is a very large number. The frequency of the event "five or six spots" can be easily derived from the column (4) of Table I. We find as the value of the relative frequency

$$\frac{106,602}{315,672} = 0.3376986;$$

we take this relative frequency, resulting from a very large number of trials, for the value of $p$. (We assume so for $p$ a value slightly higher than 1/3.)

Once $p$ is chosen, we can compute theoretical values corresponding to the observed frequencies. These theoretical values are tabulated in column (5) of Table I. Thus the columns (3) and (5) give theoretical values corresponding to the same observed numbers, but computed under different statistical hypotheses. In fact, the two statistical hypotheses differ only in the value of $p$; column (3) uses $p = 1/3$, column (5) uses the slightly higher value derived from the observations. (Column (3) can be computed before the observations, but column (5) cannot.) The differences between corresponding items of columns (5) and (4) are listed in column (6).

There is little doubt that the theoretical values in column (5) fit the observations much better than those in column (3). In absolute value, the differences in column (6) are, with just one exception, less than, or equal to, the differences in column (2) (equal in just three cases, much less in most cases). In opposition to column (2), the signs $+$ and $-$ are intermingled in column (6), so that they yield no ground to suspect a systematic deviation of the theoretical values in column (5) from the experimental data in column (4).

(3) Judged by the foregoing example, the theory of probability seems to be quite suitable for describing mass phenomena generated by such gambling devices as dice.   If it were not suitable for anything else, it would not deserve too much attention.   Let us, therefore, consider one more example.

As reported by the careful official Swiss statistical service, there were exactly 300 deliveries of triplets in Switzerland in the 30 years from 1871 to 1900.   (That is, 900 triplets were born.   In talking of deliveries, we count the mothers, not the babies.)   The number of all deliveries (some of triplets, some of twins, most of them, of course, of just one child) during the same period in the same geographical unit was 2,612,246.   Thus, we have here a mass phenomenon of considerable proportions, but the event considered, the birth of triplets, is a *rare event*.   The average number of deliveries per year is

$$2,612,246/30 = 87,075,$$

the average number of deliveries of triplets only

$$300/30 = 10.$$

Of course, the event happened more often in some years, in others less often than the average 10, and in some years exactly 10 times.   Table II gives

Table II

Triplets born in Switzerland 1871–1900.

| (1) | (2) | (3) | (4) | (5) |
|---|---|---|---|---|
| Deliveries | Years obs. | Years theor. | (2) cumul. | (3) cumul. |
| 0 | 0 | 0.00 | 0 | 0.00 |
| 1 | 0 | 0.00 | 0 | 0.00 |
| 2 | 0 | 0.09 | 0 | 0.09 |
| 3 | 1 | 0.21 | 1 | 0.30 |
| 4 | 0 | 0.57 | 1 | 0.87 |
| 5 | 1 | 1.14 | 2 | 2.01 |
| 6 | 1 | 1.89 | 3 | 3.90 |
| 7 | 5 | 2.70 | 8 | 6.60 |
| 8 | 1 | 3.39 | 9 | 9.99 |
| 9 | 4 | 3.75 | 13 | 13.74 |
| 10 | 4 | 3.75 | 17 | 17.49 |
| 11 | 4 | 3.42 | 21 | 20.91 |
| 12 | 3 | 2.85 | 24 | 23.76 |
| 13 | 2 | 2.16 | 26 | 25.92 |
| 14 | 1 | 1.59 | 27 | 27.51 |
| 15 | 2 | 1.02 | 29 | 28.53 |
| 16 | 0 | 0.66 | 29 | 29.19 |
| 17 | 1 | 0.39 | 30 | 29.58 |
| 18 | 0 | 0.21 | 30 | 29.79 |
| 19 | 0 | 0.12 | 30 | 29.91 |

the relevant details in column (2). We see there (in the row that has 10 in the first column) that there were in the period considered exactly 4 years in which exactly 10 deliveries of triplets took place. As the same column (2) shows, no year in the period had less than 3 such deliveries, none had more than 17, and each of these extreme numbers, 3 and 17, turned up in just one year.

The numbers of column (2) seem to be dispersed in some haphazard manner. It is interesting to note that the calculus of probability is able to match the irregular looking observed numbers in column (2) by theoretical numbers following a simple law; see column (3). The agreement of columns (2) and (3), judged by inspection, does not seem to be bad; the difference between the two numbers, the observed and the theoretical, is less than 1 in absolute value, except in two cases. Yet in these two cases (the rows with 7 and 8 in the first column) the difference is greater than 2 in absolute value.

There is a device that allows us to judge a little better the agreement of the two series of numbers. The column (4) of Table II contains the numbers of column (2) "cumulatively." For example, consider the row that has 7 in column (1); it has 5 in column (2) and 8 in column (4). Now

$$8 = 0 + 0 + 0 + 1 + 0 + 1 + 1 + 5;$$

that is, 8 is the sum, or the "accumulation", of all numbers in column (2) up to the number 5, inclusively, in the respective row. (In other words, 8 is the number of those years of the period in which the number of deliveries of triplets did *not exceed* 7.) Column (5) contains the numbers of column (3) "cumulatively", and so the columns (4) and (5) are analogously derived from the observed numbers in column (2) and the theoretical numbers in column (3), respectively. The agreement between columns (4) and (5) looks excellent; the difference is less than 1 in absolute value except in just one case, where it is still less than 2.

**6. Explanation of phenomena.** Ideas connected with the concept of probability play a rôle in the explanation of phenomena, and that is true of phenomena dealt with by any science, from physics to the social sciences. We consider two examples.

(1) Gregor Mendel (1822–1884), experimenting with the cross-breeding of plants, became the founder of a new science, genetics. Mendel was, by the way, an abbot in Moravia, and carried out his experiments in the garden of his monastery. His discovery, although very important, is very simple. To understand it we need only the description of one experiment and an intuitive notion of probability. To make things still easier, we shall not discuss one of Mendel's own experiments, but an experiment carried out by one of his followers.[6]

Of two closely related plants (different species of the same genus) one has

---

[6] By Correns; see W. Johannsen, *Elemente der exakten Erblichkeitslehre*, Jena 1909, p. 371.

white flowers and the other rather dark red flowers. The two plants are so closely related that they can fertilize each other. The seeds resulting from such crossing develop into hybrid plants which have an intermediate character: the hybrids have pink flowers. (In fig. 14.3 red is indicated by more, pink by less, shading.) If the hybrid plants are allowed to become self-fertilized, the resulting seeds develop into a third generation of plants in which all three kinds are represented: there are plants with white, plants with pink, and plants with red flowers. Fig. 14.3 represents schematically the relations between the three subsequent generations.

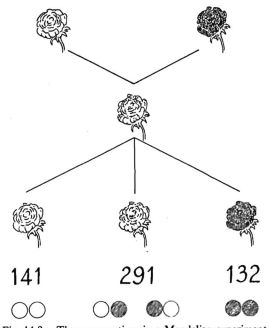

### 141          291          132

Fig. 14.3.   Three generations in a Mendelian experiment.

Yet the most striking feature of the phenomenon is the numerical proportion in which the three different kinds of plants of the third generation are produced. In the experiment described, 564 plants of the third generation have been observed. Among them, those two kinds of plants that resemble one or the other grandparental plant were about equally numerous: there were 141 plants with white flowers and 132 plants with red flowers in the third generation. Yet the plants resembling the hybrid parental plants were more numerous: there were 291 plants with pink flowers in the third generation. We can conveniently survey these numbers in fig. 14.3. We easily notice that these numbers given by the experiment are approximately in a simple proportion:

141 : 291 : 132 almost as 1 : 2 : 1.

This simple proportion invites a simple explanation.

Let us begin at the beginning. The experiment began with the crossing of two different kinds of plants. Any flowering plant arises from the union of two reproductive cells (an ovule and a grain of pollen). The pink-flowering hybrids of the second generation arose from two reproductive cells of different extraction. As the pink-flowering plants of the third generation are similar to those of the second generation, it is natural to assume that they were similarly produced, by two reproductive cells of *different* kinds. This leads us to suppose that the pink-flowering hybrids of the second generation *have* two different kinds of reproductive cells. Supposing this, however, we may perceive a possibility of explaining the mixed offspring. In fact, let us see more clearly what would happen if the pink-flowering hybrids of the second generation actually *had* two different kinds of reproductive cells, which we may call "white" and "red" cells. When two such cells are combined, the combination can be white with white, or red with red, or one color with the other, and these three different combinations *could* explain the three different kinds of plants in the third generation; see fig. 14.3.

After this remark, it should not be difficult to explain the numerical proportions. The deviation of the actually observed proportion 141 : 291 : 132 from the simple proportion 1 : 2 : 1 appears as random. That is, it looks like the deviation of observed frequencies from underlying probabilities. This leads us to wondering what the probabilities of the two kinds of cells are, or in which proportion the "white" and "red" cells are produced. As there are about as many white-flowering as red-flowering plants in the third generation, we can hardly refrain from trying the simplest thing: let us assume that the "white" and "red" reproductive cells are produced in equal numbers by the pink-flowering plants. Finally, we are almost driven to compare the random encounter of two reproductive cells with the random drawing of two balls, and so we arrive at the following simple problem.

There are two bags containing white and red balls, and no balls of any other color. Each bag contains just as many white balls as red balls. With both hands, I draw from both bags, one ball from each. Find the probability for drawing two white balls, two balls of different colors, and two red balls. As it is easily seen (cf. sect. 3 (5)), the required probabilities are

$$\frac{1}{4}, \quad \frac{2}{4}, \quad \frac{1}{4},$$

respectively. We perceive now a simple reason for the proportion 1 : 2 : 1 that seems to underlie the observed numbers, and so doing we come very close to Mendel's essential concepts.

(2) The concept of random mass phenomena plays an important rôle in physics. In order to illustrate this rôle, we consider the velocity of chemical reactions.

Relatively crude observations are sufficient to suggest that the speed of a chemical change depends on the concentration of the reacting substances. (By concentration of a substance we mean its amount in unit volume.) This dependence of the chemical reaction velocity on the concentration of the reactants was soon recognized, but the discovery of the mathematical form of the dependence came much later.    An important particular case was noticed by Wilhelmy in 1850, and the general law was discovered by two Norwegian chemists, Guldberg and Waage, in 1867.    We now outline, in a particular case and as simply as we can, some of the considerations that led Guldberg and Waage to their discovery.

We consider a bimolecular reaction.    That is, two different substances, A and B, participate in the reaction which consists in the combination of one molecule of the first substance A with one molecule of the second substance B. The substances A and B are dissolved in water, and the chemical change takes place in this solution.    The substances resulting from the reaction do not participate further in the chemical action; they are inactive in one way or another.    For example, they may be insoluble in water and deposited in solid form.

The solution in which the reaction takes place consists of a very great number of molecules.    According to the ideas of the physicists (the kinetic theory of matter) these molecules are in violent motion, traveling at various speeds, some at very high speed, and colliding now and then.    If a molecule A collides with a molecule B, the two may get so involved that they exchange some of their atoms: the chemical reaction in which we are interested consists of such an exchange, we imagine.    Perhaps it is necessary for such an exchange that the molecules should collide at a very high speed, or that they should be disposed in a favorable position with respect to each other in the moment of their collision.    At any rate, the more often it happens that a molecule A collides with a molecule B, the more chance there is for the chemical combination of two such molecules, and the higher the velocity of the chemical reaction will be.    And so we are led to the conjecture: *the reaction velocity is proportional to the number of collisions between molecules A and molecules B.*

We could not predict exactly the number of such collisions.    We have before us a random mass phenomenon like rainfall.    Remember fig. 14.2; there, too, we could not predict exactly how many raindrops would strike the subarea F.    Yet we could predict that the number of raindrops striking the subarea F would be *proportional* to the number of raindrops falling on the whole area P.    (The proportionality is approximate, and the factor of proportionality is $F/P$, as discussed toward the end of sect. 2.)    Similarly, we can predict that the number of collisions in which we are interested (between any molecule A and any molecule B) will be proportional to the number of the molecules A.    Of course, it will also be proportional to the number of the molecules B, and so finally proportional to the *product* of these two

numbers.   Yet the number of the molecules of a substance is proportional
to the concentration of that substance, and so our conjecture leads us to
the following statement: *the reaction velocity is proportional to the product of
the concentrations.*

We arrived at a particular case of the general law of chemical mass
action discovered by Guldberg and Waage.   This is the particular case
appropriate for the particular circumstances considered.   On the basis of
the law of mass action it is possible to compute the concentration of the reacting substances at any given moment and to predict the whole course of
the reaction.

**7. Judging statistical hypotheses.**   We start from an anecdote.[7]

(1) "One day in Naples the reverend Galiani saw a man from the
Basilicata who, shaking three dice in a cup, wagered to throw three sixes;
and, in fact, he got three sixes right away.   Such luck is possible, you say.
Yet the man succeeded a second time, and the bet was repeated.   He put
back the dice in the cup, three, four, five times, and each time he produced
three sixes.   'Sangue di Bacco,' exclaimed the reverend, 'the dice are loaded!'
And they were.   Yet why did the reverend use profane language?"

The reverend Galiani drew a plausible conclusion of a very important
type.   If he discovered for himself this important type of plausible inference
on the spur of the moment, his excitement is quite understandable and I,
personally, would not reproach him for his mildly profane language.

The correct thing is to treat everybody as a gentleman until there is
some definite evidence to the contrary.   Quite similarly, the correct thing is
to engage in a game of chance under the assumption that it is fairly played.
I do not doubt that the reverend did the correct thing and assumed in the
beginning that that man from the Basilicata had fair dice and used them
fairly.   Such an assumption, correctly stated in terms of probability, is a
statistical hypothesis.   A statistical hypothesis generally assumes the values
of certain probabilities.   Thus, the reverend assumed in the beginning, more
or less explicitly, that any of the dice involved will show six spots with the
probability $1/6$.   (We have here exactly the same statistical hypothesis as in
sect. 5 (1).)

The calculus of probability enables us to compute desired probabilities
from given probabilities, on the basis of a given statistical hypothesis.   Thus,
on the basis of the statistical hypothesis adopted by the reverend at the beginning, we can compute the probability for casting three sixes with three dice;
it is

$$(1/6)^3 = 1/216,$$

a pretty small probability.   The probability for repeating this feat twice,

[7] J. Bertrand, *Calcul des probabilités*, p. VII–VIII.

that is, casting three sixes at a first trial, and casting them again at the next trial, is

$$(1/216)^2 = (1/6)^6 = 1/46,656,$$

a very small probability indeed.    Yet that man from the Basilicata kept on repeating the same extraordinary thing five times.    Let us list the corresponding probabilities:

| Repetitions | Probability | |
|---|---|---|
| 1 | $1/6^3$ = | $1/216$ |
| 2 | $1/6^6$ = | $1/46,656$ |
| 3 | $1/6^9$ = | $1/10,077,696$ |
| 4 | $1/6^{12}$ = | $1/2,176,782,336$ |
| 5 | $1/6^{15}$ = | $1/470,184,984,576.$ |

Perhaps, the reverend adopted his initial assumption out of mere politeness; looking at the man from the Basilicata, he may have had his doubts about the fairness of the dice.    The reverend remained silent after the three sixes turned up twice in succession, an event that under the initial assumption should happen not much more frequently than once in fifty thousand trials.    He remained silent even longer.    Yet, as the events became more and more improbable, attained and perhaps surpassed that degree of improbability that people regard as miraculous, the reverend lost patience, drew his conclusion, rejected his initial polite assumption, and spoke out forcibly.

(2) The anecdote that we have just discussed is interesting in just one aspect: it is typical.    It shows clearly the circumstances under which we can reasonably reject a statistical hypothesis.    We draw consequences from the proposed statistical hypothesis.    Of special interest are consequences concerned with some event that appears *very improbable* from the standpoint of our statistical hypothesis; I mean an event the probability of which, computed on the basis of the statistical hypothesis, is very small.    Now, we appeal to experience: we observe a trial that can produce that allegedly improbable event.    If the event, in spite of its computed low probability, actually happens, it yields a strong *argument against* the proposed statistical hypothesis.    In fact, we find it hard to believe that anything so extremely improbable could happen.    Yet, undeniably, the thing did happen.    Then we realize that any probability is computed on the basis of some statistical hypothesis and start doubting the basis for the computation of that small probability.    And so there arises the argument against the underlying statistical hypothesis.

(3) As the reverend Galiani, we also felt obliged to reject the hypothesis of fair dice when we examined the extensive observations related in sect. 5 (1);

our reasons to reject it, however, were not quite as sharp as his.    Could we find better reasons in the light of the foregoing discussion?

Here are the facts: 315,672 attempts to cast five or six spots with a dice produced 106,602 successes; see sect. 5 (2).    *If* all dice cast were fair, the probability of a success would be 1/3.    Therefore, we should expect about

$$315,672/3 = 105,224$$

successes in 315,672 trials.    Thus, the observed number deviates from the expected number

$$106,602 - 105,224 = 1,378$$

units.    Does such a deviation speak for or against the hypothesis of fair dice? Should we regard the deviation 1,378 as small or large?    Is the probability of such a deviation high or low?

The last question seems to be the sensible question.    Yet we still need a sensible interpretation of the short, but important, word "such."    We shall reject the statistical hypothesis if the probability that we are about to compute turns out to be low.    Yet the probability that the deviation should be exactly equal to 1,378 units is very small anyhow—even the probability of a deviation exactly equal to 0 would be very small.    Therefore, we have to take into account *all the deviations of the same absolute value as, or of larger absolute value than, the observed deviation* 1,378.    And so our judgment depends on the solution of the following problem:    *Given that the probability of a success is 1/3 and that the trials are independent, find the probability that in* 315,672 *trials the number of successes should be either more than* 106,601 *or less than* 103,847.

With a little knowledge of the calculus of probability we find that the required probability is approximately

$$0.0000001983;$$

this means less than two chances in ten million.    That is, an event has occurred that looks extremely improbable, *if* the statistical hypothesis is accepted that underlies the computation of probability.    We find it hard to believe that such an improbable event actually occurred, and so the underlying hypothesis of fair dice appears extremely unlikely.    Already in sect. 5 (1) we saw a good reason to reject the hypothesis of fair dice, but now we see a still better, more distinct, reason to reject it.

(4) The actual occurrence of an event to which a certain statistical hypothesis attributes a small probability is an argument against that hypothesis, and the smaller the probability, the stronger is the argument.

In order to visualize this essential point, let us consider the sequence

$$\frac{1}{10}, \quad \frac{1}{100}, \quad \frac{1}{1,000}, \quad \frac{1}{10,000}, \quad \ldots$$

A statistical hypothesis implies that the probability of a certain event is 1/10.

The event happens. Should we reject the hypothesis? Under usual circumstances, most of us would not feel entitled to reject it; the argument against the hypothesis does not appear yet strong enough. If another event happens to which the statistical hypothesis attributes the probability 1/100, the urge to reject the hypothesis becomes stronger. If the alleged probability is 1/1,000, yet the event happens nevertheless the case against the hypothesis is still stronger. If the statistical hypothesis attributes the probability

$$\frac{1}{1,000,000,000}$$

to the event, or one chance in a billion, yet the event happens nevertheless, almost everybody would regard the hypothesis as hopelessly discredited, although there is no logical necessity to reject the hypothesis just at this point. If, however, the sequence proceeds without interruption so that events happen one after the other to which the statistical hypothesis attributes probabilities steadily decreasing to 0, for each reasonable person arrives sooner or later the critical moment in which he feels justified in rejecting the hypothesis, rendered untenable by its increasingly improbable consequences. And just this point is neatly suggested by the story of the reverend Galiani. The probability of the first throw of three sixes was 1/216; of the sequence of five throws of three sixes each, 1/470,184,984,576.

The foregoing discussion is of special importance for us if we adopt the standpoint that the theory of probability is a part of the study of nature. Any natural science must recur to observations. Therefore it must adopt rules that specify somehow the circumstances under which its statements are confirmed or confuted by experience. We have done just this for the theory of probability. We described certain circumstances under which we can reasonably consider a statistical hypothesis as practically refuted by the observations. On the other hand, if a statistical hypothesis survives several opportunities of refutation, we may consider it as corroborated to a certain extent.

(5) Probability, as defined in sect. 2, is the theoretical value of long range relative frequency. The foregoing gave us an opportunity to realize a few things. First, such a theoretical value depends, of course, on our theory, on our initial assumptions, on the statistical hypothesis adopted. Second, such a theoretical value may be very different from the actual value.

A suitable notation may help us to clarify our ideas. Let $P$ be the probability of an event $E$ computed on the basis of a certain statistical hypothesis $H$. Then $P$ depends both on $E$ and on $H$. (In fact, we could use, instead of $P$, the more explicit symbol $P(E, H)$ that emphasizes the dependence of $P$ on $E$ and $H$.)

In some of the foregoing applications we took the hypothesis $H$ for granted (at least for the moment) and, computing $P$ on the basis of $H$, we tried to predict the observable frequency of the event $E$. Yet, in the present section, we proceeded in another direction. Having observed the event $E$, we computed $P$ on the basis of the statistical hypothesis $H$ and, in view of the value of $P$ obtained, we tried to judge the reliability of the hypothesis $H$. We perceive here a new aspect of $P$. The *smaller* $P$ is, the more we feel inclined to reject the hypothesis $H$, and the *more unlikely* the hypothesis $H$ appears to us: $P$ indicates the likelihood of the hypothesis $H$. We shall say henceforward that $P$ is the *likelihood of the statistical hypothesis $H$*, judged in view of the fact that the event $E$ has been observed.

This terminology, which agrees essentially with the usage of statisticians, emphasizes a certain aspect of the dependence of $P$ on the event $E$ and the statistical hypothesis $H$. Our original terminology lays the stress on the complementary aspect of the same dependence: $P$ is the *probability of the event $E$*, computed on the basis of the statistical hypothesis $H$.

Some practice in the use of this double terminology is needed to convince us that its advantages sufficiently outweigh its dangers.

**8. Choosing between statistical hypotheses.** The following example may provide a first orientation to the applications of the theory of probability in statistical research.

(1) A consumer buys a certain article from the producer in large lots. The consumer is a big consumer, a large merchandizing or industrial firm, or a government agency. The producer is also big and manufactures the article in question on a large scale. The article can be a nail, or a knob, or anything manufactured; an interesting example is a fuze, used for firing explosives in ammunition or in blasting operations. The article has to meet certain specifications. For example, the nail should not be longer than 2.04 inches nor shorter than 1.96 inches, its thickness is similarly specified, and perhaps also its minimum breaking strength; the burning time of the fuze is specified, and so on. An article that does not meet the specifications is considered as *defective*. Even the most carefully manufactured lot may contain a small fraction of defectives. Therefore the lot has to be inspected before it passes from the producer to the consumer. The lot may be fully inspected, that is, each article in the lot may be tested whether it meets the agreed specifications. Such a full inspection would be impractical for a lot of 10,000 nails and it would be preposterous for a lot of fuzes even if the lot is small; in order to measure its burning time, you have to destroy the fuze and there is not much point in destroying the whole lot by inspecting it. Therefore in many cases instead of inspecting the whole lot before acceptance, only a relatively small sample is taken from the lot. A simple procedure of such acceptance sampling is characterized by the following rule.

"Take a random sample of $n$ articles from the submitted lot of $N$ articles. Test each article in the sample. If the number of defectives in the sample does not exceed a certain agreed number $c$, the so-called *acceptance number*, the consumer accepts the lot, but he rejects it, and the producer takes it back, if there are more defectives than $c$ in the sample."

The results obtained by this rule depend on chance.   By chance, the fraction of defectives in the sample can be much lower or much higher than in the whole lot.   If the sample turns out to be better than the lot, chance works against the consumer, and it works against the producer if the sample turns out to be worse than the lot.   In spite of these risks, some such procedure appears necessary, and the rule formulated may be quite reasonable.   We have to find out how the procedure works, how its result depends on the quality of the submitted lot.   And so we are led to formulate the following problem:  *Given p, the probability that an article chosen at random in the submitted lot is defective, find a, the probability that the lot will be accepted.*

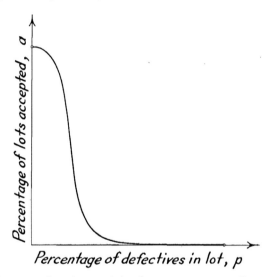

*Percentage of defectives in lot, p*

Fig. 14.4.   Operating characteristic of an acceptance sampling procedure.

In the most important practical cases $N$, the size of the lot, is large even in comparison with $n$, the size of the sample.   In such cases we may assume that $N$ is infinite; we lose little in precision and gain much in simplicity.   Assuming $N = \infty$, we easily find that

$$a = (1 - p)^n + \binom{n}{1}p(1 - p)^{n-1} + \binom{n}{2}p^2(1 - p)^{n-2}$$

$$+ \ldots + \binom{n}{c}p^c(1 - p)^{n-c}.$$

We take this expression of $a$, the probability of acceptance, for granted and we concentrate on discussing some of its practical implications.

We graph $a$ as function of $p$; see fig. 14.4.   If we graphed $100a$ as function of $100p$, the form of the curve would be the same.   Now, $100p$ is the percentage of defective articles in the lot submitted.   On the other hand, if several lots with the same percentage of defectives were subjected to the same inspection procedure, the relative frequency of acceptance, that is, the ratio of accepted lots to submitted lots, would be close to $a$.    Therefore, in the long run, $100a$ will be the percentage of the lots accepted among the lots submitted. This explains the labeling of the axes in fig. 14.4.    The curve in fig. 14.4 allows us to survey how the procedure operates on lots of various quality, and so it is appropriately called the *operating characteristic*.

Judged by its effects, does the procedure appear reasonable?    This is the question that we wish to consider.

If there are no defectives in the lot, there should be no chance for rejecting it.    In fact, if $p = 0$ our formula yields $a = 1$, as it should.    If there are only defectives in the lot, there should be no chance for accepting it.    In fact, if $p = 1$ our formula yields $a = 0$, as it should.    Both extreme points of the operating characteristic curve are obviously reasonable.

If the number of defectives increases, the chances of acceptance should diminish.    In fact, differentiating with a little skill, we easily find the surprisingly simple expression

$$\frac{da}{dp} = -(n - c) \binom{n}{c} p^c (1 - p)^{n-1-c}$$

which is always negative.    Therefore, the operating characteristic is necessarily a falling curve, as represented in fig. 14.4, which is again as it should be.

The absolute value of the derivative, or $-da/dp$, has also a certain practical significance.    The change $dp$ of the abscissa represents a change in the quality of the lot.    The change $da$ of the ordinate represents a change in the chances of acceptance, due to the change in quality.    The larger is the ratio of these chances $da/dp$ in absolute value, the sharper is the distinction made by the procedure between two slightly different lots.    Especially, the point at which $da/dp$ attains its maximum absolute value may be appropriately called the "point of sharpest discrimination."    This point is easily recognized in the graph: it is the point of inflexion, if there is one, and otherwise the left-hand extremity of the curve.    (Its abscissa is $p = c/(n - 1)$.)

(2) The rule appears sensible also from another standpoint.    It has a certain flexibility.    By choosing $n$, the size of the sample, and $c$, the acceptance number, we can adapt the rule to concrete requirements.    Both the consumer and the producer require protection against the risks inherent in sampling.    A bad lot may sometimes yield a good sample and a good lot a bad sample, and so there are two kinds of risks: the sampling procedure may accept a bad lot or reject a good lot.    The consumer is against accepting bad lots and the producer is against rejecting good lots.    Still, both kinds of undesirable decisions are bound to happen now and then and the only thing that we can reasonably demand is that they should not happen too often. This demand leads to concrete problems such as the following.

"Determine the sample size and the acceptance number so that there should be less than once chance in ten that a lot with 5% defectives is accepted and there should be less than five chances in a hundred that a lot with only 2% defectives is rejected."

In this problem, there are two unknowns, the sample size $n$ and the acceptance number $c$. The condition of the problem requires the following two inequalities:

$$a > 0.95 \text{ when } p = 0.02,$$

$$a < 0.1 \text{ when } p = 0.05.$$

It is possible to satisfy these two simultaneous conditions, but it takes considerable numerical work to find the lowest sample size $n$ and the corresponding acceptance number $c$ for which the required inequalities hold.

We shall not discuss the numerical work. We are much more concerned here with visualizing the problem than with solving it. Let us therefore look a little further into its background. As we said already, both the acceptance of a bad lot and the rejection of a good lot are undesirable, the first from the consumer's viewpoint and the second from the producer's viewpoint. Yet the two undesirable possibilities may not be equally undesirable and the interests of consumer and producer may be not quite so sharply opposed. The acceptance of a bad lot is not quite in the interest of the producer; it may damage his reputation. Yet the rejection of a good lot may be very much against the interests of the consumer; he may need the articles urgently and the rejection may cause considerable delay. Moreover, repeated rejection of good lots, or even the danger of such rejection, may raise the price. If the interests of both parties are taken into account, the rejection of a good lot may be still less desirable than the acceptance of a bad lot. Seen against this background, it appears understandable that the conditions of our problem afford more protection against the rejection of the better quality than against the acceptance of the worse quality. (Only 5 chances in a hundred are allowed for the first undesirable event, but 10 chances in a hundred for the second.)

(3) The problem discussed under (2) admits another, somewhat different, interpretation.

The producer's lawyer affirms that there are no more than 2% defectives in the lot. Yet the consumer's lawyer contends that there are at least 5% defectives in the lot. For some reason (it may be a lot of fuzes) a full inspection is out of the question; therefore some sampling procedure has to decide between the two contentions. For this purpose the procedure outlined under (1) with the numerical data given in (2) can be appropriately used.

In fact, the conflicting contentions of the two lawyers suggest a fiction. We may pretend that there are exactly two possibilities with respect to the lot: the percentage of defectives in the lot is either exactly 2% or exactly

5%. Of course, nobody believes such a fiction, but the statistician may find it convenient: it restricts his task to a decision between two clear and simple alternatives. If the parties agree that the rejection of a lot with 2% defectives is less desirable than the acceptance of a lot with 5% defectives, the statistician may reasonably adopt the procedure outlined in (1) with the numerical data prescribed in (2). Whether the statistician's choice will satisfy the lawyers or the philosophers, I do not venture to say, but it certainly has a clear relation to the facts of the case. The statistician's rule, applied to a great number of analogous cases, accepts a good lot (with 2% defectives) about 950 times out of 1,000 and rejects it only about 50 times, but the rule rejects a bad lot (with 5% defectives) about 900 times out of 1,000 and accepts it only about 100 times. That is, the statistician's rule, which is based on sampling, cannot be expected to give the right decision each time, but it can reasonably be expected to give the right decision in an assignable percentage of cases *in the long run*.

(4) To give an adequate idea of what the statisticians are doing on the basis of just one example is, of course, a hopeless undertaking. Yet on the basis of the foregoing example we can obtain an idea of the statistician's task which, although very incomplete, is not very much distorted: the statistician designs rules of the same nature as the rule of acceptance sampling procedure outlined in (1) and considered in relation to numerical data in (2). We may understand the statistician's task if we have understood the nature of the rules he designs. Therefore, we have to formulate in general terms what seems to be essential in our particular rule; I mean the rule discussed in the foregoing sub-sections (1), (2), and (3).

Our rule prescribes a choice between two courses of action, acceptance and rejection. Yet the aspect of the problem considered under (3) is more suitable for generalization. There we considered a *choice between two statistical hypotheses*. (They were "this random sample is taken from a large lot with 2% defectives" and "this random sample is taken from a large lot with 5% defectives.") Any reasonable choice should be made with due regard to past experience and future consequences. In fact, our rule is designed with regard to both.

According to our rule, the choice depends upon a set of clearly specified observations (the testing of *n* articles and the number of defectives detected among the *n* articles tested). These observations constitute the relevant experience on which the choice is based. As our rule prefers a hypothesis to another on the basis of observations, it can claim to be named an *inductive* rule.

Our rule is designed with a view to probable consequences. The statistician cannot predict the consequences of any single application of the rule. He forecasts merely how the rule will work *in the long run*. If the choice prescribed by the rule is tried many times in such and such circumstances, it will lead to such and such result in such and such percentage of the trials, in the long run.    Our rule is designed with a view to *long range consequences.*

To sum up, our *rule is designed to choose between statistical hypotheses, is based on a specified set of observations, and aims at long range consequences.*    If we may regard our rule as sufficiently typical, we have an idea what the statisticians are doing: they are designing rules of this kind.

(In fact, they try to devise "best" rules of this kind.    For example, they wish to render the chances of such and such undesirable effect a minimum, being given the size of the sample, on which the work and expense of the observations depend.)

(5) Taking a random sample from a lot is an important operation in statistical research.    There is another problem about this operation that we have to discuss here.    We keep our foregoing notation in stating the problem.

*In a very large lot, 100p percent of the articles is defective.    In order to obtain some information about p, we take a sample of n articles from the lot, among which we find m defective articles.    On the basis of this observation, which value should we reasonably attribute to p?*

There is an obvious answer, suggested by the definition of probability itself.    Yet the problem is important and deserves to be examined from various angles.

Our observation yields some information about $p$.    Especially, if $m$ happens to be different from 0, we conclude that $p$ is different from 0. Similarly, if $m$ is less than $n$, we conclude that $p$ is less than 1.    Yet in any case $p$ remains unknown and all values between 0 and 1 are eligible for $p$. If we attribute one of these values to $p$, we make a guess, we adopt a conjecture, we choose a statistical hypothesis.

Let us think of the consequences of our choice before we choose.    If we have a value for $p$, we can compute the probability of the event the observation of which is our essential datum.    I mean the probability for finding exactly $m$ defective articles in a random sample of $n$ articles.    Let us call this probability $P$.    Then

$$P = \binom{n}{m} p^m (1-p)^{n-m}.$$

The value of $P$ depends on $p$, varies with $p$, can be greater or less.    If, however, this probability $P$ of an observed event is very small, we should reject the underlying statistical hypothesis.    It would be silly to choose such an unlikely hypothesis that has to be rejected right away.    Therefore let us choose the least unlikely hypothesis, the one for which the danger of rejection is least.    That is, let us choose the value of $p$ for which $P$ is *as great as possible.*

Now, if $P$ is a maximum, log $P$ is also a maximum and, therefore,

$$\frac{d \log P}{dp} = \frac{m}{p} - \frac{n-m}{1-p} = 0.$$

This equation yields

$$p = \frac{m}{n}.$$

And so, after some consideration, we made the choice that we were tempted to make from the outset: as a reasonable approximation to $p$, the underlying probability, we choose $m/n$, the observed relative frequency.

Yet our consideration was not a mere detour. We can learn a lot from this consideration.

Let us begin by examining the rôle of $P$. This $P$ is the probability of a certain observed event $E$ ($m$ defectives in a sample of size $n$). This probability is computed on the basis of the statistical hypothesis $H_p$ that $100p$ is the percentage of defectives in the lot. The probability $P$ varies with the hypothesis $H_p$ (with the value of $p$). The smaller $P$, the less acceptable, the less likely appears $H_p$. Thus we are led to consider $P$ as indicating the *likelihood* of the hypothesis $H_p$. This term "likelihood" has been introduced before (in sect. 7 (5)), in the same meaning, but now we may see the reasons for its introduction more clearly.

Let us emphasize that we choose among the various admissible statistical hypotheses $H_p$ (with $0 \leq p \leq 1$) the one for which $P$, the likelihood of $H_p$, is as great as possible. Behind this choice there is a principle, appropriately called the *principle of maximum likelihood*, that guides the statistician also in other cases, less obvious than our case.

**9. Judging non-statistical conjectures.** We consider several examples in order to illustrate the same fundamental situation from several angles.

(1) The other day I made the acquaintance of a certain Mr. Morgenstern. This name is not very usual, but not unknown to me. There was a German author Morgenstern for whose nonsense poetry I have a great liking. And, Oh yes, my cousin who lives in Atlanta, Georgia, recently began work in the offices of Mark Morgenstern & Co., consulting engineers.

At the beginning I had no thoughts about Mr. Morgenstern. After a while, however, I hear that he is in the engineering business. Then other pieces of information leaked out. I hear that the first name of my new acquaintance is Mark, and that his place of business is Atlanta, Georgia. At this stage it is very difficult not to believe that this Mr. Morgenstern is the employer of my cousin. I ask Mr. Morgenstern directly and find that it is so.

This trivial little story is quite instructive. (It is based, by the way, on actual experience, but the names are changed, of course, and also some irrelevant circumstances.) That two different persons should have exactly the same last name is not improbable, provided that the name is very common such as Jones or Smith. It is more improbable that two different persons have the same first and last name, especially, when it is an uncommon name, such as Mark Morgenstern. That two different persons have the same profession, or the same large town as residence, is not improbable. Yet it is very improbable that two different persons taken at random should have the same unusual name, the same home town, and the same occupation. A chance coincidence was hard to believe and so my conjecture about my recent acquaintance Mr. Morgenstern was quite reasonable. It turned out to be correct, but this has really little to do with the merits of the case. My conjecture was reasonable, defensible, justifiable on the basis of the probabilities considered. Even if my conjecture had turned out incorrect, I would have no reason to be ashamed of it.

In this example, no numerical value was given for the probability decisively connected with the problem, but a rough estimate for it could be obtained with some trouble.

(2) Two friends who met unexpectedly decided to write a postcard to a third friend. Yet they were not quite sure about the address. Both remembered the city (it was Paris) and the street (it was Boulevard Raspail) but they were both uncertain about the number. "Wait," said one of the friends, "let us think about the number without talking, and each of us will write down the number when he thinks that he has got it." This proposal was accepted and it turned out that both remembered the same number: 79 Boulevard Raspail. They put this address on the postcard which eventually reached the third friend. The address was correct.

Yet what was the reason for adopting the number 79? By not talking to each other, the two friends made their memories work independently. They both knew that Boulevard Raspail is long enough to have buildings numbered at least up to 100. Therefore, it seems reasonable to assume that the probability for a chance coincidence of the two numbers is not superior to $1/100$. Yet this probability is small, and so the hypothesis of a chance coincidence appears unlikely. Hence the confidence in the number 79.

(3) According to the statement of the bank, the balance of my checking account was \$331.49 at the end of the past month. I compute my balance for the same date on the basis of my notes and find the same amount. After this agreement of the two computations I am satisfied that the amount in which they both agree is correct. Is this certain? By no means. Although both computations arrived at the same result, the result could be wrong and the agreement may be due to chance. Is that likely?

The amount, expressed in cents, is a number with five digits. If the las digit was chosen at random, it could just as well be 0 or 1 or 2, . . . or 8 as 9 and so the probability that the last digit should be 9 is just $1/10$. The same is true for each of the other figures. In fact, if all figures were chosen at random, the number could be any one of the following:

$$000.00, \ 000.01, \ 000.02, \ \ldots \ 999.99$$

I have here obviously 100,000 numbers. If that assemblage of five figures, 33149, was produced in some purely random way, all such assemblages could equally well arise.   And, as there are 100,000 such assemblages, the probability that any one given in advance should be produced is

$$\frac{1}{100,000} = \left(\frac{1}{10}\right)^5 = 10^{-5}$$

Now, $10^{-5} = 0.00001$ is a very small probability.   If, trying to produce an effect with such a small probability, somebody manages to succeed at the very first trial, the outcome may easily appear as miraculous.   I am, however, not inclined to believe that there is anything miraculous about my modest bank account.    A chance coincidence is hard to believe and so I am driven to the conclusion that the agreement of the two computations is due to the correctness of the result.   Ordinary normal people generally think so in similar circumstances and after the foregoing considerations this kind of belief appears rather reasonable.

(4) To which language is English more closely related, to Hungarian or to Polish ?   Very little linguistic knowledge is enough to answer this question, but it is certainly more fun to obtain the answer by your own means than to accept it on the authority of some book.   Here is a common sense access to the answer.

Both the form and the meaning of the words change in the course of history.   We can understand the changes of form if we realize that the same language is differently pronounced in different regions, and we can understand the changes of meaning if we realize that the meaning of words is not rigidly fixed, but shifting, and changes with the context.   In the second respect, however, there is one conspicuous exception: the meaning of the numerals one, two, three, . . . certainly cannot shift by imperceptible degrees.   This is a good reason to suspect that the numerals do not change their meaning in the course of linguistic history.   Let us, therefore, base a first comparison of the languages in question on the numerals alone.   Table III lists the first ten numerals in English, Polish, Hungarian, and seven other modern European languages.   Only languages that use the Roman alphabet are considered (this accounts for the absence of Russian and modern Greek). Certain diacritical marks (accents, cedillas) which are unknown in English are omitted (in Swedish, German, Polish, and Hungarian).

Looking at Table III and observing how the same numeral is spelled in different languages, we readily perceive various similarities and coincidences. The first five languages (English, Swedish, Danish, Dutch, and German) seem to be pretty similar to each other, and the next three languages (French, Spanish, and Italian) appear to be in even closer agreement; so we have two groups, one consisting of five languages, the other of three.   Yet even these two groups appear to be somehow related; observe the coinciding spelling of 3 in Swedish, Danish and Italian, or that of 6 in English and French.

Table III.   Numerals in ten languages.

| English | Swedish | Danish | Dutch | German | French | Spanish | Italian | Polish | Hungarian |
|---|---|---|---|---|---|---|---|---|---|
| one | en | en | een | ein | un | uno | uno | jedem | egy |
| two | tva | to | twee | zwei | deux | dos | due | dwa | ketto |
| three | tre | tre | drie | drei | trois | tres | tre | trzy | harom |
| four | fyra | fire | vier | vier | quatre | cuatro | quattro | cztery | negy |
| five | fem | fem | vijf | funf | cinq | cinco | cinque | piec | ot |
| six | sex | seks | zes | sechs | six | seis | sei | szesc | hat |
| seven | sju | syv | zeven | sieben | sept | siete | sette | siedem | het |
| eight | atta | otte | acht | acht | huit | ocho | otto | osiem | nyolc |
| nine | nio | ni | negen | neun | neuf | nueve | nove | dziewiec | kilenc |
| ten | tio | ti | tien | zehn | dix | diez | dieci | dziesiec | tiz |

Polish seems to be closer to one group in some respects, and to the other in other respects; compare 2 in Swedish and Polish, 7 in Spanish and Polish. Yet Hungarian shows no such coincidences with any of the nine other languages. These observations lead to the impression that Hungarian has little relation to the other nine languages which are all in some way related to each other. Especially, and this is the answer to our initial question, English seems to be definitely closer related to Polish than to Hungarian.

Yet there are several objections. A first objection is that "similarity" and "agreement" are vague words; we should say more precisely what we mean. This objection points in the right direction. Following its suggestion, we sacrifice a part of our evidence in order to render the remaining part more precise. We consider only the *initials* of the numerals listed in Table III. We compare two numerals expressing the same number in two different languages; we call them "concordant" if they have the same initial, and "discordant" if the initials are different. Table IV contains the number of concordant cases for each pair of languages. For instance, the

Table IV.  Concordant initials of numerals in ten languages.

| E | 8 | 8 | 3 | 4 | 4 | 4 | 4 | 3 | 1 | 39 |
|---|---|---|---|---|---|---|---|---|---|---|
| Sw | 9 | 5 | 6 | 4 | 4 | 4 | 3 | 2 | | 45 |
| Da | | 4 | 5 | 4 | 5 | 5 | 4 | 2 | | 46 |
| Du | | | 5 | 1 | 1 | 1 | 0 | 2 | | 22 |
| G | | | | 3 | 3 | 3 | 2 | 1 | | 32 |
| F | | | | | 8 | 9 | 5 | 0 | | 38 |
| Sp | | | | | | 9 | 7 | 0 | | 41 |
| I | | | | | | | 6 | 0 | | 41 |
| P | | | | | | | | 0 | | 30 |
| H | | | | | | | | | | 8 |

number 7, in the same row as the letters "Sp" and in the same column as the letter "P" indicates that Spanish and Polish have exactly seven concordant numerals out of the possible 10 cases. The reader should check this and a few other entries of Table IV. The last column of Table IV shows how many concordant cases each language has with the other nine languages altogether. This last column shows pretty clearly the isolated position of Hungarian: it has only 8 concordant cases altogether whereas the number of concordant cases varies between 22 and 46 for the other nine languages.

Yet, perhaps, any definite conclusion from such data is rash: those coincidences of initials may be due to chance. This objection is easy to raise, but not so easy to answer. Chance could enter the picture through various channels. There may be an element of chance due to the fact that the correspondence between letters and pronunciation is by no means rigid. This is true even of a single language (especially of English). *A fortiori*, the same letter is often pretty differently pronounced in different languages and, on the other hand, different letters are sometimes very similarly pronounced. We have to admit that the coincidences observed are not free from some random element. Yet the question is: Is it *probable* that such coincidences as we have observed are due to mere chance?

If we wish to answer this question precisely, numerically, we have to adopt some precise, numerically definite statistical hypothesis and draw consequences from it which can be confronted with the observations. Yet the choice of a suitable hypothesis is not too obvious. We consider here two different statistical hypotheses.

I. There are two bags. Each bag contains 26 balls, each ball is marked with a letter of the alphabet, and different balls in the same bag are differently marked. With both hands, I draw simultaneously from both bags, one ball from each. The two letters so drawn may coincide or not; their coincidence is likened to the coincidence of the initials of the same numeral written in two different languages (and non-coincidence is likened to non-coincidence). The probability of a coincidence is 1/26.

II. The coincidence of the initials of the same numeral written in two different languages is again likened to the coincidence of two letters drawn simultaneously from two different bags and, again, both bags are filled in the same way with balls marked with letters. Yet now each of the bags contains 100 balls and each letter of the alphabet is used to mark as many different balls in the bag as there are numerals in Table III having that letter as initial. The probability of a coincidence is found to be 0.0948.

On both hypotheses, the comparison of the ten first numerals is likened to ten independent drawings of the same nature.

We can compare both hypotheses with the observations if we compute suitable probabilities. Tables V and VI contain the relevant material.

Table V compares the relative frequencies actually found with the probabilities computed. Columns (2) and (3) of Table V refer to all 45 pairs of languages considered in Table IV. Columns (4) and (5) of Table V refer only to 9 pairs, formed by Hungarian matched with the remaining nine languages. For the sake of concreteness, let us focus on the line that deals with 6 or more coincidences ($n = 6$). Such coincidences turned up in

Table V. Absolute and relative frequencies, and probabilities, for $n$ or more coincidences of initials

| (1) | (2) | (3) | (4) | (5) | (6) | (7) |
|-----|-----|-----|-----|-----|-----|-----|
| | | Frequencies | | | Probabilities | |
| $n$ | 10 languages | | 9 lang. v. Hu. | | Hyp. II | Hyp. I |
| 0 | 45 | 1.000 | 9 | 1.000 | 1.000000 | 1.000000 |
| 1 | 40 | 0.889 | 5 | 0.556 | 0.630644 | 0.324436 |
| 2 | 35 | 0.778 | 3 | 0.333 | 0.243824 | 0.054210 |
| 3 | 31 | 0.689 | 0 | 0.000 | 0.061524 | 0.005569 |
| 4 | 25 | 0.556 | 0 | 0.000 | 0.010612 | 0.000381 |
| 5 | 15 | 0.333 | 0 | 0.000 | 0.001281 | 0.000018 |
| 6 | 9 | 0.200 | 0 | 0.000 | 0.000108 | 0.000001 |
| 7 | 7 | 0.156 | 0 | 0.000 | 0.000006 | 0.000000 |

9 out of 45 cases as column (2) shows. Therefore, the observed relative frequency of 6 or more coincidences is $9/45 = 0.2$, whereas this many coincidences have only a little more than one chance in ten thousand to happen on hypothesis II, and only one chance in a million on hypothesis I; see columns (6) and (7), respectively. Similar remarks apply to the other lines of Table V: what has been actually observed appears as extremely improbable on either hypothesis, so there are strong grounds to reject both hypotheses. Yet columns (4) and (5) present a different picture: the coincidences observed are somewhat improbable on hypothesis I, but they appear as quite usual and normal from the standpoint of hypothesis II. The

Table VI. Total number of coincidences, observed and theoretical (Hypothesis II).

| | Coincidences | | Deviations | |
|---|---|---|---|---|
| | Observed | Expected | Actual | Standard |
| 10 languages | 171 | 42.66 | 128.34 | 7.60 |
| 9 lang. v. Hu. | 8 | 8.53 | − 0.53 | 2.78 |

impression gained from Table V is corroborated by Table VI: if we consider all 45 pairs of languages, the actually observed total number of coincidences exceeds tremendously what we have to expect on the basis of hypothesis II, yet the expected and observed numbers agree closely if we consider only the 9 pairs in which Hungarian is matched with the other 9 languages. (On hypothesis I, we have considerably stronger disagreement in both cases.)

In short there is no obvious interpretation of "chance" that would permit us to make chance responsible for all the coincidences observable in Table III; there are too many of them. Yet we can quite reasonably make chance responsible for the coincidences between Hungarian and the other languages. The explanation that Hungarian is unrelated to the other languages which are all related to each other has been vindicated.

The point is that this explanation has been vindicated, thanks to the consideration of probabilities, by so *few observations*. The explanation itself is supported by an overwhelming array of philological evidence.

(5) From appropriate observations (with telescope and spectroscope) we can conclude that certain elements found in the crust of our globe are also present in the sun and in certain stars. This conclusion is based on a physical law discovered by G. Kirchhoff almost a century ago (which says roughly that a luminous vapor absorbs precisely the same kind of light that it emits). Yet the conclusion appeals also to probabilities, and this is the point with which we are concerned here; we shall reduce the physical part of the argument to a schematic outline.

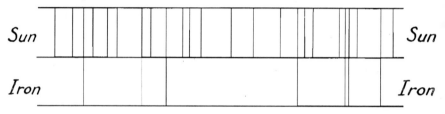

Fig. 14.5.   Coincidences.

Using suitable apparatus (a prism or a diffraction grating) we can detect a sequence of lines in the light of the sun (in the solar spectrum). We can detect a sequence of lines also in the light emitted by certain substances, such as iron, vaporized at high temperature in the laboratory. (In fact, the lines in the spectrum of the sun, the Fraunhofer lines, are dark, and the lines in the spectrum of iron are bright.) Kirchhoff examined 60 iron lines and found that each of these lines coincides with some solar line. (See the rough schematic fig. 14.5 or *Encyclopaedia Britannica*, 14th edition, vol. 21, fig. 3 on plate I facing p. 560.) These coincidences are fully understandable if we assume that there is iron in the sun. (More exactly, these coincidences follow from Kirchhoff's law on emission and absorption if we assume that in the atmosphere of the sun there is iron vapor that absorbs some of the light emitted by the central part of the sun glowing at some still higher temperature.) Yet, perhaps (here is again that ever-present objection) these coincidences are due to chance.

The objection deserves serious consideration. In fact, no physical observation is absolutely precise. Two lines which we regard as coincident could be different in reality and just by chance so close to each other that, with the limited precision of our observations, we might fail to recognize their difference. We have to concede that any observed coincidence may be only an apparent coincidence and there may be, in fact, a small difference. Yet let us ask a question: Is it *probable* that each of the 60 coincidences observed springs from a random difference so small that it failed to be detected by the means of observation employed?

Kirchhoff, who registered the observed lines on an (arbitrary) centimeter scale, estimated that he could not have failed to recognize a difference that exceeded 1/2 millimeter on his scale. On this scale the average distance between two adjacent lines of the solar spectrum was about 2 millimeters. If the 60 lines of iron were thrown into this picture at random, independently from each other, what would be the probability that each falls closer to some solar line than 1/2 millimeter?

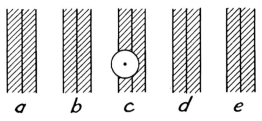

Fig. 14.6.   Equidistant lines.

We bring this question nearer to its solution by formulating an equivalent question in a more familiar domain. Parallel lines are drawn on the floor; the average distance between two adjacent lines is 2 inches. We throw a coin on the floor 60 times. If the diameter of the coin is 1 inch, what is the probability that the coin covers a line each time?

In this last formulation, the question is easy to answer. Assume first that the lines on the floor are equidistant (as in fig. 14.6) so that the distance from each line to the next is 2 inches. If the coin covers a line, the center of the coin is at most at 1/2 inch distance from the line and, therefore, this center lies somewhere in a strip 1 inch wide that is bisected by the line (shaded in

Fig. 14.7.   Lines at irregular distances.

fig. 14.6). Obviously, the probability that the coin cast on the floor should cover a line is 1/2. The probability that the coin, cast on the floor 60 times, should cover some line each time, is $(1/2)^{60}$.

Assume now that the lines on the floor are not equidistant; the average distance between two adjacent lines is still supposed to be 2 inches. We imagine that the lines, which were equidistant originally, came into their present position by being shifted successively. If a line (as line $b$ in fig. 14.7) is shifted so that its distance from its next neighbor remains more than 1 inch, the chances of the coin for covering some line remain unchanged. If however, the line is so shifted (as line $d$ in fig. 14.7) that its distance from the next line becomes less than 1 inch, the two (shaded) attached strips overlap and the chances of the coin to cover a line are diminished. Therefore, the required probability is less than $(1/2)^{60}$.

To sum up, if the iron lines were thrown by blind chance into the solar spectrum, the probability of the 60 coincidences observed by Kirchhoff would be less than $2^{-60}$ and so less than $10^{-18}$ or

$$\frac{1}{1,000,000,000,000,000,000}.$$

"This probability" says Kirchhoff, whom we quoted already in the motto prefixed to this chapter "is rendered still smaller by the fact that the brighter a given iron line is seen to be, the darker, as a rule, does the corresponding solar line appear.   Hence this coincidence must be produced by some cause, and a cause can be assigned which affords a perfect explanation of the observed facts."

(6) The following example is not based on actual observation, but it illustrates a frequently arising, typically important situation.

An extremely dangerous disease has been treated in the same locality by two different methods which we shall distinguish as the "old treatment" and the "new treatment."   Of the 9 patients who have been given the old treatment 6 died and only 3 survived, whereas of the 11 patients who received the new treatment only 2 died and 9 survived.   The twenty cases are clearly displayed in Table VII.

Table VII.   Four Place Correlation Table.

| Patients | Died | Survived | Total |
|---|---|---|---|
| Old treatment | 6 | 3 | 9 |
| New treatment | 2 | 9 | 11 |
| Total | 8 | 12 | 20 |

A first glance at this table may give us the impression that the observations listed speak strongly in favor of the new treatment.   The relative frequency of fatal cases is

6/9  or 67%   with the old treatment,

2/11 or 18%   with the new treatment.

On second thoughts, however, we may wonder whether the observed numbers are large enough to give us any reasonable degree of confidence in the percentages just computed, 67% and 18%.   Still, the fact remains that the number of fatal cases was much lower with the new treatment.   Such a low mortality, however, could be due to chance. *How easily can chance produce such a result?*

This last question seems to be the right question. Yet, at any rate, the question must be put more precisely before it can be answered. We have to explain the precise meaning in which we used the words "chance" and "such." The word "chance" will be explained if we assimilate the present case to some suitable game of chance. A fair interpretation of the words "such a result" seems to be the following: we consider all outcomes in which the number of fatalities with the second treatment is *not higher than that actually observed*. Thus, we may be led eventually to the following formulation. *There are two players, Mr. Oldman and Mr. Newman, and 20 cards, of which 8 are black and 12 are red. The cards are dealt so that Mr. Oldman receives 9 cards and Mr. Newman receives 11 cards. What is the probability that Mr. Newman receives 2 or less black cards?*

This formulation expresses as simply and as sharply as possible the contention that we have to examine: the difference between the old and the new treatment does not really matter, does not really influence the mortality, and the observed outcome is due to mere chance.

The required probability turns out to be

$$\frac{335}{8398} = 0.0399 \sim \frac{1}{25}$$

That is, an outcome that appears to be as favorable to the new treatment, as the observed outcome, or even more favorable, will be produced by chance about once in 25 trials. And so the numerical evidence for the superiority of the new treatment above the old cannot be simply dismissed, but is certainly not very strong.

In order to see clearly in these matters, let us give a moment's consideration to a situation in which the numerical data would lead us to a probability 1/10,000 instead of 1/25. Such data would make very hard to believe that the observed difference in mortality is due to mere chance but, of course, they would not prove right away the superiority of the new treatment. The data would furnish a pretty strong argument for the existence of *some non-random difference* between the two kinds of cases. What the nature of this difference actually is, the numbers cannot say. If only young or vigorous people received the new treatment, and only elderly or weak people the old treatment, the argument in favor of the medical superiority of one treatment above the other would be extremely weak.

(7) I think that the reader has noticed a certain parallelism between the six preceding examples of this section. Now this parallelism may be ripe to be brought into the open and formulated in general terms. Yet let us follow as far as possible the example of the naturalist who carefully compares the relevant details, rather than the example of those philosophers who rely mainly on words. We went into considerable detail in discussing our examples; if we do not take into account the relevant particulars carefully, our labor is lost.

In each example there is a *coincidence* and an *explanation*.   (Name, surname, occupation and home town of a person I met coincide with those of a person I heard of.   Explanation: the two persons are identical.—Two numbers, remembered or computed by two different persons, coincide.   Explanation: the number, arrived at by two persons working independently, is correct.— The initials of several couples of numerals, designating the same number in two different languages, coincide.   Explanation: the two languages are related.—The bright lines in the spectrum of iron, observed in laboratory experiments, coincide with certain dark lines in the spectrum of the sun. Explanation: there is iron vapor in the atmosphere of the sun.—A new treatment of a disease coincides with lower mortality.   Explanation: the new treatment is more effective.)

Contrasting with these specific explanations, the nature of which varies with the nature of the example, there is another explanation which can be stated in the same terms in all examples: the observed coincidences are due to chance.

The specific explanations are not groundless, some of them are reasonably convincing, but none of them is logically necessary or rigidly proven. Therefore the situation is fundamentally the same in each example: there are two rival conjectures, a specific conjecture, and the "universally applicable" hypothesis of "randomness" which attributes the coincidences to chance.

Yet, if we look at it more closely, we perceive that the "hypothesis of randomness" is vague.   The statement "this effect is due to chance" is ambiguous, since chance can operate according to different schemes.   If we wish to obtain some more definite indication from it, we have to make the hypothesis of randomness more precise, more specific, express it in terms of probability, in short, we have to raise it to the rank of a *statistical hypothesis*.

In everyday matters we usually do not take the trouble to state a statistical hypothesis with precision or to compute its likelihood numerically.   Yet we may take a first step in this direction (as in example (1)) or go even a little further (as in examples (2) or (3)).   In scientific questions, however, we should clearly formulate the statistical hypothesis involved and follow it up to a numerical estimate of its likelihood, as in examples (5) and (6).

In the transition from the general and therefore somewhat diffuse idea of randomness to a specific statistical hypothesis we have to make a choice. There are cases in which we scarcely notice this choice, since we can perceive just one statistical hypothesis that is simple enough and fits the case reasonably well; in such a case the hypothesis chosen appears "natural" (as in examples (3), (5) and (6)).   In other cases the choice is quite noticeable; we do not see immediately a statistical hypothesis that would be simple enough and fit the case somewhat "realistically," so we choose after more or less hesitation (as in example (4)).

Eventually there are two rival conjectures facing each other: a non-statistical, let us say "physical," conjecture $Ph$ and a statistical hypothesis $St$. Now, a certain event $E$ has been observed.   This event $E$ is related both to $Ph$ and to $St$, and is so related that its happening could influence our choice between the two rival conjectures $Ph$ and $St$.   If the physical conjecture $Ph$ is true, $E$ appears as easily explicable, its happening is easily understandable. In the clearest cases (as in example (5)) $E$ is implied by $Ph$, is a consequence of $Ph$.   On the other hand, from the standpoint of the statistical hypothesis $St$, the event $E$ appears as a "coincidence" the probability $p$ of which can be computed on the basis of the hypothesis $St$.   If the probability $p$ of $E$ turns out to be low, the happening of the event $E$ is not easily explicable by "chance," that is by the statistical hypothesis $St$; this weakens our confidence in $St$ and, accordingly, strengthens our confidence in $Ph$.   On the other hand, if the probability $p$ of the observed event $E$ is high, $E$ may appear as explicable by chance, that is, by the statistical hypothesis $St$; this strengthens somewhat our confidence in $St$ and accordingly weakens our confidence in $Ph$.

It should be noticed that the foregoing is in agreement with what we said about rival conjectures in sect. 13.12 and adds some precision to the pattern of plausible reasoning discussed in sect. 12.3.

The omnipresent hypothesis of randomness is an alternative to any other kind of explanation.   This seems to be deeply rooted in human nature. "Was it intention or accident?"   "Is there an assignable cause or merely chance coincidence?"   Some question of this kind occurs in almost every debate or deliberation, in trivial gossip and in the law courts, in everyday matters and in science.

**10. Judging mathematical conjectures.**   We compare some examples treated in foregoing chapters with each other and with those treated in the foregoing section.

(1) Let us remember the story of a remarkable discovery told in sect. 2.6. Euler investigated the infinite series

$$1 + \frac{1}{4} + \frac{1}{9} + \frac{1}{16} + \frac{1}{25} + \cdots + \frac{1}{n^2} + \cdots .$$

First he found various transformations of this series.   Then, using one of these transformations, he obtained an approximate numerical value for the sum of the series, the value 1.644934.   Finally, by a novel and daring procedure, he guessed that the sum of the series is $\pi^2/6$.   Euler felt himself that his procedure was daring, even objectionable, yet he had a good reason to trust his discovery: the value found by numerical computation, 1.644934, coincided, as far as it went, with the value guessed

$$\frac{\pi^2}{6} = 1.64493406 \ldots .$$

And so Euler was confident.   Yet was this confidence reasonable?   Such a coincidence may be due to chance.

In fact, it is not outright impossible that such a coincidence is due to chance, yet there is just one chance in ten million for such a coincidence to happen: the probability that chance, interpreted in a simple manner, should produce such a coincidence of seven decimals is $10^{-7}$; cf. sect. 9 (3) and ex. 11. And so we should not blame Euler that he rejected the explanation by chance coincidence and stuck to his guess $\pi^2/6$. He proved his guess ultimately. Yet we need not insist here on the fact that it has been proved: with or without confirmation, Euler's guess was, in itself, not only brilliant but also reasonable.

(2) Let us look again at sect. 3.1 and especially at fig. 3.1 which displays nine polyhedra. For each of these polyhedra we determined $F$, $V$, and $E$, that is, the number of faces, vertices, and edges, respectively, and listed the numbers found in a table (Vol. I, p. 36). Then we observed a regularity: throughout the table

$$F + V = E + 2.$$

It seemed to us improbable that such a persistent regularity should be mere coincidence, and so we were led to conjecture that the relation observed in nine cases is generally true.

There is a point in this reasoning that could be made more precise: what is the probability of such a coincidence? To answer this question, we have to propose a definite statistical hypothesis. I was not able to think of one that fits the case perfectly, but there is one that has some bearing on the situation. Let me state it in setting $F - 1 = X$, $V - 1 = Y$, $E = Z$. With this change of notation, the conjectural relation obtains the form $X + Y = Z$.

We have three bags, each of which contains $n$ balls, numbered $1, 2, 3, \ldots n$. We draw one ball from each bag and let $X$, $Y$, and $Z$ denote the number from the first, the second, and the third bag, respectively. What is the probability that we should find the relation

$$X + Y = Z$$

between the three numbers $X$, $Y$, and $Z$, produced by chance?

It is understood that the three drawings are mutually independent. With this proviso the probability required is determined and we easily find that it is equal to

$$\frac{n - 1}{2n^2}.$$

Let us apply this to our example. Let us focus on the moment when we succeed in verifying the hypothetical relation for a new polyhedron. For example, after the nine polyhedra that we examined initially (in sect. 3.1) we took up the case of the icosahedron (in sect. 3.2). For the icosahedron, as we found, $F = 20$, $V = 12$, $E = 30$, and so, in fact

$$(F - 1) + (V - 1) = 19 + 11 = 30 = E.$$

Is this merely a random coincidence? We apply our formula, taking $n = 30$ (we certainly could not make $n$ less than 30) and find that such an event has the probability

$$\frac{29}{2 \times 30^2} = \frac{29}{1800} = 0.016111;$$

that is, it has a little less prospect than 1 chance in 60. We may hesitate whether we should, or should not, ascribe the verification of the conjectured relation to mere chance. Yet if we succeed in verifying it for another polyhedron, with $F$, $V$, $E$ about as large as for the icosahedron, and we are inclined to regard the two verifications as mutually independent, we face an event (the joint verification in both cases) with a probability less than $(1/60)^2$; this event has less chance to happen than 1 in 3600 and is, therefore, even harder to explain by chance. If the verifications continue without interruption, there comes a moment, sooner or later, when we feel obliged to reject the explanation by chance.

(3) In the foregoing example we should not stress too much the numerical values of the probabilities that we computed. To realize that the probability steadily decreases as verification follows verification may be more helpful in guiding our judgment than the numerical values computed. At any rate, there are cases in which it would be hard to offer a fitting statistical hypothesis and so it is not possible to compute the probabilities involved numerically, yet the calculus of probability still yields helpful suggestions.

In sect. 4.8 we compared two conjectures concerning the sum of two squares. Let us call them conjecture $A$ and conjecture $B$, respectively. Conjecture $A$ (that we have discovered at the end of sect. 4.6) asserts a remarkable rule that precisely determines in how many ways an integer of a certain form can be represented as a sum of four odd squares. Conjecture $B$ (Bachet's conjecture) asserts that any integer can be represented as the sum of four squares in one or more ways. Each of the two conjectures offers a prediction about the sum of four squares, but the prediction offered by $A$ is more precise than that offered by $B$. Just to stress this point, let us consider for a moment a quite unbelievable assumption. Let us assume that we know from some (mysterious) source that, in a certain case, the number of representations has an equal chance to have any one of the $r + 1$ values 0, 1, 2, ... $r$, and cannot have a value exceeding $r$, which is a quite large number (and this should hold both under the circumstances specified in $A$ and under those specified in $B$—a rather preposterous assumption). Now, $A$ predicts that the number of representations has a definite value; $B$

predicts that it is greater than 0. Therefore, the probability that $A$ turns out to be true in that assumed case is $1/(r+1)$, whereas the probability that $B$ turns out to be true is $r/(r+1)$. In fact, both $A$ and $B$ turn out to be true in that case, both conjectures are verified, and the question arises which verification yields the stronger evidence. In view of what we have just discussed, it is much more difficult to attribute the verification of $A$ to chance, than the verification of $B$. By virtue of this circumstance (in accordance with all similar examples discussed in this chapter) the verification of the more precise prediction $A$ should carry more weight than the verification of the less precise prediction $B$. In sect. 4.8 we arrived at the same view without any explicit consideration of probabilities.

## EXAMPLES AND COMMENTS ON CHAPTER XIV

### First Part

Each example in this first part begins with a reference to some section or subsection of this chapter and supplies formulas or derivations omitted in the text. The solutions require some knowledge of the calculus of probability.

**1.** [Sect. 3 (3)]  Accept the scheme of sect. 3 (3) for representing the succession of rainy and rainless days. Say "sunny" instead of "rainless," for the sake of convenience, and let $r_r$, $s_r$, $r_s$, and $s_s$ denote probabilities,

$r_r$ for a rainy day after a rainy day,
$s_r$ for a sunny day after a rainy day,
$r_s$ for a rainy day after a sunny day, and
$s_s$ for a sunny day after a sunny day.

(a) Show that $r_r - r_s = s_s - s_r$.

(b) It was said that "a rainy day follows a rainy day more easily than a rainless day." What does this mean precisely?

**2.** [Sect. 3 (4)]  It was said that "each letter tends to be unlike the foregoing letter." What does this mean precisely?

**3.** [Sect. 5 (1)]  Find the general expression for the numbers in column (3) of Table I.

**4.** [Sect. 5 (2)]  Find the general expression for the numbers in column (5) of Table I.

**5.** [Sect. 5 (3)]  (a) Find the general expression for numbers in column (3) of Table II.  (b) In order to detect a systematic deviation, if there is one, examine the differences of corresponding entries (on the same row) of columns (4) and (5); list the signs.

**6.** [Sect. 7 (1)]  If a trial consists in casting three fair dice and a success consists in casting six spots with each dice, what is the probability of $n$ successes in $n$ trials?

**7.** [Sect. 7 (2)] Among the various events reported in the story of the Reverend Galiani told in sect. 7 (1), which one constitutes the strongest argument against the hypothesis of fair dice?

**8.** [Sect. 7 (3)] (a) Write down the formula that leads to the numerical value $1.983 \cdot 10^{-7}$.

(b) The probability of a success is $1/3$. Find the probability that $315672$ trials yield precisely $315672/3$ successes.

**9.** [Sect. 8 (1)] The expression given for $a$ is a sum. Each term of this sum is, in fact, a probability: for what?

**10.** [Sect. 8 (1)] Find the abscissa of the point of inflection of the curve represented by fig. 14.4.

**11.** [Sect. 9 (3)] Given a number of $n$ figures. A sequence of $n$ figures is produced at random, perhaps by a monkey playing with the keys of an adding machine. What is the probability that the sequence so produced should coincide with the given number? [Is the answer mathematically determined?]

**12.** [Sect. 9 (4)] Explain the computation of the probability 0.0948.

**13.** [Sect. 9 (4)] Find the general expression for the numbers (a) in column (6), (b) in column (7), of Table V.

**14.** [Sect. 9 (4)] Explain the computation of the expected numbers of coincidences in Table VI: (a) 42.66, (b) 8.53.

**15.** [Sect. 9 (4)] Explain the computation of the standard deviation 2.78 in the last row and last column of Table VI.

**16.** [Sect. 9 (5)] Why $(1/2)^{60}$?

**17.** [Sect. 9 (6)] Explain the computation of the probability 0.0399. [Generalize.]

**18.** [Sect. 10 (2)] Derive the expression $(n-1)/2n^2$ for the required probability.

## Second Part

**19.** *On the concept of probability.* Sect. 2 does not define what probability "is," it merely tries to explain what probability aims at describing: the "long range" relative frequency, the "final stable" relative frequency, or the relative frequency in a "very long" series of observations. How long such a series is supposed to be, was not stated. This is an omission.

Yet such omissions are not infrequent in the sciences. Take the oldest physical science, mechanics, and the definition of velocity in non-uniform, rectilinear motion: velocity is the space described by the moving point in a certain interval of time, divided by the length of that interval, provided that the interval is "very short." How short such an interval is supposed to be is not stated.

Practically, you take the interval of time measured as short, or the statistical series observed as long, as your means of observation allow you. Theoretically you may pass to the limit. The physicists, in defining velocity, let the interval of time tend to zero. R. von Mises, in defining probability, lets the length of the statistical series tend to infinity.

**20.** *How not to interpret the frequency concept of probability.* The D. Tel. shook his head as he finished examining the patient. (D. Tel. means doctor of teleopathy; although strenuously opposed by the medical profession, the practice of teleopathy has been legally recognized in the fifty-third state of the union.) "You have a very serious disease," said the D. Tel. "Of ten people who have got this disease only one survives." As the patient was sufficiently scared by this information, the D. Tel. went on. "But you are lucky. You will survive, because you came to me. I have already had nine patients who all died of it."

Perhaps the D. Tel. meant it. His grandfather was a sailor whose ship was hit by a shell in a naval engagement. The sailor stuck his head through the hole torn by the shell in the hull of the ship and felt protected "because," he reasoned, "it is very improbable that a shell will hit the same spot twice."

**21.** An official, charged to supervise an election in a certain locality, found 30 fake registrations among the 38 that he examined the first morning. A daily paper declares that at least 99% of the registrations in that locality are correct and above suspicion. How does the daily's assertion stand up in the light of the official's observation?

**22.** In the window of a watchmaker's shop there are four cuckoo clocks, all going. Three clocks out of the four are less than two minutes apart: can you rely on the time that they show? There is a natural conjecture: the clocks were originally set on time, but they are not very precise (they are just cuckoo clocks) and one is out of order. If this is so, you could rely on the time shown by three. Yet there is a rival conjecture, of course: those three clocks agree by mere chance. What is the probability of such an event?

**23.** If $a$, $b$, $c$, $d$, $e$, and $f$ are integers chosen at random, not exceeding in absolute value a given positive integer $n$, what is the probability that the system

$$ax + by = e$$
$$cx + dy = f$$

of two equations with two unknowns has just one solution?

**24.** *Probability and the solution of problems.* In a crossword puzzle one unknown word with 5 letters is crossed by two unknown words with four letters each.   You guess that the unknown 5 letter word is TOWER and then you have the situation indicated by the following diagram:

In order to test your guess, you would like to find one or the other four letter word crossing the conjectural TOWER.   One of the crossing words could verify the o, the other the E.   Which verification would carry more weight? And why?

**25.** *Regular and Irregular.*   Compare the two columns of numbers:

| I | II |
|------|------|
| 1005 | 1004 |
| 1033 | 1038 |
| 1075 | 1072 |
| 1106 | 1106 |
| 1132 | 1139 |
| 1179 | 1173 |
| 1205 | 1206 |
| 1231 | 1239 |
| 1274 | 1271 |
| 1301 | 1303 |

One of these two columns is "regular," the other "irregular."   The regular column contains ten successive mantissas from a four-place table of common logarithms.   The numbers of the irregular column agree with the corresponding numbers of the regular column in the first three digits, yet the fourth digits could be the work of an unreliable computer: they have been chosen "at random."   Which is which?   [Point out an orderly procedure to distinguish the regular from the irregular.]

**26.** *The fundamental rules of the Calculus of Probability.*   In computing probabilities we may visualize the set of possible cases and see intuitively that none is privileged among them, or we may proceed according to rules. It is important for the beginner to realize that he can arrive at the same result by these two different paths.   The rules are particularly important when we regard the theory of probability as a purely mathematical theory.

The rules will be important in the next chapter.    For all these reasons, let us introduce here the fundamental rules of the calculus of probability, using the bag and the balls;[8] cf. sect. 3.

The bag contains $p$ balls.    Some of the balls are marked with an $A$, others with a $B$, some with both letters, some are not marked at all.    (There are $p$ possible cases and two "properties," or "events," $A$ and $B$.)    Let us write $\bar{A}$ for the absence of $A$ or "non-$A$."    (We take — as the sign of negation, but place this sign on the top of the letter, not before it.)    There are four possibilities, four categories of balls.

The ball has $A$, but has not $B$.    We denote this category by $A\bar{B}$ and the number of such balls by $a$.

The ball has $B$, but has not $A$.    We denote this category by $\bar{A}B$ and the number of such balls by $b$.

The ball has both $A$ and $B$.    We denote this category by $AB$ and the number of such balls (common to $A$ and $B$) by $c$.

The ball has neither $A$ nor $B$.    We denote this category by $\bar{A}\bar{B}$ and the number of such balls (different from those having $A$ or $B$) by $d$.

Therefore, obviously,

$$a + b + c + d = p.$$

We let $\Pr\{A\}$ stand as abbreviation for the probability of $A$, and $\Pr\{B\}$ for that of $B$.    With this notation, we have obviously

$$\Pr\{A\} = \frac{a + c}{p}, \qquad \Pr\{B\} = \frac{b + c}{p}.$$

Let $\Pr\{AB\}$ stand for the "probability of $A$ and $B$," that is, the probability for the joint appearance of $A$ and $B$.    Obviously

$$\Pr\{AB\} = \frac{c}{p}.$$

Let $\Pr\{A \text{ or } B\}$ stand for the probability of obtaining $A$, or $B$, or both $A$ and $B$.[9]    Obviously

$$\Pr\{A \text{ or } B\} = \frac{a + b + c}{p}.$$

---

[8] We follow H. Poincaré, *Calcul des probabilités*, p. 35–39.

[9] The little word "or" has two meanings, which are not sufficiently distinguished by the English language, or by the other modern European languages.    (They are, however, somewhat distinguished in Latin.)    We may use "or" "exclusively" or "inclusively."    "You may go to the beach or to the movies" (not to both) is *exclusive* "or" (in Latin "aut").    "You may go the beach or have a lot of candy" is *inclusive* "or" if you mean "one or the other or both."    In legal or financial documents inclusive "or" is rendered as "and/or" (in Latin "vel").    In $\Pr\{A \text{ or } B\}$ we mean the *inclusive* "or."

We readily see that

$$\Pr\{A\} + \Pr\{B\} = \Pr\{AB\} + \Pr\{A \text{ or } B\}$$

and hence follows our first fundamental rule (the "or" rule):

(1)            $$\Pr\{A \text{ or } B\} = \Pr\{A\} + \Pr\{B\} - \Pr\{AB\}.$$

We wish now to define the *conditional* probability $\Pr\{A/B\}$, in words: probability of $A$ *if* $B$ (granted $B$, posito $B$, on the condition $B$, on the hypothesis $B$, . . . ). Also this probability is intended to represent a long range relative frequency. We draw from the bag, repeatedly, one ball each time, replacing the ball drawn before drawing the next, as described at length in sect. 2 (1). Yet we *take into account only the balls having a B*. If among the first $n$ such balls drawn, there are $m$ balls that also have an $A$, $m/n$ is the relative frequency that should be approximately, when $n$ is sufficiently large, equal to $\Pr\{A/B\}$. It appears rather obvious that

$$\Pr\{A/B\} = \frac{c}{b + c}.$$

In fact, there are $c$ balls having $A$ among the $b + c$ balls having $B$; also the reasoning of sect. 2 (1) may be repeated; from a certain viewpoint, we could regard the expression of $\Pr\{A/B\}$ also as a definition. At any rate, we easily find, comparing the expressions of the probabilities involved, that

$$\Pr\{A/B\} = \Pr\{AB\}/\Pr\{B\}.$$

Interchanging $A$ and $B$, we find the second fundamental rule (the "and" rule):

(2)            $$\Pr\{AB\} = \Pr\{A\} \Pr\{B/A\} = \Pr\{B\} \Pr\{A/B\}.$$

We can derive many other rules from (1) and (2). Observing that

$$\Pr\{A \text{ or } \bar{A}\} = 1, \qquad \Pr\{A\bar{A}\} = 0,$$

we obtain from (1), by substituting $\bar{A}$ for $B$, that

(3)            $$\Pr\{A\} + \Pr\{\bar{A}\} = 1,$$

what we could see also directly, of course. Similarly, since

$$\Pr\{AB \text{ or } \bar{A}B\} = \Pr\{B\}, \qquad \Pr\{(AB)(\bar{A}B)\} = 0,$$

we obtain from (1), by substituting $AB$ for $A$ and $\bar{A}B$ for $B$, that

(4)            $$\Pr\{B\} = \Pr\{AB\} + \Pr\{\bar{A}B\}.$$

We note here the following generalization of (2):

(5)        $\Pr\{AB/H\} = \Pr\{A/H\} \Pr\{B/HA\} = \Pr\{B/H\} \Pr\{A/HB\}.$

We can also visualize (5) by using the bag and the balls.

**27. Independence.** We call two events independent of each other, if the happening (or not happening) of one has no influence on the chances of the other. Disregard, however, for the moment this informal definition and consider the two following formal definitions.

(I)  $A$ is called *independent of B* if

$$\Pr\{A/B\} = \Pr\{A/\bar{B}\}.$$

(II)  $A$ and $B$ are called *mutually independent* if

$$\Pr\{A/B\} = \Pr\{A/\bar{B}\} = \Pr\{A\}, \qquad \Pr\{B/A\} = \Pr\{B/\bar{A}\} = \Pr\{B\}.$$

(I)     (II)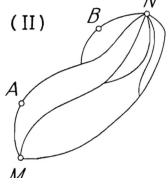

Fig. 14.8. Two systems of roads from the city $M$ to the city $N$, with an essential difference.

Obviously, if $A$ and $B$ are mutually independent, $A$ is independent of $B$. Using the rules of ex. 26, prove the theorem: *If none of the probabilities* $\Pr\{A\}$, $\Pr\{B\}$, $\Pr\{\bar{A}\}$, $\Pr\{\bar{B}\}$ *vanishes and any one of the two events $A$ and $B$ is independent of the other, they are mutually independent.*

**28.** Compare sect. 3 (5) with ex. 27.

**29.** A car traveling from the city $M$ to the city $N$ may pass through the town $A$ and also through the town $B$. This is true of both systems of roads, (I) and (II), shown in fig. 14.8. Answer the following questions (a), (b), and (c) first in assuming that (I) represents the full system of roads between $M$ and $N$, then in assuming the same thing about (II).

(a) Let $A$ stand for the event that a car traveling from $M$ to $N$ passes through the town $A$, and $B$ for the event that it passes through $B$. Assume (for both systems, (I) and (II)) that the three roads starting from $M$ are equally well frequented (have the same probability) and also that the roads ending in $N$ (there are 2 in (I), 6 in (II)) are equally well frequented. Find the probabilities $\Pr\{A\}$, $\Pr\{A/B\}$, $\Pr\{A/\bar{B}\}$, $\Pr\{B\}$, $\Pr\{B/A\}$, $\Pr\{B/\bar{A}\}$.

(b) Find $\Pr\{AB\}$ using the rule (2) of ex. 26.

(c) Verify that

$$\Pr\{A\} = \Pr\{B\}\,\Pr\{A/B\} + \Pr\{\bar{B}\}\,\Pr\{A/\bar{B}\},$$

$$\Pr\{B\} = \Pr\{A\}\,\Pr\{B/A\} + \Pr\{\bar{A}\}\,\Pr\{B/\bar{A}\}.$$

(d) What do you regard as the most important difference between (I) and (II)?

**30.** *Permutations from probability.* To decide the order in which the $n$ participants should show their skill in an athletic contest, the name of each is written on a slip of paper, and then the $n$ slips are drawn from a hat, one after the other, at random. What is the probability that the $n$ names should appear in alphabetical order?

We present two solutions, and draw a conclusion from comparing them.

(1) Call $E_1$ the event that the slip drawn first is also alphabetically the first, $E_2$ the event that the slip drawn in the second place is also alphabetically the second, and so forth. The desired probability is

$$\Pr\{E_1 E_2 E_3 \ldots E_n\} =$$

$$= \Pr\{E_1\}\,\Pr\{E_2/E_1\}\,\Pr\{E_3/E_1 E_2\} \ldots \Pr\{E_n/E_1 \ldots E_{n-1}\}$$

$$= \frac{1}{n} \cdot \frac{1}{n-1} \cdot \frac{1}{n-2} \ldots \frac{1}{1} \cdot$$

In fact, we obtain the first transformation by applying the rules (2) and (5) of ex. 26, and the second transformation by observing that there are $n$ possible cases for $E_1$, $n-1$ for $E_2$ after $E_1$, $n-2$ for $E_3$ after $E_1$ and $E_2$, and so forth, whereas, for each of these events, there is just one favorable case.

(2) Call $P_n$ the number of all the possible orderings (permutations, linear arrangements, ...) of $n$ distinct objects. The $n$ names can come out from the hat in $P_n$ ways, no one of these $P_n$ possible cases appears as more privileged than the others, and among these $P_n$ cases just one is favorable (the alphabetical order). Therefore, the desired probability is $1/P_n$.

(3) The results derived under (1) and (2) must be equal. Equating them, we evaluate $P_n$:

$$P_n = 1 \cdot 2 \cdot 3 \ldots n = n!.$$

**31.** *Combinations from probability.* Mrs. Smith bought $n$ eggs, not realizing that $r$ of these eggs are rotten. She needs $r$ eggs, and chooses as many among her $n$ eggs at random. What is the probability that all $r$ eggs chosen are rotten?

As in ex. 30 we present two solutions, and draw a conclusion from comparing them.

(1) Call $E_1$ the event that the first egg opened by Mrs. Smith is rotten, $E_2$ the event that the second egg is rotten, and so forth. The desired probability is

$$\Pr\{E_1\,E_2\,E_3\ldots E_r\}$$

$$= \Pr\{E_1\}\,\Pr\{E_2/E_1\}\,\Pr\{E_3/E_1E_2\}\ldots\Pr\{E_r/E_1\ldots E_{r-1}\}$$

$$= \frac{r}{n}\cdot\frac{r-1}{n-1}\cdot\frac{r-2}{n-2}\ldots\frac{1}{n-r+1}.$$

The first transformation is obtained by rules (2) and (5) of ex. 26, the second from the consideration of possible and favorable cases for $E_1$, for $E_2$ after $E_1$, and so on.

(2) We have a set of $n$ distinct objects. Any $r$ objects chosen among these $n$ objects form a subset of size $r$ of the given set of size $n$: call $C_r^n$ the number of all such subsets. (Usually $C_r^n$ is called the number of "combinations" of $r$ things selected from among $n$ things.) In the case of Mrs. Smith's eggs, there are $C_r^n$ possible cases, no one more privileged than the others, and among these $C_r^n$ cases just one is favorable (if getting rotten eggs is "favorable"). Hence the desired probability is $1/C_r^n$.

(3) Comparing (1) and (2), we evaluate $C_r^n$:

$$C_r^n = \frac{n(n-1)\ldots(n-r+1)}{1\cdot 2\ \ldots\ r} = \frac{n!}{r!(n-r)!} = \binom{n}{r}.$$

**32.** *The choice of a rival statistical conjecture: an example.* One person withdrew \$875 from his savings account on a certain date, and another person received \$875 two days later. The coincidence of these two amounts, one withdrawn, the other received, may be regarded as circumstantial evidence, as an indication that a crime has been committed; cf. ex. 13.6. If the jury finds it too hard to believe that this coincidence is due to mere chance, a conviction may result. Hence the problem: what is the probability of such a coincidence? The less the probability is, the more difficult it is to attribute the coincidence to chance, and the stronger is the case against the defendants.

Yet we cannot compute a probability numerically without assuming some definite statistical hypothesis. Which hypothesis should we assume? In a serious case we should give serious thought to such a question. Let us survey a few possibilities.

(1) As the number 875 has three digits, we may regard the positive integers with not more than three digits as admissible, and we may regard them as equally admissible. The probability that two such integers, chosen at random, independently from each other, should coincide, is obviously

1/999.   This probability is pretty small—but is the assumption that underlies·
its computation reasonable?

(2) As 875 has less than five digits, we could regard all positive integers
with less than five digits as equally admissible.   This leads to the probability
1/9999 for the coincidence.   This probability is very small indeed, but our
assumption is far-fetched, even frivolous.

(3) If the point appears as important, the court can order inspection of
the books of the bank or summon one of its competent officials to testify.
And so it has been ascertained that immediately before the withdrawal of
that sum $875 the amount $2581.48 was deposited on the account.   In
possession of this relevant information we may regard as possible and equally
admissible cases the sums 1, 2, 3, . . . 2581 that could have been withdrawn
from the account.   Just one of these cases, 875, has to be termed favorable
and so we are led to the probability 1/2581 for the coincidence.   This is a
small probability, but our assumption may seem reasonably realistic.

(4) We could have considered not only withdrawals in dollars, but also
withdrawals in dollars and cents, such as $875.31.   If we consider all such
cases as equally admissible, the probability for a coincidence becomes
1/258148.   This is a very small probability, but our assumption may appear
less realistic: withdrawals in dollars and cents such as $875.31 are more
usual from a checking account than from a savings account.

(5) On the contrary, one could argue that the amounts withdrawn from
a savings account are usually "round" amounts, divisible by 100, or 50, or 25.
Now, 875 is divisible by 25.   If we regard only multiples of 25 as admissible,
and equally admissible, the probability in question becomes 1/103.

Of course, we could imagine still other and more complicated ways to
compute the probability, but we should not insist unduly on such a transparent example.   The example served its purpose if the reader can see by
now the following two points.

(a) Although some of the five assumptions discussed may seem more
acceptable than others, no one is conspicuously superior to the others, and
there is little hope to find an assumption that would be satisfactory in every
respect and could be regarded as the best.

(b) Each of the five assumptions considered attributes a rather small
probability to the coincidence actually observed, and so, on the whole, our
consideration upholds the common sense view: "It is hard to believe that
this coincidence is due to mere chance."

**33.** *The choice of a rival statistical conjecture: general remarks.*   Let us try to
learn something more general from the particular example considered
(ex. 32).   Let us reconsider the general situation discussed in sect. 14.9 (7).
An event $E$ has occurred and has been observed.   Concerning this event,
there are two rival conjectures facing each other: a "physical" conjecture $P$,
and a statistical hypothesis $H$.   If we accept the physical conjecture $P$, $E$ is

easily and not unreasonably explicable.    If we accept the statistical hypothesis $H$, we can compute the probability $p$ for the happening of such an event as E.    If $p$ is "small," we may be induced to reject the statistical hypothesis $H$.    At any rate, the smallness of $p$ weakens our confidence in $H$ and therefore strengthens somewhat our confidence in the rival conjecture $P$.

Yet ex. 32 makes us aware that the quality of the statistical hypothesis $H$ plays a role in the described reasoning.    The statistical hypothesis $H$ may appear as unnatural, inappropriate, far-fetched, frivolous, cheap, *unreliable* from the start.    Or $H$ may appear as natural, appropriate, realistic, reasonable, *reliable* in itself.

Now $p$, the probability of the event $E$ computed on the basis of the hypothesis $H$, may be so small that we reject $H$: a rival of $P$ drops out of the race. This increases the prospects of $P$—but it may increase them a lot or only a little: this depends on the quality of the rival.    If the statistical hypothesis $H$ appeared to us originally as appropriate and reliable, $H$ was a dangerous rival and its fall strengthens $P$ appreciably.    If, however, $H$ appeared to us as inappropriate and unreliable from the start, $H$ was a weak rival; its fall is not surprising and strengthens $P$ very little.

Being given a clear statistical hypothesis $H$, the probability $p$ of the event $E$ is determined, and the statistician can compute it.    Yet the statistician's customer, who may be a biologist, or a psychologist, or a businessman, or any other non-statistician, has to decide what this numerical value of $p$ means in his case.    He has to decide how small a $p$ is enough to reject or weaken the statistical hypothesis $H$.    Yet the customer is usually not even directly interested in the statistical hypothesis $H$: he is primarily concerned with the rival "physical" conjecture $P$.    And he has to decide how much weight the rejection or weakening of $H$ has in strengthening $P$.    This latter decision obviously cannot depend on the numerical value of $p$ alone: it certainly depends on the choice of $H$.

I am afraid that the statistician's customer who wishes to make use of the numerical value $p$ furnished by the statistician, without realizing the import of the statistical hypothesis $H$ for his problem, just deceives himself.    He can scarcely realize the import of $H$ if he does not realize that his physical conjecture $P$ could be also confronted with statistical hypotheses different from $H$.    Cf. ex. 15.5.

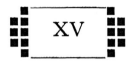

# XV

# THE CALCULUS OF PROBABILITY AND THE LOGIC OF PLAUSIBLE REASONING

*It is difficult to estimate the probability of the results of induction.*—
LAPLACE[1]

*We know that the probability of a well-established induction is great,
but, when we are asked to name its degree, we cannot. Common sense tells
us that some inductive arguments are stronger than others, and that some are
very strong. But how much stronger or how strong we cannot express.*—
JOHN MAYNARD KEYNES[2]

1. **Rules of plausible reasoning.** In the foregoing three chapters we
collected patterns of plausible reasoning. Do these "patterns" constitute
"rules" of plausible reasoning? How far and in which way are they binding,
authoritative, imperative? There is a certain danger of losing ourselves in
purely verbal explanations. Therefore I wish to consider the question more
concretely, even a little personally.

(1) I remember a conversation on invention and plausible reasoning. It
happened long ago. I talked with a friend who was much older than myself
and could look back on a distinguished record of discoveries, inventions,
and successful professional work. As he talked on plausible reasoning and
invention, he doubtless knew what he was talking about. He maintained
with unusual warmth and force of conviction that invention and plausible
reasoning have no rules. Hunches and guesses, he said, depend on experience
and intuition, but not on rules: there are no rules, there can be no rules,
there should be no rules, and if there were some rules, they were useless
anyway. I maintained the contrary—a conversation is uninteresting if
there is no difference of opinion—yet I felt the strength of his position. My
friend was a surgeon. A wrong decision of a surgeon may cost a life and

---

[1] Essai philosophique sur les probabilités; see *Oeuvres complètes de Laplace*, vol. 7,
p. CXXXIX.

[2] *A treatise on probability*, p. 259.

sometimes, when a patient suddenly starts bleeding or suffocating, the right decision must come in a second.   I understood that people who have to make such responsible quick decisions have no use for rules.   The time is too short to apply a rule properly, and any set pattern could misguide you; what you need is intense concentration upon the situation before you.   And so people come to distrust "rules" and to rely on their "intuition" or "experience" or "intuition-and-experience."

In the case of my friend, there was still something else, perhaps.   He was a little on the domineering side.   He hated to relinquish power.   He felt, perhaps, that acknowledging a rule is like delegating a part of his authority to a machine, and so he was against it.

Let us note: distrust in rules of reasoning may come naturally to intelligent people.

(2) Two people presented with the same evidence may judge it very differently.   Two jurors who sat through the same proceedings may disagree: one thinks that the evidence introduced is sufficient proof against the defendant and the other thinks that it is not.   Such disagreement may have thousand different grounds: people may be moved in opposite directions by fears, hopes, prejudices and sympathies, or by personal differences. Perhaps, one of the jurors is stupid and the other is clever, or one slept through the proceedings and the other listened intently. Yet the personal differences underlying the disagreement may be more subtle.   Perhaps both jurors are honest and reasonably unprejudiced, both followed the proceedings with attention, and both are intelligent, but in a different way.   The first juror may be a better observer of demeanor.   He observes the facial expressions of the witnesses, the tics of the defendant; he notices when an answer is haltingly given; he is impressed by quick motions of the eyes and little gestures of the hands.   The other juror may be a less skillful observer of facial expressions, but a better judge of social relations:   he understands better the milieu and the circumstances of the people involved in the case. Seeing the same things with different eyes, honestly and not unintelligently, the two jurors come to opposite conclusions.

Let us not neglect the obvious and let us note:   two people presented with the same evidence may honestly disagree.

(3) My friend and I are both interested in the conjecture $A$.   (This friend is a mathematician, and $A$ is a mathematical conjecture.)   We both know that $A$ implies $B$.   And now we find that $B$, this consequence of $A$, is true. We agree, as we honestly have to agree, that this verification of its consequence $B$ is evidence in favor of the conjecture $A$, but we disagree about the value, or weight, of this evidence.   One of us asserts that this verification adds very little to the credibility of $A$, and the other asserts that it adds a lot.

This disagreement would be understandable if we were very unequally familiar with the subject and one of us knew many more formerly verified consequences than the other.   Yet this is not so.   We know about the same consequences of $A$ verified in the past.  We agree even that there is little analogy between the just verified $B$ and those formerly verified consequences. We agree also, as we honestly should, that this circumstance renders the evidence for $A$ stronger.  Yet one of us says "just a little stronger," the other says "a lot stronger," and we disagree.

We both suspected, even a short while ago, that $B$ is false and it came to us as a surprise that $B$ is true.  In fact, from the standpoint of a rather natural assumption (or statistical hypothesis) $B$ appears pretty improbable.  We both perceive that this circumstance renders the evidence for $A$ stronger. Yet one of us says "just a little stronger," the other says "a lot stronger," and we keep on disagreeing.

We are both perfectly honest, I think, and our disagreement is not merely a matter of temperament.  We disagree because his *background* is different from mine.  Although we had about the same scientific training, we developed in different directions.  His work led him to distrust the hypothesis $A$. He hopes, perhaps, that one day he will be able to refute that conjecture $A$. As to myself, I do not dare to hope that I shall prove $A$ one day.  Yet I must confess that I would like to prove $A$.  In fact, it is my ambition to prove $A$, but I do not wish to fool myself into the illusion that I shall ever be able to prove $A$.  Such an incompletely avowed hope may influence my judgement, my evaluation of the weight of the evidence.  Yet I may have other grounds besides: still more obscure, scarcely formulated, inarticulate grounds.  And my friend may have some grounds too that he did not yet confess to himself. At any rate, such differences in our backgrounds may explain the situation: we disagree concerning the strength of the evidence, although we agree in all the clearly recognizable points that should influence the strength of the evidence in an impersonal judgement according to universally accepted reasonable standards.

Let us note: two persons presented with the same evidence and applying the same patterns of plausible inference may honestly disagree.

(4) We tried to see plausible reasoning at work, concretely, in the behavior of people facing concrete problems.  We have now, I hope, a somewhat clearer idea in which way our patterns are "binding," how far they can be regarded as "rules."

Yet there are other approaches to explore.  Formal Logic and the Calculus of Probability have clear strict rules which appear somehow related to our patterns.  What is the nature of this relation?  This is the question that we shall discuss in the next sections.

**2. An aspect of demonstrative reasoning.**  A comparison of plausible reasoning with demonstrative reasoning may be useful at this stage.  Yet of course the aspect of plausible reasoning at which we have arrived cannot stand comparison with the highly sophisticated stage at which the theory of demonstrative reasoning has arrived at this date, after a development of more than two thousand years of which the last fifty were particularly

crowded.   A  comparison  with  a  more  primitive  aspect  of  demonstrative
reasoning  may  be  more  helpful.   Let  us  put  ourselves,  more  or  less,  in  the
position  of  a  contemporary  of  Aristotle.

Aristotle  noticed  that  reasoning  conforms  to  certain  *patterns*.   He  observed,
I  imagine,  such  patterns  in  philosophical  or  political  or  legal  or  everyday
arguments,  recognized  the  patterns  as  they  occurred,  extracted  and  formulated
them.   These  patterns  are  the  syllogisms.   The  examples  by  which  Aristotle
finds  necessary  to  support  his  syllogisms  seem  to  bear  witness  to  the  idea  that
he  discovered  his  syllogisms  by  a  sort  of  induction—and  how  could  he  have
discovered  them  otherwise?   At  any  rate,  the  idea  that  the  syllogisms  may
have  been  discovered  inductively  brings  them  a  little  nearer  to  our  patterns
of  plausible  reasoning.

Instead  of  the  "subsumptive"  syllogism,  so  dear  to  Aristotle  and  still
dearer  to  his  scholastic  followers,  let  us  consider  the  "modus  tollens"  of  the
"hypothetical"  syllogism,  which  we  have  already  considered  in  sect.  12.1:

$$A \text{ implies } B$$
$$B \text{ false}$$
$$\overline{\phantom{A \text{ implies } B}}$$
$$A \text{ false}$$

Even  from  a  quite  primitive  standpoint,  we  can  see  various  remarkable
features  in  this  pattern  of  reasoning:  it  is  *impersonal, universal, self-sufficient,*
and  *definitive*.

(1)  By  using  the  word  *impersonal*  we  emphasize  that  the  validity  of  the
reasoning  does  not  depend  on  the  personality  of  the  reasoner,  on  his  mood,
or  taste,  or  class,  or  creed,  or  color.

(2)  By  using  the  word  *universal*,  we  emphasize  that  the  statements  considered  (denoted  by  $A$  and  $B$)  need  not  belong  to  this  or  that  particular
field  of  knowledge,  to  mathematics  or  physics,  to  law  or  logic,  but  can
belong  to  any  one  of  these  fields,  or  to  any  field  whatsoever.   They  can  be
concerned  with  any  sufficiently  clear  object  of  human  thought:  the  syllogistic
conclusion  applies  to  all  such  objects.

(3)  In  order  to  understand  the  next  point,  we  should  realize  that  our
knowledge  and  our  reasonable  beliefs  can  be  changed  by  new  information.
Yet  there  is  something  unchangeable  in  the  syllogism  considered.   Having
once  accepted  the  premises  we  cannot  avoid  accepting  the  conclusion.   At
some  later  date,  we  may  receive  new  information  about  the  matters  involved
in  our  syllogistic  reasoning.   If,  however,  this  information  does  not  change
our  acceptance  of  the  premises,  it  cannot  reasonably  change  our  acceptance
of  the  conclusion.   The  inference  of  a  demonstrative  syllogism  requires
nothing  from  outside,  is  independent  of  anything  not  mentioned  explicitly
in  the  premises.   In  this  sense,  the  syllogism  is  *self-sufficient*:  nothing  is  needed
beyond  the  premises  to  validate  the  conclusion  and  nothing  can  invalidate  it
if  the  premises  remain  solid.

This "self-sufficiency" or "autarky" of the syllogism is, perhaps, its most noteworthy feature. Let us quote Aristotle himself: "A syllogism is a discourse in which, certain things being stated, something other than what is stated follows of necessity from their being so. I mean by the last phrase that they produce the consequence, and by this, that no further term is required from without in order to make the consequence necessary."

(4) If the premises are unquestionably certain, we can "detach" the conclusion from the syllogism. That is, if you know for certain both that "$A$ implies $B$" and that "$B$ is false," you may forget about these premises and just keep the conclusion "$A$ is false" as your *definitive* mental possession.

We have examined just one out of the several kinds of syllogisms, but the syllogism examined is typical: also the other syllogisms are impersonal, universal, self-sufficient, and definitive. And these features foreshadow the general character of demonstrative reasoning.

**3. A corresponding aspect of plausible reasoning.** Let us compare the pattern of demonstrative reasoning (the "modus tollens") discussed in the foregoing section with the pattern of plausible reasoning introduced in sect. 12.1:

$$A \text{ implies } B$$
$$\underline{B \text{ true}}$$
$$A \text{ more credible}$$

Between these two patterns, the "demonstrative" and the "plausible," there is a certain outward similarity. (The demonstrative pattern is traditional, and the other has been fashioned after it, of course.) Yet let us compare them more thoroughly.

Both patterns have the same first premise

$$A \text{ implies } B.$$

The second premises

$$B \text{ false} \qquad\qquad B \text{ true}$$

are just opposite, but they are equally clear and definite; they are on the same logical level. Yet there is a great difference between the two conclusions

$$A \text{ false} \qquad\qquad A \text{ more credible}$$

These conclusions are on different logical levels. The conclusion of the demonstrative pattern is on the same level as the premises, but the conclusion of our pattern of plausible reasoning is of a different nature, less sharp, less fully expressed.

The plausible conclusion is comparable to a force which has *direction and magnitude*. This conclusion pushes us in a certain direction: $A$ becomes *more* credible. This conclusion has also a certain strength: $A$ may become *much more* credible or *just a little more* credible. The conclusion is not fully expressed and is not fully supported by the premises. *The direction is expressed and is implied by the premises, the strength is not.* For any reasonable person, the premises involve that $A$ becomes more credible (certainly not less credible) but my friend and I may disagree *how much* more credible $A$ becomes. *The direction is impersonal, the strength may be personal.* My friend and I may honestly disagree about the weight of the conclusion, since our temperaments, our backgrounds, and our unstated reasons may be different. Yet the strength of the conclusion matters. If two jurors judge differently the strength of a conclusion, one may be for acquittal and the other against it. If two scientists judge differently the strength of a conclusion, one may be for undertaking a certain experiment and the other against it.

The conclusion of our pattern of plausible inference appears as *one-sided* when compared with the actual beliefs and acts of the reasoning persons: it expresses merely one aspect, and neglects others. If we realize this, the nature of plausible reasoning may seem less baffling and elusive. At any rate we are now better prepared to compare the patterns, the demonstrative and the plausible, point by point. Each of the following subsections refers to the correspondingly numbered subsection of the foregoing sect. 2.

(1) When we are reasoning in accordance with our pattern of plausible inference, we conform to a principle: the verification of a consequence strengthens the conjecture. This principle seems to be generally recognized, independently of personal differences and idiosyncrasies. Thus our pattern appears as *impersonal.*

We pay, however, a price for such "impersonality." Our pattern succeeds in being impersonal because it is one-sided, restricted to one aspect of the plausible inference. When we raise the question "*How much* is the conjecture strengthened by the verification of this consequence?" we open the door to personal differences.

(2) We took pains to show by many examples in the foregoing chapters that we naturally follow our pattern of plausible inference in dealing with mathematical conjectures. The underlying principle is generally recognized in the natural sciences, and it is implicitly admitted in the law courts, and in everyday life. The verification of a consequence is regarded as reasonable evidence for a conjecture in any domain. Thus our pattern appears as *universal.*

We pay, however, a price for such "universality." Our pattern succeeds in being universal because it is one-sided, restricted to one aspect of plausible inference. The universality becomes blurred when we raise the question "What is the weight of such evidence?" In order to judge the weight of the evidence, you have to be familiar with the domain; in order to judge the weight with assurance, you have to be an expert in the domain. Yet you cannot be familiar with all domains, and you can still less be an expert in all domains. And so everyone of us will notice soon enough that there are practical limits to the universality of plausible inference.

(3) As far as it is expressed, the plausible conclusion is supported by the premises. On the basis of the evidence supplied by the premises it is reasonable to place more confidence in $A$. At some later date, however, we may receive new information which, without changing our reliance on the premises, may change our opinion about $A$: we may find $A$ less credible, or we may even succeed in proving $A$ false.

This does not constitute an objection against the pattern of reasoning: as far as the evidence expressed in the premises goes, the conclusion is justified. A verdict of the jury may condemn the innocent or acquit the criminal. Yet such injustice of the verdict may be justifiable: on the basis of the available evidence no better verdict was possible. Such is the nature of plausible inference, and so our pattern of plausible reasoning may be termed *self-sufficient*.

Yet this kind of self-sufficiency or "autarky" does not mean durability. Moreover, the weight of the evidence, which is not mentioned in the conclusion of our (one-sided) pattern, but is important nevertheless, depends on things which are not mentioned in the premises. The strength of the conclusion (not its direction) requires things outside the premises.

(4) We cannot "detach" the conclusion of our pattern of plausible reasoning. "$A$ is rendered more credible" is meaningless without reference to the premises that explain by which circumstances it was rendered so. Referred to the premises, the plausible conclusion makes perfectly good sense and is perfectly reasonable, but it may diminish in value as time goes by, although the premises remain intact. The plausible conclusion may be very valuable in the moment when it emerges, but the advance of knowledge is likely to depreciate it: its importance is only momentary, transitory, ephemeral, *provisional*.

In short, our pattern of plausible reasoning is one-sided and leaves an ample margin for disagreement in things that matter. Yet, at the price of such one-sidedness, it manages to be impersonal and universal, even self-sufficient in a way. Still, it cannot escape being merely provisional.

It would be foolish to deplore that in several respects our pattern of plausible reasoning falls short of the perfection of demonstrative reasoning. On the contrary, we should feel a little satisfaction that we succeeded in clarifying somewhat a difference that we may have suspected from the beginning.

From the outset it was clear that the two kinds of reasoning have different tasks. From the outset they appeared very different: demonstrative reasoning as definite, final, "machinelike"; and plausible reasoning as vague, provisional, specifically "human." Now we may see the difference a little more distinctly. In opposition to demonstrative inference, plausible inference leaves indeterminate a highly relevant point: the "strength" or the "weight" of the conclusion. This weight may depend not only on clarified grounds such as those expressed in the premises, but also on unclarified unexpressed grounds somewhere in the background of the person who draws the conclusion. A person has a background, a machine has not. Indeed, you can build a machine to draw demonstrative conclusions for you, but I think you can never build a machine that will draw plausible inferences.

**4. An aspect of the calculus of probability. Difficulties.** A highly important step in the construction of a physical theory is its *formulation in mathematical terms.* We come to a point in our investigation where we should undertake such a step; we should formulate our views on plausible reasoning in mathematical terms.

No attempt at formulating a theory of plausible reasoning can disregard a historical fact: the calculus of probability was considered by Laplace and by many other eminent scientists as the appropriate expression of the rules of plausible inference. There are some grounds for this opinion and some objections against it. We begin by considering some of the difficulties.

We wish to use the calculus of probability to render more precise our views on plausible reasoning. Yet we could have some misgivings about such a procedure, for we have seen in the foregoing chapter that the calculus of probability is a (quite acceptable) theory of random mass phenomena. How could the calculus of probability be both the theory of mass phenomena and the logic of plausible inference?

This is not a strong objection; there is no real difficulty. The calculus of probability *could* be both things, *could* have two interpretations. In fact, a mathematical theory may have several different interpretations. The same differential equation (Laplace's equation) describes the steady irrotational flow of an incompressible non-viscous fluid and the distribution of forces in an electrostatic field. The same equation describes also the steady flow of heat, the steady flow of electricity, the diffusion of a salt dissolved in water under appropriate conditions, and still other phenomena. And so it is not excluded *a priori* that the same mathematical theory may serve two purposes. Perhaps, we may use the calculus of probability both in describing random mass phenomena and in systematizing our rules of plausible inference.

It is important, however, to distinguish clearly between these two interpretations. Thus, we may use the symbol $\Pr\{A\}$ (see ex. 14.26) in both interpretations, but only with certain safeguards, and we must understand quite clearly both meanings of the symbol, and see the difference between the two meanings.

In the foregoing chapter on random mass phenomena, we considered some kind of event $A$ such as the birth of a boy, or the fall of a raindrop in some specified location, or the casting of a specified number of spots with a die, and so on. We used the symbol $\Pr\{A\}$ to denote the probability of the event $A$, that is, the theoretical value of the long range relative frequency of the event $A$.

In the present chapter, however, we have to deal with plausible reasoning. We consider some conjecture $A$, and we are concerned with the reliability of this conjecture $A$, the strength of the evidence in favor of $A$, our confidence in $A$, the degree of credence we should give to $A$, in short the *credibility of the conjecture $A$*. We shall take the symbol $\Pr\{A\}$ to denote the credibility of $A$.

Thus, in the present chapter we shall use the symbol $\Pr\{A\}$ in its second meaning as "credibility" unless we explicitly state the contrary. Such use of the symbol is not objectionable, but we have to discuss carefully the concept of credibility if we do not wish to expose ourselves to grave objections.

First, there is an ambiguity to avoid. The symbol $\Pr\{A\}$ should represent the credibility of $A$, or the strength of the evidence for the conjecture $A$. Such evidence is strong if it is convincing. It is convincing if it convinces somebody. Yet we did not say whom it should convince: you, or me, or Mr. Smith, or Mrs. Jones, or whom? The strength of the evidence could also be conceived *impersonally*. If we conceive it so, the degree of belief that you or me or any other person may happen to have in a proposed conjecture is irrelevant, but what matters is the *degree of reasonable belief* that anyone of us *should* have. We did not say yet, and we have still to decide, in what exact sense we should use the term "credibility of $A$" and the corresponding symbol $\Pr\{A\}$.

There is another difficulty. The magnitudes considered by the physicists such as "mass," "electric charge," or "reaction velocity" have an *operational* definition; the physicist knows exactly which operations he has to perform if he wishes to ascertain the magnitude of an electric charge, for example. The definition of "long range relative frequency," although in some way less distinct than that of an electric charge, is still operational; it suggests definite operations that we can undertake to obtain an approximate numerical value of such a frequency. The trouble with the concept of the "credibility of a conjecture" is that we do not know any operational definition for it. What is the credibility of the conjecture that Mr. Jones is unfaithful? This credibility may have at this moment a definite value in the mind of Mrs. Jones (a negligibly small value, we hope) but we do not know how to determine that value numerically. What is the credibility of the law of universal gravitation judged on the basis of the observations reported in the first edition of Newton's *Principia*? This question could be of high interest to some of us. (Not to Mrs. Jones, perhaps, but to Laplace or Keynes if they were still alive—see the quotations prefixed to this chapter.) But nobody dared to propose a definite numerical value for such a credibility.

We have still to give a suitable interpretation of the term "credibility of the conjecture $A$," and to the corresponding symbol $\Pr\{A\}$. This interpretation must be such that the difficulty of an operational definition does not interfere with it. Moreover, and this is the main thing, this interpretation should enable us to view the rules of plausible reasoning systematically and realistically.

**5. An aspect of the calculus of probability. An attempt.** You have just been introduced to Mr. Anybody and you have to say a few words to him. You two are really complete strangers to each other and so your conversation may be cautious. Still you cannot help touching upon various assertions such as "It will rain tomorrow," "The next Big Game will be won by The Blues," "Corporation So-and-so will pay a higher dividend next year," "Mrs. Somebody whose divorce is the talk of the town was unfaithful," "Polio is caused by a virus," or any other assertions, $A$, $B$, $C$, $D$, $E$, .... Mr. Anybody attaches to the assertion $A$ a definite degree of credence $\Pr\{A\}$. If you are very clever, you can feel after some time spent with Mr. Anybody whether $\Pr\{A\}$ is low or high. Yet however clever you are, I cannot believe that you are able to ascribe a definite numerical value to $\Pr\{A\}$, the credibility of the statement $A$ in the eyes of Mr. Anybody. (Although it would be interesting: the values of $\Pr\{A\}$, $\Pr\{B\}$, $\Pr\{C\}$, ... could sharply characterize Mr. Anybody's personality.)

Let us be realistic and acknowledge the impossibility of a task that is obviously beyond our means: let us regard $\Pr\{A\}$, the credibility of the conjecture $A$ in the eyes of Mr. Anybody, as a definite positive fraction

$$0 < \Pr\{A\} < 1$$

the numerical value of which, however, we *do not know*. And let us treat similarly $\Pr\{B\}$, $\Pr\{C\}$, ... if $B$, $C$, ... are conjectures, that is, clearly formulated (perhaps mathematical) statements of which, however, Mr. Anybody does not know at this time whether they are true or false. If, however, $A$ is true and Mr. Anybody knows it, we set $\Pr\{A\} = 1$. If $A$ is false and Mr. Anybody knows it, we set $\Pr\{A\} = 0$.

It seems to me that our ignorance of the numerical values of $\Pr\{A\}$, $\Pr\{B\}$, ... cannot really hurt us. In fact, we are not concerned here with the personal opinions of Mr. Anybody. We are concerned with impersonal and universal rules of plausible inference. We wish to know, in the first place, whether there are such rules at all, and then we wish to know whether or not the calculus of probability does disclose such rules (as Laplace and others maintained). For the moment we rather hope that there are such rules, and that Mr. Anybody, as a sensible person, reasons according to such rules whatever degrees of credence $\Pr\{A\}$, $\Pr\{B\}$, $\Pr\{C\}$, ... he may attach at this moment to the statements $A$, $B$, $C$, ... discussed. And so I cannot see why our ignorance of the numerical values of $\Pr\{A\}$, $\Pr\{B\}$, $\Pr\{C\}$, ... should hurt us.

Let us attempt, therefore, to apply the rules of the calculus of probability to the credibilities $\Pr\{A\}$, $\Pr\{B\}$, $\Pr\{C\}$, ... as interpreted: positive fractions measuring degrees of confidence of that mythical or idealized person, Mr. Anybody. We wish to see whether, in doing so, we can elicit anything that can be reasonably interpreted as an impersonal and universal rule of plausible reasoning.   Our attempt may fail, of course, but I cannot see at this moment why it should fail, and so I am cautiously hopeful.[3]

**6. Examining a consequence.**   Mr. Anybody is investigating a certain conjecture $A$.   This conjecture $A$ is clearly formulated, but Mr. Anybody does not know whether $A$ is true or not and wants badly to find out which is the case: is $A$ true or is it false?   He notices a certain consequence $B$ of $A$.   He is satisfied that

$$A \text{ implies } B.$$

Yet he does not know whether $B$ is true or false, and sometimes, when he is tired of investigating $A$, he thinks of switching to the investigation of $B$.

We have considered this situation many times, and now we wish to reconsider it in the light of the calculus of probability.   We wish to pay due attention to three credibilities: $\Pr\{A\}$, $\Pr\{B\}$, and $\Pr\{A/B\}$.   Mr. Anybody knows very well that $A$ and $B$ are unproved and unrefuted, but he believes in them to a certain degree and this degree is expressed by $\Pr\{A\}$ and $\Pr\{B\}$, respectively.   Also $\Pr\{A/B\}$, the degree of credence that he could place in $A$ if he knew that $B$ is true, plays an important role in his deliberations.

We have no means of assigning a numerical value to any of these credibilities, although we can sometimes imagine in which direction a change in the state of knowledge of Mr. Anybody would change the value of one or the other.   At any rate, the calculus of probability yields a relation between them.

In fact, by one of the fundamental theorems on probability (see ex. 14.26 (2))

$$\Pr\{A\} \Pr\{B/A\} = \Pr\{B\} \Pr\{A/B\}.$$

Yet now, since $A$ implies $B$, $B$ must be true if $A$ is true, and so

$$\Pr\{B/A\} = 1.$$

Hence we obtain

(I)                     $$\Pr\{A\} = \Pr\{B\} \Pr\{A/B\}.$$

Let us visualize the contents of this equation.

(1) Mr. Anybody decided to investigate the consequence $B$ of his conjecture $A$.   He did not succeed yet in bringing this investigation to a conclusion.   Yet sometimes he saw indications that $B$ may be true, and sometimes

---

[3] That the Calculus of Probability should deal primarily with degrees of belief (confidence, confirmation, certitude, . . .) and not with more or less idealized relative frequencies, is the opinion of many authors among which I quote only two: J. M. Keynes, *A treatise on probability* (cf. especially p. 34, 66, 160), and B. de Finetti, La prévision, ses lois logiques, ses sources subjectives, *Annales de l'Institut Henri Poincaré*, vol. 7 (1937) p. 1–68.   There is no space to explain my differences from these authors, or the differences between them, but I wish to express my thanks to both.   The standpoint here adopted is similar to, but not quite the same as, that of my previous paper: Heuristic reasoning and the theory of probability, *American Math. Monthly*, vol. 48 (1941) p. 450–465.

indications to the contrary.   His confidence in $B$, which we call $\Pr\{B\}$, rose and fell accordingly.   Yet he did not observe anything that would have changed his views about the relation between $A$ and $B$ or about $\Pr\{A/B\}$.   How did all this influence $\Pr\{A\}$, his confidence in $A$?

Equation (I) shows that $\Pr\{A\}$ *changes in the same direction as* $\Pr\{B\}$, provided that $\Pr\{A/B\}$ remains unchanged.   This agrees with our former remarks, especially with those in sect. 13.6.   (Observe, that we consider only the *direction* of the change that we can sometimes ascertain, and not its *amount* that we can never know precisely.)

(2) Mr. Anybody succeeded in proving $B$, which is a consequence of the conjecture $A$ that he originally investigated.   Before proving $B$, he had some confidence in $B$, the degree of which we have represented by $\Pr\{B\}$; he had also some confidence in $A$, of degree $\Pr\{A\}$.   He sometimes considered $\Pr\{A/B\}$, the confidence he could place in $A$ after a proof of $B$.   After the proof of $B$, his confidence in $B$ attains the maximum value 1, and his confidence in $A$ becomes, of course, $\Pr\{A/B\}$.   (Substituting 1 for $\Pr\{B\}$ in equation (I), we are led formally to the new value of the credibility of $A$.) We suppose here that his views about the relation between $A$ and $B$ and his evaluation of $\Pr\{A/B\}$ remain unchanged.

Observing that $0 < \Pr\{B\} < 1$, we derive from equation (I) the inequality

(II)                         $$\Pr\{A\} < \Pr\{A/B\}.$$

Now, $\Pr\{A\}$ and $\Pr\{A/B\}$ represent the credibility of $A$ before and after the proof of $B$, respectively.   Therefore inequality (II) is the formal expression of a principle with which we met so often: *the verification of a consequence renders a conjecture more credible*;   cf. sect. 12.1, for instance.

(3) Yet we can learn still more from equation (I) which we write in the form

(III)                         $$\Pr\{A/B\} = \frac{\Pr\{A\}}{\Pr\{B\}}.$$

The left-hand side, the credibility of $A$ after the verification of $B$, is expressed in terms of the confidence that the researcher had in $A$ and $B$, respectively, before such verification.   Let us compare various cases of successful verification of a consequence.   These cases have one circumstance in common: the same confidence $\Pr\{A\}$ was placed in the conjecture $A$ (at which the investigation aims) before the verification of its consequence $B$. Yet these cases differ in another respect: the consequence $B$ (which was eventually verified) was expected with more confidence in some cases and with less confidence in others.   That is, we regard $\Pr\{A\}$ as constant and $\Pr\{B\}$ as variable.   How does the variation of $\Pr\{B\}$ influence the weight of the evidence resulting from the verification of the consequence $B$?

Let us pay due attention to the extreme cases. Since $B$ is a consequence of $A$, $B$ is certainly true when $A$ is true, and so $\Pr\{B\}$, the credibility of $B$, cannot be less than $\Pr\{A\}$, the credibility of $A$. On the other hand no credibility can exceed certainty: $\Pr\{B\}$ cannot be greater than 1. We have determined the bounds between which $\Pr\{B\}$ is contained:

$$\Pr\{A\} \leqq \Pr\{B\} < 1.$$

The lower bound is attained, when not only $A$ implies $B$, but also $B$ implies $A$, so that the two assertions $A$ and $B$ are equivalent, stand and fall together, in which case they are, of course, equally credible. The upper bound 1 cannot be really attained: if it were attained, $B$ would be certain before investigation, and we have not included this case in our consideration. Yet the upper bound can be approached: $B$ can be almost certain before examination. How does the evidence resulting from the verification of $B$ change when $\Pr\{B\}$ varies between its extreme bounds?

The evidence is stronger, when $\Pr\{A/B\}$, the new confidence in $A$ resulting from the verification of the consequence $B$, is greater. It is visible from the relation (III) that

as $\Pr\{B\}$ decreases from 1 to $\Pr\{A\}$

$\Pr\{A/B\}$ increases from $\Pr\{A\}$ to 1.

This statement expresses in a new language a point that we have recognized before (sect. 12.3): *the increase of our confidence in a conjecture due to the verification of one of its consequences varies inversely as the credibility of the consequence before such verification.* The more unexpected a consequence is, the more weight its verification carries. The verification of the most surprising consequence is the most convincing, whereas the verification of a consequence that we did not doubt much anyway ($\Pr\{B\}$ almost 1) has little value as evidence.

(4) The situation just discussed can be viewed from another standpoint. Using some simple rules of the calculus of probability (cf. ex. 14.26 formulas (4), (2), (3), in this order) we obtain

$$\Pr\{B\} = \Pr\{AB\} + \Pr\{\bar{A}B\}$$
$$= \Pr\{A\}\,\Pr\{B/A\} + \Pr\{\bar{A}\}\,\Pr\{B/\bar{A}\}$$
$$= \Pr\{A\} + [1 - \Pr\{A\}]\,\Pr\{B/\bar{A}\}.$$

In passing to the last line, we also used that $\Pr\{B/A\} = 1$, which expresses that $B$ is a consequence of $A$. In substituting for $\Pr\{B\}$ the value just derived, we obtain from (III)

$$(\text{IV}) \qquad \Pr\{A/B\} = \frac{\Pr\{A\}}{\Pr\{A\} + [1 - \Pr\{A\}]\,\Pr\{B/\bar{A}\}}.$$

Let us assume as before that $\Pr\{A\}$ is constant; that is, let us survey various cases in which the confidence in $A$, the conjecture examined, was the same before testing the consequence $B$ of $A$.  Yet $\Pr\{A/B\}$, the credibility of $A$ after the verification of $B$, still depends on $\Pr\{B/\bar{A}\}$, the credibility of $B$ (before verification, of course) viewed under the assumption that $A$ is not true.  And $\Pr\{B/\bar{A}\}$ can vary; it can take, in fact, any value between 0 and 1. Now, by our formula (III),

as $\Pr\{B/\bar{A}\}$ decreases from 1 to 0

$\Pr\{A/B\}$ increases from $\Pr\{A\}$ to 1.

This statement expresses in a new language a point that we have discussed before (sect. 13.10).  Let us look at the extreme cases.  If $B$ without $A$ is hardly credible ($\Pr\{B/\bar{A}\}$ almost 0) the verification of the consequence $B$ brings the conjecture $A$ close to certainty.  On the other hand, the verification of a consequence $B$ that we would scarcely doubt even if $A$ were false ($\Pr\{B/\bar{A}\}$ almost 1) adds little to our confidence in $A$.

**7. Examining a possible ground.**  After the broad and cautious discussion in the foregoing section we can proceed a little faster in surveying similar situations.

Here is such a situation: the aim of our research is a certain conjecture $A$. We notice a possible ground for $A$, that is, a proposition $B$ from which $A$ would follow:

A is implied by B.

We start investigating $B$.  If we succeeded in proving $B$, $A$ would also be proved.  Yet $B$ turns out to be false.  How does the disproof of $B$ affect our confidence in $A$?

Let the calculus of probability answer this question.  Since $A$ is implied by $B$

$$\Pr\{A/B\} = 1.$$

Let us combine this with some basic formulas (see ex. 14.26 (4), (2), (3)):

$$\Pr\{A\} = \Pr\{AB\} + \Pr\{A\bar{B}\}$$
$$= \Pr\{B\}\Pr\{A/B\} + \Pr\{\bar{B}\}\Pr\{A/\bar{B}\}$$
$$= \Pr\{B\} + (1 - \Pr\{B\})\Pr\{A/\bar{B}\}.$$

We obtain hence that

(I) $$\Pr\{A/\bar{B}\} = \frac{\Pr\{A\} - \Pr\{B\}}{1 - \Pr\{B\}}.$$

The left-hand side represents the credibility of $A$ after $B$ (which is a possible ground for $A$) has been refuted.  The right-hand side refers to the situation before the refutation of $B$. By the way, this right-hand side can be transformed so that equation (I) appears in the form

$$\Pr\{A/\bar{B}\} = \Pr\{A\} - \Pr\{B\}\frac{1 - \Pr\{A\}}{1 - \Pr\{B\}}$$

and hence we see that

(II)                    $\Pr\{A/\bar{B}\} < \Pr\{A\}.$

Both sides of this inequality represent the credibility of the conjecture $A$, the left-hand side after the refutation of $B$, the right-hand side before this refutation. Therefore, inequality (II) expresses a rule: *our confidence in a conjecture can only diminish when a possible ground for the conjecture has been exploded.* (Cf. sect. 13.2.)

Yet we can learn more from equation (I). Let us consider $\Pr\{A\}$ as constant and $\Pr\{B\}$ as variable. That is, let us survey various cases which differ in one important respect: in the degree of our confidence in $B$. Our confidence in $B$ may be very small, but it cannot be arbitrarily large: it can never exceed our confidence in $A$, since if $B$ is true $A$ is also true. (Yet $A$ could be still true even if $B$ is false.) And so, we determined the extreme values between which $\Pr\{B\}$ can vary:

$$0 < \Pr\{B\} \leqq \Pr\{A\}.$$

We see from equation (I) that

as $\Pr\{B\}$ increases from 0 to $\Pr\{A\}$

$\Pr\{A/\bar{B}\}$ diminishes from $\Pr\{A\}$ to 0.

That is, *the more confidence we placed in a possible ground of our conjecture, the greater will be the loss of faith in our conjecture when that possible ground is refuted.*

**8. Examining a conflicting conjecture.** We consider now another situation: we examine two conflicting conjectures, $A$ and $B$. When we say that $A$ conflicts with $B$ or

$A$ is incompatible with $B$

we mean that the truth of one of them implies the falsity of the other. We are, in fact, primarily concerned with $A$ and we have started investigating $B$ because we thought that the investigation of $B$ could shed some light on $A$. In fact, a proof of $B$ would disprove $A$. Yet we succeeded in disproving $B$. How does this result affect our confidence in $A$?

Let the calculus of probability give the answer. Let us begin by expressing in the language of this calculus that $A$ and $B$ are incompatible. This means in other words that $A$ and $B$ cannot both be true, and so

$$\Pr\{AB\} = 0.$$

Now we conclude from our basic formulas (cf. ex. 14.26 (4), (2), (3))

$$\Pr\{A\} = \Pr\{AB\} + \Pr\{A\bar{B}\}$$
$$= \Pr\{A\bar{B}\}$$
$$= \Pr\{\bar{B}\}\,\Pr\{A/\bar{B}\}$$
$$= (1 - \Pr\{B\})\,\Pr\{A/\bar{B}\}$$

which yields finally

(I) $$\Pr\{A/\bar{B}\} = \frac{\Pr\{A\}}{1 - \Pr\{B\}}.$$

Equation (I) obviously implies the inequality:

(II) $$\Pr\{A/\bar{B}\} > \Pr\{A\}.$$

The left-hand side refers to the situation after the refutation of $B$, the right-hand side to the situation before this refutation. Therefore, we can read (II) as follows: *our confidence in a conjecture can only increase when an incompatible rival conjecture has been exploded.* (Cf. sect. 13.3.)

Yet we can learn more from equation (I). Let us consider $\Pr\{A\}$ as constant and $\Pr\{B\}$ as variable. Let us determine the bounds between which $\Pr\{B\}$ may vary. Of course, $\Pr\{B\}$ can be arbitrarily small. Yet $\Pr\{B\}$ cannot be arbitrarily large; in fact it can never exceed $\Pr\{\bar{A}\}$. If $B$ is correct, $\bar{A}$ is *a fortiori* correct. Since $\Pr\{\bar{A}\}$ is equal to $1 - \Pr\{A\}$

$$0 < \Pr\{B\} \leqq 1 - \Pr\{A\}.$$

We see from equation (I) that

as $\Pr\{B\}$ increases from 0 to $1 - \Pr\{A\}$

$\Pr\{A/\bar{B}\}$ increases from $\Pr\{A\}$ to 1.

That is, *the more confidence we placed in an incompatible rival of our conjecture, the greater will be the gain of faith in our conjecture when that rival is refuted.*

**9. Examining several consequences in succession.** We consider now the following important situation: the aim of our research is a certain conjecture $A$. For the moment, we do not see how we could decide whether $A$ is true or not. Yet we see several consequences $B_1, B_2, B_3, \ldots$ of $A$: $A$ implies $B_1$, $A$ implies $B_2$, $A$ implies $B_3$, .... The consequences $B_1, B_2, B_3, \ldots$ are more accessible than $A$ itself and we settle down to examine them one after the other. (This is the typical procedure of the natural sciences: we have no means of examining a general law $A$ in itself, and therefore we examine it by testing several consequences $B_1, B_2, B_3, \ldots$) We have already examined the consequences $B_1, B_2, \ldots B_n$, and we have succeeded in verifying all of them: $B_1, B_2, \ldots B_n$ turned out to be correct. Now we are testing the next consequence $B_{n+1}$: how will the outcome affect our confidence in $A$?

In order to see the answer in the light of the calculus of probability, we start from a general rule of this calculus (see ex. 14.26 (5)):

$$\Pr\{A/H\}\,\Pr\{B/HA\} = \Pr\{B/H\}\,\Pr\{A/HB\}.$$

We set $B = B_{n+1}$. Now, since $B_{n+1}$ is a consequence of $A$,

$$\Pr\{B/HA\} = \Pr\{B_{n+1}/HA\} = 1$$

and so we find that

$$\Pr\{A/H\} = \Pr\{B_{n+1}/H\}\,\Pr\{A/HB_{n+1}\}.$$

We set $H = B_1 B_2 \ldots B_n$ and obtain the decisive formula:

(I) $\qquad \Pr\{A/B_1 \ldots B_n\} = \Pr\{B_{n+1}/B_1 \ldots B_n\}\,\Pr\{A/B_1 \ldots B_n B_{n+1}\}.$

In order to understand (I) correctly, we have to realize that

$$\Pr\{A/B_1 \ldots B_n\} \text{ and } \Pr\{B_{n+1}/B_1 \ldots B_n\}$$

denote the credibilities of $A$ and $B_{n+1}$, respectively, *after* $B_1$, $B_2$, $\ldots B_n$ have been verified but, of course, *before* $B_{n+1}$ has been verified;

$\Pr\{A/B_1 \ldots B_n B_{n+1}\}$ denotes the credibility of $A$ after the verification of its $n + 1$ consequences, $B_1$, $B_2$, $\ldots B_n$ and $B_{n+1}$.

We have to keep these meanings in mind, and then we can read (I) as a precise and pregnant proposition on inductive reasoning.

Let us first focus our attention on $\Pr\{B_{n+1}/B_1 \ldots B_n\}$; the value of this credibility will be in most cases less than 1, and equal to 1 only if the correctness of $B_1$, $B_2$, $\ldots B_n$ renders the correctness of $B_{n+1}$ certain, that is, if $B_1, B_2, \ldots$ and $B_n$ jointly imply $B_{n+1}$. If this is *not* the case, we can derive from (I) the inequality:

(II) $\qquad \Pr\{A/B_1 \ldots B_n\} < \Pr\{A/B_1 \ldots B_n B_{n+1}\}.$

That is, *the verification of a new consequence enhances our confidence in the conjecture, unless the new consequence is implied by formerly verified consequences.*

Let us write equation (I) in the form:

(III) $\qquad \Pr\{A/B_1 \ldots B_n B_{n+1}\} = \dfrac{\Pr\{A/B_1 \ldots B_n\}}{\Pr\{B_{n+1}/B_1 \ldots B_n\}}.$

The left-hand side refers to the situation after the confirmation of $B_{n+1}$; the right-hand side refers to the situation before this confirmation. Let us regard the relation of $A$ to $B_1$, $B_2$, $\ldots B_n$ as fixed, but the relation of $B_{n+1}$ to $B_1$, $B_2$, $\ldots B_n$ as variable. Then we can read (III) as follows: *the increase in our confidence brought about by the confirmation of a new consequence* (or the weight of the evidence furnished by this confirmation) *varies inversely as the credibility of the new consequence, appraised* (before its confirmation, of course) *in the light of the previously verified consequences.*

We can express this same rule in other words. When we start testing the consequence $B_{n+1}$ of our conjecture $A$, we face the possibility that $B_{n+1}$ will turn out false in which case $A$ will be exploded. In view of the formerly verified consequences $B_1, B_2, \ldots B_n$ the chance for exploding $A$ by disproving $B_{n+1}$ appears strong when $\Pr\{B_{n+1}/B_1 \ldots B_n\}$ is small. Therefore we can read (III) as follows: *the consequence that, judged in the light of the preceding verifications, stands the better chance of refuting the proposed conjecture, will disclose the stronger inductive evidence if it is confirmed in spite of the forebodings.* Still shorter: "more danger, more honor." If a conjecture escapes the danger of refutation it shall be esteemed in proportion to the risk involved.

From the very beginning of our discussion we have considered the inductive evidence supplied by the successive verification of several consequences of a proposed conjecture. The extreme cases were the most conspicuous. Let us survey them once more (adding just a little color) and let us focus the moment when, having verified the consequences $B_1, B_2, \ldots B_n$ of a conjecture $A$, we start scrutinizing a new consequence $B_{n+1}$.

The new consequence under scrutiny, $B_{n+1}$, may appear "little different" from the formerly verified consequences $B_1, B_2, \ldots B_n$. Such a case is not too exciting. We confidently expect (by analogy, presumably) that $B_{n+1}$ will be verified like the other consequences (that is, $\Pr\{B_{n+1}/B_1 \ldots B_n\}$ is close to 1, its maximum). We scarcely expect that the investigation of $B_{n+1}$ will disclose some very new aspect or that it will upset the conjecture $A$, but also, when $B_{n+1}$ is finally verified, the gain in evidence for $A$ is not much.

On the other hand, the new consequence under scrutiny, $B_{n+1}$, may appear as "very different" from the formerly verified consequences $B_1, B_2, \ldots B_n$. Such a case may be exciting. Analogy with $B_1, B_2, \ldots B_n$ gives us little reason to expect that $B_{n+1}$ will be verified ($\Pr\{B_{n+1}/B_1 \ldots B_n\}$ is close to its minimum). We realize that the investigation of $B_{n+1}$ risks upsetting the conjecture $A$, but it has also a chance to disclose some new aspect, and when $B_{n+1}$ is eventually verified, the gain in evidence for $A$ may be considerable.

The reader should review some of our former examples and discussions. (Cf. sect. 3.1–3.7, ch. VI, sect. 10.1, sect. 12.2, sect. 13.11, and several other passages.) After due comparison, equation (III) of the present section may appear to him as the most concise and precise expression of the principle involved. At any rate, if he can see the bearing of equation (III) on some of our examples, he has taken a good step towards clarifying his ideas about an important subject.

**10. On circumstantial evidence.** We now consider a situation that we encountered in dealing with reasoning in judicial matters: we are examining a conjecture $A$. (This conjecture $A$ may be an accusation advanced by the prosecution.) We (the jury) have to find out whether $A$ is true or not. A circumstance $B$ is submitted (by the prosecution) which is so connected with the conjecture $A$ that

$B$ with $A$ is more credible than without $A$.

In the course of the proceedings this circumstance $B$ is so strongly confirmed that we may regard it as a proven fact.    (Perhaps $B$ was not even challenged by the defense.)    How does all this affect our belief in $A$?

Let the calculus of probability answer this question.    The essential assumption concerning the connection between $A$ and $B$ is expressed by the inequality

(I) $$\Pr\{B/A\} > \Pr\{B/\bar{A}\}.$$

By basic formulas of the theory of probability (cf. ex. 14.26)

$$\Pr\{A\}\,\Pr\{B/A\} = \Pr\{B\}\,\Pr\{A/B\},$$

$$\Pr\{B\} = \Pr\{A\}\,\Pr\{B/A\} + (1 - \Pr\{A\})\,\Pr\{B/\bar{A}\}.$$

Combining these, we obtain

(II) $$\frac{\Pr\{A\}}{\Pr\{A/B\}} = \Pr\{A\} + (1 - \Pr\{A\})\,\frac{\Pr\{B/\bar{A}\}}{\Pr\{B/A\}}.$$

Using (I), we conclude from (II) that

(III) $$\Pr\{A\} < \Pr\{A/B\}.$$

Both sides of this inequality represent the credibility of the conjecture $A$, the left-hand side before the verification of the circumstance $B$, the right-hand side after the verification of $B$.    Therefore, inequality (III) expresses a rule: *if a certain circumstance is more credible with a certain conjecture than without it, the proof of that circumstance can only enhance the credibility of that conjecture.* (Cf. sect. 13.13.)

Yet we can learn more from equation (II).    Let us regard $\Pr\{A\}$ and $\Pr\{B/A\}$ as constant, but $\Pr\{B/\bar{A}\}$ as variable.    Then $\Pr\{A/B\}$ depends on $\Pr\{B/\bar{A}\}$:

as $\Pr\{B/\bar{A}\}$ decreases from $\Pr\{B/A\}$ to $0$

$\Pr\{A/B\}$ increases from $\Pr\{A\}$ to $1$.

That is, *the less credible a circumstance appears without a certain conjecture, the more will the proof of that circumstance enhance the credibility of that conjecture.*    In sect. 13.13, led by the consideration of examples, we came very close to this rule.

Strong judicial evidence results from several proved coincidences all pointing to the same conclusion; see sect. 13.13 (4).    If there are several circumstances $B_1, B_2, B_3, \ldots$ each of which is more credible with $A$ than without $A$, and they are successively proved, the evidence for $A$ increases at each step.    The amount of additional evidence resulting from a circumstance newly proved depends on various points.    A new circumstance that is very different from the circumstances previously examined (a new witness who is visibly independent of the witnesses previously examined) carries particular weight.    To express these points, we could develop formulas so related to those introduced in the present section as the formulas developed in sect. 9 are related to those introduced in sect. 6.

## EXAMPLES AND COMMENTS ON CHAPTER XV

**1.** Examine the situation discussed in sect. 13.8 by the Calculus of Probability.

**2.** Examine the pattern encountered in the solution of ex. 13.8 by the Calculus of Probability. Can you justify it?

**3.** Reexamine ex. 13.10 by the Calculus of Probability.

**4.** *Probability and credibility.* The peculiar "non-quantitative" application of the Calculus of Probability, discussed in sect. 5–10, was designed to elucidate certain patterns of plausible reasoning. These patterns were mainly suggested by heuristic reasoning about mathematical conjectures. Can we apply the Calculus of Probability to examples of other kind in the same manner?

(1) Let $A_n$ denote the conjecture that the fair die that I am about to roll will show $n$ spots ($n = 1, 2, \ldots 6$). These conjectures $A_1, A_2, \ldots A_6$ are of a kind that we did not particularly wish to examine in sect. 5–10. Yet let us try to treat them in the same manner: we consider their credibilities $\Pr\{A_1\}$, $\Pr\{A_2\}$, $\ldots \Pr\{A_6\}$ and apply the Calculus of Probability to them. As $A_1, A_2, \ldots A_6$ are mutually exclusive and exhaust all the possibilities,

$$\Pr\{A_1\} + \Pr\{A_2\} + \ldots + \Pr\{A_6\} = 1.$$

Since the die with which we are concerned is fair, its faces are interchangeable, none of the six faces is preferable to any other, and so we are compelled to assume that

$$\Pr\{A_1\} = \Pr\{A_2\} = \ldots = \Pr\{A_6\}.$$

Combining this with the foregoing equation, we obtain that

$$\Pr\{A_1\} = \Pr\{A_2\} = \ldots = \Pr\{A_6\} = 1/6,$$

and so we are led to attribute definite numerical values to the credibilities considered. Mr. Anybody (our friend from sect. 5) would be driven to the same conclusion, I think.

The credibility of the conjecture $A_1$ turned out to have the same numerical value as the probability of the event that a fair die shows one spot. Yet this is not surprising at all: we admitted the same rules and assumed the same interchangeability (or symmetry) in computing credibilities and probabilities. (The reader should not forget, of course, that credibility and probability are quite differently defined.)

(2) The foregoing argument applies to many other cases.

By three straight lines passing through its center, the surface of a circle is divided into six equal sectors. A raindrop is about to fall on the circular surface.   Let us call $A_1$ the assertion that the raindrop will fall on the first sector, $A_2$ the assertion that it will fall on the second sector, and so on.   This situation is essentially the same as that discussed under (1) and the result is, of course, also the same: each of the six assertions $A_1, A_2, \ldots A_6$ has the same credibility 1/6.

The generalization is obvious: we need not stick to the number 6 or to raindrops.   We can divide the circle into $n$ sectors and consider some other kind of random event.   We can also pass from the area of the circle to its periphery and so we arrive at the following situation: a point will be chosen on the periphery of a circle by some random agency that has no preference whatsoever for any particular points of this periphery.   Somebody conjectures that the point will lie on a certain arc.   The credibility of this conjecture is the ratio of the length of the arc to the length of the whole periphery.

From the circle we can pass to the sphere.   Let us assume, for the sake of simplicity, that the Earth is a perfect sphere and that the meteorites striking the Earth have no preference for any particular direction.   Somebody conjectures that the next meteorite falling on the Earth will strike a certain region.   What is the credibility of this conjecture?   The mathematical treatment may be more complicated if we carry it through in minute detail, but the result is just as intuitive as for the circle: the desired credibility is the ratio of the area of that region to the area of the whole spherical surface.

We need not enter upon further analogous, or more general, situations. Let us look, however, at two particular cases.

The credibility of the conjecture that a point chosen at random on the periphery of a circle will lie within a distance of one degree from a given point of this periphery is 1/180.

The credibility of the conjecture that the next meteorite falling on the Earth will strike within a distance of one degree from the center of New York City is 0.00007615.   (This is the numerical ratio of a small spherical cap to the whole spherical surface.)

The circle and the sphere possess high symmetry: a suitable rotation that does not change the position of the figure as a whole can transport a point of the figure from any given position on the figure to any other given position. This symmetry is not enough in itself to justify the preceding results.   In deriving them we have to *assume* "physical" symmetry, the interchangeability of any two points of the geometric figure in relation to the physical agency under consideration.

The computed credibilities cannot claim much novelty: they coincide with corresponding probabilities, known for a long time, and successfully applied to a variety of mass phenomena.

The computed credibilities may appear as incapable of interfering with the credibilities with which we are primarily concerned.   Of this, however, we should not feel too sure.

(3) Kepler knew only six planets revolving around the sun and even devised a geometric argument why there *should* be exactly six planets; see sect.11.5. Yet the telescopes did not listen to his argument. In 1781, about 150 years after Kepler's death, the astronomer Herschel observed a slowly moving star and supposed it to be a comet; but it turned out to be a seventh planet (Uranus) revolving beyond the orbit of Saturn. In the years 1801–1806 four small planets (Ceres, Pallas, Juno, and Vesta) revolving between the orbits of Mars and Jupiter were similarly discovered. (Hundreds of such minor planets have been discovered later.)

On the basis of Newton's theory, the astronomers tried to compute the motions of these planets. They did not succeed very well with the planet Uranus; the differences between theory and observation seemed to exceed the admissible limits of error. Some astronomers suspected that these deviations may be due to the attraction of a planet revolving beyond Uranus' orbit, and the French astronomer Leverrier investigated this conjecture more thoroughly than his colleagues. Examining the various explanations proposed, he found that there is just one that could account for the observed irregularities in Uranus' motion: the existence of an ultra-Uranian planet. He tried to compute the orbit of such a hypothetical planet from the irregularities of Uranus. Finally Leverrier succeeded in assigning a definite position in the sky to the hypothetical planet. He wrote about it to another astronomer whose observatory was the best equipped to examine that portion of the sky. The letter arrived on the 23rd of September 1846 and in the evening of the same day a new planet was found within one degree of the spot indicated by Leverrier. It was a large ultra-Uranian planet that had approximately the mass and orbit predicted by Leverrier.

(4) The theory that renders possible such an extraordinary prediction must be a wonderful theory. This may be our first impression. Let us try to clarify this impression by the Calculus of Probability.

$T$ stands for the theory that underlies the astronomical computations: it is Newton's theory, consisting of his laws of mechanics and his law of gravitational attraction.

$N$ stands for Leverrier's assertion: at the date given, there will be a new planet, having such and such a mass and such and such an orbit (approximately), in the neighborhood of such and such a point of the sky. More precisely, let $N$ stand for that part of Leverrier's assertion that has been *verified* by subsequent observations.

$\Pr\{T\}$ denotes the degree of confidence placed in the theory $T$ on the basis of all facts known by the astronomers before the date of the discovery of the new planet.

$\Pr\{T/N\}$ denotes the degree of confidence due to the same theory $T$ when the verification of Leverrier's prediction $N$ has been added to the facts known before.

We think that the verification of Leverrier's prediction adds to the credit of the theory $T$, and so we suspect that $\Pr\{T/N\}$ is greater than $\Pr\{T\}$. In fact, we find (by ex. 14.26 (2)) that

$$\frac{\Pr\{T/N\}}{\Pr\{T\}} = \frac{\Pr\{N/T\}}{\Pr\{N\}}.$$

$\Pr\{N\}$ is the credibility of Leverrier's assertion $N$ "in itself," that is, without reference to the truth or falsity of the theory $T$; the underlying state of knowledge is the same as assumed in the evaluation of $\Pr\{T\}$. This $\Pr\{N\}$ must be extremely small: if we ignore Newton's theory $T$ what ground could we have to suspect that there is an ultra-Uranian planet of such and such precise properties so near to a given point of the sky?

$\Pr\{N/T\}$ is the credibility of Leverrier's assertion $N$ in the light of Leverrier's computation based on the theory $T$, and so it is a concept very different from $\Pr\{N\}$. Perhaps Leverrier did not show quite conclusively that the existence of an ultra-Uranian planet with such and such properties is the only explanation compatible with the theory $T$. Yet he came pretty near to showing just this, and so $\Pr\{N/T\}$ cannot be too far from certainty.

To sum up, we may regard $\Pr\{N/T\}$ as a sizable fraction, not very far from 1, and $\Pr\{N\}$ as a very small fraction, approaching 0. Yet if we think so, the ratio $\Pr\{N/T\}/\Pr\{N\}$ appears as very large, and so must appear the ratio $\Pr\{T/N\}/\Pr\{T\}$ which has the same value as the foregoing, by the above equation. Therefore, $\Pr\{T/N\}$, our confidence in the theory $T$ after the verification of Leverrier's prediction appears as very much greater than $\Pr\{T\}$, our confidence in the same theory before this verification.

(5) The foregoing consideration remains within the limits to which we cautiously restricted ourselves in sect. 5. Yet let us discard caution for a moment and indulge in some adventurous rough estimates.

Leverrier's prediction $N$ covers many details of which we pick out just one: the new planet will be near such and such a spot of the sky. It was found, in fact, within one degree from the indicated point (at a distance 52′). Yet the probability that a point chosen at random on the sphere should be within one degree from a preassigned spot can be computed under simple assumptions, as we stated above, under (2). We find that $\Pr\{N\}$ is much smaller than 0.00007615. We can scarcely regard $\Pr\{N/T\}$ as exactly equal to 1, yet we may surmise that, in setting

$$\Pr\{N\} = 0.00007615, \qquad \Pr\{N/T\} = 1,$$

we overestimate much more the first, than the second, credibility. And so we arrive at the inequality

$$\frac{\Pr\{T/N\}}{\Pr\{T\}} > \frac{1}{0.00007615} = 13131.$$

Of course, such an estimate is questionable. There may be reasons of analogy, quite independent of Newton's theory $T$, suggesting that a new planet has more chance to be near the plane of the Earth's orbit than far from it.    If we think so, we should substitute for 0.00007615 a greater fraction but one that is still less than

$$1/180 = 0.005556;$$

cf. under (2).

Such estimates could be argued endlessly.    For example, since $\Pr\{T/N\}$ is certainly less than 1, the inequality proposed implies that

$$\Pr\{T\} < 0.00007615.$$

We may be tempted to regard this as a refutation of the proposed inequality, as a "reductio ad absurdum."    In fact, Newton's theory $T$ could be regarded as firmly established in 1846, even before the discovery of Neptune, and so it could appear as absurd to attribute to $T$ such a low credibility.    I do not think, however, that we are obliged to regard a credibility $10^{-5}$ as low in such a case: we could think that logical certainty, to which we ascribe the credibility 1, is incomparably more than the reliance that we can place in the best established inductive generalization, the credibility of which we could even consider as infinitesimal; cf. ex. 8.

After all this, we may find it safer to return to the standpoint of sect. 5–10, to which we essentially adhered under (4).    Represent to yourself qualitatively how a change in this or that component of the situation would influence your confidence, but do not commit yourself to any quantitative estimate.

5. *Likelihood and credibility.*    We may have a conjecture about probabilities.    For example, you may conjecture that the die that you hold in your hands is ideally fair, that is, each face has the same probability 1/6. Of course, this conjecture is hard to believe.    Or you may conjecture that each face of that die has a probability between 0.16 and 0.17, which would be more credible.    A conjecture about probabilities is a statistical hypothesis. It often happens that we have only two obvious rival conjectures: a "physical" conjecture $P$ and a statistical hypothesis $H$; cf. sect. 14.9 (7) and ex. 14.33. In such a case we may be seriously concerned with $\Pr\{H\}$, the credibility of the statistical hypothesis $H$.

A statistical hypothesis $H$ is appropriately tested by statistical observation. Let $E$ denote the prediction that the statistical observation will yield such and such a result.    Let us consider the credibility $\Pr\{E/H\}$ and let us assume that this *credibility has a numerical value, equal to the probability* that an event of the kind predicted by $E$ will happen, computed on the basis of the statistical hypothesis $H$.    As we have seen in ex. 4, certain (quite natural) assumptions of interchangeability, or symmetry (included in the statistical hypothesis $H$) may even oblige us to equate credibility with probability.

This credibility, or probability, $\Pr\{E/H\}$ can be viewed from two different standpoints as we have discussed in sect. 14.7 (5); cf. also sect. 14.8 (5).    On the one hand $\Pr\{E/H\}$ is the probability of an event of the kind predicted by

$E$, computed on the basis of the statistical hypothesis $H$. On the other hand, if such an event actually happens and is observed, we are inclined to think that $H$ is the less likely the less the numerical value of $\Pr\{E/H\}$ is, and for this reason we call $\Pr\{E/H\}$ the *likelihood* of the statistical hypothesis $H$, judged in view of the fact that the event predicted by $E$ has actually happened. Cf. sect. 14.7 (5), also sect. 14.8 (5).

Now it follows from ex. 14.26 (2) that

$$\Pr\{H/E\} = \frac{\Pr\{E/H\}\,\Pr\{H\}}{\Pr\{E\}}.$$

In this equation, $\Pr\{E/H\}$ is not only a credibility, but also a probability, and has a definite numerical value. Yet $\Pr\{H/E\}$, $\Pr\{H\}$, $\Pr\{E\}$ are only credibilities and are not supposed to possess definite numerical values.[4] Especially both $\Pr\{H\}$ and $\Pr\{H/E\}$ denote the credibility of the same statistical hypothesis $H$, but the first before, and the second after, the observation of the event predicted by $E$. Let us restate the above equation in a less conventional form, emphasizing one of the aspects of $\Pr\{E/H\}$:

$$\text{Credibility after event} = \frac{\text{Likelihood} \times \text{Credibility before event}}{\text{Credibility of event}}.$$

Having observed the event predicted by $E$, we face a decision: should we reject the statistical hypothesis $H$, and accept the rival physical conjecture $P$, or what should we do? Our decision should be based on the latest information, and so on $\Pr\{H/E\}$, the credibility of the statistical hypothesis after the observation of the event. Of this credibility $\Pr\{H/E\}$ the likelihood $\Pr\{E/H\}$ is a factor: it is the most important factor, perhaps, because it possesses a numerical value, computable by a clear and familiar procedure, but it is still just a factor, not the full expression of the credibility.

The likelihood is an important indication, but not everything. The statistician may wisely restrict himself to the computation of the likelihood, but the statistician's customer may act unwisely if he neglects the other factors. He should carefully weigh $\Pr\{H\}$, the credibility of the statistical hypothesis $H$ before the event: we mean, in fact, this $\Pr\{H\}$ when we talk about the "appropriateness," or "realism," of $H$. Cf. ex. 14.33.

**6.** *Laplace's attempt to link induction with probability.* A bag contains black and white balls in unknown proportions; $m$ white and $n$ black balls have been successively drawn and replaced. What is the probability that $m' + n'$ subsequent drawings will yield $m'$ white and $n'$ black balls?

A particular case of this apparently harmless problem about the bag and the balls can be interpreted as the fundamental problem about inductive inference reduced to its simplest expression. In fact, let us consider the case

---

[4] This is the usual situation. Only exceptionally is the statistician in a position to ascribe a numerical value to $\Pr\{H\}$.

where $n = n' = 0$.  We drew $m$ balls from the bag of unknown composition, and all balls drawn turned out to be white.  We may assimilate this situation to the naturalist's situation who tested $m$ consequences of a conjecture and found all $m$ observations concordant with the conjecture.  The naturalist plans more observations.  What is the probability that his next $m'$ observations will also turn out to be concordant with the conjecture?  This question may be construed as the particular case $n = n' = 0$ of the proposed problem about the bag and the balls.

In this problem there is an obscure and puzzling point: the proportion of the black balls to the white balls in the bag is unknown.  Yet in view of the intended interpretation, this point appears essential: the naturalist cannot know the "inner workings" of nature.  He knows only what he has observed, and we know only that the bag yielded, up to this moment, so and so many white balls in so and so many drawings.

Common sense would suggest that we cannot compute the required probability if nothing is known about the composition of the bag; a problem without sufficient data is insoluble.  Yet Laplace gave a solution—a controversial solution.  How did he manage to arrive at a solution at all?

Laplace introduces a new principle to compensate for the lack of data; this principle is controversial.  "When the probability of a simple event is unknown, we may suppose all possible values of this probability between 0 and 1 as equally likely," says Laplace.[5]  "This is the equal distribution of ignorance," sneer his opponents.

Once Laplace's controversial principle is admitted, the derivation of the solution is straightforward; we need not consider it here.  Its result is: if $m$ drawings yielded only white balls, the probability that $m'$ subsequent drawings should also yield only white balls is

$$\frac{m + 1}{m + m' + 1}.$$

Let us call this statement the "General Rule of Succession."  The best known particular case is concerned with $m' = 1$: if $m$ drawings yielded only white balls, the probability that also the next drawing should yield a white ball is

$$\frac{m + 1}{m + 2}.$$

Let us call this the "Special Rule of Succession."[6]

Are these rules acceptable if we interpret "white balls" as "concordant observations of the same nature" and "probability" as "degree of reasonable confidence"?  This question is to the point and we shall discuss it.

---

[5] *Oeuvres complètes*, vol. 7, p. XVII.
[6] This differs somewhat from the usual terminology.  Cf. J. M. Keynes, *A treatise on probability*, p. 372–383.

(1) Let us reconsider our first example of inductive reasoning. Goldbach's conjecture asserts that, from $6 = 3 + 3$ onward, any even integer is the sum of two odd primes. The table in sect. 1.3 verifies this conjecture up to 30. Having verified it up to 30, we expect with more or less confidence that it will also be verified in the next case, for 32. The Special Rule of Succession may be interpreted to mean that, having verified Goldbach's conjecture in the first $m$ cases, we are entitled to expect its verification in the next case with the probability

$$\frac{m+1}{m+2} = 1 - \frac{1}{m+2}.$$

Let us realize what this means. As $m$ increases the probability also increases: in fact, the more cases have been verified in the past, the more confidently we expect the conjecture to be verified in the next case. If $m$ tends to $\infty$, the probability tends to 1: we could hope to approach certainty closer and closer by collecting more and more verifications. Let us now consider the difference of two probabilities, one corresponding to $m + 1$, the other to $m$, previous verifications:

$$\frac{m+2}{m+3} - \frac{m+1}{m+2} = \frac{1}{(m+2)(m+3)}.$$

This difference decreases as $m$ increases: it is true that each new verification adds to our confidence, but it adds less and less as it comes after more and more previous similar verifications. (The similarity of the verifications is essential at this point; cf. sect. 12.2.)

Let us take up now the General Rule of Succession. It could be interpreted to mean this: having verified Goldbach's conjecture in the first $m$ cases, we are entitled to expect its verification in the next $m'$ cases with the probability

$$\frac{m+1}{m+m'+1}.$$

If we keep $m$ fixed, but let $m'$ increase, this probability decreases: in fact, the further we try to predict the future on the basis of past observation, the less confidently we can predict it. If $m'$ increases indefinitely, the probability tends to 0. In fact, the verification for all values of $m'$ would mean that Goldbach's conjecture is true. Obviously, on the basis of a given number $m$ of observations we cannot assert that the conjecture is true. The rule seems to imply a stronger statement: on the basis of $m$ observations we cannot even attribute a probability different from 0 to Goldbach's conjecture, and such a strong statement may point in the right direction.

(2) Up to this point the Rule of Succession looks rather respectable. Yet let us look at it more concretely. Let us attribute numerical values to $m$ and let us not neglect everyday situations. It will be enough to consider the Special Rule of Succession.

I tested the even numbers 6, 8, 10, . . . 24, and found that each of them is a sum of two odd primes. The Rule says that I should expect with the probability 11/12 that also 26 is a sum of two odd primes.

In a foreign city where I scarcely understood the language, I ate in a restaurant with strong misgivings. Yet after 10 meals taken there I felt no ill effects and so I went quite confidently to the restaurant the eleventh time. The Rule said that the probability that I would not be poisoned by my next meal was 11/12.

A boy is 10 years old today. The Rule says that, having lived 10 years, he has the probability 11/12 to live one more year. The boy's grandfather attained 70. The Rule says that he has the probability 71/72 to live one more year.

These applications look silly, but none is sillier than the following due to Laplace himself. "Assume," he says, "that history goes back 5,000 years, that is, 1,826,213 days. The sun rose each day, and so you can bet 1,826,214 against 1 that the sun will rise again tomorrow."[7] I would certainly be careful not to offer such a bet to a Norwegian colleague who could arrange air transportation for both of us to some place within the Arctic circle.

Yet the rule can beat even this absurdity. Let us apply it to the case $m = 0$: the Rule's derivation is as valid for this case as for any other case. Yet for $m = 0$ the Rule asserts that any conjecture without any verification has the probability 1/2. Anybody can invent examples to show that such an assertion is monstrous. (By the way, it is also self-contradictory.)

(3) Our discussion was long. In more guarded language it would have been still longer, and it could be prolonged, but here is the long and the short of it: the Rule of Succession may look wise if we avoid numerical values, but it certainly looks foolish if we get down to numerical values. Perhaps, this points to a moral: in applying the Calculus of Probability to plausible reasoning, avoid numerical values on principle. At any rate, this is the standpoint advocated in this chapter.

7. *Why not quantitative?* This chapter advances a thesis: The Calculus of Probability should be applied to plausible reasoning, but only qualitatively. But there is a strong temptation to apply it quantitatively, and so we have to examine a few more relevant points.

(1) *Non-comparable.* There is some evidence that Goldbach's conjecture concerning the sum of two odd primes is correct; see sect. 1.2–1.3. There is some evidence that Norsemen landed on the American mainland a few hundred years before Columbus. Which evidence is stronger?

This seems to be a very silly question indeed. What could be the purpose of comparing two such disparate cases? And who should compare them? To judge the evidence competently you should be an expert. In one case the evidence should be judged by a mathematician expert in the Theory of

---

[7] *Oeuvres complètes*, vol. 7, p. XVII.

Numbers.    In the other case the evidence should be judged by a historian expert in Old Icelandic.    There is scarcely a person who would be expert in both.    However this may be, our apparently silly question points out a possibility. It could be that there is no reasonable decision, no reasonable way to say which evidence is stronger than the other.    This possibility is so important that it deserves a name.    If there is no reasonable way to decide which evidence is stronger, $E_1$ or $E_2$, let us call $E_1$ *non-comparable* with $E_2$.    We could find in the foregoing chapters several examples that suggest more clearly and more convincingly than the example from which we started here that one evidence may not be comparable with another.    See sect. 4.8.

(2) *Comparable.*    After having pointed out the possibility that evidences may be non-comparable, let us survey now a few cases in which evidences are obviously comparable.

Let $E_1$ denote the evidence for Goldbach's conjecture (sect. 1.2) resulting from its verification for all even numbers up to 1,000.    Let $E_2$ denote the evidence for the same conjecture resulting from its verification up to 2,000. Obviously, the evidence $E_2$ is stronger than the evidence $E_1$.

Now, let us change the notation, and let $E_1$ denote the present evidence for the landing of Norsemen on the American continent before Columbus. Let $E_2$ denote what this evidence would grow into, if somebody discovered, say, a burial place somewhere on the coast of Labrador containing shields and swords similar to those preserved elsewhere from the Viking age. Obviously, the evidence $E_2$ would be stronger than the evidence $E_1$.

Let us consider one more, somewhat subtler, case.    Let now $E_1$ denote the evidence for a conjecture $A$ resulting from the verification of one of its consequences $B$.    After the consequence $B$ has been verified, somebody remarks that $B$ is very improbable in itself.    (This remark could be quite precise;    the probability of $B$, computed on the basis of a simple and apparently appropriate statistical hypothesis, could be very low.)    This remark changes the evidence for the conjecture $A$ into $E_2$.    The evidence $E_2$ is stronger than the evidence $E_1$.    (We have said this already, perhaps a trifle less sharply, in sect. 12.3.)

In all three examples, we obtain the evidence $E_2$ from the evidence $E_1$ by adding some relevant observation.    Yet, if there is no such simple relation between $E_1$ and $E_2$, how could we decide which one is stronger?    This question brings even closer the possibility that evidences may be non-comparable.

(3) *Comparable, but still not quantitative.*    In the foregoing, subsection (2), we have seen cases in which an evidence $E_2$ may be reasonably held stronger than an evidence $E_1$.    Yet *how much stronger?*    It seems to me that there is no reasonable answer to this question in any of the foregoing cases.    And so we still remain on a qualitative level.

(4) *How would it look?*   In sect. 4 we started taking the symbol $\Pr\{A\}$ to denote the credibility of a conjecture $A$.   In the subsequent sections of this chapter we tried to get along without giving any determinate numerical value to $\Pr\{A\}$: herein consists the "qualitative" standpoint that this book advocates.   The "quantitative" standpoint would consist in giving to $\Pr\{A\}$ a definite numerical value whenever the Calculus of Probability is applied to plausible reasoning about the conjecture $A$.

The burden of proof falls squarely on those who champion a quantitative application of the Calculus of Probability to plausible reasoning.   All that they have to do is to produce a class of non-trivial conjectures $A$ for which the credibility $\Pr\{A\}$ can be computed by a clear method that leads to acceptable results at least in some cases.[8]

Nobody has yet proposed a clear and convincing method for computing credibilities in non-trivial cases, and if we visualize concrete situations in which a sound estimate of credibilities is important (as we have done), we can easily perceive that any attribution of definite numerical values to credibilities is in great danger of looking foolish.

Credibilities that have definite numerical values are comparable: two numbers are either equal, or one of them is greater than the other.   Yet, after the discussion under (1) and (2), we may find it hard to accept that any two conjectures should be comparable in credibility.   Take the two conjectures with which we began our discussion: Goldbach's conjecture and that historical conjecture about the discovery of America.   If we attributed numerical values to their credibilities, the strength of the evidence speaking for one could be compared with that for the other: yet such a comparison looks futile and foolish.

(5) *Would it be worthwhile?*   There is another point to consider which is independent of the foregoing discussion.   The weight of a plausible argument may be extremely important, but such importance is provisional, ephemeral, transient: would it be worthwhile to fasten a numerical value on something so transitory?

What is the credibility of Newton's law of gravitation, judged in the light of the facts collected in the first edition of the Principia?   Let us imagine for a moment that there is a method to evaluate numerically such a credibility. Yet we should not imagine for a moment that the evaluation could be easy: in view of the complexity of the facts and their interrelations, the evaluation must be delicate and the numerical computation long.   Would it be worthwhile to undertake it?   Perhaps, for us, in view of the historical and philosophical importance of Newton's discovery.   Yet scarcely for Newton and his contemporaries: instead of computing the credibility of the theory, they could have, with the same effort, changed the credibility by developing

---

[8] Attributing a numerical value to a credibility on the basis of some assumption of interchangeability or symmetry (as we did in ex. 4) should be regarded here as trivial: something more, and more novel, is required to justify quantitative credibilities.

the theory, and multiplying the observations. It looks preposterous to devote ten years to the computation of a degree of credibility that is valid only a second.

**8.** *Infinitesimal credibilities?*   A new particular case of some number-theoretic conjecture (such as Goldbach's conjecture; cf. sect. 1.2–1.3) has been verified. Such verification must be considered as increasing the weight of the evidence, or the credibility of the conjecture. Yet no amount of such verifications can prove the conjecture. We may even feel that no amount of such verifications can bring the conjecture measurably nearer to a proof. (Cf. ex. 4 (5), ex. 6 (1).) Such feelings may suggest the introduction of infinitesimals into the Calculus of Probability.

Infinitesimals can be quite clearly handled by modern mathematics. We consider "quantities" $a$, $b$, . . . represented by "formal power series":

$$a = a_0 + a_1\varepsilon + a_2\varepsilon^2 + a_3\varepsilon^3 + \cdots ;$$

$a_0, a_1, a_2, \ldots$ are real numbers, $\varepsilon$ is an indeterminate, no attention is paid to convergence. There is an algebra of such quantities; there are rules (familiar from the theory of convergent power series) according to which such formal power series can be added, subtracted, and multiplied; even division by $a$ can be performed if $a_0 \neq 0$. We call $a$ the "quantity zero" or 0 when $a_0, a_1, a_2, \ldots$ all vanish. Two quantities are equal if their difference is 0. The quantity $a$ reduces to the number $a_0$ when $a_1, a_2, a_3, \ldots$ all vanish.

We say that $a$ is *positive* if the *first* non-vanishing number in the sequence $a_0, a_1, a_2, \ldots$ is positive. From this definition, we easily derive two basic rules:

Either $a$ is 0 or $a$ is positive or $-a$ is positive, and these three possibilities are mutually exclusive.

The sum and product of two positive quantities are positive.

We say that $a > b$ if $a - b$ is positive. From these definitions it follows that any positive *number* is greater than $\varepsilon$, (that is, the formal power series $0 + \varepsilon + 0 \cdot \varepsilon^2 + 0 \cdot \varepsilon^3 + \ldots$) although $\varepsilon$ is positive. And so $\varepsilon$ is an "actual" infinitesimal.

We may contemplate now some form of the Calculus of Probability in which the probabilities (or credibilities) are not necessarily numbers, but quantities $a$ of the kind just introduced, subject to the condition $0 \leq a \leq 1$. In such a calculus the credibility of Goldbach's conjecture after 1,000,000, or any other number, of verifications could still be infinitesimal, that is, a quantity $a$ with $a_0 = 0$.

I refrain from comments on the prospects of such a plan. At any rate, it discloses a possibility of viewing non-numerical credibilities from another angle.

**9.** *Rules of admissibility.*   I wish the reader to draw his own conclusions and to form his own opinion. Therefore I postponed the expression of my opinion on a crucial question to this last comment of the chapter. I mean the question suggested by the first section of the chapter: Has plausible reasoning rules of any kind?

It is rather obvious that plausible reasoning has no rules of the same kind as demonstrative reasoning. A demonstration has been proposed. If it is presented in sufficiently small steps, the validity of each step can be tested by a rule of formal logic. If all steps conform to the rules, the demonstration is valid, but it is invalid if there is a step violating the rules. Thus, the rules of demonstrative logic are decisive: they can decide whether a proposed demonstrative argument is binding or not. The patterns of plausible reasoning that we have collected cannot achieve such a thing. A plausible argument has been proposed. Each step of it intends to render a certain conjecture more credible and does so following some accepted pattern. Having followed each step of the whole argument, you are not bound to trust the conjecture to any definite degree.

Yet there are rules of different kinds. Logical rules are very different from legal rules. A law court should listen to all parties concerned, but it should not listen to irrelevancies. Therefore a law court should have powers to exclude irrelevant matters from its proceedings, and such powers are regulated by *rules of admissibility*. Without some rules of this kind, there could be no orderly administration of justice: the court could not restrict an unscrupulous counsel who, by unfair or irrelevant questions, could wear down adverse witnesses, the opposition, the jury, and the judge, or drag out the proceedings indefinitely.

The patterns of plausible reasoning collected in the foregoing can be regarded as *rules of admissibility in scientific discussion*. You are by no means obliged to give a definite degree of credence to a conjecture if some of its consequences have been verified. Yet if the conjecture is discussed, it is certainly admissible to mention such verifications and it is fair and reasonable to listen to them. Our patterns register various points concerning such verifications that could reasonably influence the weight of the evidence (as analogy, or lack of analogy, with former verifications, etc.). It is fair and reasonable to admit the discussion of such points too. In collecting these patterns the author's intention was to list those general points that, *according to the usage of good scientists, are admissible in a scientific discussion*, with a view to reasonably influencing the credibility of the conjecture discussed.

In a trial by jury, the powers of the court are divided between the jury and the presiding judge. This division of powers (as conceived by certain legal authorities and accepted to some extent by the judicial practice of certain states and countries) is of great interest for us. The jury and the judge have different functions, they decide different questions. Questions concerning the admissibility of evidence are answered by the judge, questions concerning the credibility of the evidence admitted are answered by the jury. It is for the judge to decide which evidence deserves consideration by the jury.

It is for the jury to determine whether the submitted evidence is of sufficient weight. In deciding which matters are, and which are not, worthy of consideration by the jury, the judge has to know and to respect precedent, the court's usage, and the formulated rules of admissibility. In weighing the evidence submitted, the juror, who had possibly no legal training at all, has to rely on his own natural lights.

In short, the powers of the court are so divided between the judge and the jury, as the power of judging a proposed conjecture is divided in each of us between impersonal rules and personal good sense. The judge plays the role of the rules, the jury that of your personal discernment. It is for the impersonal rules of plausible reasoning to decide which kind of evidence deserves consideration. Yet it is for your personal good sense to decide whether the particular piece of evidence just submitted has sufficient weight or not.

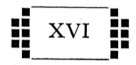

# XVI

## PLAUSIBLE REASONING IN
## INVENTION AND INSTRUCTION

*Words consist of letters of the alphabet, sentences of words which can be found in the dictionary, and books of sentences which may be found also in other authors. Yet if the things I say are consistent and so connected that they follow from each other, you can as well blame me for having borrowed my sentences from others as for having borrowed my words from the dictionary.*
---DESCARTES[1]

**1. Object of the present chapter.** The examples in the first part of this work and the discussions in the foregoing chapters of the second part elucidated somewhat, I hope, the role of plausible reasoning in the discovery of mathematical facts. Yet the mathematician does not only guess; he also has problems to solve, and he has to prove the facts that he guessed. What is the role of plausible reasoning in the discovery of the solution or in the invention of the proof? This is the question to be discussed in the present chapter. And, by the way, this is the question that attracted the author who, primarily concerned with the methods of problem-solving, was eventually led to the subject of the present book.

The subject of plausible reasoning is subtle and elusive and so is the subject of methods of solution. It was perhaps appropriate to defer the question that combines two such delicate subjects till the last chapter. The following treatment will be brief; the main purpose is to point out the connection with the matters previously discussed. A more ample treatment would fit into another book on methods of problem-solving.

**2. The story of a little discovery.** The solution of any simple but not merely routine mathematical problem may bring you some of the tension and the triumph of a discovery. Let us look at the following example: *Construct a quadrilateral, being given a, b, c, and d, its four sides, and ε, the angle included by the opposite sides a and c.*

---
[1] *Oeuvres,* edited by Adam and Tannery, vol. 10, 1908, p. 204.

The data of the problem are exhibited in fig. 16.1: four lines and one angle, fragments of a figure torn to pieces that we should reassemble to satisfy all the requirements laid down in the problem. It is understood that the sides $a$, $b$, $c$, and $d$ follow each other in this order around the desired quadrilateral so that $a$ is opposite to $c$, and $b$ to $d$. The angle $\varepsilon$, included by the opposite sides $a$ and $c$ is *not* one of the four angles of the quadrilateral.

Let us ask a few of the usual questions that may bring the problem closer to us.

*Are the data sufficient to determine the unknown?* The four sides alone would be obviously insufficient to determine the quadrilateral: four sticks attached by flexible joints at their respective ends form an articulated quadrilateral which is movable, deformable, not rigid, has no determinate shape. Yet if one of its four angles is fixed, the articulated quadrilateral cannot move any more: the quadrilateral is determined by its four sides and one of its angles. We may guess that it is also determined by four sides and some other angle, and so the data of our problem appear sufficient.

*Draw a figure.* We draw fig. 16.2 which displays all five data so assembled as they should be assembled according to the conditions of the proposed problem. Presumably, we have to *use all the data.*

It can happen that we get stuck at this point and no useful idea occurs to us for a while. In fact, fig. 16.2 looks clumsy. The sides $a$, $b$, $c$, $d$ are at their proper places, of course, but the situation of the angle $\varepsilon$ appears unfortunate. This angle is one of our data, we have to use it. Yet how can we use it if it is located so far away, at such an unusual place?

An experienced problem-solver would try to *redraw* the figure: he would try to place that angle $\varepsilon$ somewhere else. He may hit so upon fig. 16.3 where the angle $\varepsilon$ is between the side $a$ and a parallel to the side $c$ drawn through an endpoint of $a$. Fig. 16.3 looks more promising than the obvious fig. 16.2.

Why does fig. 16.3 look promising? Even good students who feel pretty sure that fig. 16.3 is promising may be unable to answer this question clearly.

"It looks good to me."

"The data are more tightly fitted together."

Only a student who is exceptionally talented or experienced will be able to give a full explanation: "In fig. 16.2 the angle is located in a triangle. Yet this triangle is not suitable for construction: only two data are known, $\varepsilon$ and $b$. As located in fig. 16.3, the angle $\varepsilon$ has more chance to be fitted into a suitable triangle. This is desirable, since, usually, this kind of construction is reduced to the construction of a triangle from suitable data."

The general idea behind the last answer seems to be: any feature in which the present situation recognizably agrees with successful past situations seems promising.

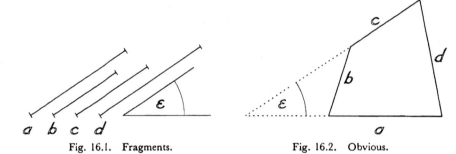

Fig. 16.1.  Fragments.                    Fig. 16.2.  Obvious.

Fig. 16.3.  Warmer.

Fig. 16.4.  Hot!

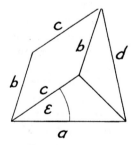

Fig. 16.5.  Done!

However this may be, fig. 16.3 lives up to our expectations. In fact, the angle ε does fit into a triangle that we can easily construct (from the sides *a* and *c* and the included angle ε, see fig. 16.4). When this triangle has been completed, the solution is quite close. In fact, to the triangle just completed another triangle is attached that we can construct (from sides *b*, *d*, and a previously constructed side, see fig. 16.5). Having drawn both triangles, we complete the required construction by drawing the parallelogram with sides *b* and *c*.

For most problem-solvers the decisive step in the foregoing solution, represented by a succession of five figures, is the step from fig. 16.2 to fig. 16.3; a longer period of hesitation may precede this step. Yet, once the angle ε is favorably located, the progress of the solution from fig. 16.3 through fig. 16.4 to the final fig. 16.5 may be quite rapid.

The foregoing solution may reveal one or two points which play a rôle also in more momentous discoveries.

**3. The process of solution.** Solving a problem is an extremely complex process. No description or theory of this process can exhaust its manifold aspects, any description or theory of it is bound to be incomplete, schematic, highly simplified. I wish to point out the place of plausible reasoning in this complex process, and I shall choose the simplest description I am able to find in which this place can be recognizably located. And even the beginning of such a simple description will suffice here.

(1) *Setting a problem to yourself.* A problem becomes a problem for you when you propose it to yourself. A problem is not yet your problem just because you are supposed to solve it in an examination. If you wish that somebody would come and tell you the answer, I suspect that you did not yet set that problem to yourself in earnest. But if you are anxious to find the answer yourself, by your own means, then you have made the problem really yours, you are serious about it.

Setting a problem to yourself is the beginning of the solution, the essential first move in the game. It is a move in the nature of a decision.

(2) *Selective attention.* You need not tell me that you have set that problem to yourself, you need not tell it to yourself; your whole behavior will show that you did. Your mind becomes selective; it becomes more accessible to anything that appears to be connected with the problem, and less accessible to anything that seems unconnected. You eagerly seize upon any recollection, remark, suggestion, or fact that could help you to solve your problem, and you shut the door upon other things. When the door is so tightly shut that even the most urgent appeals of the external world fail to reach you, people say that you are absorbed.

(3) *Registering the pace of progress.* There is another thing that shows that you are seriously engaged in your problem; you become sensitive. You keenly feel the pace of your progress; you are elated when it is rapid, you are depressed when it is slow. Whatever comes to your mind is quickly

sized up: "It looks good," "It could help," or "No good," "No help." Such judgments are, of course, not infallible. (Although they seem to be more often correct than not, especially with talented or experienced people.) At any rate, such judgments and feelings are important for you personally; they guide your effort.

(4) *Where plausible reasoning comes in.* Let us see somewhat more concretely a typical situation.

You try to attain the solution in a certain direction, along a certain line. (For example, in trying to solve the geometrical problem of sect. 2 you reject fig. 16.2 and attempt to work with the more hopeful fig. 16.3.) You may feel quite keenly that you work in the right direction, that you follow a promising line of approach, that you are on the scent. You may feel so, by the way, without formulating your feeling in words. Or even if you say something such as, "It looks good," you do not take the trouble to analyze your confidence, you do not ask, "Why does it look good?" You are just too busy following up the scent.

Yet you may have bad luck. You run into difficulties, you do not make much progress, nothing new occurs to you and then you start doubting: "Was it a good start? Is this the right direction?" And then you may begin to analyze your feeling: "The direction looked quite plausible— but why is it plausible?" Then you may start debating with yourself, and some more distinct reasons may occur to you:

"The situation is not so bad. I could bring in a triangle. People always bring in triangles in such problems."

"It was probably the right start, after all. It looks like the right solution. What do I need for a solution with this kind of problem? Such a point—and I have it. And that kind of point—I have it too. And . . . "

It would be interesting to see more distinctly how people are reasoning in such a situation—in fact, it is our main purpose to see just that. Yet we need at least one more example to broaden our observational basis.

**4. Deus ex machina.**[2] The next example has to be a little less simple than that of sect. 2. It will be given in sect. 6 after some preparations in this section and the next. Sect. 6 will bring a proof, presented in a manner that contrasts with the usual manner of presentation. To emphasize the contrast, let us first see the proof as it would be presented in a (more advanced) textbook or in a mathematical periodical.

A mathematical book or lecture should be, first of all, correct and unambiguous. Still, we know from painful experience that a perfectly unambiguous and correct exposition can be far from satisfactory and may appear uninspiring, tiresome, or disappointing, even if the subject-matter presented is interesting in itself. The most conspicuous blemish of an

---

[2] Sect. 4, 5, and 6 reproduce, except for slight changes, parts of my paper "With, or without, motivation?" *American Mathematical Monthly*, v. 56, 1949, p. 684–691.

otherwise acceptable presentation is the "deus ex machina." Before further comments, I wish to give a concrete example. Let us look at the proof of the following not quite elementary theorem.[3]

*If the terms of the sequence* $a_1, a_2, a_3, \ldots$ *are non-negative real numbers, not all equal to 0, then*

$$\sum_1^\infty (a_1 a_2 a_3 \ldots a_n)^{1/n} < e \sum_1^\infty a_n.$$

*Proof.* Define the numbers $c_1, c_2, c_3, \ldots$ by

$$c_1 c_2 c_3 \ldots c_n = (n+1)^n$$

for $n = 1, 2, 3, \ldots$ . We use this definition, then the inequality between the arithmetic and the geometric means (sect. 8.6), and finally the fact that the sequence defining $e$, the general term of which is $[(k+1)/k]^k$, is increasing. We obtain

$$\sum_1^\infty (a_1 a_2 \ldots a_n)^{1/n} = \sum_1^\infty \frac{(a_1 c_1 a_2 c_2 \ldots a_n c_n)^{1/n}}{n+1}$$

$$\leqq \sum_1^\infty \frac{a_1 c_1 + a_2 c_2 + \ldots + a_n c_n}{n(n+1)}$$

$$= \sum_{k=1}^\infty a_k c_k \sum_{n=k}^\infty \frac{1}{n(n+1)}$$

(d)

$$= \sum_{k=1}^\infty a_k c_k \sum_{n=k}^\infty \left(\frac{1}{n} - \frac{1}{n+1}\right)$$

$$= \sum_{k=1}^\infty a_k \frac{(k+1)^k}{k^{k-1}} \frac{1}{k}$$

$$< e \sum_{k=1}^\infty a_k.$$

**5. Heuristic justification.** The crucial point of the derivation (d) is the definition of the sequence $c_1, c_2, c_3, \ldots$ . This point appears right at the beginning without any preparation, as a typical "deus ex machina." What is the objection to it?

"It appears as a rabbit pulled out of a hat."

"It pops up from nowhere. It looks so arbitrary. It has no visible motive or purpose."

---

[3] I may be excused if I choose an example from my own work. See G. Pólya, Proof of an inequality, *Proceedings of the London Mathematical Society* (2), v. 24, 1925, p. LVII. The theorem proved is due to T. Carleman.

"I hate to walk in the dark.    I hate to take a step when I cannot see any reason why it should bring me nearer to the goal."

"Perhaps the author knows the purpose of this step, but I do not, and therefore I cannot follow him with confidence."

"Look here, I am not here just to admire you.    I wish to learn how to do problems by myself.    Yet I cannot see how it was humanly possible to hit upon your . . . definition.    So what can I learn here?    How could I find such a . . . definition by myself?"

"This step is not trivial.    It seems crucial.    If I could see that it has some chances of success, or see some plausible provisional justification for it, then I could also imagine how it was invented and, at any rate, I could follow the subsequent reasoning with more confidence and more under-standing."

The first answers are not very explicit, the later ones are better, and the last is the best.    It reveals that an intelligent reader or listener desires two things:

First, to see that the present step of the argument is correct.

Second, to see that the present step is appropriate.

A step of a mathematical argument is appropriate if it is essentially connected with the purpose, if it brings us nearer to the goal.    It is not enough, however, that a step *is* appropriate: it should *appear so* to the reader. If the step is simple, just a trivial, routine step, the reader can easily imagine how it could be connected with the aim of the argument.    If the order of presentation is very carefully planned, the context may suggest the connection of the step with the aim.    If, however, the step is visibly important, but its connection with the aim is not visible at all, it appears as a "deus ex machina" and the intelligent reader or listener is understandably dis-appointed.

In our example, the definition of $c_n$ appears as a "deus ex machina."    Yet this step is certainly appropriate.    In fact, the argument based on this definition proves the proposed theorem, and proves it rather quickly and clearly.    The trouble is that the step in question, although vindicated in the end, does not appear as justified from the start.

Yet how could the author justify it from the start?    The complete justifica-tion takes some time; it is supplied by the full proof.    What is needed is, not a complete, but an *incomplete* justification, a *plausible provisional ground*, just a hint that the step has some chances of success, in short, some *heuristic justification*.

**6. The story of another discovery.**    It is almost unnecessary to remind the reader that the best stories are not true.    They must contain, however, some essential elements of truth, otherwise they would not be any good. The following is a somewhat "rationalized" presentation of the steps that led me to the proof given in sect. 4.    That is, the heuristic justification of the successive steps is suitably emphasized.

The theorem proved in sect. 4 is surprising in itself. We would be less surprised if we knew how it was discovered. We are led to it naturally in trying to prove the following: *If the series with positive terms*

$$a_1 + a_2 + a_3 + \ldots + a_n + \ldots$$

*is convergent, the series*

$$a_1 + (a_1a_2)^{1/2} + (a_1a_2a_3)^{1/3} + \ldots + (a_1a_2a_3 \ldots a_n)^{1/n} + \ldots$$

*is also convergent.* I shall try to emphasize some motives which may help us to find the proof.

(1) *A suitable known theorem.* It is natural to begin with the usual questions. *What is the hypothesis?* We assume that the series $\Sigma a_n$ converges—that its partial sums remain bounded—that

$$a_1 + a_2 + \ldots + a_n \text{ is not large.}$$

*What is the conclusion?* We wish to prove that the series $\Sigma(a_1, a_2 \ldots a_n)^{1/n}$ converges—that

$$(a_1a_2 \ldots a_n)^{1/n} \text{ is small.}$$

*Do you know a theorem that could be useful?* What we need is some relation between the sum of $n$ positive quantities and their geometric mean. *Have you seen something of this kind before?* If you ever have heard of the inequality between the arithmetic and the geometric means, it has a good chance to occur to you at this juncture:

$$(ag) \qquad (a_1a_2 \ldots a_n)^{1/n} \leq \frac{a_1 + a_2 + \ldots + a_n}{n}.$$

This inequality shows that $(a_1a_2 \ldots a_n)^{1/n}$ is small when $a_1 + a_2 + \ldots + a_n$ is not large. It has so many contacts with our problem that we can hardly resist the temptation of applying it:

$$(a) \qquad \sum_{n=1}^{\infty} (a_1 a_2 \ldots a_n)^{1/n} \leq \sum_{n=1}^{\infty} \frac{a_1 + a_2 + \ldots + a_n}{n}$$

$$= \sum_{k=1}^{\infty} a_k \sum_{n=k}^{\infty} \frac{1}{n}$$

—complete failure! The series $\Sigma 1/n$ is divergent, the last line of (a) is meaningless.

(2) *Learning from failure.* It is difficult to admit that our plan was wrong. We would like to believe that at least some part of it was right. The useful questions are: *What was wrong with our plan? Which part of it could we save?* The series $a_1 + a_2 + \ldots + a_n + \ldots$ converges. Therefore, $a_n$ is small when $n$ is large. Yet the two sides of the inequality (ag) are different when $a_1, a_2, \ldots a_n$ are not all equal, and they may be very different when $a_1, a_2, \ldots a_n$ are very unequal. In our case, $a_1$ is much larger than $a_n$, and so there may be a considerable gap between the two sides of (ag). This is

probably the reason that our application of (ag) turned out to be insufficient.

(3) *Modifying the approach.* The mistake was to apply the inequality (ag) to the quantities

$$a_1, a_2, a_3, \ldots a_n$$

which are too unequal. Why not apply it to some related quantities which have more chance to be equal? We could try

$$1a_1, 2a_2, 3a_3, \ldots na_n.$$

This may be the idea! We may introduce such increasing compensating factors as 1, 2, 3, . . . *n*. We should, however, not commit ourselves more than necessary, we should reserve ourselves some freedom of action. We should consider perhaps, more generally, the quantities

$$1^\lambda a_1, 2^\lambda a_2, 3^\lambda a_3, \ldots n^\lambda a_n.$$

We could leave $\lambda$ *indeterminate* for the moment, and choose the most advantageous value later. This plan has so many good features that it seems ripe for action:

(b)
$$\sum_1^\infty (a_1 a_2 \ldots a_n)^{1/n} = \sum_1^\infty \frac{(a_1 1^\lambda \cdot a_2 2^\lambda \ldots a_n n^\lambda)^{1/n}}{(1 \cdot 2 \ldots n)^{\lambda/n}}$$

$$\leqq \sum_{n=1}^\infty \frac{a_1 1^\lambda + a_2 2^\lambda + \ldots + a_n n^\lambda}{n(n!)^{\lambda/n}}$$

$$= \sum_{k=1}^\infty a_k k^\lambda \sum_{n=k}^\infty \frac{1}{n(n!)^{\lambda/n}}.$$

We run into difficulties. We cannot evaluate the last sum. Even if we recall various relevant tricks, we are still obliged to work with "crude equations" (notation $\approx$, instead of $=$):

$$(n!)^{1/n} \approx ne^{-1},$$

$$\sum_{n=k}^\infty \frac{1}{n(n!)^{\lambda/n}} \approx e^\lambda \sum_{n=k}^\infty n^{-1-\lambda}$$

$$\approx e^\lambda \int_k^\infty x^{-1-\lambda} dx$$

$$= e^\lambda \lambda^{-1} k^{-\lambda}.$$

Introducing this into the last line of (b) we come very close to proving

(b')
$$\sum_1^\infty (a_1 a_2 \ldots a_n)^{1/n} \leqq C \sum_1^\infty a_k$$

where $C$ is some constant, perhaps $e^\lambda \lambda^{-1}$. Such an inequality would, of course, prove the theorem in view.

(4) *Looking back* at the foregoing reasoning we are led to repeat the question: "Which value of $\lambda$ is the most advantageous?" Probably the $\lambda$ that makes $e^{\lambda}\lambda^{-1}$ a minimum. We can find this value by differential calculus:

$$\lambda = 1.$$

This suggests strongly that the most obvious choice is the most advantageous: the compensating factor multiplying $a_n$ should be $n^1 = n$, or some quantity not very different from $n$ when $n$ is large. This may lead to the simple value $C = e$ in (b').

(5) *More flexibility.* We left $\lambda$ indeterminate in our foregoing reasoning (b). This gave our plan a certain *flexibility*: the value of $\lambda$ remained at our disposal. Why not give our plan still more flexibility? We could leave the compensating factor that multiplies $a_n$ quite indeterminate; we call it $c_n$, and we will dispose of its value later, when we shall see more clearly what we need. We embark upon this further modification of our original approach:

(c)
$$\sum_1^\infty (a_1 a_2 \ldots a_n)^{1/n} = \sum_{n=1}^\infty \frac{(a_1 c_1 \cdot a_2 c_2 \ldots a_n c_n)^{1/n}}{(c_1 c_2 \ldots c_n)^{1/n}}$$
$$\leq \sum_{n=1}^\infty \frac{a_1 c_1 + a_2 c_2 + \ldots + a_n c_n}{n(c_1 c_2 \ldots c_n)^{1/n}}$$
$$= \sum_{k=1}^\infty a_k c_k \sum_{n=k}^\infty \frac{1}{n(c_1 c_2 \ldots c_n)^{1/n}}.$$

How should we choose $c_n$? This is the crucial question and we can no longer postpone the answer.

First, we see easily that a factor of proportionality must remain arbitrary. In fact, the sequence $c c_1, c c_2, \ldots, c c_n, \ldots$ leads to the same consequences as $c_1, c_2, \ldots, c_n, \ldots$.

Second, our foregoing work suggests that both $c_n$ and $(c_1 c_2 \ldots c_n)^{1/n}$ should be asymptotically proportional to $n$:

$$c_n \sim Kn, \quad (c_1 c_2 \ldots c_n)^{1/n} \sim e^{-1} Kn = K'n.$$

Third, it is most desirable that we should be able to effect the summation

$$\sum_{n=k}^{n=\infty} \frac{1}{n(c_1 c_2 \ldots c_n)^{1/n}}.$$

At this point, we need whatever previous knowledge we have about simple series. If we are familiar with the series

$$\sum \frac{1}{n(n+1)} = \sum \left( \frac{1}{n} - \frac{1}{n+1} \right)$$

it has a good chance to occur to us at this juncture. This series has the property that its sum has a simple expression not only from $n = 1$ to $n = \infty$, but also from $n = k$ to $n = \infty$—a great advantage! This series suggests the choice

$$(c_1 c_2 \ldots c_n)^{1/n} = n + 1.$$

Now, visibly $n + 1 \sim n$ for large $n$—a good sign! What about $c_n$ itself? As

$$c_1 c_2 \ldots c_{n-1} c_n = (n + 1)^n, \qquad c_1 c_2 \ldots c_{n-1} = n^{n-1},$$

$$c_n = \frac{(n + 1)^n}{n^{n-1}} = \left(1 + \frac{1}{n}\right)^n n \sim en;$$

the asymptotic proportionality with $n$ is a good sign. And the number $e$ arises—a very good sign!

We choose this $c_n$ and, after this choice, we take up again the derivation (d) in sect. 4 with more confidence than before.

Now, we may understand how it was humanly possible to discover that definition of $c_n$ which appeared in sect. 4 as a "deus ex machina." The derivation (d) became also more understandable. It appears now as the last, and the only successful, attempt in a chain of consecutive trials, (a), (b), (c), and (d). And the origin of the theorem itself is elucidated. We see now how it was possible to discover the rôle of the number $e$ which appeared so surprising at the outset.

7. **Some typical indications.** We have seen two examples in the foregoing. We examined first a "problem to find" (in sect. 2) then a "problem to prove" (in sect. 6.)[4] A much greater variety of examples were needed to illustrate properly the rôle of plausible reasoning in devising a plan of the solution. At any rate, from our examples we can disentangle a few typical circumstances indicative of the worth of a plan. In dealing with other circumstances of this kind we shall appeal to whatever experience the reader had in solving mathematical problems.

In enumerating such indicative circumstances, we shall not attempt completeness. In some cases we shall find it necessary to distinguish between problems to find and problems to prove. In such cases we shall give two parallel formulations, and give the formulation relative to problems to find first.

We consider a situation in which plausible reasoning comes naturally to the problem-solver. You are engaged in an exciting problem. You have conceived a plan of the solution, but somehow you do not like it quite. You have your doubts, you are not quite convinced that your plan is workable. In debating this matter with yourself, you are, in fact, examining a conjecture:

A. *This plan of the solution will work.*

Several pros and cons may occur to you as you examine your plan from various angles. Here are some conspicuous typical indications that may speak for the conjecture $A$.

---

4 For this terminology, see *How to Solve It*, p. 141–144.

$B_1$. *This plan takes all the data into account.*
This formulation applies to problems to find. There is a parallel formulation that applies to problems to prove: *this plan takes into account all the parts of the hypothesis.* For example, fig. 16.3 combines all the data, and this is a good sign. Also fig. 16.2 contains all the data, but there is a difference between the two figures that the next points may clarify.

$B_2$. *This plan provides for a connection between the data and the unknown.*
There is a parallel formulation concerned with problems to prove: *this plan provides for a connection between the hypothesis and the conclusion.* For example, in sect. 6 (1), the inequality between the arithmetic and geometric means promised to create a connection between the hypothesis and the conclusion, and this promise moved us to work with that inequality. Fig. 16.3 appears to provide for a closer connection, and so it appears more hopeful, than fig. 16.2.

$B_3$. *This plan has features that are often useful in solving problems of this kind.*
For example, the plan starting from fig. 16.3 introduces at a more mature stage (fig. 16.4) the construction of a triangle. This is a good sign, since problems of geometric construction are often reduced to the construction of triangles.

$B_4$. *This plan is similar to one that succeeded in solving an analogous problem.*

$B_5$. *This plan succeeded in solving a particular case of the problem.*
For example, you have a plan to solve a difficult problem concerned with an arbitrary closed curve. Carrying through the plan seems to involve a lot of work and so you hesitate. Yet you observe that in the particular case when the closed curve is a circle your plan works and yields the right result. This is a good sign, and you feel encouraged.

$B_6$. *This plan succeeded in solving a part of the problem* (in finding some of the unknowns, or in proving a weaker conclusion).
This list is by no means exhaustive. There are still other typical indications and signs, but we need not list them here. At any rate, it would be useless to list them without proper illustration.[5]

**8. Induction in invention.** The problem-solver's conjecture $A$ (that his plan of the solution will work) may be supported by one, or two, or more indications $B_1$, $B_2$, $B_3$, ... (of the kind listed in the foregoing sect. 7). Such indications may occur to the problem-solver successively, each indication intensifying his confidence in his plan. Our foregoing discussions lead us to compare such a problem-solving process with the inductive process in which an investigator, examining a conjecture $A$, succeeds in verifying several consequences $B_1$, $B_2$, $B_3$, ... in succession. We may also compare it with the legal procedure in which the jury examining an accusation $A$, takes

---

[5] Cf. *How to Solve It*, p. 212–221.

cognizance of several corroborating circumstances $B_1$, $B_2$, $B_3$, . . . in succession. We should not naïvely expect the identity of the three processes, but we should examine their similarity or dissimilarity with an open mind.

(1) When the problem-solver debates his plan of the solution with himself, this plan is usually more "fluid" than "rigid," it is more felt than formulated. In fact it would be foolish of the problem-solver to fix his plan prematurely. A wise problem-solver does not commit himself to a rigid plan. Even at a later stage, when the plan is riper, he keeps it ready for modification, he leaves it a certain flexibility, he reckons with unforeseen difficulties to which he might be obliged to adapt his plan. Therefore, when the problem-solver investigates the workability of his plan, he examines a changeable, sometimes a fleeting, object.

On the other hand, the conjectures that the mathematician or the naturalist investigates are usually pretty determinate: they are clearly formulated, or at least reasonably close to a clear formulation. Also the jury has a pretty determinate conjecture to examine: an indictment, the terms of which have been carefully laid down by the prosecution.

Let us note this striking difference that separates the problem-solver's investigation of the workability of his plan from the inductive investigation of a mathematical or physical conjecture, or from the judicial investigation of a charge: it is the difference between a changeable, or fleeting, and a determinate, relatively well defined object.

(2) The proceedings and acts of a court of justice are laid down in the record. The conjecture examined by the naturalist, and the evidence gathered for or against it, are also destined for a permanent record. Not so the problem-solver's conjecture concerning the workability of his scheme, or the signs speaking for or against it: their importance is ephemeral. They are extremely important as long as they guide the problem-solver's decisions. Yet, when the problem-solver's work enters a new phase, the plan itself may change, and then the indications speaking for or against it lose almost all interest. At the end, when the solution is attained and the problem is done, all such accessories are cast away. The final form of the solution may be recorded, yet the changing plans and the arguments for or against them are mostly or entirely forgotten. The building erected remains in view, but the scaffoldings, which were necessary to erect it, are removed.

Let us note this aspect of the difference between an inductive, or judicial, investigation on the one hand, and the problem-solver's appraisal of the prospects of his plan on the other: one is, the other is not, for permanent record.

(3) The conjecture $A$ and the indications $B_1$, $B_2$, . . . listed in sect. 7 can be interpreted with a certain latitude. After the foregoing remarks (under (1) and (2)) we should not expect that a sharply defined interpretation will be too often applicable. Still, there is some advantage in beginning with

such an interpretation. We consider the problem-solver's conjecture $A$ and an indication $B$ supporting it, stated as follows:

A. This plan of the solution will work in its present form.

B. This plan of the solution takes into account all the data.

In order to describe the situation more precisely, we add: *It is known that each of the data is necessary.* If this is so

$$A \text{ implies } B.$$

In fact, if the plan should work and yield the correct solution, it must use all the data, each of which is necessary to the solution.

Now it is important to visualize clearly the situation: $A$ is a conjecture in which the problem-solver is naturally interested, $B$ is a statement that may, or may not, be true. Let us examine both possibilities.

(4) If all the data are necessary to the solution, but our plan of the solution does not take into account all the data, our plan, in its present form, cannot work. (It could work in a modified form.) That is, if $B$ is false, $A$ must also be false.

Now it is important to observe that we could reach this conclusion also by formal reasoning. In fact, we have followed here a classical elementary pattern of reasoning (already quoted in sect. 12.1) the "modus tollens" of the so-called hypothetical syllogism:

$$A \text{ implies } B$$
$$B \text{ false}$$
$$\overline{\phantom{AAAA}}$$
$$A \text{ false}$$

(5) If, however, our plan of the solution does take into account all the data, it is natural to regard this circumstance as a good sign, as a favorable omen, as a forecast that our plan might work. (I imagine the problem-solver's relief when he notices that a datum that at first seemed to be neglected by his plan is used by it after all, and used in a neat manner, too.) In short, if $B$ is true, $A$ becomes more credible.

Now it is important to observe that, in fact, we could have reached this conclusion by simply following our fundamental inductive pattern:

$$A \text{ implies } B$$
$$B \text{ true}$$
$$\overline{\phantom{AAAA}}$$
$$A \text{ more credible}$$

(6) We consider now another situation. It is similar to, but clearly different from, the situation explained under (3) and discussed under (4) and (5). Again we are concerned with the problem-solver's conjecture $A$

and an indication *B* supporting it. Yet the situation is now different (less sharply defined). *A* and *B* have the meanings:

*A*. This plan of the solution will work (in a modified form, perhaps).

*B*. This plan of the solution takes into account all the data.

In order to characterize the situation more fully, we have to add: *We strongly suspect, although we do not definitely know, that all the data are necessary.*

As above, *A* is a conjecture in which the problem-solver has a stake, and *B* is a statement that may, or may not, be true. We have to examine both possibilities.

If *B* turns out to be false, there is an argument against *A*, but it is not quite decisive. As *B* is false, our plan does not take into account all the data; nevertheless, we may stick to our plan (if we have some strong, although unexpressed, ground for it). There may be some (not yet clarified) ground to hope that a modification of our plan will take care of all the data eventually. Also the doubt that all the data may not be necessary could have some little influence.

If *B* turns out to be true, we can take this circumstance for an encouraging sign. In fact, even if *A* does not imply *B*, and so *B* is not absolutely certain with *A*, it may be that still

*B* with *A* is readily credible,

*B* without *A* is less readily credible.

In such a case the verification of *B* still counts as a sort of circumstantial evidence for *A*. (Cf. sect. 13.13 (5).)

(7) In the foregoing subsections (3), (4), (5), and (6), we discussed the indication listed under $B_1$ in sect. 7. (We called it just *B*.) The discussion of the other indications listed in sect. 7 (under $B_2$, $B_3$, . . . $B_6$) would disclose a similar picture.

As we have seen, *A* may imply $B_1$, but even if this is not so and $B_1$ is not necessarily associated with *A*, the chances may be strong that *A* will be accompanied by $B_1$. The relation of *A* to $B_2$ (or $B_3$, or $B_4$, . . . ) is of the same nature. If the problem-solver's plan is any good, it must create some connection between the data and the unknown (or between the hypothesis and the conclusion); cf. $B_2$. It is not absolutely necessary that the solution should be similar to the solution of some formerly solved similar problem, yet the chances are usually pretty strong that this should be so; cf. $B_3$, $B_4$. If the plan works for the whole problem it must work, of course, for any particular case or any portion of the problem; cf. $B_5$, $B_6$.

Therefore, if we are in doubt about *A*, but succeed in observing $B_1$ or $B_2$ or $B_3$ . . . , we can reasonably regard our observation as some kind of inductive or circumstantial evidence for *A*, as an indication in favor of the problem-solver's conjecture that his plan will work.

(8) If, in spite of much work, the naturalist succeeds in verifying only a few not too surprising consequences of his conjecture, he may be moved to drop it. If too little evidence is submitted against the defendant, the court may drop the case. If, after a long and strenuous effort, only a few weak indications in favor of his plan have occurred to the problem-solver, he may be moved to modify his plan radically, or even drop it altogether.

On the other hand, if several consequences are verified, several pieces of evidence against the defendant submitted, several indications observed, the case for the naturalist's conjecture, for the prosecution, or for the problem-solver's plan may be greatly strengthened. Yet even more important than the number may be the variety. Consequences that are very different from each other, witnesses who are obviously independent, indications that come from different sides, count more heavily. (Cf. sect. 12.2, 13.11, 15.9.)

(9) In spite of such similarities, there is a considerable difference. The naturalist's task is to gather as much evidence as he can for or against his conjecture. The court's task is to examine all the relevant evidence submitted. Yet it is not the problem-solver's task to collect as much evidence as he can for or against his plan of the solution, or to debate such evidence to the bitter end: his task is to solve the problem, by any means, in following this plan of the solution, or any other plan, or no plan.

Even a faulty plan may serve the problem-solver's purpose. To solve his problem, he has to mobilize and organize the relevant parts of his past experience. Working with a faulty plan, but with genuine effort, the problem-solver may stir up some relevant item which otherwise would have remained hidden and unawakened in the background; this may give him a new departure. In problem-solving a bad plan is frequently useful; it may lead to a better plan.

(10) Two persons presented with the same evidence may honestly disagree, even if they are relying on the same patterns of plausible reasoning. Their backgrounds may be different. My unexpressed, inarticulate grounds, my whole background may influence my evaluation of experimental or judicial evidence. They may influence still more my evaluation of the indications for or against my plan of the solution, and this is not unreasonable. It is reasonable that, working at the solution of a problem, I should attach more weight than under other conditions to the promptings of my background and less weight to distinctly formulated grounds: to stir up relevant material hidden somewhere in the background is the thing I am working for.

Still, it seems to me that one of the principal assets of a seasoned problem-solver is that he is able to deal astutely with indications for or against the workability of his plan, as a well-trained naturalist deals with experimental evidence, or an experienced lawyer with legal evidence.

**9. A few words to the teacher.** Mathematics has many aspects. To many students, I am afraid, mathematics appears as a set of rigid rules, some of which you should learn by heart before the final examinations, and all of which you may forget afterwards.   To some instructors, mathematics appears as a system of rigorous proofs which, however, you should refrain from presenting in class, but instead present some more popular although inconclusive talk of which you are somewhat ashamed.   To a mathematician, who is active in research, mathematics may appear sometimes as a guessing game: you have to guess a mathematical theorem before you prove it, you have to guess the idea of the proof before you carry through the details.

To a philosopher with a somewhat open mind all intelligent acquisition of knowledge should appear sometimes as a guessing game, I think.   In science as in everyday life, when faced by a new situation, we start out with some guess.   Our first guess may fall wide of the mark, but we try it and, according to the degree of success, we modify it more or less.   Eventually, after several trials and several modifications, pushed by observations and led by analogy, we may arrive at a more satisfactory guess.   The layman does not find it surprising that the naturalist works in this way.   The knowledge of the naturalist may be better ordered with a view to selecting the appropriate analogies, his observations may be more purposeful and more careful, he may give more fancy names to his guesses and call them "tentative generalizations," but the naturalist adapts his mind to a new situation by guessing like the common man.   And the layman is not surprised to hear that the naturalist is guessing like himself.   It may appear a little more surprising to the layman that the mathematician is also guessing.   The result of the mathematician's creative work is demonstrative reasoning, a proof, but the proof is discovered by plausible reasoning, by guessing.

If this is so, and I believe that this is so, there should be a place for guessing in the teaching of mathematics.   Instruction should prepare for, or at least give a little taste of, invention.   At all events, the instruction should not suppress the germs of invention in the student.   A student who is somewhat interested in the problem discussed in class *expects* a certain kind of solution. If the student is intelligent, he foresees the solution to some extent: the result may look thus and so, and there is a chance that it may be obtained by such and such a procedure.   The teacher should try to realize what the students might expect, he should find out what they do expect, he should point out what they should reasonably expect.   If the student is less intelligent and especially if he is bored, he is likely to produce wild and irresponsible guesses. The teacher should show that guesses in the mathematical domain may be reasonable, respectable, responsible.   I address myself to teachers of mathematics of all grades and say: *Let us teach guessing!*

I did not say that we should neglect proving.   On the contrary, we should teach both proving and guessing, both kinds of reasoning, demonstrative and plausible.   More valuable than any particular mathematical fact or trick, theorem, or technique, is for the student to learn two things:

First, to distinguish a valid demonstration from an invalid attempt, a proof from a guess.

Second, to distinguish a more reasonable guess from a less reasonable guess.

There are special cases in which it is more important to teach guessing than proving. Take the teaching of calculus to engineering students. (I have a long and varied experience of this kind of teaching.) Engineers need mathematics, quite a few of them have a healthy interest in mathematics, but they are not trained to understand ε-proofs, they have no time for ε-proofs, they are not interested in ε-proofs. To teach them the rules of calculus as a dogma imposed from above would not be educational. To pretend that your proof is complete when, in fact, it is not, would not be honest. Confess calmly that your proofs are incomplete, but give respectable plausible grounds for the incompletely proved results, from examples and analogy. Then you need not be ashamed of fake proofs, and some students may remember your teaching after the examinations. On the basis of a long experience, I would say that talented students of engineering are usually more accessible to and more grateful for well-presented plausible grounds than for strict proofs.

I said that it is desirable to teach guessing, but not that it is easy to teach it. There is no foolproof method for guessing, and therefore there cannot be any foolproof method to teach guessing. I may have said a few foolish things in the foregoing, perhaps, but I have avoided the most foolish thing, I hope, which would be to pretend that I have an infallible method to teach guessing.

Still, it is not impossible to teach guessing. I hope that some of the examples explained at length and some of the exercises proposed in the foregoing will serve as useful suggestions. These suggestions have the best chance to fall on fertile ground with teachers who have some real experience in problem-solving.

Take, for instance, the example treated in sect. 4 and 6. The two presentations, in sect. 4 and sect. 6, are very different. The most obvious difference is that one is short and the other long. The most essential difference is that one gives proofs and the other plausibilities. One is designed to check the *demonstrative conclusions* justifying the successive steps. The other is arranged to give some insight into the *heuristic motives* of certain steps. The demonstrative presentation follows the accepted manner, usual since Euclid; the heuristic presentation is extremely unusual in print. Yet an alert teacher can use both manners of exposition. In fact, he could construct, if need be, a third presentation that is between the two with due regard to the available time, to the interest of his students, to all the conditions under which he works.[6]

---

[6] For a presentation intermediate between sect. 4 and sect. 6 see G. H. Hardy, J. E. Littlewood, and G. Pólya, *Inequalities*, p. 249–250.

This book is principally addressed to students desiring to develop their own ability, and to readers curious to learn about plausible reasoning and its not so banal relations to mathematics. The interests of the teacher are not neglected, I hope, but they are met rather indirectly than directly. I hope to fill that gap some day. In the meantime, I reiterate my hope that this book, as it is, may be useful to some teachers, at least to those teachers who have some genuine experience in problem-solving. The trouble is that there are so few teachers of mathematics who have such experience. And even the best School of Education has not yet succeeded in producing the marvellous teacher who has such an excellent training in teaching methods that he can make his students understand even those things that he does not understand himself.

### EXAMPLES AND COMMENTS ON CHAPTER XVI

1. *To the teacher: some types of problems.* This book is designed to serve various categories of readers: those who wish to understand guessing, those who wish to learn guessing, and those who wish to teach guessing. The reader of the last category is seldom directly addressed, but an alert teacher could learn something both from the examples presented in this book and from the manner of presentation. He could see, for instance, that there are manners of proposing a problem very different from the usual manner. I wish to point out a few types of problems, the "guess-and-prove" problems, the "test-consequences" problems, the "you-may-guess-wrong" problems, and the "small-scale-theory" problems. All these types could be used by an alert teacher to challenge his more intelligent students and to relieve the monotony of routine problems which fill the textbooks.

*Guess and prove.* Mathematical facts are first guessed and then proved, and almost every passage in this book endeavors to show that such is the normal procedure. If the learning of mathematics has anything to do with the discovery of mathematics, the student must be given some opportunity to do problems in which he first guesses and then proves some mathematical fact on an appropriate level. Still, the usual textbooks do not offer such an opportunity: ex. 1.2, 5.1, 5.2, 7.1–7.6 (and many others) do.

*Test consequences.* Philosophers and non-philosophers disagree about almost everything touching induction, but there is little doubt that the most usual inductive procedure consists in examining a general statement by testing its particular consequences. This inductive procedure is of daily use in mathematical research, and could be of daily use in the classroom with real benefit to the students. See sect. 12.2 and ex. 12.3–12.6. Cf. ex. 6.

*You may guess wrong.* You should acquire some experience in guessing. You should know from personal contact with the real thing that guesses may be respectable, that guesses may go wrong, and that even your own quite respectable guesses may go wrong. For such experience, solve ex. 11.1–11.12.

*Small scale theory.* On almost every page of this book some relatively elementary problem is discussed so that the discussion should shed some light upon a point that may arise in connection with other, not so elementary, problems. There is a reason to prefer such "small scale research": a less elementary problem could show the point in question on a more impressive scale, but it would demand a much longer explanation and much more preliminary knowledge. It is not too easy to "reduce the scale": elementary problems that show clearly enough the relevant features of plausible, or inventive, reasoning may be hard to find. It is also possible, but still less easy, to devise elementary problems to illustrate the scientist's activity in constructing a theory. The following ex. 2, 3, and 4 offer such "small-scale-theory problems"; ex. 5 and 6 are somewhat similar.

**2.** A quadrilateral is cut into four triangles by its two diagonals. We call two of these triangles "opposite" if they have a common vertex but no common side. Prove the following statements:

(a) The product of the areas of two opposite triangles is equal to the product of the areas of the other two opposite triangles.

(b) The quadrilateral is a trapezoid if, and only if, there are two opposite triangles equal in area.

(c) The quadrilateral is a parallelogram if, and only if, all four triangles are equal in area.

**3.** (a) Prove the following theorem: A point lies inside an equilatera triangle and has the distances $x$, $y$, and $z$ from the three sides, respectively; $h$ is the altitude of the triangle. Then $x + y + z = h$.

(b) State precisely and prove the analogous theorem in solid geometry concerning the distances of an inner point from the four faces of a regular tetrahedron.

(c) Generalize both theorems so that they should apply to any point in the plane or space, respectively (and not only to points inside the triangle or tetrahedron). Give precise statements, and also proofs.

**4.** Consider the following four propositions, (I)–(IV), which are not necessarily true.

(I) If a polygon inscribed in a circle is equilateral it is also equiangular.

(II) If a polygon inscribed in a circle is equiangular it is also equilateral.

(III) If a polygon circumscribed about a circle is equilateral it is also equiangular.

(IV) If a polygon circumscribed about a circle is equiangular it is also equilateral.

(a) State which of the four propositions are true and which are false, giving a proof of your statement in each case.

(b) If, instead of general polygons, we should consider only quadrilaterals which of the four propositions are true and which are false?

(c) What about pentagons?

(d) Could you guess, or perhaps even prove, more comprehensive statements? They should explain your observations (b) and (c).

**5.** Let $\alpha$, $\beta$, and $\gamma$ denote the angles of a triangle. Show that

(a) $\sin \alpha + \sin \beta + \sin \gamma = 4 \cos \dfrac{\alpha}{2} \cos \dfrac{\beta}{2} \cos \dfrac{\gamma}{2}$,

(b) $\sin 2\alpha + \sin 2\beta + \sin 2\gamma = 4 \sin \alpha \sin \beta \sin \gamma$,

(c) $\sin 4\alpha + \sin 4\beta + \sin 4\gamma = -4 \sin 2\alpha \sin 2\beta \sin 2\gamma$.

**6.** Consider the frustum of a right pyramid with square base. Call "midsection" the intersection of the frustum with a plane parallel to the base and the top and at the same distance from both. Call "intermediate rectangle" the rectangle of which one side is equal to a side of the base and the other side is equal to a side of the top.

Four different friends of yours agree that the volume of the frustum equals the altitude multiplied by a certain area, but they disagree and make four different proposals regarding this area:

(I) the midsection

(II) the average of the base and the top

(III) the average of the base, the top, and the midsection

(IV) the average of the base, the top, and the intermediate rectangle.

Let $h$ be the altitude of the frustum, $a$ the side of its base, and $b$ the side of its top. Express each of the four proposed rules in mathematical notation, decide whether it is right or wrong, and prove your answer.

**7.** *Qui nimium probat, nihil probat.* That is, if you prove too much, you prove nothing. I cannot tell in which sense the inventor of this classical saying intended to use it, but I wish to explain a meaning of the sentence that I found extremely helpful when I started doing some work in mathematics, and ever since. The sentence reminds me of one of the most useful signs by which we can judge the workability of a plan of the solution.

Here is the situation: you wish to prove a proposition. This proposition consists of a conclusion and a hypothesis which has several clauses, and you know that each one of these clauses is necessary to the conclusion, that is, none of them can be discarded without rendering the conclusion of the proposition invalid. You have conceived a plan of the proof, and you are weighing the chances of your plan. If your plan does not bring into play all the clauses, you have to modify your plan or reject it: if it would work as it is and prove the conclusion, although it leaves aside this or that clause of the hypothesis, it would prove too much, that is, something false, and so it would prove nothing.

I said that your plan should bring into play those clauses. I mean that mere lip-service is not enough, just mentioning them does not do: your plan should provide for essential use of each clause in the proof. The framework intended to support the conclusion cannot stand up unless it has a solid foothold in each clause of the hypothesis.

It may be very difficult to devise a plan that duly brings into play all the clauses of the hypothesis. Therefore, if a plan promises to catch all those clauses, we greet it with relief: there is an excellent sign, a strong indication that the plan may work.

For corresponding remarks on the solution of "problems to find" all the data of which are necessary see sect. 8 (3), (4), (5).

If you prefer a French sentence to a Latin saying, here is one: "La mariée est trop belle"; the bride looks too good. I do not think that I need to amplify this; after the foregoing the reader can picture all the details to himself.

**8.** *Proximity and credibility.* How far away is the solution? How much remains to be done? Such questions lie heavily on the mind of the student who has to finish his task within a set time, but they are present in the mind of every problem-solver.

(1) We are even able to answer such questions to a certain extent, not precisely, of course, but rather correctly on the average, I am inclined to believe.

For instance, let us look back at figs. 16.1–16.5 and the process of solution that they represent. The problem-solver may feel that fig. 16.3 is much closer to the solution than fig. 16.2; and he may feel that the solution is within easy reach once he has arrived at fig. 16.4.

In judging the proximity of the solution, we may rely on inarticulate feelings, or on more distinct signs. Any sign that indicates that our plan of the solution might work may be interpreted also as a sign of progress toward the solution, and help us to estimate the distance that we still have to go.

(2) Let us consider the solution of a "problem to prove." The aim is to prove (or disprove) a certain theorem. The problem-solver may trust, or distrust, the theorem at which he works. Yet, if he is any good as a problem-solver, he should be prepared to revise his beliefs. And so the questions, "Is the theorem credible? How credible is it?" are ever present in his mind, although sometimes more, sometimes less, in the foreground. If anything new comes in sight, he has two questions: "Does it render the theorem more, or less, credible?" "Does it bring the solution nearer or not?" His attention may be, of course, so absorbed by the new fact that he finds no time to formulate either question in words. He may also give the answer to himself without words. Even if he says "Good sign" or "Bad sign," he will scarcely take the trouble to express at length what he means. Is it a sign of the *proximity* of the solution, or a sign of the *credibility* of the theorem? Yet, if he is a good problem-solver, he knows well enough the important difference between proximity and credibility, and this difference will show up in his work, in his handling of the problem.

(3) Recalling a name once known, but now forgotten, is a task that is simpler than but somewhat similar to a mathematical problem. We can often observe people trying to recall a name, and we could learn a few interesting things from such observations.

In a conversation, your friend wants to tell you a name (the name of a shop, of an acquaintance, or of an author, perhaps). He gets stuck and you hear him say: "Just a minute and I shall get it," or "Wait a little, it may take a short while till I recall it," or "How stupid of me, but I am not able to recall it now, although I am sure it will occur to me in a few hours, perhaps tomorrow morning." Obviously, your friend tries to judge the proximity of that name, he tries to measure a sort of "psychological distance." I would surmise that his predictions will turn out about right, that his evaluation of that "psychological distance" is roughly correct.

With a view to a possible comparison with problem-solving, it is also interesting to note that a person who is not able to recall a name fully may be able to recall it partially or, better expressed perhaps, he may be able to recall certain features of the name. You may hear your friend say: "The name is not Battenberg—he is not the husband of the Queen, after all—but it is a German name, of three syllables, very similar to Battenberg." And (I have observed such cases) your friend may be completely right in all these particulars, although the correct name will occur to him only a few days later.

Quite similarly, a mathematician, although he has not yet solved his problem, may foresee certain features of the solution quite reliably.

Those aspects of problem-solving that are the most interesting for the future mathematician or the teacher are not readily accessible to the usual methods of experimental psychology. Perhaps, recalling a name, a process in some respects similar to solving a mathematical problem, could be brought more easily within the scope of psychological experiments.

9. *Numerical computation and plausible reasoning.* Although numbers are often regarded as symbols of the highest attainable certainty, the results of numerical computation are by no means certain; they are only plausible. Numerical computation depends on plausible reasoning in many ways.

(1) You have to do a long numerical computation. The final result is attained in a sequence of steps. You have a very good chance to do correctly any single step, yet there are many steps, there is a possibility of a mistake at each step, and the final result may be wrong. How can you protect yourself from error?

Compute the desired number twice by procedures as different as possible. If the two computations yield different results, it is certain that at least one of them is wrong, but both may be wrong. If the two computations agree, it is by no means certain that the result twice obtained is correct, but it may be correct, and the agreement is an indication of its correctness. The weight of this indication depends on the difference between the two procedures used.

For example, the weight is very small if, having done the computation once, you repeat it immediately afterwards, without any change in the method: with the first computation still fresh in your mind and in your fingers, you can easily stumble a second time at the same place where you stumbled the first time. To repeat the computation after a while is a little better, to let another person do it the second time is considerably better, to do the second computation by a very different method is still better.

In fact, if two quite different procedures arrive at the same result, we have only two obvious conjectures: the result may be correct, or the agreement may be due to chance. If the probability of an agreement by mere chance is very small, the second of the two rival conjectures is correspondingly unlikely, we are inclined to reject it, and so we are disposed to place more confidence in the first conjecture, that is, we may rely more on the correctness of the result.

The more the procedures of the two computations differ, the more realistic is the simplest evaluation of the probability of their agreement: the probability is $10^{-n}$ that the two computations arrive by mere chance at the same $n$ figures; cf. sect. 14.9 (3), ex. 14.11, but also ex. 14.32, and ex. 12.

In large-scale computation it is good practice to render a chance coincidence still more improbable by the introduction of *multiple controls*. Two computations are performed by methods as different as conveniently possible, yet so that they should agree, if correct, not only in the final result, but also in several intermediate results. As the agreement increases, it becomes increasingly difficult to attribute it to mere chance although, of course, chance can never be fully excluded and the result can never be fully guaranteed.

(2) In the foregoing we tacitly assumed that the two computations of which we compare the results are known to be strictly equivalent theoretically. Yet in applied mathematics we often have to work with approximations, and we may compare numerical results which need not agree completely even if all arithmetical operations involved are faultless; we just hope that they will agree "roughly." Moreover, the theory of the method of approximation with which we work may be very imperfectly known. Under such circumstances, of course, the scope of plausible reasoning is even wider and such reasoning is more hazardous. Cf. ex. 11.23.

(3) Two mathematicians, $A$ and $B$, investigated the same set of nine combinatorial problems. We need not know the subject of these problems (they deal with the hypercube in four dimensions) but it is important to know that they are arranged in order of increasing difficulty. The first two problems are trivial, the third problem is easy, the fourth is less easy, then they become increasingly more complex, and the last problem is the hardest.

Both $A$ and $B$ solved the problems, but their results did not quite agree. Here are their solutions for the nine problems:

$$A: 1 \quad 1 \quad 4 \quad 6 \quad 19 \quad 27 \quad 47 \quad 55 \quad 78$$
$$B: 1 \quad 1 \quad 4 \quad 6 \quad 19 \quad 27 \quad 50 \quad 56 \quad 74.$$

That is, $A$ and $B$ agree about the first six problems, which are easier, but disagree concerning the last three problems, which are harder. In fact, they followed very different methods.

$A$ attacked each of the nine problems independently of the others. His method for each problem is somewhat different, and as he proceeds to harder problems, also his method becomes increasingly complex.

$B$ attacked the problems by a uniform method. His work consists of two parts. The first part, which is more difficult, is a common preparation for the solution of all nine problems. The second part, which is more routine, applies the result of the first part to each single problem according to a uniform rule. Treated with $B$'s method, the problems appear to differ in difficulty much less than with $A$'s method.

It seems to me that the situation described gives us a reasonable ground to trust $B$'s solution more than $A$'s solution.

Since the two methods, which are very different, agree with regard to the first six problems, and these problems are easier anyhow, there is good ground to believe that the solution of these problems is correct. About the solution of the first three problems there is no doubt.

Since the result of the first part of $B$'s work is verified by its consequences in 3 cases out of 9, and is presumably verified in 3 more cases (it is neither verified nor refuted in the 3 remaining cases) there is good ground to trust this result.

If, however, the first part of $B$'s work were correct (as it presumably is) he could only err in the second, more routine, part in dealing with the last three problems. Yet $A$ had the greatest difficulty in dealing with them. And so $A$ appears to have more chances of error than $B$.[7]

The case just discussed is rather special, but it shows that there are further possibilities in exploring the patterns of plausible reasoning. For example, it may be a rewarding task to express the plausible argument just presented in formulas of the Calculus of Probability as fittingly as possible.

**10.** You have to add a column of ten six-place numbers, beginning so, for instance:

$$1 \quad 5 \quad 9 \quad 6 \quad . \quad 0 \quad 3$$
$$1 \quad 6 \quad 4 \quad 6 \quad . \quad 0 \quad 7$$
$$1 \quad 7 \quad 8 \quad 1 \quad . \quad 1 \quad 0$$
$$\cdot \quad \cdot \quad \cdot \quad \cdot \quad \cdot \quad \cdot \quad \cdot$$

Describe various procedures to do it.

---

[7] See G. Pólya, Sur les types des propositions composées, *Journal of Symbolic Logic*, vol. 5 (1940), p. 98–103.

**11.** Call "elementary step" the addition of a one-digit number that you see written to a two-digit number that you have in mind; include, however, the possibility (which renders the step easier) that the second number is also written, or that it has one digit only. How many elementary steps are required to perform the addition mentioned in ex. 10 in the most usual manner?

**12.** In doing a computation, you obtain first two nine-place numbers, and then the final result as their difference, which turns out to be a three-place number. Another procedure of computation obtains the same three-place result as the difference of two seven-place numbers. Compute the probability that such an agreement is due to chance, using the formula given in ex. 9 (1).

**13.** *Formal demonstration and plausible reasoning.* You have to do a long numerical computation. The final result is attained in a sequence of steps and must be correct if each step is correct. Each single step (as the addition 3 + 7 or the multiplication 3 × 7) is so simple and familiar that you cannot slip under somewhat favorable circumstances, when your attention is "undivided." Still, like everyone else, you are liable to make mistakes in a computation. After having performed the successive steps quite carefully, you should not trust the final result without checking it.

You go through a lengthy mathematical demonstration. The demonstration is supposed to be decomposed into steps each of which you can check perfectly, and the final conclusion must be correct if each step is correct. Yet you may make mistakes like everybody else. After having checked the successive steps quite carefully, can you trust the final conclusion? Not more, and perhaps less, than the final result of a long computation.

In fact, a mathematician who has checked the details of a demonstration step by step and has found each step in order may be still dissatisfied. He needs something more to satisfy himself than the correctness of each detail. What?

He wants to *understand* the demonstration. After having struggled through the proof step by step, he takes still more trouble: he reviews, reworks, reformulates, and rearranges the steps till he succeeds in grouping the details into an understandable whole. Only then does he start trusting the proof.

I would not dare to analyze what constitutes "understanding." Some people say that it is based on "intuition" and they credit intuition with perceiving the whole and grouping the details into a well-arranged harmonious whole. I would not dare to contradict this, although I have some misgivings.[8] Yet I wish to call attention to a point strongly suggested by the examples and discussions in this book.

---

[8] The meaning of intuition and its rôle in grouping the details is usually not too well explained. It is remarkable, however, that Descartes, with whom the modern usage of the term "intuition" originates, explains both points rather impressively, in the Third and the Seventh of his Rules for the Direction of the Mind. See *Oeuvres*, edited by Adam and Tannery, vol. 10, p. 368–370 and 387–388.

Some practice may convince us that analogy and particular cases can be helpful both in finding and in understanding mathematical demonstrations. The general plan, or considerable parts, of a proof may be suggested or clarified by analogy. Particular cases may suggest a proof (see, for instance, sect. 3.17); on the other hand, we may test an already formulated proof by observing how it works in a familiar or critical particular case. Yet analogy and particular cases are the most abundant sources of plausible argument: perhaps, they not only help to shape the demonstrative argument and to render it more understandable, but also add to our *confidence* in it. And so we are led to suspect that a good part of *our reliance on demonstrative reasoning may come from plausible reasoning*.

# SOLUTIONS

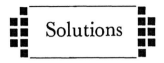

## Solutions

**1.** The resemblance to the table of sect. 3.12 (partitions of space) can be regarded as closer. All the tables mentioned, except that in sect. 3.1 that lists polyhedra, are concerned with inductive evidence supporting a proposition $A$ of a particular nature: $A$ asserts that a certain statement $S_n$, the meaning of which depends on a variable integer $n$, is true for $n = 1, 2, 3, \ldots$. Inductive examination of such a proposition $A$ naturally proceeds in a certain order: we test $S_1$ first, then $S_2$, then $S_3$, and so on. This order is visible in the arrangement of the tables. Yet if we examine a proposition concerned with polyhedra, as in sect. 3.1, there is no such "natural" order. We may begin our investigation with the tetrahedron which, from various viewpoints, may be regarded as the "simplest" polyhedron. Yet which polyhedron should we examine next? There is no convincing ground to regard some polyhedron as the "second in simplicity," another as the third, and so on.

**2.** If the cases 101 and 301 appear "further removed" from the already verified cases 1, 2, 3, . . . 20 than the cases 21 and 22 (and in Euler's investigation they do appear so), it seems reasonable (and it is certainly in harmony with the pattern introduced in sect. 2) to attach more weight to the verification of the cases 101 and 301 than to that of the cases 21 and 22.

**3.** (1) $c = p$: the triangle degenerates into a straight line; $A = 0$.

(2) $c > p$ makes $A$ imaginary: there is no triangle with $c > p$.

(3) $a = b = c$: the triangle is equilateral and $A^2 = 3a^4/16$, which is correct.

(4) $a^2 = b^2 + c^2$: the triangle is a right triangle and

$$16A^2 = (a + b + c)(b + c - a)(a - b + c)(a + b - c)$$
$$= [(b + c)^2 - a^2][a^2 - (b - c)^2]$$
$$= (2bc)^2$$

or $A^2 = b^2c^2/4$, which is correct.

(5) $b = c = (h^2 + a^2/4)^{1/2}$: the triangle is isosceles, with height $h$, and

$$16A^2 = (a + 2b)(2b - a)a^2$$
$$= (4b^2 - a^2)a^2$$
$$= 4h^2a^2,$$

which is correct.

(6) The dimension is correct.

(7) The expression for $A^2$ is symmetric in the three sides $a$, $b$, and $c$, as it should be.

**4.** (1) $d = 0$: the quadrilateral becomes a triangle, the asserted formula reduces to Heron's formula, ex. 3.

(2) $d = p$: the quadrilateral degenerates into a straight line; $A = 0$.

(3) $d > p$ makes $A$ imaginary: there is no quadrilateral with $d > p$.

(4) $a = b = c = d$ yields $A^2 = a^4$, which is correct for a square, but incorrect (too large) for a rhombus: the square is inscribable in a circle, the rhombus is not.

(5) $c = a$, $d = b$ yields $A^2 = (ab)^2$ which is correct for a rectangle but incorrect (too large) for an oblique parallelogram: the rectangle is inscribable in a circle, the oblique parallelogram is not.

(6) In the foregoing cases (4) and (5) the asserted formula attributes a value too large to the area of non-inscribable quadrilaterals: this agrees with sect. 10.5 (2) and 10.6 (3).

(7) The asserted formula gives the dimension of $A$ correctly.

(8) According to the asserted formula $A$ is symmetric in the four sides $a$, $b$, $c$, and $d$: an inscribed quadrilateral remains inscribed in the same circle if two neighboring sides are interchanged. (Consider the four isosceles triangles with common vertex at the center of the circle of which the bases are the four sides.)

These remarks do not prove the proposed formula, of course, but they render it very plausible, according to the pattern exhibited in sect. 2. For a proof see ex. 8.41.

**5.** (1) $a = b = c = e = f = g$: the tetrahedron is regular; $V = 2^{1/2}a^3/12$.

(2) $e^2 = b^2 + c^2$, $f^2 = c^2 + a^2$, $g^2 = a^2 + b^2$: the tetrahedron is "tri-rectangle," that is, the three edges $a$, $b$, and $c$, starting from the same vertex, are perpendicular to each other; $V = abc/6$.

(3) $e = 0$, $b = c$, $f = g$: the tetrahedron collapses, becomes a plane figure, a triangle; $V = 0$.

(4) $e = a$, $f = b$, $g^2 = c^2 = a^2 + b^2$: the tetrahedron collapses, becomes a rectangle with sides $a$ and $b$; $V = 0$.

(5) A particular case more extended than (4): the tetrahedron becomes a plane quadrilateral, with sides $a$, $\overline{b}$, $e$, $f$ and diagonals $c$, $g$. Then $V = 0$ yields a relation between the sides and the diagonals of a general quadrilateral which also can be verified directly, although less easily.

(6) The dimension is correct.

(7) The expression of $V$ is *not* symmetric in all six edges, but need not be: the three edges $a$, $b$, $c$ which start from the same vertex do not play the same rôle as the three edges $e$, $f$, $g$ which include a triangle (a face). We can transform, however, the proposed expression into the following:

$$144V^2 = a^2e^2 (b^2 + f^2 + c^2 + g^2 - a^2 - e^2)$$
$$+ b^2f^2 (c^2 + g^2 + a^2 + e^2 - b^2 - f^2)$$
$$+ c^2g^2 (a^2 + e^2 + b^2 + f^2 - c^2 - g^2)$$
$$- e^2f^2g^2 - e^2b^2c^2 - a^2f^2c^2 - a^2b^2g^2.$$

The first three lines correspond to the three pairs of opposite edges, the four terms in the last line to the four faces of the tetrahedron: the new algebraic form exhibits all the symmetry (interchangeability) of the data due to the geometric configuration. By the way, the new form (the correctness of the algebraic manipulation that yielded it) can also be checked by the particular cases (1), (2), (3), (4).

**6.** (1) We find for

$$s_1, \quad s_3, \quad s_5, \quad q, \quad r$$

the numerical values

$$6, \quad 36, \quad 276, \quad 11, \quad 6$$

respectively, which verify both formulas.

(2) More generally, let

$$a + b + c = s_1 = p = 0$$

Then

$$s_3 = -3ab(a + b),$$
$$s_5 = -5ab(a + b)(a^2 + ab + b^2),$$
$$q = -a^2 - ab - b^2,$$
$$r = -ab(a + b)$$

and both formulas are verified since

$$-\frac{3s_5}{5s_3} = -a^2 - ab - b^2, \quad \frac{s_3}{3} = -ab(a + b).$$

(3) More generally, let $c = 0$. A longer straightforward computation verifies both formulas.

(4) More generally, let $b + c = 0$. Then $s_1 = a$, $s_3 = a^3$, $s_5 = a^5$: denominator *and* numerator vanish in the expressions proposed for $q$ and $r$.

For proof and generalization see *Journal des math. pures et appliquées*, ser. 9, vol. 31 (1952) p. 37–47.

**7.** The statement $B_4$ (with $r = h = 0$) is contained as a particular case both in $B_2$ ($r = 0$) and in $B_3$ ($h = 0$). Hence, if either $B_2$ or $B_3$ is true, $B_4$ must be true too. If we have this clearly in mind, the verification of $B_4$, coming after that of $B_2$ or $B_3$, does not convey to us new information. And, where there is no new information, there can be no new evidence, I should think. Still, it is desirable to observe $B_4$; it rounds off the picture.

**8.** Closer attention to the derivation in No. 9–13 of Euler's memoir in sect. 6.2 shows that $C_1^*, C_2^*, \ldots C_{20}^*$ follow from $C_1, C_2, \ldots C_{20}$ mathematically: therefore, the verification of $C_1^*, C_2^*, \ldots C_{20}^*$ did not really yield new information or evidence, but that of $C_{101}^*$ and $C_{301}^*$ did.

**9.** The pattern agrees essentially with the pattern that will be introduced in ex. 11; see comments there.

**10.** If $8n + 3 = w^2 + 2p$, the integer $w$ is necessarily odd. Therefore, $w^2$ is of the form $8n + 1$ and so $p$ of the form $4n + 1$. Euler proved that

$$p = u^2 + v^2;$$

of the two integers $u$ and $v$, one must be odd and the other even. Hence

$$2p = 2u^2 + 2v^2 = (u + v)^2 + (u - v)^2.$$

Now, $w$, $u + v$, and $u - v$ are odd. Let

$$w = 2x - 1, \; u + v = 2y - 1, \; u - v = 2z - 1$$

and we obtain

$$8n + 3 = (2x - 1)^2 + (2y - 1)^2 + (2z - 1)^2$$

or

$$n = \frac{x^2 - x}{2} + \frac{y^2 - y}{2} + \frac{z^2 - z}{2}.$$

**13.** Yes, it should, in harmony with the pattern introduced in sect. 6. No solution: **11, 12, 14**.

### SOLUTIONS, CHAPTER XIII

**1.** The following pattern is generally applicable:

$$A \text{ implies } B$$
$$B \text{ false}$$
$$\overline{\phantom{A \text{ implies } B}}$$
$$A \text{ false}$$

Apply it in substituting non-$B$ for $B$. You obtain

$$A \text{ implies non-}B$$
$$\text{non-}B \text{ false}$$
$$\overline{\phantom{A \text{ implies non-}B}}$$
$$A \text{ false}$$

We have noted in sect. 4 (5) the equivalence:

"$A$ implies $B$"   eq.   "$A$ incompatible with non-$B$".

Also this equivalence is generally applicable. Apply it in substituting non-$B$ for $B$. You obtain

"$A$ implies non-$B$"   eq.   "$A$ incompatible with $B$"

We took here for granted that the negation of non-$B$ is $B$ (since the negation of $B$ is non-$B$) and, therefore, we substituted $B$ for non-(non-$B$). In substituting $B$ for $A$ in the equivalence noted at the end of sect. 4 (3), we obtain also

"non-$B$ false"   eq.   "$B$ true".

Substituting for the two premises of the last pattern the corresponding equivalent statements, displayed on the right-hand sides of the two foregoing equivalences, we obtain:

$$A \text{ incompatible with } B$$
$$B \text{ true}$$
$$\overline{\qquad\qquad\qquad}$$
$$A \text{ false}$$

This is, in fact, the demonstrative pattern of sect. 3.

**2.** Assume that the following pattern is generally applicable:

$$A \text{ implies } B$$
$$B \text{ true}$$
$$\overline{\qquad\qquad\qquad}$$
$$A \text{ more credible}$$

Apply it in substituting non-$A$ for $A$ and non-$B$ for $B$. You obtain:

$$\text{non-}A \text{ implies non-}B$$
$$\text{non-}B \text{ true}$$
$$\overline{\qquad\qquad\qquad}$$
$$\text{non-}A \text{ more credible}$$

Collect the following three equivalences:

"non-$A$ implies non-$B$"   eq.   "$B$ implies $A$"

"non-$B$ true"   eq.   "$B$ false"

"non-$A$ more credible"   eq.   "$A$ less credible".

The first has been derived in sect. 4 (5). The second has been mentioned in sect. 4 (3), in another notation, with $A$ for $B$. The third has been stated (invented just for the present purpose) in sect. 5. Substitute for the premises and the conclusion of the last considered pattern the three corresponding equivalent statements just displayed. You obtain:

$$B \text{ implies } A$$
$$B \text{ false}$$
$$\overline{A \text{ less credible}}$$

Except for a slight change in the wording (or notation) this is, in fact, the heuristic pattern introduced in sect. 2.

3. Start from the same pattern as in ex. 2. Substitute non-$B$ for $B$ (as in ex. 1). You obtain so:

$$A \text{ implies non-}B$$
$$\text{non-}B \text{ true}$$
$$\overline{A \text{ more credible}}$$

Collect the following two equivalences:

"$A$ implies non-$B$"   eq.   "$A$ incompatible with $B$"

"non-$B$ true"   eq.   "$B$ false".

The first has been derived in ex. 1. The second is given, except for notation ($A$ instead of $B$) in sect. 4 (3). Substitute for the two premises of the pattern considered the equivalent statements just displayed. You obtain

$$A \text{ incompatible with } B$$
$$B \text{ false}$$
$$\overline{A \text{ more credible}}$$

This is, in fact, the heuristic pattern of sect. 3.

4. (a) At any rate $A$ implies $B$, where the statement $B$ is defined as follows: $B$. The letters in which the required nine-letter word is crossed by other words of the puzzle are chosen among the letters of the word TIREDNESS.

Let us interpret $B$ as restricted to the two places filled in the proposed diagram (end and third from the end). We should distinguish between two cases.

If we regard the solutions for the two crossing words as final, we found that $B$ is true and so we verified a consequence of the conjecture $A$. Therefore, we consider $A$ as more credible according to the fundamental inductive pattern (sect. 12.1).

If, however, we regard the solutions for the crossing words only as tentative, we rendered $B$ only more credible. Therefore the shaded version of the fundamental inductive pattern, defined in sect. 6, is appropriate and, of course, the evidence for $A$ is weaker than in the former case.

(b) DISSENTER.

5. The pattern is that of sect. 13 (5):

The Factum Probans is readily credible or understandable under assumption of the Factum Probandum.

The Factum Probans is (much) less readily credible or understandable without the assumption of the Factum Probandum.

The Factum Probans itself is proved.

This renders the Factum Probandum more credible.

The presentation that we are considering here appears to be more appropriate to a court case, but the presentation of sect. 10 is more suitable to show the connection with the most usual form of inductive reasoning in the physical sciences or in mathematical research.

6. We have to treat the charge as a conjecture:
A. The down payment for the official's car came from the contractor's pocket.
We have to regard as a fact:
B. The withdrawal from the contractor's account of an amount ($875) equal to the down payment on the official's car, the date of the withdrawal preceding the date of the payment by two days.
B with A is much more readily understandable than B without A: if the withdrawal was not connected with the following payment, the exact coincidence of the amounts and the near coincidence of the dates has to be ascribed to mere chance. Such chance is not impossible, but improbable. The strength of the evidence hinges on this point. The pattern of sect. 13 (5) seems to fit very well.

7. Let us call poor Mrs. White and Mr. Black the "defendants." (They cannot answer charges, but at least they cannot be cross-examined by Mrs. Green.) Mrs. Green's accusation, stripped from her pious circumlocutions, is, of course,
A. The defendants live in double adultery.
We accept as a fact
B. The defendants had a long conversation in the obscurity over the fence.
This fact, unfortunately, yields some circumstantial evidence for A according to a reasonable pattern (sect. 13 (5)) if the following two premises are accepted:

B with A readily credible,

B without A less readily credible.

There is no use trying to shake the faith of Suburbia in the first premise, I am afraid. Yet even some Suburbans may see that, on that famous evening, the defendants might well have discussed the lease in which both had a legitimate interest. This makes $B$ about as credible without $A$ as with $A$, knocks out the second premise, and explains away the alleged circumstantial evidence. This argument may be of no avail against Mrs. Green's gossip, although it seems to me reasonable and typical. In arguing against some piece of circumstantial evidence, lawyers very often try to knock out just that second premise of the pattern in a "rebuttal."

**8.** We have to consider another contention or conjecture:

$B$. The defendant was well acquainted with the victim three years before the crime.

It would be too much to say that $A$ is implied by $B$, but a weaker statement in this direction is obviously justified:

$A$ is rendered more credible by $B$.

Now $C$ does not prove $B$, but certainly renders $B$ more credible.

From the two displayed premises we are tempted to draw the conclusion: $A$ more credible.

This seems to suggest a new pattern:

$$A \text{ more credible with } B$$
$$B \text{ more credible}$$
$$\overline{A \text{ more credible}}$$

The first premise of this pattern is weaker than the corresponding premise of the pattern in line 2, column (2) of Table I, the second premise is the same: the conclusion must also be weaker. [See, however, ex. 15.2.]

By the way, the size of the firm matters: if the firm is small, $B$ is rendered much more credible than in the opposite case.

No solution: Ex. **9** through **20**.

### SOLUTIONS, CHAPTER XIV

**1.** (a) Follows from $r_r + s_r = r_s + s_s = 1$.

(b) $r_r - r_s = s_s - s_r > 0$.

**2.** With the notation of ex. 1 appropriately adapted, $r_r - r_s = s_s - s_r < 0$.

**3.** $N \binom{n}{s} p^s q^{n-s}$, with $N = 26306$, $n = 12$, $p = 1/3$, $q = 2/3$ where $s$ is the number in the same row in column (1).

**4.** Same expression as in the solution of ex. 3 with the same numerical values for $N$ and $n$ and the same meaning of $s$, but $p = 0.3376986$, $q = 1 - p$.

**5.** (a) $N \mu^s e^{-\mu}/s!$ with $N = 30$, $\mu = 10$, and $s$ the corresponding entry in column (1).

(b) $- - - + + - - + - - - + + + - + - + + +$ .

**6.** $6^{-3n}$.

**7.** The compound event that consists in casting six spots with each of the three dice five times in uninterrupted succession and has the probability $6^{-15}$ on the hypothesis of fair dice: $n = 5$ in ex. 6.

**8.** (a) $2 \int\limits_{\alpha}^{\infty} y\, dx = 1.983 \cdot 10^{-7}$

where $y = (2\pi)^{-1/2}\, e^{-x^2/2}$, $\alpha = 1377.5\ (pqn)^{-1/2}$,

$$p = 1/3, \qquad q = 2/3, \qquad n = 315672.$$

(b) $\int\limits_{\alpha}^{\beta} y\, dx = 1.506 \cdot 10^{-3}$

with $\beta = -\alpha = 0.5\ (pqn)^{-1/2}$ and $y$, $p$, $q$, and $n$ have the same meaning as under (a). To compute the numerical values as exactly as they are given here the simplest current tables of the probability integral are not sufficient.

**9.** For finding no defective article in the sample, just one defective, precisely two defectives, . . . precisely $c$ defectives, respectively.

**10.** $d^2a/dp^2 = 0$ if the derivative of log $(da/dp)$ vanishes, which yields the equation

$$\frac{c}{p} - \frac{n-1-c}{1-p} = 0$$

and hence the value given at the end of sect. 8 (1).

**11.** The required probability is $10^{-n}$, *provided* that we accept one or the other of the following assumptions:

(I) All possible sequences of $n$ figures are equally probable. (There are $10^n$ such sequences.)

(II) The various figures in the sequence are mutually independent, and for each figure the ten possible cases 0, 1, 2, . . . 9 are equally probable. (Apply the rule of sect. 3 (5) repeatedly.)

Both assumptions look "natural," but no such assumption is logically binding: the answer $10^{-n}$, although *strongly suggested*, is *not* mathematically determined.

**12.** Suppose that $l$ different letters can be drawn from a bag, with probabilities $p_1, p_2, \ldots p_l$, respectively. We have two such bags and we pick out a letter from each: the probability of a coincidence is $p_1^2 + p_2^2 + \ldots + p_l^2$. In the case of Hypothesis II, $l = 17$ and $p_1, \ldots p_{17}$ can be found by actual count.

**13.** In both cases

$$\sum_{k=n}^{10} \binom{10}{k} p^k q^{10-k} = 1 - \sum_{k=0}^{n-1} \binom{10}{k} p^k q^{10-k}$$

where $q = 1 - p$; (a) $p = 0.0948$, (b) $p = 1/26 = 0.03846$.

**14.** In both cases $np$ with $p = 0.0948$; (a) $n = 450$, (b) $n = 90$.

**15.** $(npq)^{1/2}$ for $n = 90$, $p = 0.0948$, $q = 1 - p$. The computation of the standard deviation 7.60 is based on a formula which is not found in the textbooks. With the notation of ex. 12 set

$$p = p_1^2 + p_2^2 + \ldots + p_l^2,$$

$$p' = p_1^3 + p_2^3 + \ldots + p_l^3,$$

$$\sigma^2 = n(n-1) \, w[p(1-p) + 2(n-2)(p' - p^2)]/2.$$

Then $n = w = 10$, $p = 0.0948$, $p' = 0.01165$ yield $\sigma = 7.60$.

**16.** We assume that the 60 trials with the coin are independent and apply the rule introduced in sect. 3 (5) repeatedly.

**17.** Generalize the proposed numerical table

| $s$ | $r$ | $n$ |
|-----|-----|-----|
| $s'$ | $r'$ | $n'$ |
| $S$ | $R$ | $N$ |

and interpret it as follows. There are $N = R + S = n + n'$ cards, among which $R = r + r'$ cards are red and $S = s + s'$ cards are black. The cards are distributed at random between two players; one receives $n$ cards, and the other $n'$ cards. What is the probability that the first player receives $r$ red cards and $s$ black cards, and the other player $r'$ red cards and $s'$ black cards? (Of course, $r + s = n$, $r' + s' = n'$.) The answer is, as is well known,

$$\frac{\binom{R}{r}\binom{S}{s}}{\binom{N}{n}} = \binom{n}{r} \frac{R(R-1)\ldots(R-r+1)\,S(S-1)\ldots(S-s+1)}{N(N-1)\ldots(N-n+1)}$$

$$= \frac{1}{N!} \frac{R!\,S!\,n!\,n'!}{r!\,s!\,r'!\,s'!}.$$

Only four quantities out of the 9 contained in the table can be arbitrarily given, the values of the remaining 5 follow from the relations written above. We take the numbers $n = 9$, $n' = 11$, and $S = 8$ as given (the number of patients receiving each treatment and the total number of fatal cases) from which $N = 20$ and $R = 12$ follow. Yet we take for $s' = 2, 1, 0$ in succession (number of fatalities with the second treatment $\leq 2$). From the formula, we compute the probability in each of these three cases (not more than 2 fatalities with the new treatment, 2, 1 or 0 black cards to Mr. Newman) and adding the probabilities for these mutually exclusive events we find:

$$\frac{12!\,8!\,9!\,11!}{20!}\left[\frac{1}{3!\,6!\,9!\,2!}+\frac{1}{2!\,7!\,10!\,1!}+\frac{1}{1!\,8!\,11!\,0!}\right]$$

$$=\left[\binom{12}{9}\binom{8}{2}+\binom{12}{10}\binom{8}{1}+\binom{12}{11}\binom{8}{0}\right]\bigg/\binom{20}{11}=\frac{335}{8398}.$$

**18.** By an obvious extension of the reasoning of sect. 3 (5) (three-dimensional analogue of fig. 14.2) the number of possible cases is $n^3$. All the favorable cases, that is, all the admissible solutions of the equation $Z = X + Y$ can be enumerated as follows:

$$2 = 1 + 1$$

$$3 = 1 + 2 = 2 + 1$$

$$4 = 1 + 3 = 2 + 2 = 3 + 1$$

$$\cdot \qquad \cdot \qquad \cdot \qquad \cdot$$

$$n = 1 + (n - 1) = 2 + (n - 2) = \ldots = (n - 1) + 1 \; .$$

Hence the number of favorable cases is

$$1 + 2 + 3 + \ldots + (n - 1) = n(n - 1)/2$$

and the required probability

$$\frac{n(n - 1)/2}{n^3} = \frac{n - 1}{2n^2}.$$

**21.** The probability that a sample of 38 from an infinite population contains 30 or more defectives is

$$\sum_{i=30}^{38}\binom{38}{i}p^i(1 - p)^{38-i} < \binom{38}{8}p^{30}(1 - p)^8\sum_{i=0}^{\infty}\left(\frac{p}{1 - p}\right)^i$$

$$= \binom{38}{8}\frac{(1 - p)^9}{1 - 2p}p^{30}$$

$$\sim 4.61 \times 10^{-23}$$

provided that $100p\% = 1\%$ of the population is defective.  The probability estimated is the likelihood of the daily's assertion in the light of the official's observation, computed under the simplest assumptions.

**22.** The probability required is

$$4 \left\{ \left( \frac{2}{12 \times 60} \right)^3 + 3 \left( \frac{2}{12 \times 60} \right)^2 \left( 1 - \frac{2}{12 \times 60} \right) \right\} \sim 0.0000924.$$

(Any one of the four clocks could be the one among the three agreeing clocks that shows the earliest time. The possibility that all four clocks show times less than 2 minutes apart accounts for the first term in the curly brackets.)

**23.** The integer $a$ takes one of the values

$$-n, \dots , -2, -1, 0, 1, 2 \dots n.$$

We assume that these $2n + 1$ values are equally probable, we make the corresponding assumption for $b$, $c$, $d$, $e$, and $f$, and assume also that $a$, $b$, $c$, $d$, $e$, and $f$ are mutually independent. Only now, after having given a precise meaning to "chosen at random," can we proceed to solve the problem.

There is just one solution if, and only if, $ad - bc \neq 0$. We can, and do, neglect $e$ and $f$: there are $(2n + 1)^4$ possible cases. We count *unfavorable* cases, distinguishing two possibilities.

(I) $a = 0$. Then $bc = 0$, $d$ arbitrary, and there are $(2n + 2n + 1)(2n + 1)$ cases.

(II) $a \neq 0$. Then $a$ can take $2n$ values, $b$ and $c$ are arbitrary, and $d$ is uniquely determined by $a$, $b$, and $c$: there are $2n(2n + 1)^2$ cases.

The probability required is

$$1 - \frac{(4n + 1)(2n + 1) + 2n(2n + 1)^2}{(2n + 1)^4}$$

$$= 1 - \frac{4n^2 + 6n + 1}{8n^3 + 12n^2 + (6n + 1)} > 1 - \frac{1}{2n}$$

and so it tends to 1 as $n$ tends to $\infty$. This gives another precise meaning to the statement: "a system of two equations with two unknowns has *in general* just one solution." Cf. sect. 11.3, ex. 11.16.

**24.** The verification of the O gives more confidence in TOWER than the verification of the E. Since O is less frequent than E, the occurrence of O in the crossing word is less easily interpreted as chance coincidence than that of the E.

**25.** Tabulating the differences between the successive numbers of each column, and then again the differences of the differences (the so-called "second differences") we obtain:

| I | | II | |
|---|---|---|---|
| 1005 | | 1004 | |
| | 28 | | 34 |
| 1033 | + 14 | 1038 | 0 |
| | 42 | | 34 |
| 1075 | — 11 | 1072 | 0 |
| | 31 | | 34 |
| 1106 | — 5 | 1106 | — 1 |
| | 26 | | 33 |
| 1132 | + 21 | 1139 | + 1 |
| | 47 | | 34 |
| 1179 | — 21 | 1173 | — 1 |
| | 26 | | 33 |
| 1205 | 0 | 1206 | 0 |
| | 26 | | 33 |
| 1231 | + 17 | 1239 | — 1 |
| | 43 | | 32 |
| 1274 | — 16 | 1271 | 0 |
| | 27 | | 32 |
| 1301 | | 1303 | |

The first differences show it clearly enough that II is regular and I is not. Yet the second differences are still more suggestive: II shows a minimum of irregularity due to the unavoidable rounding errors, but the second differences in I vary in sign and are quite large. Such "differencing" is an important operation in checking the construction of numerical tables. There are two remarks.

(1) In the table of a function $f(x)$ the first differences are connected with $f'(x)$, and the second differences with $f''(x)$, by the mean value theorem. This provides us with an opportunity to check the differences.

(2) The last decimals in I are, in fact, the first ten decimals of $\pi$ in reverse order. The view that the successive decimals of $\pi$ behave *as if* they were produced by chance, has been expressed many times with many variations.

**27.** Assume that $A$ is independent of $B$:

(6)                     $\Pr\{A/B\} = \Pr\{A/\bar{B}\}.$

(We begin with (6) to avoid interference with the numbering in ex. 26.)

Using (4), (2), (6), and (3) (in this order) we obtain

(7)
$$\Pr\{A\} = \Pr\{AB\} + \Pr\{A\bar{B}\}$$
$$= \Pr\{B\} \Pr\{A/B\} + \Pr\{\bar{B}\} \Pr\{A/\bar{B}\}$$
$$= \Pr\{A/B\} (\Pr\{B\} + \Pr\{\bar{B}\})$$
$$= \Pr\{A/B\}.$$

From (2), (7), and $\Pr\{A\} \neq 0$ follows that

(8)
$$\Pr\{A\} \Pr\{B/A\} = \Pr\{B\} \Pr\{A\}$$
$$\Pr\{B/A\} = \Pr\{B\}.$$

Using again (4) and (2), as in (7), and using also (8), (3), and $\Pr\{\bar{A}\} \neq 0$, we find that

(9)
$$\Pr\{B\} = \Pr\{A\} \Pr\{B/A\} + \Pr\{\bar{A}\} \Pr\{B/\bar{A}\}$$
$$(1 - \Pr\{A\}) \Pr\{B\} = \Pr\{\bar{A}\} \Pr\{B/\bar{A}\}$$
$$\Pr\{B\} = \Pr\{B/\bar{A}\};$$

(6), (7), (8), and (9) show the required conclusion.

**28.** If $A$ and $B$ are mutually independent.

$$\Pr\{AB\} = \Pr\{A\} \Pr\{B\}.$$

This follows from rule (2) of ex. 26 and definition (II) of ex. 27.

**29.** (a)   $\Pr\{A\}$,   $\Pr\{A/B\}$,   $\Pr\{A/\bar{B}\}$,   $\Pr\{B\}$,   $\Pr\{B/A\}$,   $\Pr\{B/\bar{A}\}$

(I)   $\frac{1}{3}$,   $\frac{1}{3}$,   $\frac{1}{3}$,   $\frac{1}{2}$,   $\frac{1}{2}$,   $\frac{1}{2}$

(II)   $\frac{1}{3}$,   1,   $\frac{1}{5}$,   $\frac{1}{6}$,   $\frac{1}{2}$,   0

(b)    $\Pr\{AB\} = \Pr\{A\} \Pr\{B/A\} = \Pr\{B\} \Pr\{A/B\}$

(I)     $\frac{1}{3} \cdot \frac{1}{2} = \frac{1}{2} \cdot \frac{1}{3}$

(II)    $\frac{1}{3} \cdot \frac{1}{2} = \frac{1}{6} \cdot 1$ .

(c) The formulas follow generally from ex. 26 (4), (2).    Numerically

$$\Pr\{A\} = \Pr\{B\} \Pr\{A/B\} + \Pr\{\bar{B}\} \Pr\{A/\bar{B}\},$$

(I)   $\frac{1}{3} = \frac{1}{2} \cdot \frac{1}{3} + \frac{1}{2} \cdot \frac{1}{3}$ ,

(II)   $\frac{1}{3} = \frac{1}{6} \cdot 1 + \frac{5}{6} \cdot \frac{1}{5}$ .

$$\Pr\{B\} = \Pr\{A\}\,\Pr\{B/A\} + \Pr\{\bar{A}\}\,\Pr\{B/\bar{A}\},$$

(I) $\quad \tfrac{1}{2} = \tfrac{1}{3} \cdot \tfrac{1}{2} + \tfrac{2}{3} \cdot \tfrac{1}{2}$ ,

(II) $\quad \tfrac{1}{6} = \tfrac{1}{3} \cdot \tfrac{1}{2} + \tfrac{2}{3} \cdot 0$ .

(d) With (I), $A$ and $B$ are mutually independent, with (II) they are not.

No solution: **19, 20, 26, 30, 31, 32, 33.**

## SOLUTIONS, CHAPTER XV

**1.** Since $H$ implies both $A$ and $B$,

$$\Pr\{A/H\} = 1, \qquad \Pr\{B/H\} = 1.$$

Therefore, by rule (2) of ex. 14.26,

$$\Pr\{H\} = \Pr\{A\}\,\Pr\{H/A\}, \qquad \Pr\{H\} = \Pr\{B\}\,\Pr\{H/B\}.$$

Eliminating $\Pr\{H\}$, we obtain

$$\Pr\{A\}\,\Pr\{H/A\} = \Pr\{B\}\,\Pr\{H/B\}.$$

If we regard the relation between $H$ and $A$, and also the relation between $H$ and $B$, as unchanged, and, therefore, $\Pr\{H/A\}$ and $\Pr\{H/B\}$ as constants, and let $\Pr\{B\}$ increase, also $\Pr\{A\}$ will increase by virtue of our last equation.

**2.** By the formulas ex. 14.26 (2), (3), (4)

$$\Pr\{A\} = \Pr\{A/\bar{B}\} + \Pr\{B\}\,(\Pr\{A/B\} - \Pr\{A/\bar{B}\}).$$

We take $\Pr\{A/B\}$ and $\Pr\{A/\bar{B}\}$ as given, but let $\Pr\{B\}$ increase. (This can be quite naturally visualized in the case considered in ex. 13.8.) The increase of $\Pr\{B\}$ implies a corresponding increase of $\Pr\{A\}$ *if*

(*) $$\Pr\{A/B\} > \Pr\{A/\bar{B}\}.$$

In the concrete case of ex. 13.8 this inequality seems acceptable. Yet the statement of the pattern given in the solution of ex. 13.8 appears unacceptable. Here is a statement which looks better and is certainly in line with the formula:

$A$ more credible with $B$ than without $B$

$B$ (becomes) more credible

$A$ (becomes) more credible

If $A$ is a consequence of $B$, $\Pr\{A/B\} = 1$, (*) is certainly correct, and the pattern becomes that in sect. 13.7, Table I, line 2, column 2.

**3.** Review the few cases in which we have represented patterns of plausible reasoning by formulas of the Calculus of Probability: sect. 6–10 and ex. 1–2.

The monotonicity and continuity asserted by ex. 13.10 is obvious in these cases from the following simple mathematical fact: If $a$, $b$, $c$, $d$, $x_1$, $x_2$ are real constants, $ad - bc \neq 0$, and the function $y$ of $x$, defined by

$$y = \frac{ax + b}{cx + d},$$

has the property that $0 \leq y \leq 1$ for $x_1 < x < x_2$, then $y$ is strictly monotonic and continuous for $x_1 \leq x \leq x_2$. With respect to the generality of the remarks of ex. 13.10 we should observe this: if $y$ is not a linear fractional function but, more generally, a rational function of $x$, it is still necessarily a continuous function of $x$, except at points where it becomes infinite, but no more necessarily a monotonic function.

No solution: **4, 5, 6, 7, 8, 9.**

## SOLUTIONS, CHAPTER XVI

**2.** [Stanford 1951] The quadrilateral has to be convex. Let us call I, II, III, IV the triangles into which it is divided by its diagonals, (I), (II), (III), (IV) the areas of the four triangles, respectively, and $p$, $q$, $r$, $s$ the lengths of the four straight lines drawn from the intersection of the diagonals to the four vertices of the quadrilateral. Name and number in "cyclic order" so that the side of length $p$ is common to IV and I, $q$ to I and II, $r$ to II and III, $s$ to III and IV; I is opposite to III, II to IV; $p + r$ is the length of one diagonal, $q + s$ that of the other. Let $p$ and $q$ include the angle $\alpha$. Then

$$2(\mathrm{I}) = pq \sin \alpha, \qquad 2(\mathrm{II}) = qr \sin \alpha,$$

$$2(\mathrm{III}) = rs \sin \alpha, \qquad 2(\mathrm{IV}) = sp \sin \alpha.$$

Hence

(a) $(\mathrm{I})\,(\mathrm{III}) = (\mathrm{II})\,(\mathrm{IV})$.

(b) The base of I is parallel to that of III if, and only if,

$$p/q = r/s \qquad \text{or} \qquad (\mathrm{II}) = (\mathrm{IV}).$$

(c) The quadrilateral is a parallelogram if, and only if,

$$p = r, \quad q = s \quad \text{or} \quad (\mathrm{I}) = (\mathrm{II}) = (\mathrm{III}) = (\mathrm{IV}).$$

**3.** [Stanford 1949] (a) Let $a$ be a side of the equilateral triangle. Joining the point inside it to its three vertices, you divide it into three triangles with areas that added together give the whole area: $ax/2 + ay/2 + az/2 = ah/2$. Divide by $a/2$. See fig. 8.8.

(b) A point inside a regular tetrahedron with altitude $h$ has the distances $x, y, z$, and $w$ from the four faces, respectively. Then $x + y + z + w = h$. The proof is analogous: divide the regular tetrahedron into four tetrahedra.

(c)  The relation remains valid in both cases (a) and (b) for outside points, provided that the distances $x, y, z$ (and $w$) are taken with the proper sign: $+$ when a spectator placed in the point sees the side (face) from inside, $-$ when he sees it from outside.  The proof is essentially the same.

4. [Stanford 1946]  (a) (I) and (IV) are generally true, but (II) and (III) are not; see (b).

(b) (II) and (III) are false: the rectangle and the rhombus are counter-examples, respectively.

(c) (II) and (III) are true for pentagons.

(d) (II) and (III) are true for polygons with an odd number of sides, 3 or 5 or 7 . . . , as it follows from (II') and (III').

(II') If a polygon inscribed in a circle is equiangular, any two sides separated by just one intervening side are equal.  Therefore, if the number of sides is even, equal to $2m$, either all $2m$ sides are equal, or $m$ sides are equal to $a$ and the $m$ remaining sides to $b$, $a \neq b$, and no two sides having a vertex in common are equal.

(III') If a polygon circumscribed about a circle is equilateral, any two angles separated by just one intervening angle are equal.  Therefore, if the number of angles is even, equal to $2m$, either all $2m$ angles are equal, or $m$ angles are equal to $\alpha$ and the $m$ remaining angles to $\beta$, $\alpha \neq \beta$, and no two angles having a side in common are equal.

To prove (I) (II') (III') (IV) join the center of the circle to the vertices of the polygon, draw perpendiculars from the center to the sides, and pick out congruent triangles.

5. [Stanford 1947]  The relation between (a) and (b) appears similar to that between (b) and (c); the similarity of the left-hand sides appears more clearly than that of the right-hand sides.  Hence it is natural to look for some transition from (a) to (b) that could also serve as a transition from (b) to (c).  There is such a transition: if

$$\alpha + \beta + \gamma = \pi$$

$$2\alpha + 2\beta + 2\gamma = 2\pi \neq \pi, \text{ but}$$

$$(\pi - 2\alpha) + (\pi - 2\beta) + (\pi - 2\gamma) = \pi.$$

Taking for granted that (a) holds for any three angles $\alpha$, $\beta$, $\gamma$ with sum $\pi$, we obtain (b) by substituting for these angles $\pi - 2\alpha$, $\pi - 2\beta$, $\pi - 2\gamma$, respectively, and we pass from (b) to (c) by using the same substitution.

It remains to verify (a) which can be done in many ways, for instance as follows.  Substituting $2u$, $2v$, and $\pi - 2u - 2v$ for $\alpha$, $\beta$, and $\gamma$, respectively, we transform (a) into

$$\sin u \cos u + \sin v \cos v = [2 \cos u \cos v - \cos (u + v)] \sin (u + v).$$

Use the addition theorems of cosine and sine.

**6.** [Stanford 1952] According to the four proposals, the volume of the frustum would be

(I) $[(a + b)/2]^2h$

(II) $[(a^2 + b^2)/2]h$

(III) $[a^2 + b^2 + (a + b)^2/4]h/3$

(IV) $[a^2 + b^2 + ab]h/3$,

respectively. If $b = a$, the frustum becomes a prism with volume $a^2h$: (I) (II) (III) (IV) agree in yielding the correct result. If $b = 0$ the frustum becomes a pyramid with volume $a^2h/3$: only (IV) yields this and so (I) (II) (III) which yield different values must be incorrect. That (IV) is generally correct, must still be proved; see the textbooks.

**10.** (1) Doing the addition with paper and pencil in the most usual manner, you begin at the top of the last column and proceed downward; in the given example the first two steps are $3 + 7 = 10$, $10 + 0 = 10$. Then you proceed downward in the next column, and so on. (2) You take the columns in the same order as before, but you start at the bottom and proceed upward in each. (3) You take the columns again in the same order but do each column twice, first downward, then upward; you note the result when, working downward, you reach the bottom, and check it by working upward. (4) Add first the 1st, 3rd, 5th, . . . number, then the 2nd, 4th, 6th, . . . number, and finally the two sums obtained; this requires some extra writing. (5) Do the addition once by hand, and once by a machine. And so on.

**11.** 59.

**12.** $10^{-9}$. (Not $10^{-3}$ or $10^{-7}$.)

No solution: **1, 7, 8, 9, 13.**

# BIBLIOGRAPHY

## I. CLASSICS

EUCLID, *Elements*. The inexpensive shortened edition in Everyman's Library is sufficient here. "EUCLID III, 7" refers to Proposition 7 of Book III of the Elements.

DESCARTES, *Oeuvres*, edited by Charles Adam and Paul Tannery. The work "Regulae ad Directionem Ingenii," vol. 10, pp. 359–469, is of especial interest.

EULER, *Opera Omnia*, edited by the "Societas scientiarum naturalium Helvetica."

LAPLACE, *Oeuvres complètes*. The "Introduction" of vol. 7, pp. V–CLIII, also separately printed (and better known) under the title "Essai philosophique sur les probabilites" is of especial interest.

## II. SOME BOOKS OF SIMILAR TENDENCY

R. COURANT and H. ROBBINS, *What is mathematics?*

H. RADEMACHER and O. TOEPLITZ, *Von Zahlen und Figuren.*

O. TOEPLITZ, *Die Entwicklung der Infinitesimalrechnung;* of especial interest.

## III. RELATED FORMER WORK OF THE AUTHOR

*Books*

1. *Aufgaben und Lehrsätze aus der Analysis*, 2 volumes, Berlin, 1925. Jointly with G. Szegö. Reprinted New York, 1945.

2. *How to Solve It*, Princeton, 1945. The 5th printing, 1948, is slightly enlarged.

3. Wahrscheinlichkeitsrechnung, Fehlerausgleichung, Statistik. From *Abderhalden's Handbuch der biologischen Arbeitsmethoden*, Abt. V, Teil 2, pp. 669–758.

*Papers*

1. Geometrische Darstellung einer Gedankenkette. *Schweizerische Pädagogische Zeitschrift*, 1919, 11 pp.

2. Wie sucht man die Lösung mathematischer Aufgaben? *Zeitschrift für mathematischen und naturwissenschaftlichen Unterricht*, v. 63, 1932, pp. 159–169.

3. Wie sucht man die Lösung mathematischer Aufgaben? *Acta Psychologica*, v. 4, 1938, pp. 113–170.

4. Heuristic reasoning and the theory of probability. *American Mathematical Monthly*, v. 48, 1941, pp. 450–465.

5. On Patterns of Plausible Inference. *Courant Anniversary Volume*, 1948, pp. 277–288.

6. Generalization, Specialization, Analogy. *American Mathematical Monthly*, v. 55, 1948, pp. 241–243.

7. Preliminary remarks on a logic of plausible inference. *Dialectica*, v. 3, 1949, pp. 28–35.

8. With, or without, motivation? *American Mathematical Monthly*, v. 56, 1949, pp. 684–691.

9. Let us teach guessing. *Etudes de Philosophie des Sciences, en hommage à Ferdinand Gonseth*, 1950, pp. 147–154. Editions du Griffon, Neuchatel, Switzerland.

10. On plausible reasoning. *Proceedings of the International Congress of Mathematicians*, 1950, v. 1, pp. 739–747.

## IV. PROBLEMS

Among the examples proposed for solution there are some taken from the *William Lowell Putnam Mathematical Competition* or the *Stanford University Competitive Examination in Mathematics*. This fact is indicated at the beginning of the solution with the year in which the problem was proposed as "Putnam 1948" or "Stanford 1946." The problems of the Putnam Examination are published yearly in the *American Mathematical Monthly* and most Stanford examinations have been published there too.

CPSIA information can be obtained at www.ICGtesting.com
Printed in the USA
LVOW10*1718010514

384026LV00004B/50/P